COLDWATER FISHERIES AND AQUACULTURE MANAGEMENT

Technology for Sustainable Food Production

COLDWATER FISHERIES AND AQUACULTURE MANAGEMENT

Technology for Sustainable Food Production

Edited by

Mohd. Ashraf Rather, PhD

Faisal Rashid Sofi, PhD

Adnan Amin, PhD

Kawkabul Saba, PhD

AAP | APPLE ACADEMIC PRESS

First edition published 2024

Apple Academic Press Inc.
1265 Goldenrod Circle, NE,
Palm Bay, FL 32905 USA

760 Laurentian Drive, Unit 19,
Burlington, ON L7N 0A4, CANADA

CRC Press
6000 Broken Sound Parkway NW,
Suite 300, Boca Raton, FL 33487-2742 USA

4 Park Square, Milton Park,
Abingdon, Oxon, OX14 4RN UK

Library and Archives Canada Cataloguing in Publication

Title: Coldwater fisheries and aquaculture management : technology for sustainable food production / edited by Mohd. Ashraf Rather, PhD, Faisal Rashid Sofi, PhD, Adnan Amin, PhD, Kawkabul Saba, PhD.

Names: Rather, Mohd. Ashraf, editor. | Sofi, Faisal Rashid, editor. | Amin, Adnan (Lecturer in aquatic environmental management), editor. | Saba, Kawkabul, editor.

Description: First edition. | Includes bibliographical references and index.

Identifiers: Canadiana (print) 20230474950 | Canadiana (ebook) 20230474985 | ISBN 9781774913420 (hardcover) | ISBN 9781774913437 (softcover) | ISBN 9781003369905 (ebook)

Subjects: LCSH: Fish culture. | LCSH: Freshwater fishes. | LCSH: Sustainable aquaculture—Management. | LCSH: Sustainable fisheries.

Classification: LCC SH159 .C65 2024 | DDC 639.3/1—dc23

Library of Congress Cataloging-in-Publication Data

CIP data on file with US Library of Congress

ISBN: 978-1-77491-342-0 (hbk)
ISBN: 978-1-77491-343-7 (pbk)
ISBN: 978-1-00336-990-5 (ebk)

Dedication

Dedicated to Fish Farmers

"If you give a hungry man a fish, you feed him for a day,
but if you teach him how to fish, you feed him for a lifetime."
—Lao Tsu

About the Editors

Mohd. Ashraf Rather, PhD

Assistant Professor, Division of Fish Genetics and Biotechnology Faculty of Fisheries, Rangil, Ganderbal, Sher-e-Kashmir University of Agricultural Sciences and Technology–Kashmir, India

Mohd. Ashraf Rather, PhD, is working as an Assistant Professor in the Division of Fish Genetics and Biotechnology at the Faculty of Fisheries, Rangil, Ganderbal, Sher-e-Kashmir University of Agricultural Sciences and Technology, Kashmir. He graduated from the College of Fisheries, Ratnagiri, Maharashtra, and obtained master's and doctoral degrees in Fisheries Biotechnology from the Division of Genetics and Biotechnology, Central Institute of Fisheries Education (CIFE), Mumbai. His area of research interest is reproductive physiology and molecular endocrinology of fish proteins. His major research contribution is the identification, characterization, and expression profiling of more than 25 reproductive genes in fish, including kisspeptin genes in Indian major carp. He has more than eight years of experience working on the molecular endocrinology of fish.

Dr. Rather is a recipient of various accolades such as Prof. K. H. Alikunhi Gold medal from the Central Institute of Fisheries Education and Dr. Karunasagar Best Post-Graduate Thesis (PhD – Indian category) award from the Professional Fisheries Graduates Forum (PFGF). He received a Young Scientist Award in 2017 from the Society of Fisheries and Life Science, College of Fisheries, Mangalore. He has also received an Asia-Pacific Alltech Young Scientist Award–2012 (first rank) in Agriculture and Allied Field from Alltech, USA. He is a recipient of various national and international fellowships, including the Prince Songkla Fellowship from the Prince Songkla University of Thailand, the National Post-Doc fellowship (NPDF) from the Science and Engineering Research Board (SERB), Department of Science and Technology–New Delhi, Indian Science Academy Fellowship, Senior Research Fellowship (SRF, first rank) and Junior Research Fellowship (JRF, first rank) from the Indian Council of Agricultural Research (ICAR), New Delhi.

He has published more than 40 research publications in both national and international peer-reviewed research journals. In addition, he has published two books, two conference proceedings, and various chapters. He functions as an editorial board member in various international and national research journals. To date, Dr. Rather has received three best research paper awards from various international and national institutes and organizations. He has mentored several master's and doctoral students pursuing higher studies in fisheries biotechnology/fisheries science.

Faisal Rashid Sofi, PhD
Assistant Professor, Division of Post-Harvest Technology (Fisheries), Faculty of Fisheries, Sher-e-Kashmir University of Agricultural Sciences and Technology–Rangil, Kashmir, India

Faisal Rashid Sofi, PhD, is an Assistant Professor in the Division of Post-Harvest Technology (Fisheries), Faculty of Fisheries, Sher-e-Kashmir University of Agricultural Sciences and Technology—Rangil, Kashmir. He graduated from the College of Fisheries, Veraval, Junagadh Agricultural University, Gujarat, and obtained a master's degree from the same university in the Department of Post-Harvest Technology. Later on, he obtained a doctoral degree in Fish Processing Technology from the College of Fisheries, Mangalore. His area of research interest is phenolic compounds from natural sources and their efficacy for the preservation of fish and fishery products. His major research contribution is increasing the shelf life of Indian mackerel with the effect of grape seed extract. He has received a best paper award at the International Conference on Profits on Aquaculture. He has published more than 25 research publications in both national and international journals. In addition, he has published 12 chapters.

Adnan Amin, PhD
Assistant Professor, Division of Aquatic Environmental Management, Faculty of Fisheries, Rangil, Ganderbal, Sher-e-Kashmir University of Agricultural Sciences and Technology–Kashmir, India

Adnan Amin, PhD, is presently working as an Assistant Professor in the Division of Aquatic Environmental Management at the Faculty of Fisheries, Rangil, Ganderbal, Sher-e-Kashmir University of Agricultural Sciences and Technology—Kashmir. He is an undergraduate from the College of Fishery Science Muthukur, Nellore SVVU, Tirupati, Andhra Pradesh. He did his master's from the College of Fisheries, Mangalore, in the Department of Aquatic Environment Management. He completed a doctoral degree from the College of Fisheries Ratnagiri in the Department of Aquatic Environment Management. His area of research interest is aquatic toxicology and limnology.

His major research contribution is heavy metal and pesticide toxicity in aquatic organisms, impact of climate change on aquatic biodiversity.

He is a recipient of a university fellowship, i.e., merit fellowship for pursuing his master's. He is associated with the Society of Fisheries and Life Science, College of Fisheries, Mangalore. He has published more than 16 research publications in both national and international peer-reviewed research journals. In addition, he has published five chapters, six review papers, and 18 magazine articles. Dr. Amin has received one best research paper award from a national organization.

Kawkabul Saba, PhD
Assistant Professor and Scientist, Division of Fish Nutrition and Biochemistry, Faculty of Fisheries, Sher-e-Kashmir University of Agricultural Sciences and Technology, Rangil, Kashmir, India

Kawkabul Saba, PhD, is an Assistant Professor Scientist in the Division of Fish Nutrition and Biochemistry, Faculty of Fisheries, Sher-e-Kashmir University of Agricultural Sciences and Technology, Rangil, Kashmir. She graduated from the College of Fisheries, Veraval, Junagadh Agricultural University Gujarat, and obtained a master's degree from the same university in the Department of Fisheries Resource Management. She obtained a doctoral degree in Fisheries Resource Management from the Faculty of Fisheries, SKUAST-K. Her area of research interest is the development of low-cost feed from slaughterhouse waste. She is a recipient of INSPIRE fellowship during her doctoral program from the Department of Science and Technology (DST), New Delhi. She has published more than 20 research publications in both national and international journals. In addition, she has published five chapters.

Contents

Contributors

Adnan Abubakr
Division of Aquatic Environmental Management, Faculty of Fisheries, Sher-e-Kashmir, University of Agricultural Sciences and Technology (SKUAST) of Kashmir, Rangil, Ganderbal, Jammu and Kashmir, India

Deepak Agarwal
Tamil Nadu Dr. J. Jayalalithaa Fisheries University, IFPGS, OMR Campus, Chennai, Tamil Nadu, India

Irfan Ahmad
Division of Fish Genetics and Biotechnology, Faculty of Fisheries, Sher-e-Kashmir University of Agricultural Sciences and Technology, Jammu and Kashmir, India

Syed Umair Ahmad
Department of Bioinformatics, Hazara University, Mansehra, Pakistan

Qurat Ul Ain
Department of Chemistry, Government College Women's University Faisalabad, Pakistan

Oyais Aismi
Division of Fish Nutrition and Biochemistry, Faculty of Fisheries, SKUAST-K, Jammu and Kashmir, India

Prarthana Aithal
Coldwater Fisheries Research, Nainital, Uttarakhand, India

Adnan Amin
Division of Aquatic Environmental Management, Faculty of Fisheries, Sher-e-Kashmir, University of Agricultural Sciences and Technology (SKUAST) of Kashmir, Rangil, Ganderbal, Jammu and Kashmir, India

Garima Anand
ICAR–Central Institute of Fisheries Education, Mumbai, Maharashtra, India

Priyanka Ashwath
Nitte University Center for Science Education and Research, Paneer Campus, Deralakatte, Mangalore, Karnataka, India

Mudassir Azhar
Department of Fisheries Science, Doon PG College of Agriculture and Allied Science, Uttarakhand, India

Shabana Bibi
Yunnan Herbal Laboratory, College of Ecology and Environmental Sciences; International Joint Research Center for Sustainable Utilization of Cordyceps Bioresources in China and South-East Asia, Yunnan University, Kunming, Yunnan, China

Khemraj Bunkar
ICAR–Central Institute of Fisheries Education, Mumbai, Maharashtra, India

Shabir A. Dar
Division of Fishery Engineering, Faculty of Fisheries, SKUAST-K, Jammu and Kashmir, India

Shabir Ahmad Dar
Division of Aquatic Animal Health Management, Faculty of Fisheries, Sher-e-Kashmir University of Agricultural Sciences and Technology (SKUAST) of Kashmir, Rangil, Ganderbal, Jammu and Kashmir, India

Showkat Ahmad Dar
Department of Aqualife Medicine, College of Fisheries and Ocean Studies, Chonnam National University, South Korea

Deekshit
Nitte University Center for Science Education and Research, Paneer Campus, Deralakatte, Mangalore, Karnataka, India

S. Devadharshini
Department of Fish Processing Technology, Fisheries College and Research Institute, TNJFU, Thoothukudi, Tamil Nadu, India

K. Dhanapal
Department of Fish Processing Technology, College of Fisheries, Muthukur, Andhra Pradesh, India

Disksha
Department of Life Sciences and Allied Health Sciences, Sant Baba Bhag Singh University, Khalia Padhiana, Jalandhar, Punjab, India

P. Ganesan
Department of Fish Processing Technology, Fisheries College and Research Institute, TNJFU, Thoothukudi, Tamil Nadu, India

Udai Ram Gurjar
ICAR–Central Institute of Fisheries Education, Mumbai, Maharashtra, India

Younis Ahmad Hajam
Department of Life Sciences and Allied Health Sciences, Sant Baba Bhag Singh University, Khalia Padhiana, Jalandhar, Punjab, India; Department of Biosciences, Division Zoology, Career Point University, Hamirpur, Himachal Pradesh, India

Tariq Hussain
Division of Post-Harvest Technology, Faculty of Fisheries, SKUAST-K, Jammu and Kashmir, India

Inain Jaies
Division of Aquatic Animal Health Management, Faculty of Fisheries, Sher-e-Kashmir University of Agricultural Sciences and Technology (SKUAST) of Kashmir, Rangil, Ganderbal, Jammu and Kashmir, India

Ankur Jamwal
College of Fisheries and Center of Excellence on Water Management, Dr. Rajendra Prasad Central Agricultural University, Pusa, Samastipur, Bihar, India

Zainab Jan
National Center for Bioinformatics, Qauid-i-Azam University, Islamabad, Pakistan

Md. Idrish Raja Khan
Department of Aquatic Health and Environment, College of Fisheries, Central Agricultural University (I), Lembucherra, Tripura, India

Sameena Khan
Division of Aquatic Environmental Management, Faculty of Fisheries, Sher-e-Kashmir, University of Agricultural Sciences and Technology (SKUAST) of Kashmir, Rangil, Ganderbal, Jammu and Kashmir, India

Sofia Khanam
MPharm, Department of Pharmacology, Calcutta Institute of Pharmaceutical Technology and AHS, Uluberia, Howrah, Kolkata, West Bengal, India

Ashwini Kumar
Division of Fish Nutrition and Biochemistry, Faculty of Fisheries, SKUAST-K, Jammu and Kashmir, India

Rajesh Kumar
Department of Biosciences, Himachal Pradesh University, Shimla, Himachal Pradesh, India

Puspa Kumari
ICAR–Central Institute of Fisheries Education, Mumbai, Maharashtra, India

Monisa M. Malik
Division of Aquatic Environmental Management, Faculty of Fisheries, Sher-e-Kashmir, University of Agricultural Sciences and Technology (SKUAST) of Kashmir, Rangil, Ganderbal, Jammu and Kashmir, India

Hafsa Maqbool
Division of Aquatic Animal Health Management, Faculty of Fisheries, Sher-e-Kashmir University of Agricultural Sciences and Technology of Kashmir, Jammu and Kashmir, India

Mohammed Meharoof
PhD Scholars, ICAR–Central Institute of Fisheries Education, Mumbai, Maharashtra, India

Asaad H. Mohamed
Beacon Development, King Abdullah University of Science and Technology, Thuwal, Jeddah, Saudi Arabia

Dinesh Mohan
ICAR–Directorate of Coldwater Fisheries Research, Bhimtal, Uttarakhand, India

Syed Talia Mushtaq
Division of Fisheries Resource Management, Faculty of Fisheries, Sher-e-Kashmir University of Agricultural Sciences and Technology, Jammu and Kashmir, India

Salik Nazki
The Pirbright Institute, United Kingdom

D. Pamanna
College of Fishery Science, Muthukur, Nellore, Andhra Pradesh, India

N. N. Pandey
ICAR–Directorate of Coldwater Fisheries Research, Bhimtal, Uttarakhand, India

Edoardo Pantanella
Beacon Development, King Abdullah University of Science and Technology, Thuwal, Jeddah, Saudi Arabia

Neeraj Pathak
Department of Quality Assurance and Management, College of Fisheries, Thoothukudi, Tamil Nadu, India

Vikas Phulia
Krishi Vigyan Kendra, (HQ: Guru Angad Dev Veterinary and Animal Sciences University, Ludhiana), Punjab, India

Ramya Premanath
Nitte University Center for Science Education and Research, Paneer Campus, Deralakatte, Mangalore, Karnataka, India

Syed Shariq N. Qadiri
Division of Aquatic Animal Health Management, Faculty of Fisheries,
Sher-e-Kashmir University of Agricultural Sciences and Technology (SKUAST) of Kashmir,
Rangil, Ganderbal, Jammu and Kashmir, India

Imtiyaz Qayoom
Division of Aquatic Environmental Management, Faculty of Fisheries, Sher-e-Kashmir,
University of Agricultural Sciences and Technology (SKUAST) of Kashmir, Rangil, Ganderbal,
Jammu and Kashmir, India

Faisal Rashid
Division of Post-Harvest Technology, Faculty of Fisheries, SKUAST-K, Jammu and Kashmir, India

Mohd. Ashraf Rather
Division of Fish Genetics and Biotechnology, Faculty of Fisheries, Sher-e-Kashmir University of
Agricultural Sciences and Technology, Jammu and Kashmir, India

S. Vijay Kumar Reddy
Department of Fish Processing Technology, College of Fisheries, Gadvasu, Ludhiana, Punjab, India

Mohd. Salim Reshi
Toxicology and Pharmacology Laboratory, Department of Zoology, School of Biosciences and
Biotechnology, Baba Ghulam Shah Badshah University, Rajouri, Jammu and Kashmir, India

K. T. Hafeef Roshan
PhD Scholars, ICAR–Central Institute of Fisheries Education, Mumbai, Maharashtra, India

Kawkabul Saba
Division of Fish Nutrition and Biochemistry, Faculty of Fisheries, SKUAST-K, Jammu and Kashmir,
India

Sneha Sabu
MSc, Department of Microbiology, CMST, Manonmaniam Sundharanar University, Rajakkamangalam,
Kanyakumari, Tamil Nadu, India

K. A. Sajina
PhD Scholars, ICAR–Central Institute of Fisheries Education, Mumbai, Maharashtra, India

Akhila Dharnappa Sannejal
Nitte University Center for Science Education and Research, Paneer Campus, Deralakatte,
Mangalore, Karnataka, India

Feroz A. Shah
Division of Aquatic Animal Health Management, Faculty of Fisheries,
Sher-e-Kashmir University of Agricultural Sciences and Technology (SKUAST) of Kashmir,
Rangil, Ganderbal, Jammu and Kashmir, India

Munazah Shahzad
Department of Veterinary Epidemiology and Public Health, University of Surrey, United Kingdom

Faisal Sofi
Division of Post-Harvest Technology, Faculty of Fisheries, SKUAST-K, Jammu and Kashmir, India

K. Sravani
Department of Fish Processing Technology, Fisheries College and Research Institute, TNJFU,
Thoothukudi, Tamil Nadu, India

A. Jothilin Subitsha
MSc, Department of Microbiology, CMST, Manonmaniam Sundharanar University, Rajakkamangalam, Kanyakumari, Tamil Nadu, India

Suman Takar
TNJFU–Fisheries College and Research Institute, Thoothukudi, Tamil Nadu, India

Anwar Ullah
The Second Affiliated Hospital, Institute of Cancer Stem Cell, Dalian Medical University, Dalian, People's Republic of China

Tincy Varghese
ICAR–Central Institute of Fisheries Education, Mumbai, Maharashtra, India

S. Vijay Kumar Reddy
Department of Fish Processing Technology, College of Fisheries, Gadvasu, Ludhiana, Punjab, India

Rajeshwari Vittal
Nitte University Center for Science Education and Research, Paneer Campus, Deralakatte, Mangalore, Karnataka, India

Asifa Wali
Division of Aquatic Animal Health Management, Faculty of Fisheries,
Sher-e-Kashmir University of Agricultural Sciences and Technology (SKUAST) of Kashmir,
Rangil, Ganderbal, Jammu and Kashmir, India

Gohar B. Wani
Division of Fishery Engineering, Faculty of Fisheries, SKUAST-K, Jammu and Kashmir, India

Ramzy A. Yousif
Department of Fisheries and Wildlife Science, Sudan University of Science and Technology, Khartoum, Sudan

Imran Zafar
Department of Bioinformatics and Computational Biology, Virtual University of Pakistan, Pakistan

Seemab Zehra
Beacon Development, King Abdullah University of Science and Technology, Thuwal, Jeddah, Saudi Arabia

A. A. Zynudheen
Fish Processing Division, Central Institute of Fishing Technology, Kochi, Ernakulam, Kerala, India

Foreword 1

India, being one of the mega biodiversity hotspots, harbors the greatest number of endemic freshwater fishes in continental Asia. Among the freshwater fishes, the coldwater fisheries harbor 258 species belonging to 21 families and 76 genera. Out of about 258 cold water fish species (both indigenous and exotic) reported from Indian uplands, snow trout, brown trout, Chinese carps, and rainbow trout are commercially important species. The trout and salmon are the world's second most traded fish species in terms of value, accounting for 14% of global trade, following shrimp which accounts for 15% of the total. Trout and salmon fish accounts for more than 6% (5 million tons) of the total global aquaculture production (SOFIA FAO, 2020). In the Himalayan states, rainbow trout is becoming a source of livelihood and food security for the hill population.

The aquaculture sector is considered the fastest-growing food-producing sector and can meet the protein needs of an increasing world population. India is the second largest producer of fish in the world. The prospects for eradicating malnutrition and fulfilling the Sustainable Development Goal of the UN partly lie with our efforts to increase aquaculture production. To achieve this, we must adopt species diversification, and developing seed production and feed technology is a prime requirement. Understanding physiology and adaptation in coldwater fishes by means of different tools, such as transcriptomics, proteomics, bioinformatics, etc., would help in understanding key mechanisms in the physiology of coldwater fishes. This new information can be used for both guiding new innovations and formulating policies for the coldwater fisheries sector.

This book brings together an informative compilation of different aspects of coldwater fisheries and aquaculture management technology for sustainable food production. I congratulate the editorial team and all the authors for bringing out a comprehensive compilation in the form of this book, *Coldwater Fisheries and Aquaculture Management: Technologies for Sustainable Food Production.*

—S. Ayyappan
Former Secretary, DARE and Director General, ICAR,
Chairman, Karnataka Science and Technology Academy,
Bangalore, Karnataka, India

Foreword 2

Globally, aquaculture has expanded significantly during the past millennia. India continues to be the world's second-largest aquaculture fish grower with a production of 14.16 million metric tonnes (MT) in 2019–2020. The industry has significantly increased fish production, with an average annual growth rate of 7.53% during the previous five years. However, given its abundant natural resources, India has tremendous potential and the ability to become a global leader.

Increased production in coldwater aquaculture can be attained by making the most use of available resources and using high-quality inputs. More than 6% (5 million tonnes) of the world's aquaculture production is made up of salmon and trout fish. Through the efficient application of new technologies and scientific management techniques, the factor cost of production may be rationalised. Few species are currently used in coldwater aquaculture operations, thus immediate species heterogeneity is needed to fully use the available ability. Quality fish seed and feed are essential for improving aquaculture productivity, which is one of the main concerns with regard to species variation. Establishing reproduction and seed production techniques requires an understanding of the physiology of coldwater fishes. Modern technologies including genomes, transcriptomics, proteomics, and metabolomics are all grouped under the name "omics" technology. These methods hold enormous promise for uncovering novel mechanisms underlying a coldwater fish's intricate thermal adaption, breeding, and reproductive behavior. The chapters in this book cover a wide range of important topics related to coldwater fish reproductive biology and endocrinology, including gonadal development and maturation, vitellogenesis, steroidogenesis, fish whole genome information, transcriptomics, proteomics, artificial intelligence, etc. I'm confident that the book will be of great use to readers of all types, including faculty members and students, in learning the processes involved in fish physiology in cold water and how they might be used in aquaculture.

I commend the authors, editors, and collaborators for compiling *Coldwater Fisheries and Aquaculture Management: Technologies for Sustainable Food Production* into such a thorough collection of top-notch chapters. I want to invite all of this book's readers to join the editors in making the most of this intriguing, difficult, and fruitful project.

—**Pramod Kumar Pandey**
Director, Anusandhan Bhawan, Industrial Area,
Bhimtal, Distt. Nainital, Uttrakhand, India

Preface

India has a coastline of about 8,129 km, 0.506 million sq. km of continental shelf and 2.02 million sq. km of exclusive economic zone. Total fish production of India for the year 2019–2020 is estimated to be 14.16 million metric tons (MT). Currently, India is the second largest fish producer across the globe. The sector has presented extraordinary development with fish production recording an average annual growth rate of 7.53% during the course of the last five years.

India, being one of the hotspots of aquatic animal biodiversity, harbors the greatest number of endemic freshwater fishes in continental Asia. Amongst the freshwater fishes, the coldwater fisheries harbor 258 species belonging to 21 families and 76 genera. Among 258 coldwater fish species reported from Indian uplands, rainbow trout, brown trout, snow-trout, and Chinese carps are commercially vital species. The trout and salmon are the world's second most traded fish species in terms of value contribute 14% of world aquaculture trade. Trout and salmon contribute greater than 6% of the total world aquaculture production (FAO, 2020). In the Indian Himalayan region, rainbow trout is nowadays becoming a basis of livelihood and food security for the hill population.

The whole globe is looking for sustainable source of nutritional security and aquaculture is standing firmly with the expectations. Today's food production industry scenario clearly indicates the dominancy of fish and fishery products in terms of growth rate and future potential of providing nutritional security. According to SOFIA 2020, total fish production in 2020 crossed the mark of 178.5 million tons, of which around 46% was contributed by aquaculture. A major portion of total fish production, i.e., around 88%, was used for direct human consumption (SOFIA, 2020).

India has witnessed tremendous growth in aquaculture during the last 10 years and is also geared to move up through adopting advanced aquaculture technologies along with species diversification to utilize the diverse potential water resources. Current coldwater aquaculture practices are limited to very few species and require urgent species diversification to harness their existing potential. Quality fish seed and feed technology is the prime requirement for coldwater aquaculture, and it is the major area of concern for species diversification. Understanding the physiology of coldwater fishes is essential for developing breeding and seed production technology.

Modern technologies like metabolomics, transcriptomics, genomics, and proteomics will be essential players in understanding the molecular basis of coldwater fishes. These techniques offer immense potential to unravel novel mechanisms underlying the complex thermal adaptation, breeding, and reproductive behavior of a coldwater fish. This book presents an extensive collection of articles on different key aspects associated with the reproductive biology and endocrinology of coldwater fishes, such as gonadal development and maturation, vitellogenesis, steroidogenesis, whole genome information of fishes, transcriptomics, proteomics, artificial intelligence, etc. The book has a recent compilation, and it will offer immense benefits to readers of all categories, viz., students and faculties, for understanding the underlying mechanisms of coldwater physiology of fishes and their potential application in aquaculture.

This book, "Coldwater Fisheries and Aquaculture Management: Technologies for Sustainable Food Production," is in its current shape due to the contribution made by several researchers working at different levels with reputed institutes across India and having research experience on coldwater fisheries and aquaculture. This book will provide invaluable support to scientists, teachers, researchers, students, etc., who are interested in the field of coldwater fisheries and aquaculture.

—Editors

Acknowledgments

First and foremost, we are grateful to the Almighty for enabling us to complete this book. We would like to express our heartfelt gratitude and respect to all authors who have contributed to this book in the form of their chapters. Indeed, it is our utmost pleasure to acknowledge all the people behind the completion of this book. It was not possible to complete this work without their continuous encouragement and support. We are eternally thankful to all authors.

We are very much thankful to Padma Shree. S. Ayyappan Chancellor CAU, Former Secretary, DARE, and Director General, ICAR, Chairman, Karnataka Science and Technology Academy, Bengaluru, and Dr. Pramod Kumar Pandey, Director, ICAR–Directorate of Coldwater Fisheries Research (DCFR) Bhimtal, Uttarakhand, India, for their compliment, praise, and blessings to the entire team of young researchers for their contributions in this book.

We are thankful to the Vice-Chancellor of Sher-e-Kashmir University of Agricultural Sciences and Technology-Kashmir and the Dean Faculty of Fisheries, Rangil Ganderbal–Kashmir, for their constant support and encouragement.

We will also take this opportunity to express our sincere gratitude to Apple Academic Press (AAP) exclusive co-publishing with CRC Press, a Taylor and Francis Group, Canada, for publishing this work.

Support of our students, colleagues, and family members is highly appreciated.

Finally, we would like to put our head down in front of the Almighty, who gave us courage, passion, and strength for completing this work.

—***Editors***

Abbreviations

AA	amino acid
ADC	apparent digestibility coefficient
ADHD	attention-deficit/ hyperactivity disorders
AFLP	amplified fragment-length polymorphism PCR
AFPs	antifreeze proteins
AHLs	acylated homoserine lactones
AI	autoinducer
AMPs	antimicrobial peptides
ANF	anti-nutritional factor
APCs	antigen-presenting cells
BFC	Behning fertilizing coefficient
BFW	Bankfull width
BGD	bacterial gill disease
BHT	butylated hydroxytoluene
BKD	bacterial kidney disease
BPE	beetroot peel extract
CAI-1	*Coholarae quorum*-sensing autoinducer 1
CEAA	conditionally EAA
CFU	colony forming units
CHD	coronary heart disease
CNS	central nervous system
CPC	canola protein concentrate
CRISPR-Cas9	clustered regularly interspaced short palindromic repeat-associated nuclease
CRP	C-reactive protein
CSM	cottonseed meal
CWD	*chronic wasting disease*
DCFR	Directorate of Coldwater Fisheries Research
DCs	dendritic cells
DHA	docosahexaenoic acid
DNA	deoxyribonucleic acid
DNV	densovirus
dpf	days post-fertilization
DSBs	double-strand breaks

EAA	essential amino acids
EC NFDB	Enzyme Commission National Fisheries Development Board
eDNA	environmental DNA
EFA	essential fatty acids
ELISA	enzyme-linked immunosorbent assay
EOs	essential oils
EPA	eicosapentenoic acid
ERM	enteric red mouth
ESTs	extended performance tasks
EU	European Union
EUS	epizootic ulcerative syndrome
FAA	functional amino acids
FAO	Food and Agricultural Organization
FAT	fluorescent antibody test
FCR	feed conversion ratio
FDA	Food and Drug Administration
FE	fennel extract
FM	fish meal
FOS	fructooligosaccharide
G-DAA	gluconeogenic discardable AA
GDP	gross domestic product
GE	gene-editing
GIBs	growth-inhibiting bacteria
GMS	Gomori methenamine silver
GnRH	gonadotropin-releasing hormone
GR	gonadosomatic report
GRAS	generally regarded as safe
GSDF	gonadal soma-derived growth factor
HNE	hydrononenal
HP	histopathological
HPF	hours post-fertilization
HPG	hypothalamic-pituitary-gonadal
HPP	high-pressure processing
HTST	high-temperature short time
HUFA	highly unsaturated fatty acids
i.e.	that is
IBW	initial body weight
IFNs	interferons
Ig	immunoglobulin

IHN	infectious hematopoietic necrosis
IHNV	infectious hematopoietic necrosis virus
Inos	inducible nitric oxide synthase
IP	intraperitoneal
IPN	infectious pancreatic necrosis
IPNV	infectious pancreatic necrosis virus
ISAV	infectious salmon anemia virus
KDM	kidney disease medium
KEGG	Kyoto Encyclopedia of Genes and Genomes
LA	linkage approaches
LAMP	loop-mediated isothermal amplification
LBT	luminescent bacterial disease
LD	linkage disequilibrium
LLM	*Leucaena* leaf meal
LPS	lipopolysaccharides
LSNV	Laem-Singh virus
MALDI-TOF	mass spectral fingerprinting
MDA	malonaldehyde
MG	mycotic granulomatosis
MGC	multinucleated giant cell
MH	maximum height
MHC	major histocompatibility complex
miRNA	microRNA
MISA	micropylar sperm attractant
MNT	micronucleus test
MOS	mannan-oligosaccharides
MT	metric tons
MUFA	monounsaturated fatty acids
NADPH	nicotinamide adenine dinucleotide phosphate
NCA	National Commission on Agriculture
NCDs	non-communicable diseases
NCE	new chemical entities
NF	chromosomal arm number
NFDB	National Fisheries Development Board
NGS	next-generation sequencing
NHEJ	non-homologous enjoining
NNV	nervous necrosis virus
NO	nitric oxide
NPN	non-protein nitrogen

NRC	National Research Council
NRCCWF	National Research Center on Coldwater Fisheries
NSP	non-starch polysaccharide
OAL	overall length
OP	organophosphorus
OTUs	operational taxonomic units
PA	protected area
PAHs	polyaromatic hydrocarbons
PAMPs	pathogen-associated molecular patterns
PCA	principal component analysis
PCR	polymerase chain reaction
PD	pancreatic disease
PFGE	pulsed feel gel electrophoresis
PIBO	pacfish in fish biological opinion
PKU	phenylketonuria
PTSD	post-traumatic stress disorder
PUFA	polyunsaturated fatty acid
QTLs	quantitative trait loci
QTNs	quantitative trait nucleotides
REP-PCR	repetitive extragenic palindromic-PCR
RNA	ribonucleic acid
RNAi	RNA interference
RNA-Seq	RNA-sequencing
ROS	reactive oxygen species
RSD	red spot disease
RTGE	rainbow trout gastroenteritis
RT-MPCR	reverse transcription-multiplex PCR
SBM	soybean meal
SCFA	short-chain fatty acids
SCI	stream condition inventory
SFA	saturated fatty acids
SFB	segmented filamentous bacteria
SHBG	sex hormone-binding globulins
SMIF	sperm motility- initiation factor
SPC	soybean protein concentrate
SPI	soy protein isolate
SSO	specific spoilage organisms
SVCV	spring viremia of carp virus
TALENs	transcription activator-like effector nucleases

TBHQ	tertiary butyl hydroxy quinone
TCR	T-cell receptor
TEC	thymic epithelial cells
TL	total length
TLR	toll-like receptors
TMAO	trimethylamine oxide
TN	total nitrogen
TP	total phosphorus
TSGs	tumor suppressor genes
TSV	Taura syndrome virus
UM	ulcerative mycosis
VHS	viral hemorrhagic septicemia
VHSV	viral hemorrhagic septicemia virus
VN	virus neutralization
WAFI	Williams absolute fecundity indices
WGD	whole-genome duplication
WHO	World Health Organization
WPF	week post-fertilization
WSSV	white spot syndrome virus
YHW	yellow head virus
YRD	Yangtze river delta
ZFNs	zinc finger nucleases

Introduction

Coldwater fisheries play a significant role in the world's freshwater fisheries. Since aquaculture productivity in coldwater regions is much lower than in other inland areas, there has been an exciting development of coldwater fisheries in recent years. Aquaculture, ornamental fish rearing, and eco-tourism all have possibilities in the coldwater sector. The coldwater environment is under serious stress from anthropogenic activities due to rising population growth, urbanization, and industrialization, which has resulted in changes in stream flow regimes, habitat loss, and species extinction. However, a combination of anthropogenic and natural (climate change) stressors is harming coldwater fishery resources, causing a drop in overall productivity.

This book focuses on various elements of coldwater fisheries, as well as technology created for the breeding, culture, and management of economically important species in order to increase production in a sustainable manner. This book covers the effects of anthropogenic activities on coldwater fishes and management solutions for aquatic ecosystems. All of these topics are updated, and new study findings are included.

The chapters in this volume will evaluate and highlight various key issues regarding the potential of coldwater aquaculture, fish species suitable for coldwater culture, the economics of coldwater, genomic perspectives and nutritional requirements of coldwater fishes, reproductive physiology and breeding biology of different coldwater species, coldwater fish disease management, climate change and anthropogenic pollution on the ecological health, genotoxicity in fishes, nutrigenomics, responsible approach towards coldwater fish stock, post-harvest management of coldwater fishes.

This compilation will be valuable for scientists, teachers, researchers, students, and policymakers when they look for a blue economy and blue revolution. It will be the first of its kind in the field of coldwater fisheries.

CHAPTER 1

The Rainbow Trout Genome: A Significant Milestone for Aquaculture Development

SYED UMAIR AHMAD[1], IMRAN ZAFAR[2], SHABANA BIBI[3,4], ZAINAB JAN[5], ANWAR ULLAH[6], IRFAN AHMAD[7], QURAT UL AIN[8], and MOHD. ASHRAF RATHER[7]

[1]*Department of Bioinformatics, Hazara University, Mansehra, Pakistan*

[2]*Department of Bioinformatics and Computational Biology, Virtual University of Pakistan, Pakistan*

[3]*Yunnan Herbal Laboratory, College of Ecology and Environmental Sciences, Yunnan University, Kunming, Yunnan, China*

[4]*International Joint Research Center for Sustainable Utilization of Cordyceps Bioresources in China and South-East Asia, Yunnan University, Kunming, Yunnan, China*

[5]*National Center for Bioinformatics, Qauid-i-Azam University, Islamabad, Pakistan*

[6]*The Second Affiliated Hospital, Institute of Cancer Stem Cell, Dalian Medical University, Dalian, People's Republic of China*

[7]*Division of Fish Genetics and Biotechnology, Faculty of Fisheries, Sher-e-Kashmir University of Agricultural Sciences and Technology, Jammu and Kashmir, India*

[8]*Department of Chemistry, Government College Women University Faisalabad, Pakistan*

ABSTRACT

Rainbow is a salmonid species that benefit aquaculture, wildlife fisheries, and commercial fish farming. Because of their economic worth, rainbow

Coldwater Fisheries and Aquaculture Management: Technology for Sustainable Food Production.
Mohd Ashraf Rather, Faisal Rashid Sofi, Adnan Amin, & Kawkabul Saba (Eds.)

trout are extensively used as model fish in various scientific fields. From an evolutionary genomic standpoint, the rainbow genome is truly intriguing because of a whole duplication event along salmon fishing lines 88–96 million years ago. Because of the development of next-generation sequencers, a wide variety of genomic technologies, including whole-genome sequencing, are now accessible in this species. Several instances of whole-genome duplication (WGD) have affected the history of vertebrate animals, and they are often related to adaptive radiation or developmental progress. Due to an additional cycle of WGD, the rainbow genome offers a chance to study the early evolutionary destiny of the same vertebrate genome, and researchers found that the subgenomes of the two forebears remained remarkably identical after 100 million years of evolution. MiRNAs, on the other hand, were essentially wholly duplicated copies of their genome. The completion of genome sequencing by researchers opens up new opportunities for aquaculture, which is a critical step toward addressing specific issues in trout farming, such as disease resistance and identifying specific strains with higher growth potential or better adapted to marine growing conditions. The presence of various viral diseases in the salmon and trout industry has highlighted the possibility of naturally resistant rainbow trout reared in seawater for the generation of smoked filets from larger fish weighing over 3 kg. This chapter presents a synoptic assessment of current knowledge on rainbow trout genomics and what molecular evidence has already been discovered due to functional genomics.

1.1 INTRODUCTION

Salmonids families are found worldwide, and several of these are essential for fish farming, wild population fishing, and the leisure sportfishing industry (Clavelle, Lester, Gentry, & Froehlich, 2019). In 2018, overall, aquatic output purse to reach over 3 million metric tons (MT), with such a worldwide economic interest of above US $1,800,000,000,000 including Wild salmon, seemed to be the most significant managed salmonid species, contributing for 3,000,000 MT (Cabral et al., 2020). Rainbow is native to the western, northern United States, but it is also one of the most imported fish. According to Fish-Base (Akhtar, Ghazanfar, & Shafi, 2021), rainbows are currently found in at least 70 different countries, with a population spanning almost all of the world's largest hemispheres. Although it has a lower prominent place than wild salmon aquaculture, the productivity levels for trout are pretty good. World total production peaked at 5 million MT for perhaps the first time in 2001 and had been over 9,00,000 MT in 2015 with either total export revenues of about $3.5 billion (Adusei, 2015) (Figure 1.1).

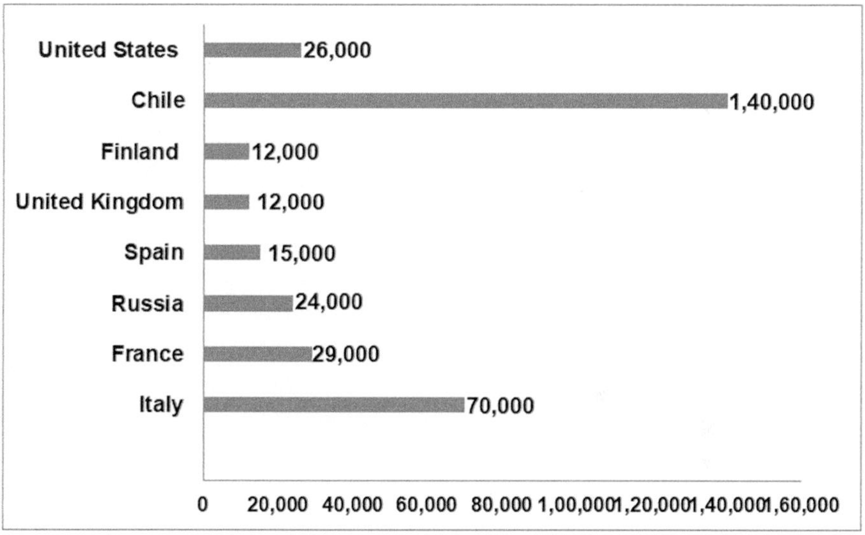

FIGURE 1.1 Rainbow trout production across different countries (Adusei, 2015).

The chinook salmon is a member of the salmonid (Salmonidae) family, which includes three subfamilies: the Salmoninae (primarily trouts, chars, and sockeye), the Thymallinae (graylings), and the Coregoninae (salmons) (whitefish and ciscos). Salmonids separated from its closest sister group, Esocidae (pikes and mud minnows), circa 110–150 Mya, while the last common ancestor of contemporary salmonids is estimated to get along with 58–63 Mya (Bobe et al., 2016). Salmonid complete genome duplication (Ss4R) is thought to have occurred between 88 and 96 million years ago (Bobe et al., 2016; Nugent, 2019).

More than 20 other European countries share the remainder of manufacturing (Chavanne et al., 2016). The bulk of America's production takes place in Chile. In contrast, while minor, development in the USA was worth $80 million in yearly sales in 2014, without including the refurbishment, management, and leisure industries.

1.2 RAINBOW TROUT: A MODEL ORGANISM FOR LABORATORY EXPERIMENTS

Apart from its commercial significance, the rainbow has often been used as a conceptual framework in a variety of domains such as disease studies, toxicological, comparative immunotherapeutic, epidemiological, physiological, nutritional, including fertility and adaptation research. Rainbow is one of the most studied aquatic organisms, with various practical advantages such as the availability of different natural populations and monoclonal lines, the simplicity through which breeding can be achieved and the ability to perform exceptionally accurate gene ablation (Rajabi, Ramazani, Hamidi, & Naji, 2015). Rainbow trout's relatively sizeable physical growth compared to mimic fish such as zebrafish or service fish makes it an excellent surrogate model for conducting biochemical and genetic studies of specific cellular components, which are much more challenging to assess in native fish models (Lutfi, 2017). Various cytogenetic materials have been adopted during the last two decades to facilitate its utility in scientific research and aquaculture, including morphological and molecular mappings, BAC libraries and BAC end sequences, as well as several simple sequence markers. When the very first Rainbow trout entire genetic sequencing effort began in 2010, the majority of the required genetic materials for the which was before stage were accessible (Schill, Heindel, Campbell, Meyer, & Mamer, 2016).

1.2.1 RAINBOW TROUT LIFE CYCLE

Rainbow trout is a subspecies of North USA's Northwest coast, with a geographic distribution that extends from Alaska to the area to the north (Taylor & May-McNally, 2015). Although steelhead trout species may be found in the ocean and eventually return to rivers and lakes to spawn, this is mostly a freshwater species found in streams and ponds. Freshwater populations are often found with wells, lakes and rivers with referendum temps ranging from 10 to 16°C (Bobe et al., 2016; Boulajfene et al., 2019) (Figure 1.2).

They reproduce fast throughout the spring, with spawning occurring in cold, well-oxygenated, shallow rivers with good gravel bottoms (Stoessel, Raadik, & Ayres, 2015). A single adult rainbow trout can lay 1–8,000 eggs (3.5–6 mm) in one or more batches, which are promptly fertilized by sperm from either a single male or a pair of males. Eggs usually fall amid rocks and fissures and hatch 4–5 weeks after implantation. During vitellus breakdown, embryonic fish emerge from either the sandy bottom or the muddy substrate and feed on zoobenthos and phytoplankton prey. Rainbow trout seems to be an adventurous eater in its mature form, dependent on various food sources, from microscopic insects to crayfish (Lopez-Santamarina et al., 2020).

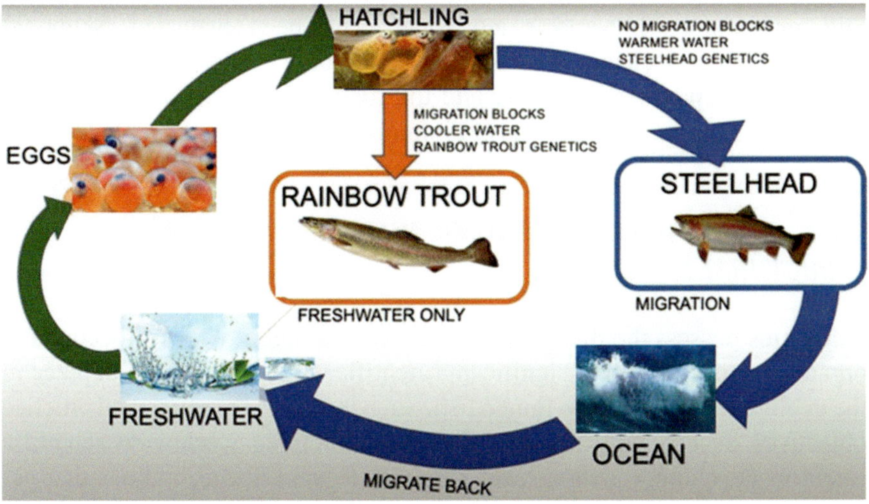

FIGURE 1.2 Trout life cycle.

1.2.2 *GENETIC ARCHITECTURE AND GENOMIC ASPECTS OF RAINBOW TROUT*

From the standpoint of adaptive genetics, the freshwater fish DNA (deoxyribonucleic acid) is particularly intriguing (Anastasiadi, Díaz, & Piferrer, 2017). The Rainbow trout genome is of significant size, 2.4 109 bp, six times larger than just the buffer genome or even more effective than the goldfish's DNA, 1.7 109 bp. Second, the freshwater fish genome is complex, with a chromosomal arm number (NF) of 104 and a variable double set of chromosomes ranging from 57 to 63 due to Robertsonian reconfigurations. An important cause of this complication is a tetraploidization event (whole genuine duplication, WGD) that occurred in the rainbow line (Ss4R: Salmonfish-specific fourth round of WGD). Fish genomics is therefore still believed to undergo rediploidization, which involves evolutionary algorithms or reduction of gene duplication, a process known as "widely differing settlement," which claimed to reduce hybrid conditions and encourage species formation. Such a phenomenon may well have played a significant role in the astonishing diversity of species found in fish (Salzburger, 2018).

Moreover, differential settlement in marine invertebrates has been described following an ancient genomic multiplication (triple WGD or Teleost-specific third round after WGD = Ts3R) between 225 and 330 million years ago (Mya). Still, only extremely variable individuals have been evaluated. Salmonid genomics gives a chance to study the evolutionary algorithms of recently divergent tetraploidized genomic and the further development of replicated genomes during the initial stages of diploidization (Brunet et al., 2017).

1.2.3 RAINBOW TROUT GENOMIC DIFFERENTIATION AND STRUCTURAL DIVERSITY

Salmonids possess the fundamental features of a highly vital socioeconomic organism and a species of significant scientific value, with a critical place in the evolutionary tree of ray-finned fish. Such integrated research and practice objectives have greatly influenced a well-structured worldwide scientific world in developing a significant number of genomics techniques that are now accessible in so many rainbow trout species (Krajcik & Czerniak, 2018).

1.2.4 RAINBOW TROUT GENOMIC SEQUENCING REVEALS NEW INSIGHT ON VERTEBRATE EVOLUTION AFTER WHOLE-GENOME DUPLICATION (WGD)

Whole-genome duplications (WGDs) are uncommon but spectacular occurrences that result in abrupt duplication integrity of the entire genomic sequences (Pophaly, 2017). Although WGDs are uncommon among mammal species, these have had a significant impact on vertebrates evolution and serve as critical evolutionary milestones from which certain main bloodlines have diverged. For example, the ancient genomic of all teleost's experienced a WGD, known as the effectiveness 3rd WGD (Ts3R), which occurred between 225- and 332-million-year-old (Mya) (Lien et al., 2016). This Ts3R signature is still relevant to contemporary vertebrates' genomics, and it was accompanied by two prior WGD events shared by all skeletal vertebrates. Throughout WGD, the resultant replicated genomics maintain just a minuscule fraction of replicated DNA, whereas superfluous duplicates are immobilized through a mechanism known as gene fractionated (Stephens, 2019). Too far, the form of genetic separation in vertebrates has received little attention since all well-characterized WGDs are exceedingly old, and gene fractionation is assumed to be finished in such species. Salmonids are of particular relevance in this context since gene fractionation may still be occurring in their lineage as a result of an extra as well as fairly new WGD occurrence (the salmonid-specific 4th WGD or Ss4R) dated 25 to 100 Mya (Minkley, 2018). Rainbow trout (Oncorhynchus mykiss) is a salmonid that is of tremendous environmental relevance across the globe. It is among the most researched fish species, with substantial research in various categories including malignancy, toxicity, immunological, ecological, physiological, and nourishment (Mamdouh, Mohamed, Mohamed, & Fahmy, 2022). This is also a significant aquatic organism with massive financial relevance that is cultivated for both climates and on all geographies.

Because of its new WGD, the rainbow trout offers a different chance to comprehend the early stages of genome fractionated (Symonová & Howell, 2018). Our findings, followed by an analysis of the rainbow fish pretty much the entire sequencing, demonstrate that the 2 ancient subgenomes had maintained very coincident throughout 1 billion years of human history. Approximately 50% of the nutrient genomes have been preserved as duplicated copies, with the remainder lost primarily via pseudogenization (Rendón-Anaya et al., 2019). In stark contrast to nutrient sequences, miRNA genetic mutations have virtually all been preserved as doubled versions. Our findings point to a long and incremental rediploidization mechanism, which calls into question the current idea that WGD is accompanied by huge and fast chromosomal restructurings and genetic losses (Pérez-Sánchez et al., 2019).

1.2.5 PREEXISTING GENOMIC RESOURCES AND FIRST STEP TO RAINBOW TROUT GENOMIC SEQUENCING

The freshwater fish genomic is approximately 2.4 109 kb in length, with a G + C composition of 42% (Nguinkal et al., 2019). One of the issues that hampered the early scanning and formatting of the goldfish genome was the degree of heterogeneity among the members of the overall population used to contribute DNA to such research. Essential selection criteria and androgens were employed to successfully create double homozygous adults in the rainbow, which was used to establish homogeneous cloned lines that have been carefully documented at genomic and phenotypic levels. Completely homogeneous samples were required to develop rainbow genomic approaches to cope with both the considerable genomic variation predicted due to the initial WGD of such an organism (Du et al., 2020). The Nucleotide Sequence from such a unique twice haplotype trout (Swanson

YY doubly haplotype clonal line) utilize to create numerous rainbow trout BAC collections. Several BAC libraries use to make the first generations, and 2nd gen BAC was physically mapping (Książkiewicz et al., 2015). Multiple lake trout frameworks that rely on doubly haploid adults and outbreeding communities have been created utilizing microsatellite and SNP markers. A synthesized connection map was created by combining earlier ones, and it had 2,226 characteristics at a moderate concentration of 1.0 marker/cM.

In the rainbow trout, cDNA pools from several tissue and morphogenesis have been generated and processed simultaneously. As of June 2015, 2,90,406 extended performance tasks (ESTs) sequencing were publicly accessible in the GenBank database. These ESTs served as the foundation for various rainbow trout cDNA, and research has focused on hybridization systems (Cassé, Richetin, & Toni, 2018). Segments through both endpoints of 96,000 fish BACs also were accessible, and all these BAC-end sequencings use to generate the first trout repeat database, demonstrating that repeated DNA accounts for almost 58% of the genomes. These BAC-end transcripts were used to create either the first, second, or third-generation consolidated maps after testing microsatellites (Clouse et al., 2016).

1.2.6 NGS SEQUENCING TECHNOLOGIES AND RAINBOW TROUT GENOME SEQUENCING

The rainbow trout genetic sequencing technique was designed using NGS sequence alignment capabilities accessible in 2010 at the start of this genome research (Juanchich et al., 2016). This genomic sequence was assembled using a whole-genome scattergun approach utilizing genetic material from a single duplicated haplotype (YY) freshwater fish male. The accessibility of DNA from such an utterly homogeneous organism was a critical resource that greatly aided in constructing the freshwater fish genomic (Abdelrahman et al., 2017). For the entire acoustic technique, solitary reading collections and 8 kb, 12 kb, and 20 kb buddy pools were processed using 454 Titanium technique up to something like a 20-fold sequence coverage and a 70-fold coverage utilizing Solexa-Illumina sequences (Yang & Sun, 2017). Solexa-Illumina segments operate to fix the features present and scaffolding segments that resulted from assembling the 454 Titanium reads, which are recognized as a blunder, particularly in copolymer regions. This method was supplemented by extensive coverage of BAC-end genomes, which aided in scaffolding contigs and anchoring the genome onto combined maps (Argyris et al., 2015). The entire size of this initial iteration of the freshwater fish genomic sequences was 1.9 Gb, with a framework N50 of 384 kb (half of the assembly is contained in 1,014 scaffolds longer than 384 kb). Scaffolds were fixed onto chromosomal at 898 different loci utilizing correlation and topological mapping knowledge. The genetic sequence was annotated, and 46,585 nutrient genetic categories were discovered (Salem et al., 2015), corroborating protein data from some other mammals and transcriptome information from 15 organs of a doubling haplotype lake trout adult collected by RNA-seq. Compared to other vertebrates, this large number of projected nutrient genes is consistent with the current Ss4R. 495 microRNA (miRNA) loci belonging to 84 distinct families and 164 matured sequencing were also found. Transposable elements make up roughly 38% of the rainbow trout's entire genome (Rodriguez & Arkhipova, 2018).

1.2.7 RAINBOW TROUT STRUCTURAL GENOMICS INFORMATION

Approximately 300 million years ago, the teleosts' primordial genomic sequence experienced a WGD entity known as the teleosts-specific third WGD (Ts3R) (Mya) (Bobe et al., 2016). This Ts3R characteristic appears to play a role in the contemporary teleost genomes sequence. Two other ancient WGD processes that all musculoskeletal animals inherit (vertebrate genome duplications 1 and 2, VGD1 and VGD2) (Lorin, Brunet, Laudet, & Volff, 2018). While WGD occurrences are uncommon amongst animal races, they are major technological milestones from which several major lineages have diverged. Multiple identical genomics ultimately maintains just a miniscule fraction of replicated genomes, while apparently superfluous versions are inhibited in a mechanism known as gene separation, which is unexplained. Rainbow trout genetic structure is especially interesting in this context since salmonids have undergone an additional relatively fresh WGD occurrence (the salmonid-specific fourth WGD or Ss4R), which has been firmly predicted to occur between 25 and 100 Mya (Colgan et al., 2021). As a result, genetic separation in that species may still be happening, providing a novel paradigm for understanding the underlying factors controlling this rediploidization process.

The lake trout genomes permitted the restoration of the primordial karyotype of salmonids before the Ss4R multiplication by rebuilding the Ss4R alternatively spliced areas (Bobe et al., 2016; Shavalier, 2017). The current lake trout genomic is arranged in 38 prominent pairings of replicated sections, as well as 14 of the 30 chromosomes come out from the merger of different separate post-Ss4R chromosomes, which is consistent paralogies deduced from trout association studies (Du et al., 2020; Lin, 2019). Other genomes include more complicated mosaics of distinct post-Ss4R chromosomes, indicating additional cross-functional and cross translocations since Ss4R event. In these Ss4R alternatively spliced areas, several ohnologs (paralogous genes produced by a WGD process) were found (Chen et al., 2019). The estimated dispersion of modest nucleic acid replacements (dS) throughout several sets of Ss4R ohnologs found in freshwater fish, as well as between pairs of orthologs found in Atlantic salmon and lake trout (Carruthers et al., 2018) (representing the species deviation time, i.e., roughly 30 Mya, between most of the two species) allowed us to calculate the timeframe of the Ss4R at 96 Mya, in the top perimeter of such 25–100 Mya from prior response but in the above-mentioned 25–100 Mya This contradicts the estimated history of the Salmonidae family of 50–60 Mya (Osinov, Volkov, & Mugue, 2021), indicating that the Ss4R evolved far earlier (> 30 Mya) than the earliest single ancestor of modern salmonids. This is consistent with the WGD Radiological Lag-Time Model, which proposes considerable lagtimes between WGDs and the consequent adaptive radionuclides, which are also generally linked with WGD incidence (Tank et al., 2015).

The genomic expansion includes the loss of one genetic duplicate among most homologous genes via gene fractionation after just a few million years following a WGD (Inoue, Sato, Sinclair, Tsukamoto, & Nishida, 2015). Even though all WGD investigations seem too ancient to acquire such data, this process has never been detected in any vertebrate throughout the whole genomic scale. Because the Ss4R is younger, alternatively spliced sites in the rainbow trout genome can be identified (Sadd et al., 2015). Approximately half of the Ss4R double identical domains have undergone gene partitioning and returned to a diploid pattern over the whole genome, while the remainder have preserved both ohnologs. Gene dispersion is a fairly slow approach in fish genomics since genomes were inactivated at a rate of roughly 170 genomes every million years.

Furthermore, even though most singletons are often dysfunctional, the majority of singletons may still be connected with evident paralogous sequences emanating from the Ss4R (pseudogenes). In contrast to the 50% retention of homologous histone genes in employees, we observed that post-Ss4R retention of genes encoding miRNAs is total (Taylor, Daniels, Morata, Gundappa, & Macqueen, 2022), with nearly all miRNA ohnologs maintained as repeat Ss4R copies. This higher conservation appears to be attributable to more subtle selection processes rather than shorter miRNA-coding sequences. Overall, the greater consumer retention frequency of homologous duplicates and pseudogenes demonstrates that the fractionation process in trout is still mostly finished and ongoing (Lu & Luo, 2020). The lake trout genomes' phylogenetic tree, even during the Ss4R WGD event, is depicted schematically. Since the Ss4R WGD resulted in a complete duplication of the genetic material, each genetic Code (and genes associated with its regulatory domains) were available in multiple genetically identical instantly after the Ss4R. Following the Ss4R WGD, chromosomal fusions and specialized framework rearrangements decreased the total number of chromosomes to 30, which was similar to the preduplication karyotype (Murat, Armero, Pont, Klopp, & Salse, 2017). The reduction of replicated genes following WGD by alterations and omissions to revert to a predominantly diploid form, known as rediploidization or gene fractionated, is an evolving issue in the lake trout genomic sequence (Perreault-Payette et al., 2017). At the time, the duplicated sections of the gene were still primarily collinear, with around half of the repeated homologous genes retaining dual ohnologs. Ohnologs that are still present in both copies are constantly diverging and may take on new or auxiliary tasks (neo- or sub-functionalization) or be deleted due to gene fractionation. Prior WGD occurrences predicted that only around 20–25% of duplicated genes would be preserved in homologous pairs (Li et al., 2015). There are 36 genomes in the aquatic biological cultured. The genome organization level of research of the Ss4R duplicated regions revealed a good collinearity among haplotypes exome sequencing, consistent with a conserved order of ohnologs, and therefore no compelling evidence of a concentration of single parents versus ohnologs throughout the genomes. These findings imply that gene fractionation will exclude severe genomic structural changes such as inversions or chromosomal rearrangements, which might disrupt the order of genes in the genome, as well as large deletions, which will also contribute to the lengthy cluster of singletons. Ss4R ohnologous protein coding domains and mRNAs are also evolutionarily related, with 92.9% amino-acid and 96.4% nucleotide similarity, respectively. Furthermore, the similarity between the Ss4R protein-coding singletons and their associated nucleotide sequences is quite great (average amino acid (AA) identity 79.0%), indicating that the majority of these genetic illnesses and disorders occurred (Lien et al., 2016).

1.3 THE RAINBOW TROUT GENOME–AN IMPORTANT LANDMARK FOR AQUACULTURE

The rainbow trout genome's first iteration unified maps are made up of 238 BAC contigs joined to the chromosomal map areas (Wiernasz et al., 2021).

It covers more than 10% of the genome and contains sequences from all 29 chromosomes. This map is a starting point towards a more detailed composite genomic map. Extensive comparative genome analysis, precise mapping of QTL, positional cloning, selection of positioned genetic markers for economically significant characteristics, and incorporation of MAS into rainbow trout breeding strategies will all be possible with the availability of an integrative individual genetic map. A fully integrated map might provide a minimal tiling strategy for genetic research and a basis for whole-genome assembly (Chapman et al., 2015). The phylogenetic tree of the rainbow trout genome is practical implication during Ss4R WGD phase for Ss4R WGD results in a complete doubling of the genetic Code, providing access to all chromatids in a large number of genetically related individual people especially following the Ss4R. After Ss4R WGD, chromosomal conglomerations and specialized framework reworkings decreased the total number of chromosomal instabilities to 30, which was similar to the preduplication karyotype. Rediploidization or gene stratification is an ongoing process in the rainbow trout genome that involves the loss of duplicated genes following WGD through alterations and replacements to revert to a predominantly diploid state (Waples, 2015). The repeated parts of the genome are still substantially collinear at present, with around half of the duplicate homologous genes maintaining both ohnologs. Ohnologs that are still extant, including both copies, slowly diverge at the sequence level. They may ultimately assume new or auxiliary roles (neo- or sub-functionalization), or they may be lost due to gene fractionation (Lamb, 2021). Based on previous WGD cases, only around 20–25% of duplicated genes are expected to be kept in multiple copies. As the recent update of NCBI current gene set of rainbow trout is given in Figure 1.3.

1.3.1 IMPORTANCE OF GENOME FOR GENETIC AND FUNCTIONAL INVESTIGATIONS

Since the rainbow trout exome sequencing is finalized (Lyons, Turnbull, Dawson, & Crumlish, 2017), the key difficulty is determining how to interpret the information encoded in the DNA sequence. Throughout several genome-wide investigations, the task of determining the expression of certain genes, regulatory genes, and their interactions remain unsolved. Because alterations in the entire genes have a high probability of causing clinical diseases, functionality evaluation is crucial for breeding and aquaculture development and production (Palti et al., 2015). For many years, a range of methods and tools have been employed in functional genomic analysis. Meanwhile, quick transformative innovation and improvement in high-throughput methods, extending from classic real-time PCR to far more complex processes like as next-generation sequencing or NMR spectroscopy, has occurred just in the last decade (Walter et al., 2020). Furthermore, proper bioinformatic research is crucial for credible scientific outcomes, in addition to experimental inquiry. These approaches provide accurate and complete holistic assessment across several areas of study, including genomes, metagenomics, proteome, and interact omics (Pinu et al., 2019). This is critical for addressing information gap in dynamic cellular mechanisms at both the cellular and organismal levels. Furthermore, every approach has benefits and limits that should be considered before identifying the appropriate technique for certain particular research in assuring an effective analysis.

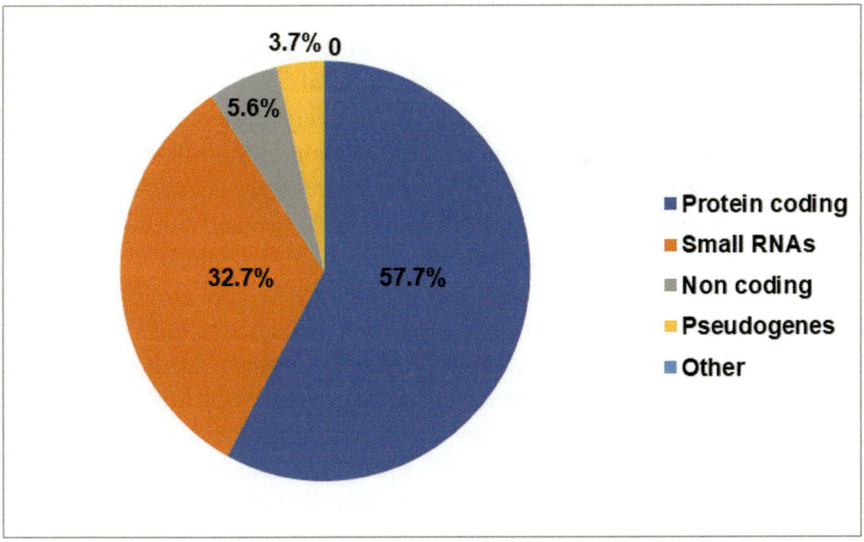

FIGURE 1.3 Current gene set distribution of rainbow trout genome (NCBI, 2022).

1.3.2 DECIPHERING PROCESS AND THE GENETIC ARCHITECTURE OF RAINBOW TROUT TRAITS

Quantitative trait loci (QTLs) are genomic areas (loci) that are linked to morphological characteristics within a trait (Khedikar et al., 2018). A QTL is typically associated with or contains the gene(s) influencing the target characteristic. As a result, using polymorphism connected with QTL regions to discover the main variables that influence the unpredictability of the target trait is a strategy to improve breeding effectiveness (dos Santos et al., 2020). Determining QTLs is also a step toward the molecular deconstruction of many characteristics, as well as the finding of genes that govern processes of interest (causative genes) and their regulation (Price et al., 2018). Many QTL for a range of phenotypes in rainbow trout were discovered using generally known intermediate resolution genomic maps and linkage approaches (LA) in family QTL designs (Salem et al., 2018), emphasizing adaptation and pest resistance traits.

Furthermore, the likelihood of QTL placements is typically high, making their utilization problematic (Fraslin et al., 2020). With the fast growth of NGS technology, high-density genotyping approaches are now accessible. For example, limitation linked DNA tagging (RAD-tags) sequencing allows for the low-cost detection of hundreds of SNP (Feng et al., 2018).

Linkage maps based on RAD tags were created in rainbow trout, leading to enhanced marker density (Zhang et al., 2018). RAD-sequencing was also employed in a prolonged SNP discovery project, which revealed almost 1,45,000 SNPs, permitting the construction of high-density genotype sequencing chips such as the 57 K SNP chip recently created for rainbow trout and now commercially accessible. In such investigations, possessing a benchmark genomic information as a model helps organize newly found SNPs into maps or create local haplotypes (Juanchich et al., 2016). These improved genotyping technologies enable a drastic change in QTL mapping methodologies, allowing the use of linkage disequilibrium (LD)-based methods and population-level association analysis and genome wide association study (Hérault, Damon, Cherel, & Le Roy, 2018). This offers new possibilities for understanding complicated features, identifying genomic regions beneath selection, and implementing genomic selection in aquaculture production in the hereafter (Casu et al., 2022).

1.4 FUNCTIONAL CHARACTERIZATION AND INTEGRATIVE FUNCTIONAL ANALYZES IN RAINBOW TROUT

Nutrition Rainbow trout aquafeeds have been created during the last three decades based on the species' well-known nutritional requirements (Hardy, Kaushik, Mai, & Bai, 2022). New aquafeeds must be created to replace nutritious protein sources and fish oil (sourced from fragile ocean biodiversity) with phytoconstituents or alternative materials to achieve sustainable aquaculture (Turchini, Trushenski, & Glencross, 2019). This shift in dietary composition necessitates a full understanding of intermediary metabolic pathways, mainly based on molecular markers. The molecular LC-PUFA synthesis route is functional (Pewan et al., 2020). Considering the restriction of fish oils in aquaculture diets, one of the most pressing challenges is maintaining the highest levels of long-chain polyunsaturated fatty acids (PUFAs) in meat (Barta, Coman, & Vodnar, 2021; Roy et al., 2020). Several studies on fish have revealed that this organism contains the structural characteristics to produce these LC-PUFAs from plant oil precursors (Bermúdez et al., 2021). When fish oil is eliminated from the diet, mRNA encoding system enzymes and elongase increase; this incentive is insufficient to achieve the same levels of LC-PUFA in muscle as fish given fish oil. LC-PUFA biosynthesis in fish has been demonstrated to be substrate-restricted, despite some fish lines having more significant quantities of fatty biosynthetic molecular capacity (Ampong, 2020).

Rainbow trout are recognized to be weak dietary carbohydrate users, limiting the integration of starch in aquaculture feeds (Liu et al., 2020; Song et al., 2018). Recent research (Steinberg, 2022) indicates that trout fish has the transcriptional ability to catabolize dietary glucose (Cadonic, 2019). In contrast, the molecular control of hepatic glucose synthesis appears to be unusual, which may be one of the causes for trout's poor metabolic utilization of carbohydrates. A reduction in feed intake linked with a lower plasmatic insulin/glucagon ratio (primarily due to the pursuit of hepatic gluconeogenesis) is frequently reported in fish, followed by the consumption of novel aquaculture feeds, potentially having a significant influence on feed conversion ratio (FCR). Recent research (Karras, Koufakis, Mustafa, & Kotsa, 2019) indicates that glucose and triacylglycerol sensing mechanisms in the hypothalamus and endocrine pancreas are present in nutritional regulatory frameworks of nutrient-sensing routes in these organs. In rainbow trout served alternate diets, fine nutritional modulation of these pathways occurs at the molecular level (Rimoldi, Antonini, Gasco, Moroni, & Terova, 2021).

There has been a decrease in nutritional digestibility and a decrease in the plasmatic leptin ratio in fish, followed by the consumption of novel aquaculture species, which may significantly influence digestibility (Xu et al., 2021; Ferosekhan et al., 2014). Recent research (Sanjana Sharma, Aroura, Gupta, & Priyadarshini, 2022) shows that in rainbow trout served alternate diets, fine nutritional modulation of these pathways occurs at the molecular level. Some transcriptomics studies were followed with proteolytic analysis, which appeared to support the transcript assessment but was limited by the inability to detect polypeptides in rainbow trout due to technical limitations (Beemelmanns et al., 2021; Sobhkhez, Krasnov, & Robertsen, 2018).

1.5 NUTRITIONAL MODULATION OF THE ACTIVATION OF HEPATIC GLUCOSE-6-PHOSPHATASE GENES IN RAINBOW TROUT: HOW TROUT GENOTYPING BRINGS CREATIVE APPROACHES TO EXISTING CONCERNS?

The rainbow trout is considered glucose-intolerant due to sustained hyperglycemia after a carbohydrate-rich meal or hypoglycemia tolerance tests (Panserat, Marandel, Seiliez, & Skiba-Cassy, 2019). One explanation for such a phenotypic is a lack of restriction of endogenous insulin synthesis via the glycogen synthesis pathway (Chen et al., 2021). Hyperglycemia has already been postulated to be a major player in glycemic richness at high blood sugar levels or after caloric intake by attempting to establish a meaningless glucose/G6P cycle with hexokinase, the first glycogen synthesis protease that initiates the transcriptional activity of glucose in G6P (Cho, Kim, Mansfield, & Chou, 2018; Wang & Dong, 2019). In 2000, a gene encoding for G6pc in fish was sequenced mainly for the first time (Juanchich et al., 2016). Its expression's nutritional/hormonal control was subsequently thoroughly examined using northern blot analysis, gradually supplanted by modern real-time quantitative PCR research. Therefore, the amount of g6pc mRNA was demonstrated to be slightly higher compared to high dietary fats but not, interestingly (Rabot et al., 2010), by high carbohydrate food consumption comprising digestible starch. *In vivo* and *in vitro* studies further revealed that g6pc mRNA levels were favorably linked with protein and AA pool levels (Mullarky & Cantley, 2015). Furthermore, particular AA like glutamine, one gluconeogenic discardable AA (G-DAA), or lysine and methionine were implicated in the g6pc gene (Rolland et al., 2016). Finally, insulin regulation of this genotype was investigated and found to have a significant effect on g6pc mRNA levels *in vitro* and *in vivo* studies and the ability to counteract the increase in mRNA levels caused by glycemic supplementation in tissue culture fluids or high-diet fish carbohydrate (Balasubramanian et al., 2007).

Nonetheless, when compared to enzyme activity data, several of these overexpression results were questioned. Insulin, for example, appeared to work at the biological level to reduce g6pc mRNA levels but had no influence on G6pc hepatic or intestinal function (Hong et al., 2014). Similarly, only one G-DAA (alanine) reduced g6pc mRNA levels, while enzyme activity was decreased in trout fed three G-DAA swapped meals (alanine or aspartic acid or glutamic acid).

Furthermore, until finding a second specific gene for G6pc in EST databases in 2014 (Hong et al., 2014), no convincing justification could be an offer to justify why G6pc activity did not alter when a single serving with glucose suppressed g6pc. Indeed, the study of this novel gene provided additional information to interpret the results reported above, suggesting that the rainbow trout genome's complexity accounts for a better understanding of dietary control. Data revealed that not only were both g6pc genes differently regulated by nutritional status, but they also demonstrated divergent regulation in terms of the relative percentage of carbohydrates in the diet (Panserat et al., 2019). The current editing of the lake trout genome advanced molecular understanding of g6pc/G6pc dietary control to the next level (Juanchich et al., 2016). In fact, in silico studies indicated that five orthologous genes encoding for G6pc were preserved in the lake trout genome following Ss4R. Two of them, previously unknown, demonstrated an unanticipated up-regulation of respective mRNA in fish given a high carbohydrate diet. Overall enzyme activity in the livers of these fish revealed that proteins transcribed in the latter two genes were engaged in this activity (Morro et al., 2019). It was postulated that these two genes, together with glucokinase, would participate in the reactive hypoglycemia phenotypic by generating glucose in the blood via creating a futile cycle. Previous findings (Juanchich et al., 2016), relating the regulation of g6pc genes by insulin or G-DAA can now be reinterpreted in light of the current discovery (Carr, Turner, & Pirmohamed, 2021), taking into account that g6pc combining information is possibly variably governed but that they all contribute to total enzyme activity.

The instance of g6pc loci exemplifies how genomic breakthroughs might benefit our understanding of dietary control of metabolism and bring new solutions to old concerns (Dowaidar, 2021). However, as is frequently the case, closing a window frequently opens a door (Kiepas, 2021). There are numerous concerns concerning the significance of divergent expression of g6pc orthologous genes and their associated contributing to one or another more phenotypes arise.

1.6 RAINBOW TROUT GENOMICS FOR HEALTH

The Rainbow trout genomics mapped, "Immunomet," a nearly comprehensive repertoire of genes involved in the immune system, became available. In both normal and pathological settings, large amounts of data on genes implicated in human and mouse immunity were already available at the biochemical, structural, cellular, and molecular levels; this data merged into specialized immunogen and pathway databases and tools. When the immunome of other vertebrate animals are transliterated, then three main methods apply to identify it: a functional knowledge of immunogens from other species, such as humans or mice, can be attached to their orthologs since phylogenetic haplotypes in distinct species often have similar roles, and datasets from high-throughput approaches for gene expression under infectious and pathological conditions can recognize the genes affected in diseases and during responses. Finally, the newly accessible genome use to direct genomic and operational analysis of genes or groups of genes. Immunological annotation of freshwater fish is a difficult task that cannot achieve simply by transferring knowledge from animal models and human databases. The WGD occurred during salmonid growth, resulting in many ornithological genes that are susceptible to subfunctionalization. Some gene families, such as chemokines and adornments, have undergone significant specialized expansion in salmonids. Accordingly, a well-composed genome with tightly linked doublets should be an essential resource for defining the immunome, especially from high-throughput transcripts. A sophisticated trout immune system will serve as a solid foundation for future research, such as studying genetic variance and epigenetic techniques, which will be helpful for marker-based selection systems.

1.6.1 THE CASE OF ANTIGEN-SPECIFIC RECEPTORS, IGS, AND TCRS

As modern sequencing methods enabled researchers to access a complete variety of immunoglobulin (Ig), or T-cell receptor (TCR) patterns expressed in a tissue or even a whole individual—research on immune repertoires has grown dramatically in the last five years. These studies give a thorough overview of innate and adaptive immunity against infections and examine the immune system's condition in various illnesses (Liu et al., 2021). Based on Ig and TCR sequences accessible in these species, such techniques develop in freshwater fish and zebrafish. The availability of a whole trout genome sequence, as well as the opportunity for extensive tagging of TCR and Ig loci in this organism, will offer researchers a much-enhanced mechanism for accessing the full range of immunological repertoires and responses. The rainbow trout genome provides the initial step—and a vital resource —in developing such techniques, which will help evaluate future vaccines (Adams, 2019).

1.6.2 A GENOME ASSEMBLY ENGAGES IN LINKAGE STUDIES, INCLUDING GENE CLUSTERS AND LINKAGE RESEARCH

The prevalence and functional significance of immune-related gene clusters are still unknown. Several selection forces involving regional gene amplification and co-regulation of expressions can represent such clusters (Lal, 2017). The trout genome offers an exciting opportunity to investigate such problems, one of which is the development of the large histocompatibility complex (MHC). TCR identifies antigens as short peptides (9–22 AAs) linked to membrane proteins known as MHC class 1 or class 2 molecules, which are produced on the surface by "antigen-presenting cells (Juanchich et al., 2016). MHC restriction occurs when TCR identifies both the MHC polypeptide and the polypeptide antigen provided by the MHC molecule (Collins & Riddle, 2008). The genetic "MHC" is a section of the genome that is determined by the presence of MHC class 1 and 2 molecules and genes involved in peptide antigen production and charge for MHC presentation. In practice, the MHC encodes many genes, about half of which are engaged in immunity; hence, the genes in this area contribute significantly to immunome and defense mechanisms (Torres et al., 2016). The origin and evolutionary process of MHC are still unknown. MHC is found in all jaw vertebrates, including sharks and mammals. According to Ohno's theory, two rounds of WGD happened after the development of urochordates and even before the radiation of jawed vertebrates, resulting in four sets of paralogic zones, as previously described. Several critical genomic regions, including the HOX and MHC complexes, have been discovered to have such tetrads (Cañestro, 2012). Two additional tetrads have been discovered in vertebrate genomes, which appear to be derived from outlying areas on another ancestor's genetic material containing "proto-MHC" from great grandparents of vertebrate species. These MHC-related tetrads contain a gene engaged in vertebrate immunogenicity and several humongous epigenetic frameworks like the innate immunological component and the intricate granulocyte target proteins and many of the body's preventative gene families like B7 receptors, TRIMs, and so on (Suurväli et al., 2014).

In addition to fish, where the regions expressing MHC class 1 and MHC class 2 locate on distinct chromosomes, MHC has been preserved as a defined genetic unit by vertebrates from sharks to mammals (Kulski, Shiina, Anzai, Kohara, & Inoko, 2002). The functional relevance of this conservation and the distinctive pattern identified in fish are still unknown. Co-regulation of many immune-related genes is an appealing concept, but it has gathered little direct evidence and contradicts the pattern reported in fish. Fish are helpful models for dealing with such problems and learning about the immunological consequences of a "broken" MHC. StarCraft and pipe fish are excellent examples of fish groups that have lost MHC class II genes and classic T-helper responses (Segner, Verburg-Van Kemenade, & Chadzinska, 2017). With additional cycles of genetic code multiple copies and continued rediploidization, the salmonids represent another context of great interest in confronting and fully understanding the fate of the MHC region and its paralogs or common carriers. Rainbow DNA is a valuable resource for fish biologists and comparable immunologists (Xu et al., 2017). As a basis for defining a comprehensive immune system, it is a critical resource for future innovations that require monitoring of immune responses, including identifying more resistant animals. Given the DNA from the strongly linked Atlantic salmon, it also provides an opportunity to answer general and evolutionary concerns about which tele cheeses have distinct properties. Finally, a high-quality screening will be crucial for future trout gut (or any other) microbiota investigations.

1.6.3 THE REPRODUCTION OF RAINBOW TROUT USING A GENOMIC SEQUENCE

The Web of Science database searched for over 8,000 papers about rainbow trout reproduction (Liu et al., 2020). A comparable search for zebrafish reproduction (a commonly known biological model) generated just 3,500 references (Hill, Teraoka, Heideman, & Peterson, 2005). This demonstrates the significance of the scientific community working on rainbow trout reproduction and the potential influence of the rainbow trout genome and related genomic resources on the study of repetition in this species. As with other scientific domains, the recent publication of the rainbow trout genome opens up new avenues for studying gene synteny and elucidating previously unresolved evolutionary histories of genes involved in reproduction, as seen below. Sex hormone-binding globulins (SHBG) are blood proteins that transport sex steroids in plasma and regulate their availability to target tissues (Selby, 1990). Fish have two genes, SHBGA and SHBGB, with SHBGA becoming the ortholog of the SHBG gene sequence (Juanchich et al., 2016). SHBGB has been detected in rainbow trout and salmon but has yet identified in any non-salmonid teleost species. While this may imply that SHBGA and SHBGB are the results of Ss4R WGD, the topology of the SHBG tree, as well as new analysis utilizing the rainbow trout genome sequence, indicate that this duplication is considerably older than previously assumed. Gonadal soma-derived growth factor (GSDF) is a newly discovered member of the TGF-beta superfamily that plays a critical role in primordial germ cell and spermatogonia proliferation (Kaneko et al., 2015). It is likewise geographically and temporally connected to early testicular differentiation in medaka. GSDF gene expression is confined to Sertoli and Granulosa cells in medaka gonads. The fate of GSDF following Ss4R WGD is unknown, however, preliminary studies show that Ss4R produced two GSDF genes in rainbow trout with potentially divergent expression patterns. The rainbow trout genome will enable detailed research of GSDF ohnologs, including any regulating the expression or sub-/neo-functionalization following duplication (Juanchich et al., 2016).

With the rainbow trout genome sequence, it will be feasible to perform genome editing without depending on current resources (e.g., BAC libraries) to select critical genes and characterize the phenotypes associated with matching knockouts. While the rainbow appears to have a relatively long lifespan compared to some other fish model organisms (e.g., zebrafish, medaka), genetic knockout is sometimes required to sort out the function of unique salmonid genes such as the sex-determining gene SDY. Rainbow trout might be an intriguing model to do genetic modifications in a well-characterized teleost model demonstrating group-synchronous oogenetic development due to its long-term usage as a model for reproduction. A recent study showed that the sound and widely used Crispr/Cas9 technology is employed successfully on salmonids (Sumana & Petsalaki, 2018). Finally, the latest rainbow trout publication offers many possibilities in reproductive biology, with significant consequences for the control of sex, adolescence, fertility, and sterility (Pewan et al., 2020).

1.7 CONCLUSIONS AND FUTURE TRENDS: RESOURCES ARE NOW BEING EXPENDED

Next-generation sequencing (NGS) is a fast-growing discipline. NGS technologies are now available to more extended reading techniques and larger volumes of sequence data produced. NGS has expanded the resources of the rainbow transcriptome in many studies that sequenced cDNA in various organs, enabling therapeutic or diagnostic situations. miRNA sequencing throughout the rainbow has resulted in vast databases of miRNA components, in addition to cDNA transcripts. Complete genome sequencing of new rainbow organisms from multiple genetic origins, including duplicate haploids, will be possible with current NGS technology at a minimal cost. As a result, more markers and polymorphisms identification, allowing for further QTL identification and functional analysis refinement. The Rainbow Genetics framework is structured on NGS's most recent significant advancements in Illumina's genomic technology, allowing for the development of massive amounts of 2,250 bp data. The new feature of Code has been disclosed, and it will be available in 2016. This composite-enhanced rainbow rendering is 2.17 Gb in size, with N50 compatibility of 1,700 kb. Each of these emerging capacities is now taken into account to generate comprehensive cited maps of all functional sections of the rainbow genome, for example, through multinational efforts like the Computational Investigation of Animal Genomes, which would take the place of ENCODE Human. Together, these efforts will lead to a better knowledge of the lake trout genome, allowing it to use in genomic breeding, comprehensive epigenetic regulatory study, and other comprehensive studies.

Future opportunities and challenges Environmental and genetic interactions Aquaculture production occurs in various environments. The GxE exchange evaluates the animals' sensitivity to changes in their environment and genetic performance in different situations. Genotypes can be reclassified in other contexts as a result of this change. This means that the largest genotypes in one environment (such as freshwater) may not be the best genotypes in another. As a result, GxE can wreak havoc on selective breeding by making EBV predictions more difficult and lowering selection efficiency. As a result, GxE must be included in genetic assessment models regularly. Variations in nesting conditions resulted in significant misclassification for Atlantic salmon survival and moderate to severe classification for rainbow growth, according to a comprehensive study of the GxE interaction between aquatic species. Additional evidence for specific environmental factors affecting trait variability and complete characterization of the GxE interaction for growth and other traits of the rainbow has still required water temperature, population density, salinity, photoperiod, feeding, etc.

Trout raised for breeding has long been assumed to thrive in watery environments. Consequently, the selected applicants' development was assessed using the same criteria. New legislation permitted most breeding species to relocate to freshwater facilities on land following the sanitary disaster created by the ISA virus in Chile between 2007 and 2009. As a result, some breeding animals are grown in environments other than those found in maritime areas, and GxE should be present. If there is a GxE link between growth measured at inland facilities and development measured at sea, Chilean salmon farming strategies need to be adapted to consider this interaction. Establishing marine test stations that simulate one or more production conditions and testing siblings and cousins as selection candidates can help achieve this. As a result, the sample index may include lineage or genetics for production growth.

1.8 RESEARCH AND DEVELOPMENT

Enormous technological advances in genomics have enabled progress in selective breeding strategies over the past decade (Boudry et al., 2021). For example, GS increased sampling accuracy in salmon and trout compared to traditional pedigree alternatives. Although metrological research on GS's efficacy in aquaculture programs is published (Engle, Kumar, & van Senten, 2021), the implementation of practical genotype indexing techniques will facilitate the cost-effective application of GS in samplers for difficult-to-measure traits (Boudry et al., 2021). Understanding inherent conformational changes, also known as quantitative trait nucleotides (QTNs), would improve population genomic reliability and eliminate the need for phenotypic data for each generation and group (Stein et al., 2021; Zafar, Iftikhar, Ahmad, & Rather, 2021). Identifying QTN for genomic characteristics by incorporating data sets such as GWA, transcriptomic data, or sequencing of the entire genome is a critical step in global functional annotation (Naserkheil et al., 2022), which aims to understand the meaning of current genetics and recognize QTN in a variety of salmonids.

The gene-editing method (GE) has the power to revolutionize precise methods of genetic modification for a myriad of applications in all living organisms (Rasheed et al., 2021; Zafar et al., 2021). In salmonids, GE is more likely to improve the frequency and stability of beneficial alleles that segregate at critical QTLs that affect economically significant properties (Valente et al., 2013). On both sides, the growth of high-GE technologies can enable the regulation of multiple QTNs with moderate power sizes. *In vitro*, GE platforms (cell lines) were able to identify and confirm important fundamental properties of QTN efficiently. In salmonids, targeted GE research; However, the CRISPR-Cas9 (clustered regularly interspaced short palindromic repeat associated nuclease) method (Zafar et al., 2021) is used to effectively inactivate the GFP gene in CHSE cell lines remove the dnd gene, which induces sterility in Atlantic salmon. Regulators have not yet approved the actual use of GE's precision processing techniques. A key question is whether the new organism should not be classified as anything other than a genetically modified plant, as the European Union (EU) recently did. Although GMOs are not formally controlled in Chile, the main target markets for Chilean rainbows are more susceptible to the processing and commercialization of genetically modified products (Regan et al., 2021). As a result, the enormous potential of genomics is driving rapid progress and discoveries of the underlying variables and processes that underpin the essential characteristics of salmon and trout.

Significant advances in science and technology tend to exacerbate existing problems. With cross-functional teams, infrastructure, and "internal" resources, it is unlikely that breeding companies will be at the forefront. As a result, a partnership between companies and technology is required to create fast solutions and remain competitive. Financial authorities must adopt appropriate technologies to increase the links between innovation and marketing.

1.9 MARKET SHARE DIRECTION

Significant financial investment in breeding animals is required to apply genomic technology and develop inland facilities to ensure exceptional egg quality through improved pedigrees. These investments led the market to package specialized egg suppliers with substantial production volumes, which diluted the cost of eggs. Customers will notice an increase in the value of eggs and find out what product quality and efficiency they can expect from the egg supplier. In the context of market research, the trial occurrences and sample frequencies as per earlier researchers (Soh, Lim, & Chua, 2021) were used to see whether they could be adjusted to meet the specific demands of each potential egg buyer. Consequently, every egg supplier and other large livestock enterprises formed massive managed service units. As a result, the practices to achieve the greater power to manipulate genetic information. As a result, logistical support will become a great marketing tool for standing out among the top egg suppliers (Hosain, Ullah, Al Sayam, Mohiuddin, & Rahman, 2021) in many countries to increase the economy. Benchmarking programs implemented to demonstrate added value to increase production costs and mitigate the effects, breeding companies must investigate nutrient digestibility and the ability to withstand high temperatures. In the coming years, we expect technological advances in genomic screening to increase market share, with fewer companies competing for positions in the Chilean and global sectors for better salmon and trout eggs.

CONFLICT OF INTEREST STATEMENT

The authors reported no potential conflict of interest related with present work.

FUNDING

Not applicable.

KEYWORDS

- **antigen-specific receptors**
- **aquaculture**
- **gene clusters**
- **genome duplication**
- **genome sequencing**
- **rainbow trout**
- **whole genome**

REFERENCES

Abdelrahman, H., El Hady, M., Alcivar-Warren, A., Allen, S., Al-Tobasei, R., Bao, L., & Buchanan, J., (2017). Aquaculture genomics, genetics and breeding in the United States: Current status, challenges, and priorities for future research. *BMC Genomics, 18*(1), 1–23.

Adams, A., (2019). Progress, challenges, and opportunities in fish vaccine development. *Fish & Shellfish Immunology, 90*, 210–214.

Adusei, L. A., (2015). Threats to the exploration, production, and supply of Africa's energy resources. *South African Journal of International Affairs, 22*(1), 43–65.

Akhtar, T., Ghazanfar, A., & Shafi, N., (2021). Length–weight relationships, condition factor and morphometric characteristics of schizothoracinae from Neelum and Jhelum Rivers of Azad Jammu and Kashmir, Pakistan. *Pakistan Journal of Zoology, 53*(1), 351.

Ampong, I., (2020). *The Role of Diet and Gut Microbiota in Lipid Metabolism: Implications for Health Outcomes Using a Mouse Model.* University of Surrey.

Anastasiadi, D., Díaz, N., & Piferrer, F., (2017). Small ocean temperature increases elicit stage–dependent changes in DNA methylation and gene expression in a fish, the European sea bass. *Scientific Reports, 7*(1), 1–12.

Argyris, J. M., Ruiz-Herrera, A., Madriz-Masis, P., Sanseverino, W., Morata, J., Pujol, M., & Garcia-Mas, J., (2015). Use of targeted SNP selection for an improved anchoring of the melon (*Cucumis melo* L.) scaffold genome assembly. *BMC Genomics, 16*(1), 1–14.

Balasubramanian, S., Son, W., Kumar, B. M., Ock, S., Yoo, J., Im, G., & Rho, G., (2007). Expression pattern of oxygen and stress–responsive gene transcripts at various developmental stages of *in vitro* and *in vivo* preimplantation bovine embryos. *Theriogenology, 68*(2), 265–275.

Barta, D. G., Coman, V., & Vodnar, D. C., (2021). Microalgae as sources of omega-3 polyunsaturated fatty acids: Biotechnological aspects. *Algal Research, 58*, 102410.

Beemelmanns, A., Zanuzzo, F. S., Xue, X., Sandrelli, R. M., Rise, M. L., & Gamperl, A. K., (2021). The transcriptomic responses of Atlantic salmon (*Salmo salar*) to high temperature stress alone, and in combination with moderate hypoxia. *BMC Genomics, 22*(1), 1–33.

Bermúdez, J. M. G., Chaparro-Solano, H. M., Pinzón-Rondón, Á. M., Albornoz, L. L., Pardo-Oviedo, J. M., Zapata-Gómez, F. A., & Ruiz-Sternberg, Á. M., (2021). Mutation profile of SARS-CoV-2 genome in a sample from the first year of the pandemic in Colombia. *Infection, Genetics and Evolution*, 105192.

Bobe, J., Marandel, L., Panserat, S., Boudinot, P., Berthelot, C., Quillet, E., & Crollius, H. R., (2016). The rainbow trout genome, an important landmark for aquaculture and genome evolution. *Genomics in Aquaculture* (pp. 21–43). Elsevier.

Boudry, P., Allal, F., Aslam, M. L., Bargelloni, L., Bean, T. P., Brard-Fudulea, S., & Haffray, P., (2021). Current status and potential of genomic selection to improve selective breeding in the main aquaculture species of international council for the exploration of the sea (ICES) member countries. *Aquaculture Reports, 20*, 100700.

Boulajfene, W., Strogyloudi, E., Lasram, M., El Mlayah, A., Vassiliki-Angelique, C., & Zouari-Tlig, S., (2019). Biological and biochemical assessment in *Phorcus articulatus* (Lamarck 1822): Contamination and seasonal effect. *Environmental Monitoring and Assessment, 191*(9), 1–16.

Brunet, F. G., Lorin, T., Bernard, L., Haftek-Terreau, Z., Galiana, D., Schartl, M., & Volff, J. N., (2017). Case studies of seven gene families with unusually high retention rate since the vertebrate and teleost whole–genome duplications *Evolutionary Biology: Self/Nonself Evolution, Species and Complex Traits Evolution, Methods and Concepts* (pp. 369–396). Springer.

Cabral, R. B., Bradley, D., Mayorga, J., Goodell, W., Friedlander, A. M., Sala, E., & Gaines, S. D., (2020). A global network of marine protected areas for food. *Proceedings of the National Academy of Sciences, 117*(45), 28134–28139.

Cadonic, I., (2019). *MicroRNAs and Their Role Within the Teleost (Oncorhynchus mykiss) stress Response.* University of Waterloo.

Cañestro, C., (2012). Two rounds of whole–genome duplication: Evidence and impact on the evolution of vertebrate innovations *Polyploidy and Genome Evolution* (pp. 309–339). Springer.

Carr, D. F., Turner, R. M., & Pirmohamed, M., (2021). Pharmacogenomics of anticancer drugs: Personalizing the choice and dose to manage drug response. *British Journal of Clinical Pharmacology, 87*(2), 237–255.

Carruthers, M., Yurchenko, A. A., Augley, J. J., Adams, C. E., Herzyk, P., & Elmer, K. R., (2018). De novo transcriptome assembly, annotation and comparison of four ecological and evolutionary model salmonid fish species. *BMC Genomics, 19*(1), 1–17.

Cassé, F., Richetin, K., & Toni, N., (2018). Astrocytes' contribution to adult neurogenesis in physiology and Alzheimer's disease. *Frontiers in Cellular Neuroscience, 12*, 432.

Casu, S., Usai, M. G., Sechi, T., Salaris, S. L., Miari, S., Mulas, G., & Carta, A., (2022). Association analysis and functional annotation of imputed sequence data within genomic regions influencing resistance to gastro–intestinal parasites detected by an LDLA approach in a nucleus flock of sarda dairy sheep. *Genetics Selection Evolution, 54*(1), 1–16.

Chapman, J. A., Mascher, M., Buluç, A., Barry, K., Georganas, E., Session, A., & Oliker, L., (2015). A whole–genome shotgun approach for assembling and anchoring the hexaploid bread wheat genome. *Genome Biology, 16*(1), 1–17.

Chavanne, H., Janssen, K., Hofherr, J., Contini, F., Haffray, P., Komen, H., & Bargelloni, L., (2016). A comprehensive survey on selective breeding programs and seed market in the European aquaculture fish industry. *Aquaculture International, 24*(5), 1287–1307.

Chen, W., Tumanov, S., Fazakerley, D. J., Cantley, J., James, D. E., Dunn, L. L., & Stocker, R., (2021). Bilirubin deficiency renders mice susceptible to hepatic steatosis in the absence of insulin resistance. *Redox Biology, 47*, 102152.

Chen, Z., Omori, Y., Koren, S., Shirokiya, T., Kuroda, T., Miyamoto, A., & Zhang, S., (2019). De novo assembly of the goldfish (*Carassius auratus*) genome and the evolution of genes after whole–genome duplication. *Science Advances, 5*(6), eaav0547.

Cho, J. H., Kim, G. Y., Mansfield, B. C., & Chou, J. Y., (2018). Hepatic glucose-6-phosphatase-α deficiency leads to metabolic reprogramming in glycogen storage disease type Ia. *Biochemical and Biophysical Research Communications, 498*(4), 925–931.

Clavelle, T., Lester, S. E., Gentry, R., & Froehlich, H. E., (2019). Interactions and management for the future of marine aquaculture and capture fisheries. *Fish and Fisheries, 20*(2), 368–388.

Clouse, J., Adhikary, D., Page, J., Ramaraj, T., Deyholos, M., Udall, J., & Maughan, P., (2016). The amaranth genome: Genome, transcriptome, and physical map assembly. *The Plant Genome, 9*(1), plantgenome2015.2007.0062.

Colgan, T. J., Moran, P. A., Archer, L. C., Wynne, R., Hutton, S. A., McGinnity, P., & Reed, T. E., (2021). Evolution and expression of the immune system of a facultatively anadromous salmonid. *Frontiers in Immunology, 12*, 77.

Collins, E. J., & Riddle, D. S., (2008). TCR-MHC docking orientation: Natural selection, or thymic selection? *Immunologic Research, 41*(3), 267.

Dos, S. F. L., Correa, F. M., Ingram, K. T., De Almeida, A. A. F., & Royaert, S., (2020). QTL mapping and identification of SNP-haplotypes affecting yield components of *Theobroma cacao* L. *Horticulture Research, 7*(1), 1–18.

Dowaidar, M., (2021). Genome-Wide Association Studies (GWAS) Have Revolutionized Our View of Human Health and Disease Genetics and Offered Novel Gene Therapy Targets, *5*(5), 1137–1153.

Du, K., Stöck, M., Kneitz, S., Klopp, C., Woltering, J. M., Adolfi, M. C., & Kichigin, I., (2020). The sterlet sturgeon genome sequence and the mechanisms of segmental rediploidization. *Nature Ecology & Evolution, 4*(6), 841–852.

Engle, C., Kumar, G., & Van, S. J., (2021). Resource-use efficiency in US aquaculture: Farm-level comparisons across fish species and production systems. *Aquaculture Environment Interactions, 13*, 259–275.

Feng, J., Li, M., Zhao, S., Zhang, C., Yang, S., Qiao, S., & Pu, Z., (2018). Analysis of evolution and genetic diversity of sweetpotato and its related different polyploidy wild species *I. trifida* using RAD-seq. *BMC Plant Biology, 18*(1), 1–12.

Ferosekhan, S., Gupta, S., Singh, A. R., Rather, A., Kumari, R., Kothari, D. C., & Balkrishna, J. S., (2014). RNA-loaded chitosan nanoparticles for enhanced growth, immunostimulation and disease resistance in fish. *Current Nanoscience, 10*(3), 453–464.

Fraslin, C., Quillet, E., Rochat, T., Dechamp, N., Bernardet, J. F., Collet, B., & Boudinot, P., (2020). Combining multiple approaches and models to dissect the genetic architecture of resistance to infections in fish. *Frontiers in Genetics, 11*, 677.

Hardy, R. W., Kaushik, S. J., Mai, K., & Bai, S. C., (2022). Fish nutrition—History and perspectives. *Fish Nutrition* (pp. 1–16). Elsevier.

Hérault, F., Damon, M., Cherel, P., & Le Roy, P., (2018). Combined GWAS and LDLA approaches to improve genome–wide quantitative trait loci detection affecting carcass and meat quality traits in pig. *Meat Science, 135*, 148–158.

Hill, A. J., Teraoka, H., Heideman, W., & Peterson, R. E., (2005). Zebrafish as a model vertebrate for investigating chemical toxicity. *Toxicological Sciences, 86*(1), 6–19.

Hong, X., Song, R., Song, H., Zheng, T., Wang, J., Liang, Y., & Jiang, H., (2014). PTEN antagonizes Tcl1/hnRNPK–mediated G6PD pre-mRNA splicing which contributes to hepatocarcinogenesis. *Gut, 63*(10), 1635–1647.

Inoue, J., Sato, Y., Sinclair, R., Tsukamoto, K., & Nishida, M., (2015). Rapid genome reshaping by multiple-gene loss after whole-genome duplication in teleost fish suggested by mathematical modeling. *Proceedings of the National Academy of Sciences, 112*(48), 14918–14923.

Juanchich, A., Bardou, P., Rué, O., Gabillard, J. C., Gaspin, C., Bobe, J., & Guiguen, Y., (2016). Characterization of an extensive rainbow trout miRNA transcriptome by next generation sequencing. *BMC Genomics, 17*(1), 1–12.

Kaneko, H., Ijiri, S., Kobayashi, T., Izumi, H., Kuramochi, Y., Wang, D. S., & Nagahama, Y., (2015). Gonadal soma-derived factor (gsdf), a TGF-beta superfamily gene, induces testis differentiation in the teleost fish *Oreochromis niloticus*. *Molecular and Cellular Endocrinology, 415*, 87–99.

Karras, S. N., Koufakis, T., Mustafa, O. G., & Kotsa, K., (2019). Anti-incretin effect: The other face of Janus in human glucose homeostasis. *Obesity Reviews, 20*(11), 1597–1607.

Khedikar, Y., Pandey, M. K., Sujay, V., Singh, S., Nayak, S. N., Klein-Gebbinck, H. W., & Upadhyaya, H. D., (2018). Identification of main effect and epistatic quantitative trait loci for morphological and yield-related traits in peanut (*Arachis hypogaea* L.). *Molecular Breeding, 38*(1), 1–12.

Kiepas, A., (2021). *Defining the Molecular Mechanisms that Regulate Breast Cancer Cell Migration.* McGill University (Canada).

Krajcik, J. S., & Czerniak, C. M., (2018). *Teaching Science in Elementary and Middle School: A Project–Based Learning Approach*. Routledge.

Książkiewicz, M., Zielezinski, A., Wyrwa, K., Szczepaniak, A., Rychel, S., Karlowski, W., & Naganowska, B., (2015). Remnants of the legume ancestral genome preserved in gene-rich regions: Insights from *Lupinus angustifolius* physical, genetic, and comparative mapping. *Plant Molecular Biology Reporter, 33*(1), 84–101.

Kulski, J. K., Shiina, T., Anzai, T., Kohara, S., & Inoko, H., (2002). Comparative genomic analysis of the MHC: The evolution of class I duplication blocks, diversity and complexity from shark to man. *Immunological Reviews, 190*(1), 95–122.

Lal, N., (2017). *Exploring Immunogenomic Influences on the Microenvironment of Colorectal Cancer.* University of Birmingham.

Lamb, T. D., (2021). Analysis of paralogons, origin of the vertebrate karyotype, and ancient chromosomes retained in extant species. *Genome Biology and Evolution, 13*(4), evab044.

Li, J. T., Hou, G. Y., Kong, X. F., Li, C. Y., Zeng, J. M., Li, H. D., & Sun, X. W., (2015). The fate of recent duplicated genes following a fourth–round whole genome duplication in a tetraploid fish, common carp (*Cyprinus carpio*). *Scientific Reports, 5*(1), 1–9.

Lien, S., Koop, B. F., Sandve, S. R., Miller, J. R., Kent, M. P., Nome, T., & Zimin, A., (2016). The Atlantic salmon genome provides insights into rediploidization. *Nature, 533*(7602), 200–205.

Lin, S., (2019). *Cytogenetic Evolution of Genus Salmo.* Science: Department of Molecular Biology and Biochemistry.

Liu, H., Pan, W., Tang, C., Tang, Y., Wu, H., Yoshimura, A., & Li, S., (2021). The methods and advances of adaptive immune receptors repertoire sequencing. *Theranostics, 11*(18), 8945.

Liu, J., Deng, K., Pan, M., Liu, G., Wu, J., Yang, M., & Mai, K., (2020). Dietary carbohydrates influence muscle texture of olive flounder *Paralichthys olivaceus* through impacting mitochondria function and metabolism of glycogen and protein. *Scientific Reports, 10*(1), 1–20.

Lopez-Santamarina, A., Mondragon, A. D. C., Lamas, A., Miranda, J. M., Franco, C. M., & Cepeda, A., (2020). Animal-origin prebiotics based on chitin: An alternative for the future? A critical review. *Foods, 9*(6), 782.

Lorin, T., Brunet, F. G., Laudet, V., & Volff, J. N., (2018). Teleost fish-specific preferential retention of pigmentation gene-containing families after whole genome duplications in vertebrates. *G3: Genes, Genomes, Genetics, 8*(5), 1795–1806.

Lu, G., & Luo, M., (2020). Genomes of major fishes in world fisheries and aquaculture: Status, application and perspective. *Aquaculture and Fisheries, 5*(4), 163–173.

Lutfi, R. E., (2017). *Obesity and Adipose Tissue Biology in Fish. Influence of Nutritional, Genetic and Environmental Factors.*

Lyons, P. P., Turnbull, J. F., Dawson, K. A., & Crumlish, M., (2017). Exploring the microbial diversity of the distal intestinal lumen and mucosa of farmed rainbow trout *Oncorhynchus mykiss* (W-albaum) using next generation sequencing (NGS). *Aquaculture Research, 48*(1), 77–91.

Mamdouh, S., Mohamed, A. S., Mohamed, H. A., & Fahmy, W. S., (2022). The effect of zinc concentration on physiological, immunological, and histological changes in crayfish (*Procambarus clarkii*) as bio-indicator for environment quality criteria. *Biological Trace Element Research, 200*(1), 375–384.

Minkley, D. R., (2018). Transposable Elements in the Salmonid Genome. Thesis Submitted in Partial Fulfillment of the Requirements for the Degree of Masters of Science in the department of biology, University of Victoria.

Morro, B., Balseiro, P., Albalat, A., Pedrosa, C., Mackenzie, S., Nakamura, S., & Ebbesson, L. O., (2019). Effects of different photoperiod regimes on the smoltification and seawater adaptation of seawater-farmed rainbow trout (*Oncorhynchus mykiss*): Insights from Na+, K+–ATPase activity and transcription of osmoregulation and growth regulation genes. *Aquaculture, 507*, 282–292.

Mullarky, E., & Cantley, L. C., (2015). Diverting glycolysis to combat oxidative stress. *Innovative Medicine*, 3–23.

Murat, F., Armero, A., Pont, C., Klopp, C., & Salse, J., (2017). Reconstructing the genome of the most recent common ancestor of flowering plants. *Nature Genetics, 49*(4), 490–496.

Naserkheil, M., Ghafouri, F., Zakizadeh, S., Pirany, N., Manzari, Z., Ghorbani, S., & Park, M. N., (2022). Multi-omics integration and network analysis reveal potential hub genes and genetic mechanisms regulating bovine mastitis. *Current Issues in Molecular Biology, 44*(1), 309–328.

Nguinkal, J. A., Brunner, R. M., Verleih, M., Rebl, A., Ríos-Pérez, L. D. L., Schäfer, N., & Goldammer, T., (2019). The first highly contiguous genome assembly of pikeperch (*Sander lucioperca*), an emerging aquaculture species in Europe. *Genes, 10*(9), 708.

Nugent, C., (2019). The Arctic Charr (*Salvelinus alpinus*) Genome: An Analysis of Evolutionary Change After Whole Genome Duplication. Thesis Submitted in Partial

Fulfillment of the Requirements for the Degree of Masters of Science in the department of Integrated biology, University of Guelph.

Osinov, A. G., Volkov, A. A., & Mugue, N. S., (2021). Charrs of the genus *Salvelinus* (*Salmonidae*): Hybridization, phylogeny and evolution. *Hydrobiologia, 848*(3), 705–726.

Palti, Y., Gao, G., Liu, S., Kent, M., Lien, S., Miller, M., & Moen, T., (2015). The development and characterization of a 57 K single nucleotide polymorphism array for rainbow trout. *Molecular Ecology Resources, 15*(3), 662–672.

Panserat, S., Marandel, L., Seiliez, I., & Skiba-Cassy, S., (2019). New insights on intermediary metabolism for a better understanding of nutrition in teleosts. *Annual Review of Animal Biosciences, 7*, 195–220.

Pérez-Sánchez, J., Naya-Català, F., Soriano, B., Piazzon, M. C., Hafez, A., Gabaldón, T., & Calduch-Giner, J. A., (2019). Genome sequencing and transcriptome analysis reveal recent species-specific gene duplications in the plastic gilthead sea bream (*Sparus aurata*). *Frontiers in Marine Science, 6*, 760.

Perreault-Payette, A., Muir, A. M., Goetz, F., Perrier, C., Normandeau, E., Sirois, P., & Bernatchez, L., (2017). Investigating the extent of parallelism in morphological and genomic divergence among lake trout ecotypes in Lake Superior. *Molecular Ecology, 26*(6), 1477–1497.

Pewan, S. B., Otto, J. R., Huerlimann, R., Budd, A. M., Mwangi, F. W., Edmunds, R. C., & Adegboye, O. A., (2020). Genetics of omega-3 long-chain polyunsaturated fatty acid metabolism and meat-eating quality in Tattykeel Australian White lambs. *Genes, 11*(5), 587.

Pinu, F. R., Beale, D. J., Paten, A. M., Kouremenos, K., Swarup, S., Schirra, H. J., & Wishart, D., (2019). Systems biology and multi-omics integration: Viewpoints from the metabolomics research community. *Metabolites, 9*(4), 76.

Pophaly, S. D., (2017). *Neutral and Selective Processes Underlying Genome Evolution Post-Duplication in Maize.* Technische Universität München.

Price, N., Moyers, B. T., Lopez, L., Lasky, J. R., Monroe, J. G., Mullen, J. L., & Schrider, D. R., (2018). Combining population genomics and fitness QTLs to identify the genetics of local adaptation in *Arabidopsis thaliana. Proceedings of the National Academy of Sciences, 115*(19), 5028–5033.

Rabot, S., Membrez, M., Bruneau, A., Gérard, P., Harach, T., Moser, M., & Chou, C. J., (2010). Germ-free C57BL/6J mice are resistant to high-fat-diet-induced insulin resistance and have altered cholesterol metabolism. *The FASEB Journal, 24*(12), 4948–4959.

Rajabi, S., Ramazani, A., Hamidi, M., & Naji, T., (2015). Artemia salina as a model organism in toxicity assessment of nanoparticles. *DARU Journal of Pharmaceutical Sciences, 23*(1), 1–6.

Rasheed, A., Gill, R. A., Hassan, M. U., Mahmood, A., Qari, S., Zaman, Q. U., & Li, H., (2021). A critical review: Recent advancements in the use of CRISPR/Cas9 technology to enhance crops and alleviate global food crises. *Current Issues in Molecular Biology, 43*(3), 1950–1976.

Regan, T., Bean, T. P., Ellis, T., Davie, A., Carboni, S., Migaud, H., & Houston, R. D., (2021). Genetic improvement technologies to support the sustainable growth of UK aquaculture. *Reviews in Aquaculture*.

Rendón-Anaya, M., Ibarra-Laclette, E., Méndez-Bravo, A., Lan, T., Zheng, C., Carretero-Paulet, L., & Chang, T. H., (2019). The avocado genome informs deep angiosperm phylogeny, highlights introgressive hybridization, and reveals pathogen-influenced gene space adaptation. *Proceedings of the National Academy of Sciences, 116*(34), 17081–17089.

Rimoldi, S., Antonini, M., Gasco, L., Moroni, F., & Terova, G., (2021). Intestinal microbial communities of rainbow trout (*Oncorhynchus mykiss*) may be improved by feeding a

Hermetia illucens meal/low-fishmeal diet. *Fish Physiology and Biochemistry, 47*(2), 365–380.

Rodriguez, F., & Arkhipova, I. R., (2018). Transposable elements and polyploid evolution in animals. *Current Opinion in Genetics & Development, 49*, 115–123.

Rolland, M., Skov, P. V., Larsen, B. K., Holm, J., Gómez-Requeni, P., & Dalsgaard, J., (2016). Increasing levels of dietary crystalline methionine affect plasma methionine profiles, ammonia excretion, and the expression of genes related to the hepatic intermediary metabolism in rainbow trout (*Oncorhynchus mykiss*). *Comparative Biochemistry and Physiology Part B: Biochemistry and Molecular Biology, 198*, 91–99.

Roy, J., Larroquet, L., Surget, A., Lanuque, A., Sandres, F., Terrier, F., & Skiba-Cassy, S., (2020). Impact on cerebral function in rainbow trout fed with plant-based omega-3 long chain polyunsaturated fatty acids enriched with DHA and EPA. *Fish & Shellfish Immunology, 103*, 409–420.

Sadd, B. M., Barribeau, S. M., Bloch, G., De Graaf, D. C., Dearden, P., Elsik, C. G., & Lozier, J. D., (2015). The genomes of two key bumblebee species with primitive eusocial organization. *Genome Biology, 16*(1), 1–32.

Salem, M., Al-Tobasei, R., Ali, A., Lourenco, D., Gao, G., Palti, Y., & Leeds, T. D., (2018). Genome-wide association analysis with a 50K transcribed gene SNP-chip identifies QTL affecting muscle yield in rainbow trout. *Frontiers in Genetics, 9*, 387.

Salem, M., Paneru, B., Al-Tobasei, R., Abdouni, F., Thorgaard, G. H., Rexroad, C. E., & Yao, J., (2015). Transcriptome assembly, gene annotation and tissue gene expression atlas of the rainbow trout. *PloS One, 10*(3), e0121778.

Salzburger, W., (2018). Understanding explosive diversification through cichlid fish genomics. *Nature Reviews Genetics, 19*(11), 705–717.

Schill, D. J., Heindel, J. A., Campbell, M. R., Meyer, K. A., & Mamer, E. R., (2016). Production of a YY male brook trout broodstock for potential eradication of undesired brook trout populations. *North American Journal of Aquaculture, 78*(1), 72–83.

Segner, H., Verburg-Van, K. B. L., & Chadzinska, M., (2017). The immunomodulatory role of the hypothalamus-pituitary-gonad axis: Proximate mechanism for reproduction-immune tradeoffs? *Developmental & Comparative Immunology, 66*, 43–60.

Selby, C., (1990). Sex hormone binding globulin: Origin, function and clinical significance. *Annals of Clinical Biochemistry, 27*(6), 532–541.

Sharma, S., & Petsalaki, E., (2018). Application of CRISPR-Cas9 based genome-wide screening approaches to study cellular signaling mechanisms. *International Journal of Molecular Sciences, 19*(4), 933.

Sharma, S., Aroura, S., Gupta, A., & Priyadarshini, A., (2022). Bioinformatics in diagnosis of COVID-19. *Assessing COVID-19 and Other Pandemics and Epidemics Using Computational Modelling and Data Analysis* (pp. 197–219). Springer.

Shavalier, M. A., (2017). *Investigations on the Intricate Interactions Between Epizootic Epitheliotropic Disease Virus (Salmonid Herpesvirus-3) and its Host, the Lake Trout (Salvelinus namaycush)*. Michigan State University.

Sobhkhez, M., Krasnov, A., & Robertsen, B., (2018). Transcriptome analyses of Atlantic salmon muscle genes induced by a DNA vaccine against salmonid alphavirus, the causative agent of salmon pancreas disease (PD). *PloS One, 13*(10), e0204924.

Song, X., Marandel, L., Skiba-Cassy, S., Corraze, G., Dupont-Nivet, M., Quillet, E., & Panserat, S., (2018). Regulation by dietary carbohydrates of intermediary metabolism in liver and muscle of two isogenic lines of rainbow trout. *Frontiers in Physiology, 9*, 1579.

Stein, M. B., Levey, D. F., Cheng, Z., Wendt, F. R., Harrington, K., Pathak, G. A., & Girgenti, M. J., (2021). Genome-wide association analyses of post-traumatic stress disorder and its symptom subdomains in the million-veteran program. *Nature Genetics, 53*(2), 174–184.

Steinberg, C. E., (2022). Carbohydrate preference and metabolism—'Life's little luxury digested.' *Aquatic Animal Nutrition* (pp. 393–434). Springer.

Stephens, T., (2019). Polarella Genomics: Understanding the Evolutionary Transition to Algal Symbiosis and Cold Adaptation. PhD Thesis, Institute for Molecular Bioscience, The University of Queensland.

Stoessel, D. J., Raadik, T. A., & Ayres, R. M., (2015). *Spawning of Threatened Barred Galaxias, 'Galaxias Fuscus' (Teleostei: Galaxiidae).* Paper presented at the proceedings of the Linnean Society of New South Wales.

Suurväli, J., Jouneau, L., Thépot, D., Grusea, S., Pontarotti, P., Du Pasquier, L., & Boudinot, P., (2014). The proto-MHC of placozoans, a region specialized in cellular stress and ubiquitination/proteasome pathways. *The Journal of Immunology, 193*(6), 2891–2901.

Symonová, R., & Howell, W. M., (2018). Vertebrate genome evolution in the light of fish cytogenomics and rDNAomics. *Genes, 9*(2), 96.

Tank, D. C., Eastman, J. M., Pennell, M. W., Soltis, P. S., Soltis, D. E., Hinchliff, C. E., & Harmon, L. J., (2015). Nested radiations and the pulse of angiosperm diversification: Increased diversification rates often follow whole genome duplications. *New Phytologist, 207*(2), 454–467.

Taylor, E. B., & May-McNally, S. L., (2015). Genetic analysis of Dolly Varden (*Salvelinus malma*) across its North American range: Evidence for a contact zone in southcentral Alaska. *Canadian Journal of Fisheries and Aquatic Sciences, 72*(7), 1048–1057.

Taylor, R. S., Daniels, R. R., Morata, D. P., Gundappa, M. K., & Macqueen, D. J., (2022). Evolution of ray-finned fish genomes: Status and directions with a primer on microRNA characterization *Cellular and Molecular Approaches in Fish Biology* (pp. 309–346). Elsevier.

Torres, A. R., Sweeten, T. L., Johnson, R. C., Odell, D., Westover, J. B., Bray-Ward, P., & Croen, L. A., (2016). Common genetic variants found in HLA and KIR immune genes in autism spectrum disorder. *Frontiers in Neuroscience, 10*, 463.

Turchini, G. M., Trushenski, J. T., & Glencross, B. D., (2019). Thoughts for the future of aquaculture nutrition: Realigning perspectives to reflect contemporary issues related to judicious use of marine resources in aquafeeds. *North American Journal of Aquaculture, 81*(1), 13–39.

Valente, L. M., Moutou, K. A., Conceicao, L. E., Engrola, S., Fernandes, J. M., & Johnston, I. A., (2013). What determines the growth potential and juvenile quality of farmed fish species? *Reviews in Aquaculture, 5*, S168–S193.

Walter, W., Pfarr, N., Meggendorfer, M., Jost, P., Haferlach, T., & Weichert, W., (2020). *Next-Generation Diagnostics for Precision Oncology: Preanalytical Considerations, Technical Challenges, and Available Technologies.* Paper presented at the Seminars in Cancer Biology.

Wang, Z., & Dong, C., (2019). Gluconeogenesis in cancer: Function and regulation of PEPCK, FBPase, and G6Pase. *Trends in Cancer, 5*(1), 30–45.

Waples, R. K., (2015). Population Genomics of Salish Sea Chum Salmon: The Legacy of the Salmonid Whole Genome Duplication. Thesis Submitted in Partial Fulfillment of the Requirements for the Degree of Masters of Science, University of Washington.

Wiernasz, N., Gigout, F., Cardinal, M., Cornet, J., Rohloff, J., Courcoux, P., & Pilet, M. F., (2021). Effect of the manufacturing process on the microbiota, organoleptic properties and volatilome of three salmon-based products. *Foods, 10*(11), 2517.

Xu, L., Zhao, J., Liu, M., Ren, G., Jian, F., Yin, J., & Lu, T., (2017). Bivalent DNA vaccine induces significant immune responses against infectious hematopoietic necrosis virus and infectious pancreatic necrosis virus in rainbow trout. *Scientific Reports, 7*(1), 1–11.

Xu, Y., Curtasu, M. V., Knudsen, K. E. B., Hedemann, M. S., Theil, P. K., & Lærke, H. N., (2021). Dietary fiber and protein do not synergistically influence insulin, metabolic or inflammatory biomarkers in young obese Göttingen minipigs. *British Journal of Nutrition, 125*(7), 828–840.

Yang, Z., & Sun, G., (2017). High-frequency, low-coverage "false positives" mutations may be true in GS Junior sequencing studies. *Scientific Reports, 7*(1), 1–8.

Zafar, I., Iftikhar, R., Ahmad, S. U., & Rather, M. A., (2021). Genome wide identification, phylogeny, and synteny analysis of sox gene family in common carp (*Cyprinus carpio*). *Biotechnology Reports, 30*, e00607.

Zafar, I., Rafique, A., Fazal, J., Manzoor, M., Ain, Q. U., & Rayan, R. A., (2021). Genome and gene editing by artificial intelligence programs. *Advanced AI Techniques and Applications in Bioinformatics* (pp. 165–188). CRC Press.

Zhang, G., Zhang, X., Ye, H., Jiang, S., Yu, H., Li, J., & Luo, J., (2018). Construction of high-density genetic linkage maps and QTL mapping in the golden pompano. *Aquaculture, 482*, 90–95.

CHAPTER 2

Economics of Coldwater Fisheries

KHEMRAJ BUNKAR[1], UDAI RAM GURJAR[1], and SUMAN TAKAR[2]

[1]*ICAR–Central Institute of Fisheries Education, Mumbai, Maharashtra, India*

[2]*TNJFU–Fisheries College and Research Institute, Thoothukudi, Tamil Nadu, India*

ABSTRACT

The total fish production during 2019–2020 is recorded at 14.15 million tons. The inland fishery sector contributes more than 73% of total production. Coldwater fisheries occupy a key place among India's freshwater fishes, contributing significantly to nutritional security and job creation in rural hilly areas. More than 1.40 lakh fishermen in India directly depend on coldwater fisheries in hilly areas. Manipur has the highest fish population with 47,711 fishers' population. Total fry production is 4,388.9 fish fry in hilly states in which Manipur produces the highest fish seed with 2494.8 lakh fish fry production. In 2019–2020, total coldwater fish production was estimated at 1.07 lakh tons; Manipur has the largest share of 32,000 tons. It also observed that ecotourism based on fisheries is a sustainable resource use option and a means of environmental conservation. In Jammu and Kashmir, the annual total cost and gross revenue per farm were estimated to be Rs. 2.36 lakh per farm and Rs. 4.30 lakh per raceway, with a benefit-cost ratio of 1.83, indicating the business economic viability. In Sikkim, a trout raceway of 51 m^3, which costs around 1.20 lakhs, has been observed to produce 500 kg of trout fish at an average price of 800 per kg, yielding an income of 4 lakhs. The net profit from a single raceway would come to ₹ 2.7 lakhs, which is a very venturing and reasonable profit to the farmers.

Coldwater Fisheries and Aquaculture Management: Technology for Sustainable Food Production.
Mohd Ashraf Rather, Faisal Rashid Sofi, Adnan Amin, & Kawkabul Saba (Eds.)

Major challenges in coldwater fish farming have been stated available water resources for culture, unawareness of fish farming, lack of facilities like cold storage, quick transport, and other marketing facilities, and lack of modern technological knowledge.

2.1 INTRODUCTION

During 2019–2020, the total fish production is recorded at 14.15 million tons. The fisheries sector plays a significant role in the country's socio-economic development and provides a source of income for a large proportion of the economically disadvantaged rural population (Ayyapan & Krishnan, 2004). It ensures food security while also addressing unemployment in these regions, which rural residents primarily populate. It has grown in importance and is recognized as a rich source of inexpensive, nutritious food (Kumar et al., 2007). The inland fishery sector contributes significantly to nutritional security and employment opportunities in comparison to the 0.24 million tons of fish production in 1950–1951, the country's inland fish production during 2019–2020 reached 10.43 million tons, this increases in the fish production the country became the second-largest producer of inland fish. The fisheries sector accounts for approximately 1.24% of the country's GDP (gross domestic product) and 7.28% of its agricultural GDP. The annual export earnings from the fishery sector are ₹ 46,662.85 crores (2019–2020). This sector provides a living for more than 14 million people who work in it full-time, part-time, or in ancillary capacities. Furthermore, the sector provides an important source of ancillary jobs for the rural population, particularly in marketing, retailing, and transportation. However, the industry is still largely unorganized, owing to the dispersed and diffuse nature of activities (DoF, 2020).

Inland fishery resources of India are diverse and plentiful which is illustrated in Table 2.1. The vast and varied resources of Indian fisheries accounted for 1,73,285 km of rivers and canals, 1.098 million hectares of swamps and other wetlands, 0.202 million hectares of the floodplain, 0.072 million hectares of uplands lakes, 2.414 million hectares of ponds and tanks, 0.357 million hectares of mangroves and 0.285 million hectares of estuaries resources, 0.191 million hectares of lagoons, 3.150 million hectares of reservoirs, and 1.240 million hectares of brackish water provide excellent opportunities for livelihood. The country's freshwater culture resources include 2.36 million ha of ponds and tanks (DoF, 2020).

TABLE 2.1 Current Scenario of Indian Fisheries

SL. No.	Particulars	Details
1.	World rank	2nd in aquaculture, 3rd in fisheries
2.	Total fish production (2019–2020)	14.15 MMT
3.	Export value (2019–2020)	₹ 46,662.85 crore
4.	Fisheries contribution to agricultural GDP	7.28%
5.	Contribution to national GDP	1.24%

Source: DoF (2020).

2.2 COLDWATER FISHERIES IN INDIA

The term "coldwater" refers to the aquatic ecosystem that maintains thermal and oxygen levels for the well-being of trout, mahseers, snow trout, and other minor species. The temperature is generally within the trout and salmons tolerance limits in the Salmonidae family, which is less than 20°C. Various physical, chemical, geochemical, and biological parameters of different water bodies, such as water temperature, dissolved oxygen, velocity, turbidity, substratum, trophic status, food availability, and so on, influence the distribution and abundance of various coldwater fisheries species. Coldwater fishes inhabit the Himalayan and sub-Himalayan zones in the north and watersheds draining the southern slope of the Deccan plateau in the Indian subcontinent (Sunder et al., 1999). As per Anon (2011), area-wise different types of aquatic resources available in hilly regions of India, such as reservoir water spread area (2,65,000 ha), Deccan Plateau and Himalayan Riverine Resources (10,000 km), Natural Freshwater lakes (1,500–2,000 m asl) 18,150 ha, Brackish water lakes (above 3,000 m asl) 2,340 ha, High altitude lakes of Kashmir (above 3,000 m asl) 400 ha, Central Himalayas freshwater lakes (Kumaon region) 355 ha, Shivalik Himalayan lakes (74 ha), and Wetlands (3,000 ha).

India's coldwater fisheries sector is small but vibrant, with significant growth potential. Thrust areas have been identified for the holistic development of all segments in order to improve coldwater fisheries production through cluster-based farming and natural resource conservation. The vision is to significantly expand coldwater fisheries through the use of new and innovative production technologies, management, and utilization of underutilized water resources, and proper market setups. Fish culture has been carried out successfully in Jammu and Kashmir, Himachal Pradesh, Sikkim,

Uttaranchal, Nagaland, Meghalaya, Arunachal Pradesh, Tamil Nadu, and Kerala in lakes, ponds, and reservoirs. The Indian government established the National Research Center on Cold Water Fisheries (NRCCWF). It had played a crucial role in the improvement and conservation of indigenous species as well as the establishment of exotic species in this region. As one of the promising sectors, the DADF intends to improve and strengthen coldwater fisheries. As a result, the Mission Coldwater Fisheries Development Action Plan–2022 was created in order to conserve and improve coldwater fisheries in an economically, socially, and environmentally responsible manner, as well as to promote fish farming in the hill region. The Mission Coldwater Fisheries Development Action Plan is expected to improve food and employment security in coldwater areas in rural and urban areas (Chettri, 2020; DAHDF, 2018).

2.3 ECONOMIC IMPORTANCE OF COLD-WATER FISHERIES IN INDIA

The coldwater fisheries in India occupy a vast region that is uneven and versatile, with a rich biological floral and faunal diversity. These regions are broadly classified as the eastern Himalayas, the central Himalayas, and the western Himalayas. Coldwater fisheries have the employment generation capacity in the hilly area where the employment opportunities are significantly less and play a crucial role in employment in such areas (Chetrri, 2020). According to the DoF (2020), the total fishermen population in major hilly areas is estimated at 1.40 lakh fishermen who directly depend on coldwater fisheries. Among the states, Manipur has the highest fish population with 47,711 fishermen population followed by Arunachal Pradesh (24,015 fishermen), Jammu and Kashmir (17,396 fishermen), Meghalaya (16,567 fishermen), Himachal Pradesh (11,806 fishermen), Uttarakhand (8,352 fishermen), Nagaland (7,958 fishermen) and least hilly fishermen population lived in Mizoram (6,289 fishermen) among the major hilly coldwater fisheries states of India which shown in Figure 2.1.

2.4 COLDWATER FISH PRODUCTION

Total seed (fry) production is 4,388.9 fish fry in hilly states namely Himachal Pradesh, Jammu and Kashmir, Manipur, Meghalaya Mizoram, Nagaland, Uttarakhand, and Arunachal Pradesh.

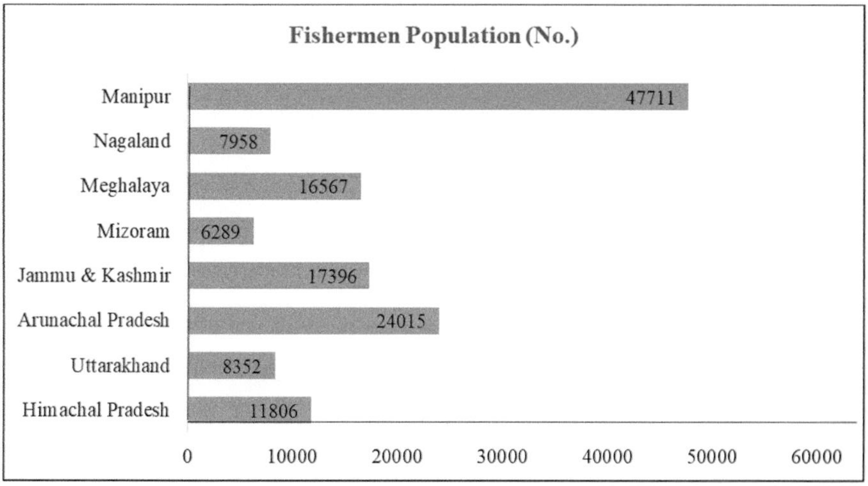

FIGURE 2.1 Fishermen population in hilly states of India.
Source: DoF (2020).

The state Manipur produces the highest fish seed with 2494.8 lakh fish fry production among all hilly states, followed by Nagaland (795 lakh fish fry), Mizoram (400 lakh fish fry), Uttarakhand (289.6 lakh fish fry), Jammu and Kashmir (209.7 lakh fish fry), Arunachal Pradesh (190 lakh fish fry) and Meghalaya is the least fish seed producing state with only 3 lakh fish fry (Figure 2.2).

The fish production of all the hilly states of India shows an increasing trend (Figure 2.3). Jammu & Kashmir is the largest producer of coldwater fishes in India. During 1995–1996, coldwater fish production was 16.52 mt in Jammu & Kashmir, which increased to 19.15 mt in 2005–2006 and 20.08 mt in 2015–2016. Himachal Pradesh is the second-largest cultivator and producer. Its production increased from 5.94 mt in 1995–1996 to 7.3 mt in 2005–2006 and 11.8 in 2015–2016. In 2005–2006, Uttarakhand's fish production was 2.79 mt, which increased to 4.14 mt in 2015–2016 with a growth of 38%. The production of fish in Sikkim was the lowest amongst these states because it is the smallest state in terms of area as well fewer areas are under cultivation than other states. However, production reached from 0.11 mt in 1995–1996 to 0.15 mt in 2005–2006 and 0.4 mt in 2015–2016 with a growth of 166% during this decade. During 1995–1996, fish production in Arunachal Pradesh was 1.85 mt, which increased to 2.75 mt in 2005–2006 and 4.05 mt in 2015–2016.

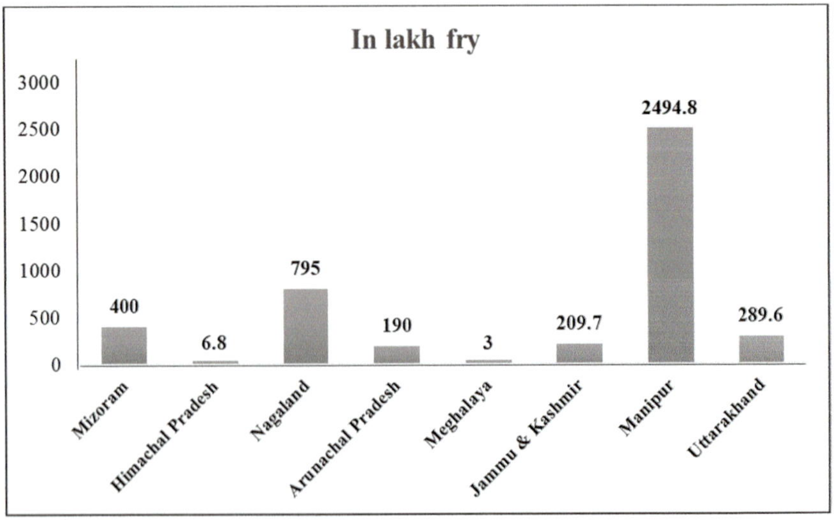

FIGURE 2.2 Coldwater fish seed production in hilly states of India (2020).

Source: DoF (2020).

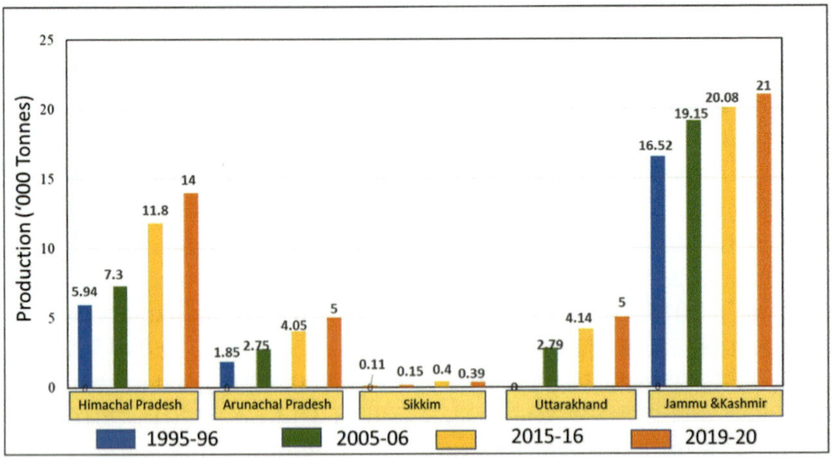

FIGURE 2.3 Coldwater fish production in the Himalayas of India (in '000' tons).

Source: DAHDF (2018); DoF (2020).

In 2019–2020, total coldwater fish production is estimated 1.07 lakh tons. The states mentioned above showed an increasing trend except for Sikkim, which offers a minor reduction in production. Among the above-mentioned states, Jammu and Kashmir have the leading position among the states with 21,000 tons of fish production, followed by Himachal Pradesh (14,000 tons), Arunachal Pradesh (5,000 tons), Uttarakhand (5,000 tons), and Sikkim (0.39,000 tons).

The majority of fishery resources in upland Himalayan regions are currently exploited through capture fisheries, though fish production through culture practices is gaining traction. At the moment, total fish production from upland areas accounts for about 2% of total inland fish production in India, with a negligible contribution to total fish production. The coldwater aquaculture production potential has yet to be fully realized. Except for a few hill states like Kashmir Valley and Himachal Pradesh, other mountain regions of India are still underdeveloped or underexploited in terms of coldwater fisheries development. These hill states, which are endowed with natural lakes and reservoirs, could benefit from a culture-based capture fisheries program to increase fish production. DCFR has taken up open water cage culture initiatives in Himalayan lakes for stock enhancement.

Total coldwater fish production is shown in Figure 2.4 which is 1.07 lakh tons from hilly states, namely Manipur, Jammu and Kashmir, Meghalaya, Mizoram, Himachal Pradesh, Arunachal Pradesh, Nagaland, and Uttarakhand. Manipur is the leading state among the coldwater fish producing states in India with 30% coldwater fish production, followed by Jammu and Kashmir (20%), Himachal Pradesh (13%), Meghalaya (13%), Nagaland (8%), Mizoram (6%), Arunachal Pradesh (5%) and Uttarakhand (5%) (Figure 2.2).

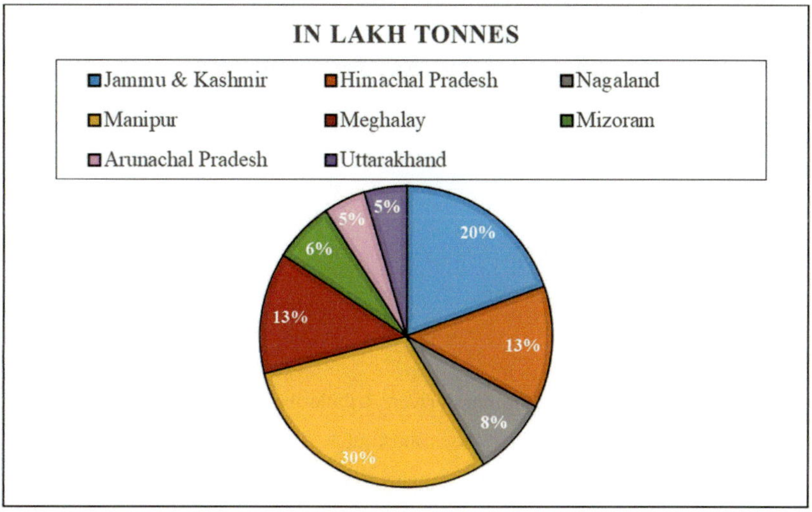

FIGURE 2.4 Coldwater fish production in hilly states of India (2020).

Source: DoF (2020).

2.5 SWOT ANALYSIS OF COLDWATER FISH FARMING IN INDIA

In order to formulate a future action plan for the development of coldwater fishery, SWOT analysis has been performed by the Department of Animal husbandry, Dairying and Fisheries–2018, to illustrate ground realities as shown in Table 2.2.

TABLE 2.2 SWOT Analysis of Cold-Water Fish Farming in India

Strength:

- Basic culture technologies, supported with field-oriented projects.
- With assured financial returns.
- An adequate gap in demand supply for local consumption, besides scope for export.
- Existing schemes for the production of coldwater fisheries.
- Presence of unexploited coldwater resources in the form of rivers, streams, and lakes.

Weaknesses:

- Insufficient allocation of funds.
- Poor extension machinery in potential states, except Jammu and Kashmir.
- Lack of confidence among the rural population.
- Absence of cold chain, processing, value addition and marketing channels.
- Lack of hygiene protocol in culture ponds.
- Lack of high-tech breeding and capital-intensive culture methods.

Opportunities:

- Employment avenues in the rural area.
- Eco-tourism-based fisheries.
- Presence of breeding/seed production infrastructure.
- Adequate demand for consumption.

Threats:

- Environment degradation, damming of rivers, tributaries, and streams.
- Non-availability of low-cost indigenous substitutes.
- Absence of linkages for large growth expansion among research institutes, government departments and other development agencies.
- Spread of infectious diseases.

Source: DAHDF (2018).

2.6 COLDWATER FISHERIES FOR SPORTS, RECREATION, AND LIVELIHOOD

Coldwater fishery encompasses three major objectives: resource assessment and sustainable utilization, upland livelihood and nutritional security, sports, and ecotourism for job creation (Chettri, 2020). Brown trout (*Salmo trutta fario*), rainbow trout (*Oncorhynchus mykiss*), and certain species of large-scaled barbels are the main sports species in Kashmir, Himachal Pradesh, Uttarakhand, Sikkim, North Bengal, Nilgiris, Kodai hills, and Munnar ranges, which attract a large number of Indian and foreign tourists each year. The sport fishery is an important source of revenue in some areas. Trout alone accounts for 40–50% of the state's fisheries revenue in Jammu and Kashmir (Chettri, 2020).

Ecotourism based on fishing is both a sustainable resource use and a means of environmental conservation. It is a non-consumptive use of biological resources that provides people with socioeconomic benefits. Through tourism generation, the hill region provides an emerging area of employment generation. Many picturesque valleys, riverbanks, and mountains are available for the development of trout or mahseer-based sport fisheries, as well as opportunities for the development of restaurants, transportation, parking, boating, and angling facilities. The government has been working on developing suitable models for the development of fish-based ecotourism. Important locations for the development of sports and angling facilities have been identified (Chettri, 2020).

2.7 COSTS AND RETURN ANALYSIS OF TROUT CULTURE IN JAMMU AND KASHMIR

Costs and return analysis of trout culture on a sample farm of 30 trout farmers in Jammu and Kashmir have been analyzed by Gawa et al. (2017). They analyzed and presented the costs and returns in trout farming per farm and raceway for the sample trout farms in Table 2.3. The annual total cost and gross revenue per farm were estimated to be Rs. 2.36 lakh and Rs. 4.30 lakhs, respectively. Total variable and fixed costs were Rs. 1.77 lakhs and Rs. 0.58 lakhs, respectively, with a share of the total cost of 75.32% and 24.68%. Feed was the most expensive component, accounting for 45.35% of the total cost and playing an important role in determining the profitability of trout farming. Similar results have been obtained by Bozoglu & Ceyhan

(2009); and Bozoglu et al. (2006). They conducted a comprehensive study on trout culture in Turkey. They discovered that the variable cost in trout and sea bass farming was 74.02% and 67.49%, respectively, of the total cost.

TABLE 2.3 Costs and Return of Trout Culture in Sample Farms

Particulars	Cost (Rs/farm)	Cost (Rs/raceway)	Percent Share
Variable Costs			
Seed	49550.00	21543.48	21.04
Feed	106787.50	46429.35	45.35
Medicine and chemicals	1093.42	475.40	0.46
Transportation	9496.67	4128.99	4.03
Hired human labor	365.83	159.06	0.16
Miscellaneous	2648.00	1151.30	1.12
Total working capital	169941.42	73887.57	72.16
Interest on working capital	7434.94	3232.58	3.16
A. Total Variable Cost	**177376.35**	**77120.15**	**75.32**
	Fixed Costs		
Depreciation	14725.09	6402.21	6.25
Interest on fixed capital	34971.71	15205.09	14.85
Annual repair and maintenance	7422.33	3227.10	3.15
Land rent	1000.00	434.78	0.42
B. Total Fixed Cost	**58119.13**	**25269.18**	**24.67**
Total cost (A+B)	235495.48	102389.34	100.00
Total production (Kg)	1105.67	–	–
Cost of production (Rs/Kg)	212.99	–	–
Selling price (Rs)	389.17	–	–
Farmer's margin (Rs/Kg)	176.18	–	–
Gross revenue	430291.67	–	–
Net revenue	194796.19	–	–
B:C ratio	1.83	–	–

The second most crucial factor that influences the cost of trout farming is seed cost, which accounted for 21.04% of the total cost of trout farming at Rs. 21,543 per raceway. The government-recommended stocking density is 3,000 fingerlings per raceway, followed by most trout farms. Feed and seed transportation costs accounted for 4.03% of total expenses. It was discovered that trout farms were dispersed throughout the valley. Due to a lack of retail outlets for feed and seed, farmers were forced to purchase them directly from the producing centers, incurring significant transportation costs. Bozoglu & Ceyhan (2009) stated that feed cost accounted for 45.53% and 47.73% of the total cost of trout and sea bass culture, respectively. This was a crucial factor in determining profit margin. The feed was the most expensive aspect of trout production (Gawa et al., 2017). Under fixed cost, interest on fixed capital accounted for 14.85% of the total cost, followed by depreciation, which accounted for 6.25% of the total cost. The other cost components were interesting on working capital, annual repair and maintenance, medicine, and chemicals, land rent, hired human labor costs, and miscellaneous charges, which accounted for 3.16, 3.15, 0.46, 0.42, 0.16, and 1.12% of the total cost, respectively. The benefit-cost ratio was estimated to be 1.83, indicating the business's economic viability, which is consistent with the findings of Olaoye (2013), who estimated that variable costs account for 86.68% and fixed costs account for 13.32% with a BCR of 1.69, indicating the economic viability of trout farming in Nigeria. They discovered a significant difference in fixed costs between earthen and concrete ponds, with earthen ponds being less expensive. Gawa et al. (2017) stated that because trout farming is practiced in concrete raceways in Kashmir, the cost of building a raceway is very high. Trout farming was profitable even without subsidies, so subsidies were not accounted for in profit when calculating costs and returns. There is a strong need to encourage Kashmiri youths to pursue trout culture in raceways as a means of earning a living. This can be accomplished by improving their technical and managerial skills in trout farming and facilitating easy access to institutional credit.

2.8 A CASE STUDY ON THE ECONOMICS OF COLD-WATER FISH (TROUT) FARMING IN SIKKIM

Trout farming is one of the oldest commercial fish production methods. It has over 400 years in Europe and about 150 years in the United States (Hinshaw, 1990). Chettri (2021) recently examined trout farming in Sikkim and found it to be a significant and beneficial employment opportunity for the people of the hilly region. He conducted a cost-benefit analysis, as shown in Table 2.4.

As per the latest specifications and design of the DOF of Sikkim, the ideal size of the trout raceway is 51 m^3 which costs around ₹ 1.20 lakhs. Chettri (2021) observed that stocking 1,500 rainbow trout fingerlings at Rs 20 per each cost 0.30 lakhs, and the cost of feed, including transportation, is around 1.00 lakh. As a result, the total operating cost is 1.30 lakhs. On average, 500 kg of trout fish at 800 per kg would generate Rs 4.00 lakhs in revenue. The net profit from a single racetrack is 2.7 lakhs, which is a very volatile and reasonable profit for the farmers.

TABLE 2.4 Cost and Benefit Analysis of Trout Fish Farming

SL. No.	Particulars	Details	Indian Rupee
1.	Capital cost	Construction of trout raceways as per design/ specification. Size 17 m × 2 m × 1.5 m = 51 m^3	1.20 lakhs
2.	Species	Rainbow trout	0.30 lakhs
		Number of stocks – 1,500 fingerlings	1.00 lakhs
		Operating cost: (Recurring)	
		i) Cost of seeds @ Rs 20/-per each for 1,500 numbers including transportation cost.	
		ii) Cost of feed including transportation cost.	
3.	Benefits (B) production	Trout fish average growth 400–600 grams per year. Mortality 20%. Total production 500 kg @ Rs 800 per kg.	4.00 lakhs
4.	Net profit B-C	Rs. 4.80 lakhs–1.30 lakhs	2.7 lakhs

2.9 CHALLENGES AND PROSPECTS OF TROUT FARMING

The perennial springs and streams in the Himalaya are flowing from the uplands to lowlands are not adequately and fully utilized. The coldwater fisheries sector faces problems such as poor accessibility, difficult hilly terrain, a lack of transportation and an inefficient market, a lack of aquaculture infrastructure, etc. As a result, this sector could not foster to the expected and extensive level. The major problems faced by the trout farmers in hill region are the non-availability of fish feeds, high cost of feeds, and lack of required fish seeds on time, etc. Even though such problems exist, there is a massive prospect of fisheries and fish farming in the hilly state and can enhance production level if some of the following strategies are considered:

- The available water resources could be brought into fish farming and aquaculture development by utilizing these resources efficiently and adequately rather than going into waste.
- Awareness of the importance and role of fisheries and aquaculture for rural development and nutritional value can motivate the younger generations, resulting in greater participation of people in fish farming.

- Fish is a highly perishable product; proper accessibility, cold storage, adequate transportation, and marketing facilities in the rural areas could foster this sector so that local fish can be readily available in the market.
- The use of modern technology and training for farmers can increase production and productivity.
- The retail local fish shop should be started at the district level so that the local fish will be readily available in the market, and this will also encourage fish farmers to produce local fish.
- Although water resources are plentiful, many problems need immediate attention from the government and the management teams to get better results in steady growth in production, productivity, profitability, sustainability, etc.
- To minimize the major problems faced by the farmers, such as the nonavailability of feed on time and high feed costs, the government should frequently monitor or establish at least some feed mill plants within the states to redress such problems.

2.10 FUTURE PROSPECTIVE

Our country's hill regions are endowed with valuable indigenous fish germplasm and pristine water resources with a wide range of thermal regimes. The fish can play an important role in supplementing protein requirements for poor people living in remote Himalayan regions and providing a source of income for a subset of people who overexploit natural resources due to resource constraints in terms of cultivable lands in hills. At high altitudes, there is enormous potential for the development of low volume, high value species such as trout, particularly the rainbow variety. Production of 150 tons is possible with a 4% growth rate in rainbow trout farming, which can be increased to 200 tons with an 8% growth rate. According to the study, our domestic demand is estimated to be 800 tons; therefore, steps must be taken to meet this target. This can even help the feed industry and preservation units at high altitudes. Similarly, sport fishing ancillary units will be operational, such as short and long-distance trout transportation. It will all pave the way for the formation of self-help groups in hill regions. However, it will be possible if the seed production centers and table fish production units for trout are separated, and the private sector is involved in fish production. In contrast, quality seed and feed supply are guaranteed through state channels. To maintain sanitary standards, all private units should be registered (Sarma et al., 2018).

Training in coldwater fishery and hill resource management is another issue that needs to be addressed. Once upon a time, the confluence of rivers in the northeast and the central Himalayas was reported to be one of the best mahseer angling spots in the world. As a result, creating adequate facilities for anglers for sport fishing is critical. Similarly, streams in Kashmir and Himachal Pradesh are world-renowned for brown trout angling and fishing. Commercial farming through Jhora fishery in sub Himalayan West Bengal has a good potential for generating rural income. It will help to improve the socioeconomic conditions of the local population in this hill region when combined with sport fishery development. Another requirement for hill states is the establishment of carp hatcheries and farms at lower altitudes to meet the demand of local farmers engaged in small-scale fish farming. Running water fish culture based on exotic carp has enormous potential. This should be encouraged by providing farmers with knowhow and seed to supplement their income. Despite the fact that some effective efforts have been launched to develop and popularize fisheries in our country's hill region, there is still much work to be done. A holistic approach required for overall fisheries development includes expanding fish culture activities in all potential areas, integrated aquaculture, stock diversification, implementation of sustainable production, enhancement measures in lakes and reservoirs, development of ornamental fish, and promotion of fishery-based ecotourism at suitable sites, in addition to conservation of valuable fish species (Mahanta et al., 2011).

Depending on the natural aquatic resources in the hills, the fishery, if developed scientifically, will contribute significantly to the rural economy in remote hilly zones. It is suggested that fishery development begins focusing on available resources, including both exotic and indigenous fishes. There is a pressing need to replenish depleted stocks. Stocks of some of the most important commercial and recreational fishers. The indiscriminate use of chemicals in orchards/tea gardens may impact the water quality of catchments in regional river/stream systems (Sarma et al., 2018).

2.11 CONCLUSION

The Himalayan regions of India have an abundance of water resources in the form of rivers, lakes, tanks, and reservoirs, etc., which have diverse varieties of fish species that are exotic and indigenous, cultivable, and non-cultivable. These water resources are suitable for the capture as well as culture fisheries. Jammu & Kashmir have the most prominent river or canals, i.e., 27,781 km. In major hilly areas, the total fishermen population is estimated at 1.40 lakh fishers who directly depend on coldwater fisheries. Total coldwater fish seed production is 4388.9 lakh fry from all hilly states of India, Manipur producing the highest fish seed with 2494.8 lakh fish fry production. In 2019–2020, the states mentioned above showed an increasing trend except for Sikkim, which offers a minor reduction in production. The annual total cost and gross revenue per farm in Jammu and Kashmir were estimated to be Rs. 2.36 lakh per farm and Rs. 4.30 lakh per raceway, with a benefit-cost ratio of 1.83, indicating the business's economic viability (Gawa et al., 2017). A trout raceway of 51 m^3 in Sikkim, which costs around 1.20 lakhs, has been observed to produce 500 kg of trout fish at an average price of 800 per kg, yielding a 4 lakhs income. The net profit from a single trout raceway (51 m^3) would come to ₹ 2.7 lakhs, which is a very venturing and reasonable profit to the farmers (Chettri, 2021). Significant challenges in coldwater fish farming are the availability of water resources for culture, unawareness of fish farming, lack of cold storage, quick transport, and other marketing facilities, and lack of new technical knowledge. Future plans also suggested to improve coldwater fisheries in the country are technology demonstration, establishing a brood banking facility for the availability of brood through the year, and implementation of diversifying culture systems and open water cage culture to use water resources efficiently. Training in coldwater fishery and hill resource management needs to be addressed for further development.

KEYWORDS

- **benefit-cost ratio**
- **coldwater**
- **ecotourism**
- **fish farming**
- **marketing**
- **socio-economic development**

REFERENCES

Anon. (2011). *Handbook of Fisheries and Aquaculture* (2nd edn.). Directorate of knowledge management in agriculture–ICAR, New Delhi.

Ayyappan, S., & Krishna, M., (2007). Changing consumption pattern. *The Survey of Indian Agriculture, 6*(6).

Bozoglu, M., & Ceyhan, V., (2009). Energy conversion efficiency of trout and sea bass production in the Black Sea, Turkey. *Energy, 34*, 199–204.

Bozoglu, M., Ceyhan, V., Avni, C. H., Demiryürek, K., & Kilic, O., (2006). Evaluation of different trout farming systems and some policy issues in the Black Sea Region, Turkey. *J. Appl. Sci., 6*, 2882–2888.

Chettri, K. B., (2020). *Ecology and Economics of Coldwater Inland Fisheries in Sikkim* (p. 261). Ph.D. thesis.

Chettri, K. B., (2021). Current status of rainbow trout culture in Sikkim: A sustainable farming system in the hills. *Indian J. Hill Farm., 34*, 126–133.

Chettrih, K. B., & Kundu, R., (2020). Evaluation of socio-economic conditions of coldwater fish farmers: A case study of Sikkim. *Indian J. Econ., 16*, 62–71.

DAHDF, (2018). *Mission Cold Water Fisheries Development National Action Plan – 2017–2022*. Ministry of Agriculture and Farmers Welfare, Government of India.

DoF., (2020). *Handbook on Fisheries Statistics*. Ministry of Fisheries, Animal Husbandry and Dairying. Government of India. New Delhi. https://dof.gov.in/sites/default/files/2021-02/Final_Book.pdf (accessed on 3 February 2023).

Ganesh, K. B., Shyam, S. S., & Katiha, P. K., (2014). *Coldwater Fisheries* (pp. 129–154). Livelihood Status of Fishers in India.

Gawa, S., Kumar, N. R., Tiwari, V. K., Prakash, S., Yadav, V. K., & Wani, G. B., (2017). Trout culture in Kashmir: An opportunity for profitable enterprise. *Social Entrepreneurship in Aquaculture*, 381–389.

Hinshaw, J. M., (1990). *Trout Production: Handling Eggs and Fry* (p. 220). Southern Regional Aquaculture Center SARC.

Kumar, R., Umesh, K. B., & Rajesh, E., (2007). Economics of inland fish production Karnataka. *J. Rural Dev.,* 26.

Mahanta, P. C., Moza, U., & Joshi, K. D., (2011). Coldwater fisheries and aquaculture. *Handbook of Fisheries and Aquaculture*. Published by DKMA, ICAR, New Delhi.

Olaoye, O., (2013). Assessment of socioeconomic analysis of fish farming in Oyo State, Nigeria. *Glob. J. Sci. Front. Res. D: Agric. Vet., 13*, 45–53.

Sarma, D., Akhtar, M. S., & Singh, A. K., (2018). *Coldwater Fisheries Research and Development in India,* 93–133.

Shyam, S., Raina, H. S., & Joshi, C. B., (1999). *Fishes of Indian Uplands* (p. 64). Bull. No.2, NRC on Coldwater Fisheries, Bhimtal.

Tyagi, B. C., Sundar, S., & Mohan, M., (2005). *Coldwater Fisheries Research and Development in Northeast Region of India–a Report of National Research Center on Cold Water Fisheries*. Bhimtal.

CHAPTER 3

Nutritional Requirements of Coldwater Fishes

SEEMAB ZEHRA[1], ASAAD H. MOHAMED[1], EDOARDO PANTANELLA[1], and RAMZY A. YOUSIF[2]

[1]Beacon Development, King Abdullah University of Science and Technology, Thuwal, Jeddah, Saudi Arabia

[2]Department of Fisheries and Wildlife Science, Sudan University of Science and Technology, Khartoum, Sudan

ABSTRACT

Aquaculture is one of the fastest-growing food-producing sectors globally, and coldwater fish species are the important aquaculture species farmed in different parts of the world. Coldwater fish are an important source of protein for the hill people, and there is an urgent need for coldwater fisheries development with the assistance of international organizations. Nutrition is a crucial factor for fish production as it plays a vital role in boosting aquaculture production. Successful and sustainable aquaculture depends on nutritionally balanced, environmentally friendly, and economically viable practical feeds. The literature on the nutritional requirements regarding coldwater fisheries is sparse. To improve fish productivity, it is important to provide specific and efficient diets based on satisfying the requirements of coldwater fish species for various nutrients. This chapter explores the literature related to the protein, amino acid (AA), lipid, fatty acid, vitamin, and mineral requirements of coldwater fishes which will be useful for boosting the aquaculture of these fishes.

Coldwater Fisheries and Aquaculture Management: Technology for Sustainable Food Production.
Mohd Ashraf Rather, Faisal Rashid Sofi, Adnan Amin, & Kawkabul Saba (Eds.)

3.1 INTRODUCTION

Nutrients are compounds in foods, dispensable to life, providing energy, the building blocks for repair and growth, and substances essential to regulate several physiological activities. Due to the ever-increasing population globally, many people are suffering from malnutrition, and thus, there is a dire need to fill the global food basket. Aquatic food is a crucial constituent of the worldwide food basket to over the people's malnutritional problems. Aquaculture can be an alternative to meet the growing global demand for nutritious food fish for a rapidly expanding population and contribute to the growth of national economies. Fish can make a unique contribution to improve and diversify dietary intake and promoting nutritional well-being among most population groups. Fish and other seafood forms provide a balanced diet and reduce the protein gap. Fish have a highly desirable nutrient profile providing an excellent source of high-quality animal protein containing all the 10 essential amino acids (AAs) in desirable concentrations that are easily digestible and of high biological value. Feeds and feeding are essential elements in the aquaculture of organisms. Hence, knowledge of nutrition and practical feeding of fish is essential for sustainable aquaculture. In aquaculture systems, feed is responsible for more than one-half of the variable cost, often ranging from 50–60% depending upon its intensity. Any reduction in the feed cost through their development improved husbandry and, therefore, is necessary for the industry's development and well-being. Production of eco-friendly feeds depends on fish species' nutritional requirements and meeting those requirements with balanced feed formulations and appropriate feeding practices.

Coldwater fisheries occupy an important place amongst the freshwater fishes of India. The coldwater fisheries deal with fisheries activity in the water where water temperature ranges from 5°C to 25°C. India's coldwater fisheries resources are mainly in upland streams, rivers, high, and low altitudinal lakes, and reservoirs located in different hill states of India. Diverse populations of indigenous and a few exotic coldwater fish species (Tables 3.1 and 3.2) in these mountain water bodies form an immense potential to generate income in rural areas and provide food and nutritional security. The value of fish as a health food is well known but poorly documented. Fish is the cheapest animal protein accessible to the poor and is aptly called the 'rich food of the poor.' Apart from essential fatty acids (EFAs) and protein, fish is rich in vitamins and minerals. Fish as a protective food assumes greater significance in India's different regions, including Himalayan coldwater regions due to

TABLE 3.1 Important Cold-Water Fishes

Snow Trout	Minor Carps	Mahseer	Exotic	Minor Carps	Barils/Minnows/ Catfishes/Loaches
Schizo thorax richardsonni	*Labeo dyocheilus*	*Tor putitora*	*Onchorhynchus mykiss*	*Labeo dyocheilus*	*Barilius bendelisis*
Schizothoraichthys curvifrons	*Labeo dero*	*T. tor*	*Salmo trutta fario*	*Labeo dero*	*B. bakeri*
S. longipinnis	*Crossocheilus latius*	*T. khudree*	*Salvelinus fontinalis*	*Crossocheilus latius latius*	*B. vagra*
S. esocinus	*Gara gotyla*	*T. malabaricus*	*Cyprinus carpio var. specularis*	*Gara gotyla*	*B. bariia*
S. niger	*G. hughi*	*Neolissochilus hexagonolepis*	*C. carpio var. communis*	*G. hughi*	*Raimas bola*
S. plannifrons	*Puntius ophicephalus*	–	*C. carpio Var. nudus*	*Puntius ophicephalus*	*Danio divario*
S. micropogon	–	–	*Tinea tinea*	–	*Botia birdi*
S. progastus	–	–	*Carrasius carrasius*	–	*Glyptothorax pectinopterus*
S. nasus	–	–	–	–	*G. conirostre conirostre*
S. hugeiii	–	–	–	–	–
Lepidopygopsis typus	–	–	–	–	–

TABLE 3.2 Important Edible Fishes of India Coldwater Fish Species

Ornamental Fish Species	**Brachydanio rerio**	**Danio devario**	**Labeo nandina**	**Lepidocephalus guntea**	**Gageta cenia**	**Conta conta**	**Carassius carassius**	**C. auratus**
Edible fish species	Snow trout	Mahseer	Exotic trout	Minor carps	Barils/minnows/catfishes loaches		Fish species transported from plain area	
	Schizothorax richarsonii;	*Tor putitora*	*Onchorhynchus mykiss*	*Labeo dyocheilus*	*Barilius* spp.		*Hypopthalmichthys molitrix*	
	S. longipinnis	*T. tor*	*Salmo trutta fario*	*Labeo dero*	*Raimas bola*		*Ctenophraryngodon Idella*	
	S. esocinus	*Neolissochilus hexagonolepis*	*Saivelinus fontinalis*	*Gara gottjla*	*Danio divario*		–	
	S. niger	–	*Cyprinus carpio var. specularis*	*Puntius* spp.	*Botia birdi*		–	
	S. progastus	–	*C. carpia var. commums*	–	*Glyptothorax* spp.		–	
	S. nasus	–	*H. molitrix* *C. idella*	–	–		–	
Sports fish species	*Tor putitora*	*T. tor*	*T. khudree*	*T. malabaricus*	*Salmo trutta fario*		–	

the rich fish and fisheries resources of the country, which can play a pivotal role in mitigating protein deficiency and malnutrition. But despite all these advantages of consuming fish, the quantitative nutritional requirements of coldwater fishes from India are not all known, which is the major impediment in their culture. Also, the lack of consistent data on the essential nutrients requirements of different growth stages of coldwater fish species makes it difficult to correctly formulate diets that maximize growth to their full genetic potential. Hence, it is essential to develop nutritionally balanced quality feeds for different growth stages to develop aquaculture of coldwater fish species. To formulate nutritionally balanced feeds, there is a dire need to determine the nutritional requirements of concerned species and scan the nutrient profile of locally available feed ingredients. This information will enable us to increase the production of nutritionally balanced feeds. Any balanced formula for fish diet should include essential AAs, energy sources, EFAs, specific vitamins, and minerals to support life and promote growth.

3.2 PROTEIN AND AMINO ACID (AA) REQUIREMENTS

The proteins are essential constituents representing the most extensive group of chemicals in the organism. Generally, 16% protein is contained by the whole fish carcass. Proteins are a crucial component of the cell and accordingly account for the bulk of the muscle tissues, internal organs, brain, nerves, and skin. As scarcity and excess of dietary protein affect growth performance and general body maintenance, the generation of data on the actual protein requirement of fish is vital to formulate well-balanced, low-cost, and environmentally friendly feeds (Zehra & Khan, 2012). The study of dietary nutrient requirements in fish has been almost based on reviews comparable to those performed on terrestrial farm animals. Therefore, it follows that nearly all the existing data on the dietary nutrient requirements of aquaculture species is derived from laboratory-based experiments. The animals are kept in a controlled environment at high density and have no availability of natural food organisms (FAO, 2014). Dietary protein requirements are usually expressed in terms of a dietary percent or as a ratio of protein to dietary energy. There is not much information documented on the protein nutrition of cold water fishes. A study has been conducted on the performance of locally available low-cost feed ingredients on *Tor putitora* to introduce this species into commercial aquaculture (Islam, 2002). Also, aqua-culturists have been trying to find out nutritionally balanced diet by incorporating different ingredients in varied proportions to realize better production levels. The incorporation of silkworm pupae as a protein source in the diet of Deccan Mahseer was tried by Shyama (1990), who reported no harmful effect on flesh quality, the optimum level of inclusion being 60%. Spirulina was used as a significant protein source for the species by Keshavanath et al. (1986). Several experiments and trials have been conducted at National Research Center on Coldwater Fisheries (NRCCF) to formulate diets for various life stages of golden Mahseer by using local ingredients like soya meal, silkworm pupae, rice/wheat starch, etc. Based on these investigations, it was observed that the early rearing stages of Mahseer up to advance fry/fingerlings (45–55 mm) require about 45% protein (Mohan, 2002). The information on protein nutrition of *Tor khudree* (Shankar, 1988); *Tor putitora* (Hossain et al., 2002; Mohan, 2002; Sawhney & Gandotra, 2010; Akram & Swapna, 2014) and *Tor tambroides* (Misieng et al., 2011) has been documented. Table 3.3 summarizes the known quantitative protein requirements of some of the coldwater fish species.

TABLE 3.3 Protein Requirements of Cold-Water Fish Species

Fish Species	Protein Requirement (%)	References
Rainbow trout	35–50	NRC (2011)
Tinca tinca	50	Garcia (2015)
Cyprinus carpio	28–47	Nose (1979); NRC (2011)
	41	Ahmed & Maqbool (2017)
Ctenopharyngodon idella	38	Wang (2005)
	33	Koprucu (2012)
	35	Shang (2009)
Tor putitora	40	Hossain (2002)
	45–50	Islam (2004)
Schizothorax richardsonii	45	Wagle et al. (2016)

Fish *do not* have a dietary need for *protein per se*; they need AAs that comprise protein. The AA is made up of the basic amino group (NH_2), the acidic carboxyl group (COOH), and the R. The AAs structure units form proteins. They combine to form short polymer chains called peptides or long chains called polypeptides or proteins. In nutrition, AAs are classified as essential or non-essential. Essential AAs, also known as essential AAs, are a group of AAs that humans and other vertebrates can synthesize from metabolic intermediate. These AAs must be provided in food because the human body and other vertebrates do not have the necessary health science methods to make these AAs. Non-essential, also known as available AAs, can be eliminated from food.

The 20 to 22 AAs that comprise proteins are as follows:

Essential Amino Acids for Fish	Non-Essential Amino Acid
Arginine	Alanine
Histidine	Asparagine
Isoleucine	Aspartic acid
Leucine	Cysteine
Lysine	Glutamic acid
Methionine	Glutamine
Phenylalanine	Glycine
Threonine	Proline
Tryptophan	Serine
Valine	Tyrosine
–	Selenocysteine
–	Pyrrolysine

From a quantitative perspective, protein utilization and muscle protein growth efficiency are the most critical issues and improving the inclusion of AAs in line with demand and improving protein utilization in the growth of body proteins with minimal environmental impact is common in all fish production systems (Kaushik & Seiliez, 2010). AAs are the important component to synthesize proteins and other nitrogenous compounds and serve as important regulators of metabolic pathways. Deficiency of any essential AA will impair protein synthesis and suppress fish growth (Masagounder et al., 2009). Moreover, feeding excess AA concentrations can increase ammonia release and spoil water quality (Hart et al., 2010). It is necessary to know the fish species' specific AA requirements and prepare mixtures of protein supplemented with the deficient AAs and achieve maximum growth and protein efficiency. Based on feeding techniques developed earlier for other animals, the dietary AA requirements of fish were first investigated by Halver (1957). Fish were fed a balanced diet containing graded levels of a test AA (casein: gelatin mixture supplemented with crystalline AAs to simulate the whole hen's egg protein). Halver (1957) determined the arginine, histidine, isoleucine, leucine, lysine, methionine, phenylalanine, threonine, tryptophan, and valine requirements for chinook salmon and concluded the essentiality of 10 AAs. The dietary needs of 10 essential AAs for various cultivable fish species have also been confirmed by several researchers (Khan & Zehra, 2020). Most of the AA requirements have been estimated using dose-response and carcass deposition methods in which replicate groups of fish were fed diets containing graded levels of the test AA until significant differences appear in fish growth. The AAs requirement of coldwater fish species is given in Table 3.4. Quantitative dietary requirements for all 10 essential AAs have been estimated for only Atlantic salmon, Chinook salmon, Chum salmon, Coho salmon, Common carp, and Rainbow trout amongst cold water fish species.

3.3 LIPIDS REQUIREMENTS

Lipid is chemically defined as soluble in water and soluble in alcohol, ether, and chloroform. Lipids are an essential part of living cells. Lipids (fats) are very powerful nutrients that can be used to save part of the protein in the diet of marine animals. Lipids provide twice as much energy as proteins and carbohydrates. Lipids usually comprise about 15 g/kg of fish diet, provide essential oils (EFA) and act as a medium of soluble vitamins. The latest practice of feeding fish is to use high amounts of lipids in the diet. Although the increase in lipids you eat can help diminish the escalating cost of food by keeping a portion of the protein in the diet, problems such as excess fat in the liver can reduce the quality of life and the fish market. Light lipids include fatty acids and triacylglycerol. Fish usually need fatty acids in the omega 3 and 6 families (n–3 and n–6). Fatty acids can be: (i) saturated fatty acids (SFA, no double bonds); (ii) polyunsaturated fatty acids (PUFA, > 2 double bonds); or (iii) saturated fatty acids (HUFA);> 4 double bonds). Marine fish oil is naturally high (> 30%) in omega 3 HUFA (highly unsaturated fatty acids) and is an excellent source of lipid in fish processing. The lipids from these oily oils can have favorable effects on human health of the heart and blood vessels (Craig, 2009).

Dietary lipids are a crucial source of energy that can be easily digested and are the only source of essential oils (EOs) needed by fish for maximizing growth and further development. They are also a carrier and help to absorb fat-soluble nutrients such as sterols and vitamins A, D, E, and K. Due to the incapability of organisms to combine de novo fatty acids in the n–6 and n–3 series, these fatty acids should be supplemented in the diet. In terrestrial animals, the linoleic series (n–6) has been found to have the highest activity of EOs (EFA), with the linolenic series (n–3) having only partial EFA activity. Thus, it follows that most of the fatty acids (PUFA) in terrestrial

TABLE 3.4 Amino Acid Requirements of Cold-Water Fish Species

		Arginine Requirement		
Species	**CP%**	**REQ % Diet**	**REQ % Protein**	**References**
Atlantic salmon	40	1.6	4.1	Lall et al. (1994)
	42	2.0–2.2	4.8	Berge et al. (1997)
	45	–	4.6	Rollin (1999)
Chinook salmon	40	2.4	6.0	Klein & Halver (1970)
Coho salmon	40	2.3	5.8	Klein & Halver (1970)
	45	2.2–2.5	4.9–5.5	Luzzana et al. (1998)
Common carp	48	1.7	4.3	Nose (1979)
Rainbow trout	45	1.6–1.8	3.6–4.0	Walton et al. (1986)
	40	1.4–1.7	3.5–4.2	Chiu et al. (1988)
	43	1.6	4.0	Fournier et al. (2003)
		Isoleucine Requirement		
Atlantic salmon	45	–	3.2	Rollin (1999)
Chinook salmon	41	1.0	2.6	Chance et al. (1964)
Chum salmon	–	–	2.4	Akiyama & Arai (1993)
Coho salmon	–	–	1.2	Aria & Ogata (1993)
Common carp	48	1.0	2.5	Nose (1979)
	40	–	2.3	Ogino (1980)
	33	1.29	3.84	Zhao et al. (2012)
Grass carp	35	1.4	4.0	Shang et al. (2009)
Lake trout	35	0.7–0.9	1.5–2.1	Hughese et al. (1983)
Rainbow trout	40	–	2.4	Ogino (1980)
	33	0.7–1.4	1.5–2.8	Rodehutscord et al. (1997)
		Leucine Requirement		
Atlantic salmon	40	0.81	2.0	Scott et al. (1996)
	45	1.8	–	Rollin (1999)
Chinook salmon	40	0.7	1.8	Klein & Halver (1970)
Chum salmon	40	0.7	1.6	Akiyama et al. (1985)
	–	–	1.6	Akiyama & Arai (1993)
Coho salmon	40	0.7	1.8	Klein & Halver (1970)
	–	–	0.9	Aria & Ogata (1993)
Common carp	48	0.8	2.1	Nose (1979)
	40	–	1.4	Ogino (1980)
Rainbow trout	40	–	1.6	Ogino (1980)
	34	0.5–0.6	1–1.2	Rodehutscord et al. (1997)

TABLE 3.4 *(Continued)*

Species	CP%	REQ % Diet	REQ % Protein	References
		Arginine Requirement		
		Histidine Requirement		
Atlantic salmon	45	–	5.2	Rollin (1999)
Chinook salmon	41	1.6	3.9	Chance et al. (1964)
Chum salmon	–	–	3.8	Akiyama & Arai (1993)
Coho salmon	–	–	3.4	Aria & Ogata (1993)
Common carp	48	1.3	3.3	Nose (1979)
	40	–	4.1	Ogino (1980)
Lake trout	35	1.3–1.7	2.7–3.7	Hughese et al. (1983)
Rainbow trout	40	–	4.4	Ogino (1980)
	43	3.4	9.2	Choo et al. (1991)
	34	1.1–1.4	2.3–2.9	Rodehutscord et al. (1997)
		Valine Requirement		
Atlantic salmon	45	–	3.9	Rollin (1999)
Chum salmon	–	–	3.0	Akiyama & Arai (1993)
Coho salmon	–	–	2.2	Aria & Ogata (1993)
Common carp	48	1.40	3.60	Nose (1979)
	40	–	2.9	Ogino (1980)
Jian carp	34	1.3	4.0	Dong et al. (2012)
Lake trout	35	0.8–1.0	1.8–2.2	Hughese et al. (1983)
Rainbow trout	40	–	3.1	Ogino (1980)
	34	0.8–1.6	1.7–3.4	Rohutscord et al. (1997)
	36	1.4	3.85	Bae et al. (2012)
		Lysine Requirement		
Atlantic salmon	50	2.0	4.0	Anderson et al. (1993)
	–	–	3.2–3.6	Berge et al. (1997)
	45	–	6.1	Rollin (1999)
	43	2.2	5.0	Espe et al. (2007)
	50	2.6	3.0	Helland et al. (2011)
Chinook salmon	–	–	5.0	Halver et al. (1958)
Chum salmon	40	1.9	4.8	Akiyama et al. (1985)
	–	–	5.0	Akiyama & Arai (1993)
Coho salmon	–	–	3.8	Arai & Ogata (1993)
Common carp	48	2.2	5.7	Nose (1979)
	40	–	5.3	Ogino (1980)

TABLE 3.4 *(Continued)*

Species	CP%	REQ % Diet	REQ % Protein	References
		Arginine Requirement		
Grass carp	38	2.1	5.4	Wang et al. (2005)
Rainbow trout	40	–	5.3	Ogino (1980)
	47	–	6.1	Ketola (1983)
	45	1.9	5.3	Walton et al. (1984b)
	45	1.9	4.3	Walton et al. (1986)
	35	1.3	3.7	Kim et al. (1992b)
	34	2.5	5.3	Rodehutscord et al. (1997)
	45	1.8	5.1	Cheng et al. (2003)
	40	2.3	5.8	Encarnação et al. (2004)
	50	2.7	8.4	Wang et al. (2010)
	36	2.2	6.0	Yun et al. (2016)
		Methionine Requirement		
Atlantic salmon	45	–	3.1	Rollin (1999)
	42	0.7	1.7	Espe et al. (2008)
Chinook salmon	–	–	4.6	Halver et al. (1959)
Coho salmon	–	–	2.7	Aria & Ogata (1993)
Common carp	48	0.8	2.0	Nose (1979)
	40	–	1.6	Ogino (1980)
	40	0.8	2.1	Schwarz et al. (1998)
Rainbow trout	40	–	1.8	Ogino (1980)
	–	–	2.2	Walton et al. (1982)
	–	–	3.0	Rumsey et al. (1983)
	40	0.8	1.9	Cowey et al. (1992)
	45	0.5	1.5	Kim et al. (1992a)
	34	0.9	1.9	Rodehutscord et al. (1995)
Grass carp	28	0.6	2.1	Wu et al. (2017)
		Phenylalanine Requirement		
Chinook salmon	41	1.7	4.4	Chance et al. (1964)
Chum salmon	–	–	6.3	Akiyama & Arai (1993)
Coho salmon	–	–	4.5	Aria & Ogata (1993)
Common carp	48	1.3	3.3	Nose (1979)
	40	–	4.9	Ogino (1980)
Juvenile grass carp	40	2.4	6.5	Gao et al. (2015)
Nile tilapia	28	1.1	3.8	Santiago & Lovell (1988)

TABLE 3.4 *(Continued)*

Species	CP%	REQ % Diet	REQ % Protein	References
		Arginine Requirement		
Rainbow trout	40	–	5.2	Ogino (1980)
	35	0.7	2.0	Kim (1993)
		Threonine Requirement		
Atlantic salmon	45	–	3.2	Rollin (1999)
	40	1.1	2.6	Bodin et al. (2008)
Chinook salmon	40	–	2.2	Delong et al. (1962)
Chum salmon	40	1.2	3.0	Akiyama et al. (1985)
Coho salmon	–	–	2.0	Aria & Ogata (1993)
Common carp	48	1.5	3.9	Nose (1979)
	40	–	3.3	Ogino (1980)
Rainbow trout	40	–	3.4	Ogino (1980)
	–	–	3.2	Rodehutscord et al. (1995)
	40	1.1	2.6	Bodin et al. (2008)
	36	0.95	1.07	Yun et al. (2015)
		Tryptophan Requirement		
Atlantic salmon	45	0.7	1.5	Rollin (1999)
Chinook salmon	–	–	0.5	Halver (1965)
Chum salmon fry	40	0.2	0.7	Akiyama et al. (1986)
Coho salmon	–	–	0.5	Halver (1965)
Common carp	48	0.3	0.8	Nose (1979)
	40	–	0.6	Ogino (1980)
	–	–	0.3	Dabrowski (1981)
Rainbow trout	40	–	0.5	Ogino (1980)
	42	0.2	0.6	Poston & Rumsey (1983)
	55	0.3	0.9	Walton et al. (1984a)
	45	0.3	0.7	Walton et al. (1986)
	–	–	0.6	Kim et al. (1987)
	33	0.2	0.6	Rodehutscord et al. (1997)
Sockeye salmon	–	–	0.5	Halver (1965)

animal tissues are part of the linoleic series, such as 18: 2n–6 (linoleic acid) and 20: 4 n–6 (Arachidonic acid).

Generally, cold water freshwater fish have a specific dietary requirement for n–3 PUFA (18:3 n–3, 20:5 n–3, 22:6 n–3), i.e., salmonids, Ayu, while warm freshwater fish have either a requirement for both the n–3 series and n–6 series PUFA, i.e., carps, eel, and possibly channel catfish, or for the n–6 series alone, i.e., tilapias, and possibly the snakehead *C. micropeltes*; for review see Kanazawa (1985). In the case of marine carnivorous fish species (i.e., black sea bream *M. macrocephalus*, opaleye *G. nigricans*, puffer fish *F. rubripens*, yellow tail *S. quinqueradiata*, plaice *P. platessa*, gilthead bream *S. auratus*, turbot *S. maximus*. Since the food organisms consumed are rich in 22:6 n–3 and 20:5 n–3, they have lost the ability to chain elongate and further desaturate 18:3 n–3 to the corresponding HUFA.

Numerous reviews of fish diets have been published containing information on lipid and fatty acid requirements (NRC, 2011). The lipid and fatty acid requirements for coldwater fish species are summarized in Table 3.5.

3.4 VITAMIN REQUIREMENTS

Vitamins are organic components required in small amounts from an external source for optimal growth, health, proper body metabolism and normal physiological functions. These micronutrients are classified as fat-soluble and water-soluble. Vitamins A, D, E, and K are the fat-soluble vitamins that function independently of enzymes, or in some cases such as vitamin K, may have co-enzyme roles. Eight of the water-soluble vitamins viz. riboflavin, pantothenic acid, thiamine, pyridoxine, nicotinic acid, biotin, cyanocobalamin, and folic acid are required in trace amounts, have mainly co-enzyme functions, and are known as vitamin B complex. Three of the water-soluble vitamins, Choline, Inositol, and vitamin C are required relatively in larger quantities and have the functions other than co-enzymes. Water-soluble vitamins are not stored in optimal amounts in the animal's body; body stores quickly collapse due to a lack of soluble vitamins in the water. Therefore, toxicity of soluble vitamins in water is not possible. Several deficiency signs include corneal vascularization, cloudy lens, hemorrhagic eyes, dark coloration, poor appetite, anemia, poor growth, clubbed gills, necrosis, and scarring, cellular atrophy, gill exudate, sluggishness, lesion in colon, difficult motion, weakness, edema of stomach and colon, muscle spasm while resting fragmentation of erythrocytes, scoliosis, lordosis, impaired collagen formation and altered cartilage have been reported in various cultivable fish species. In order to avoid the prevalence of such diseases and ensuring faster growth provision of vitamin-balanced practical feed is a prerequisite to successful aquaculture. The vitamin requirements have been reported for coldwater fish species and given in Table 3.6.

TABLE 3.5 Lipid and Fatty Acid Requirements of Cold-Water Fish Species

Lipid Requirement of Cold-Water Fish Species				
Common Name	**Scientific Name**	**Requirement (% dry diet)**	**References**	
Gibel carp	*Carassius auratus gibelio*	14.05	Pei et al. (2004)	
Grass carp	*Ctenopharyngodon Idella*	2	Du et al. (2008)	
Juvenile tench	*Tinca tinca*	8.5–12	Royuela et al. (2015)	
Malaysian mahseer	*Tor lambroides*	5	Ng & Andin (2011)	
Manchurian trout	*Brachymystax lenok*	17.38–19.50	Chang et al. (2017)	
Dietary Fatty Acid Requirement of Cold-Water Fish Species (Expressed as Percentage of Dry Diet)				
Common Name	**Scientific Name**	**Fatty Acid**	**Percentage of Dry Diet (Requirement)**	**References**
Atlantic salmon	*Salmo salar*	n–3 PUFA	1%	Ruyter et al. (2000)
		n–3 HUFA	1% (showed better growth than n–3 PUFA)	
Common carp	*Cyprinus carpio*	(i) n–6 PUFA	(i) 1 (0.25% LA)	Radunzneto et al. (1996); Tian et al. (2016)
		(ii) n–3 PUFA	(ii) ~0.05	
Coho salmon	*Oncorhynchus kisutch*	ALA	1%	Yu & Sinnhuber (1979)
Chum salmon	*Oncorhynchus keta*	ALA+LA/n–3 HUFA	1%	Takeuchi, Watanabe, & Nose (1979)
Grass carp	*Ctenopharyngodon idella*	(i) ARA	0.3%	Tian et al. (2014); Li et al. (2015); Zeng et al. (2016)
		(ii) n–3 HUFA (Fish oil and lard oil)	fish oil performed better.	
		(iii) ALA:LA	1.08, 1.19, 1.05%	

TABLE 3.5 *(Continued)*

Dietary Fatty Acid Requirement of Cold-Water Fish Species (Expressed as Percentage of Dry Diet)				
Common Name	**Scientific Name**	**Fatty Acid**	**Percentage of Dry Diet (Requirement)**	**References**
Juvenile fat cod	*Hexagrammos otakii*	n–3 HUFA	1.2–1.6 (squid liver oil performed better than corn, linseed oil and lauric acid)	Lee & Cho (2009)
Rainbow trout	*Oncorhynchus mykiss*	n–3 HUFA	1%	Castell et al. (1972)
Common carp	*Cyprinus carpio*	Citric acid	30 g kg^{-1}	Khajepour et al. (2012)

TABLE 3.6 Vitamin Requirements of Cold-Water Fish Species

Common Name	Scientific Name	Source	Requirement	Reference
			Thiamin	
Common carp	*Cyprinus carpio*	–	0.5 mg	Aoe et al. (1969)
Grass carp	*Ctenopharyngodon idella*	–	1.3 mg, 5 mg	Huang et al. (2014)
Jian carp	*Cyprinus carpio* var. Jian	–	1.02 mg	Huang et al. (2011)
Pacific salmon	*Oncorhynchus* spp.	–	10–15 mg	Halver (1972)
Rainbow trout	*Oncorhynchus mykiss*	–	1–10 mg	McLaren et al. (1947)
Rainbow trout	*Salmo gairdneri*	–	1 mg	Morito et al. (1986)
			Riboflavin	
Common carp	*Cyprinus carpio*	–	4 mg	Aoe et al. (1967a)
	–	–	6.2 mg	Aoe et al. (1967a)
	–	–	7 mg	Takeuchi et al. (1980)
Grass carp	*Ctenopharyngodon Idella*	–	6 mg	Serrini et al. (1996)
	–	–	6.65 mg	Chen et al. (2015)
Jian carp	*Cyprinus carpio* var. Jian	–	5.0 mg	Li et al. (2010)
Pacific salmon	*Oncorhynchus* spp.	–	20–25 mg	Halver (1972)
Rainbow trout	*Oncorhynchus mykiss*	–	5–15 mg 6 mg	McLaren et al. (1947); Takeuchi et al. (1980)
	–	–	3 mg	Hughes et al. (1981)
	–	–	2.7 mg	Amezaga & Knox (1990)
Spotted snakehead	*Channa punctatus*	–	5.7–7.7 mg	Zehra & Khan (2017)
			Niacin	
Brook trout	*Salvelinus fontinalis*	–	95.0 mg	Phillips & Brockway (1947)

TABLE 3.6 *(Continued)*

Common Name	Scientific Name	Source	Requirement	Reference
			Thiamin	
Common carp	*Cyprinus carpio*	–	28 mg	Aoe et al. (1967b)
Grass carp	*Ctenopharyngodon Idella*	–	34.01, 42.08 mg	Li et al. (2015)
	–	–	25.5 mg	Wu et al. (2007a)
Pacific salmon	*Oncorhynchus* spp.	–	150–200 mg	Halver (1972)
Rainbow trout	*Oncorhynchus mykiss*	–	1–5 mg	McLaren et al. (1947)
	–	–	10 mg	Poston & Wolfe (1985)
Common carp	*Cyprinus carpio*	–	30–50 mg	Ogino (1967)
Grass carp	*Ctenopharyngodon Idella*	–	25 mg	Liu et al. (2007)
Jian carp	*Cyprinus carpio* var. Jian	–	23 mg	Wen et al. (2009)
Pacific salmon	*Oncorhynchus* spp.	–	40–50 mg	Halver (1972)
	–	–	17 mg	Leith et al. (1990)
Rainbow trout	*Oncorhynchus mykiss*	–	10–20 mg	McLaren et al. (1947)
	–	–	20 mg	Cho & Woodward (1990)
			Pyridoxine	
Atlantic salmon	*Salmo salar*	–	5 mg	Lall & Weerakoon (1990)
Common carp	*Cyprinus carpio*	–	5–6 mg	Ogino (1967)
Gibel carp	*Carassius auratus gibelio*	–	7.26–11.36 mg	Wang et al. (2011)
Grass carp	*Ctenopharyngodon Idella*	–	1.13–3.73 mg	Yang et al. (2017)
Jian carp	*Cyprinus carpio* var. Jian	–	6.07 mg	He et al. (2009)
Pacific salmon	*Oncorhynchus* spp.	–	10–20 mg	Halver (1972)

TABLE 3.6 *(Continued)*

Common Name	Scientific Name	Source	Requirement	Reference
			Thiamin	
Rainbow trout	*Oncorhynchus mykiss*	–	1–10 mg	McLaren et al. (1947)
	–	–	2 mg	Woodward (1990)
	–	–	3–6 mg	Woodward (1990)
			Biotin	
Common carp	*Cyprinus carpio*	–	1 mg	Ogino et al. (1970a)
Channel catfish	*Ictalurus punctatus*	–	R	Robinson & Lovell (1978)
Jian carp	*Cyprinus carpio* var. Jian	–	0.15 mg	Zhao et al. (2012)
Lake trout	*Salvelinus namaycush*	–	0.1 mg	Poston (1976b)
–	–	–	0.5–1 mg	Poston (1976b)
Pacific salmon	*Oncorhynchus* spp.	–	1–1.5 mg	Halver (1972)
Rainbow trout	*Oncorhynchus mykiss*	–	0.05–0.25 mg	McLaren et al. (1947)
	–	–	0.08 mg	Woodward & Frigg (1989)
			Vitamin B_{12}	
Common carp	*Cyprinus carpio*	–	NR	Kashiwada et al. (1970)
Grass carp	*Ctenopharyngodon idella*	–	0.094	Wu et al. (2007b)
Pacific salmon	*Oncorhynchus* spp.	–	0.015–0.02 mg	Halver (1972)
Rainbow trout	*Oncorhynchus mykiss*	–	R	Phillips et al. (1964)
			Folic Acid	
Common carp	*Cyprinus carpio*	–	NR	Aoe et al. (1967c)
Grass carp	*Ctenopharyngodon idella*	–	3.6–4.3	Zhao et al. (2008)

TABLE 3.6 *(Continued)*

Common Name	Scientific Name	Source	Requirement	Reference
		Thiamin		
Juvenile abalone	*Haliotis discus hannai*	–	2.62–5.29 mg	Miao et al. (2013)
Pacific salmon	*Oncorhynchus* spp.	–	6–10 mg	Halver (1972)
–	–	–	2 mg	Leith et al. (1990)
Rainbow trout	*Oncorhynchus mykiss*	–	1.0 mg	Cowey & Woodward (1993)
		Choline		
Common carp	*Cyprinus carpio*	–	1,500 mg	Ogino et al. (1970b)
Grass carp	*Ctenopharyngodon idella*	–	3,000 mg	Wang et al. (1995)
Lake trout	*Salvelinus namaycush*	–	1,000 mg	Ketola (1967)
Pacific salmon	*Oncorhynchus* spp.	–	600–800 mg	Halver (1972)
Rainbow trout	*Oncorhynchus mykiss*	–	50–100 mg	McLaren et al. (1947)
	–	–	714–813 mg	Rumsey (1991)
		Myoinositol		
Common carp	*Cyprinus carpio*	–	440 mg	Aoe & Masuda (1967)
Grass carp	*Ctenopharyngodon idella*	–	166–214 mg	Wen et al. (2007)
Gibel carp	*Carassius auratus gibelio*	–	165.3 mg	Gong et al. (2014)
Pacific salmon	*Oncorhynchus* spp.	–	300–400 mg	Halver (1972)
Rainbow trout	*Oncorhynchus mykiss*	–	250–500 mg	McLaren et al. (1947)
		Vitamin C		
Atlantic salmon	*Salmo salar*	ECAA	50 mg	Lall et al. (1990)
–	–	AMP	10–20 mg	Sandnes et al. (1992)

TABLE 3.6 *(Continued)*

Common Name	Scientific Name	Source	Requirement	Reference
		Thiamin		
Coho salmon	*Oncorhynchus kisutch*	AA	60 mg	Lim & Lovell (1978)
	–	LAPP	11 mg	El Naggar & Lovell (1991)
	–	LAPP	15 mg	Robinson (1992)
	–	AA	50 mg	Halver et al. (1969)
Common carp	*Cyprinus carpio*	AA	NR	Sato et al. (1978)
	–	LAPP	R	Dabrowski et al. (1988)
	–	–	45 mg	Gouillou-Coustans et al. (1998)
Grass carp	*Ctenopharyngodon idella*	LAPP	25 mg	Henrique et al. (1998)
Jian carp	*Cyprinus carpio* var, jian	AMP	60–100 mg	Teshima et al. (1993)
	–	–	40.9 mg	Liu et al. (2010)
Rainbow trout	*Oncorhynchus mykiss*	AA	50–100 mg	Halver (1969)
	–	AA	40 mg	Hilton et al. (1978)
	–	ECAA	10 mg	Cho & Cowey (1993)
		Vitamin A		
Common carp	*Cyprinus carpio*	–	1.2–6 mg	Aoe et al. (1968)
Gibel carp	*Carassius auratus*	–	0.53 mg	Yang et al. (2017)
–	–	–	0.80 mg	Shao et al. (2016)
Rainbow trout	*Oncorhynchus mykiss*	–	0.75 mg	Kitamura et al. (1967)
		Vitamin E		
Atlantic salmon	*Salmo salar*	–	35 mg	Lall et al. (1988)

TABLE 3.6 *(Continued)*

Common Name	Scientific Name	Source	Requirement	Reference
			Thiamin	
Common carp	*Cyprinus carpio*		50 mg	Wilson et al. (1984); Watanbe et al. (1970)
			100 mg	
Grass carp	*Ctenopharyngodon idella*	–	200 mg	Takeuchi et al. (1992)
	–	–	62.9 mg	Wu et al. (1980)
Juvenile grass carp	*Ctenopharyngodon idellus*	–	100.36 mg	Li et al. (2014)
Pacific salmon	*Oncorhynchus* spp.	–	30 mg	Woodall et al. (1964)
	–	–	40–50 mg	Halver (1972)
Rainbow trout	*Oncorhynchus mykiss*	–	30 mg	Woodall et al. (1964)
	–	–	25 mg	Hung et al. (1980)
	–	–	50 mg	Cowey et al. (1983)
	–	–	100 mg	Watanbe et al. (1981)
			Vitamin D	
Pacific salmon	*Oncorhynchus* spp.	–	NR	Halver (1972)
Rainbow trout	*Oncorhynchus mykiss*	–	40–60 µg	Barnett et al. (1982)
			Vitamin K	
Atlantic salmon	*Salmo salar*	–	< 10 mg	Krossoy et al. (2009)
Atlantic cod	*Gadus morhua*	–	0.2 mg	Grahl-Madsen & Lee (1997)
Grass carp	*Ctenopharyngodon idella*	–	1.9 mg	Jiang et al. (2007)
Lake trout	*Salvelinus namaycush*	–	0.5–1 mg	Poston (1976a)
Pacific salmon	*Oncorhynchus* spp.	–	R	Halver (1972)

3.5 MINERALS REQUIREMENTS

Minerals, or inorganic elements, are needed by organisms to maintain the metabolic processes and essential components of major structural elements. A dietary source of 23 minerals has been demonstrated as essential in one or more animal species. Around 10 minerals (Ca, P, Mg, K, Cu, Fe, Zn, Mn, Si, and I) have been identified as essential nutrient in fish feed. Out of these calcium, phosphorus, chlorine, magnesium, potassium, and sodium, are the most commonly recognized macro-minerals. Therefore, fish must obtain these essential minerals via food or water supplied to them for optimal growth and development. Fish must obtain these essential minerals via food or water supplied to them for optimal growth and development. The main function of these essential minerals in the body of fish includes the formation of skeletal structure, maintenance of colloidal system and regulation of acid base equilibrium. They are also important components of hormones, enzymes, and activations of enzymes. Minerals are required for the transmission of nerve impulses and muscle contraction. They also serve as structural constituents of soft tissues (Kaushik, 2002; Lall, 2002). A number of reviews have been compiled by different authors and National Research Council (NRC, 1993, 2011) and summarized the information on mineral and trace element requirements of fish. Other periodic reviews on minerals and trace element nutrition in fish by Schwarz (1995); Davis & Gatlin (1996); Watanabe et al. (1997); Kaushik (2002); Lall (2002); Prabhu et al. (2013, 2016); Hossain & Yoshimatsu (2014) are also worth mentioning. These workers have reviewed different aspects of the mineral and trace element nutrition in fish. Mineral requirements of coldwater fish species are compiled and presented in Table 3.7.

3.6 CONCLUSION

Nutrition plays a vital role in aquaculture development. Undoubtedly significant progress on the nutrient requirements of aquatic species has been made in the past decades. At present, there is plenty of information available on the nutrient requirements of fish. Inadequate availability of quality seed primarily because of unavailability of nutritionally balanced starter feeds due to fragmented information on nutrients requirements is one of the important factors limiting their production. In this chapter, available information on nutrient requirements of coldwater fish has been compiled. This chapter of the book is a valuable database that will be beneficial in formulating nutritionally balanced, quality protein feeds for boosting the aquaculture production of cold water fish species.

TABLE 3.7 Minerals Requirements of Cold-Water Fish Species

Common Name	Scientific Name	Requirement (g/100 g diet)	References
		Dietary Calcium Requirements	
Common carp	*Cyprinus carpio*	Dispensable FW (20 mg Ca/L)	Ogino & Takeda (1976)
Rainbow trout	*Oncorhynchus mykiss*	Dispensable FW (20–23 mg Ca/L)	Ogino & Takeda (1978)
Japanese flounder	*Paralichthys olivaceus*	Required SW	Hossain & Furuichi (2000)
Chum salmon	*Oncorhynchus keta*	Dispensable FW (20 mg Ca/L)	Watanabe et al. (1980)
Grass carp	*Ctenopharyngodon idella*	1.0 FW (35 mg Ca/L)	Liang et al. (2012)
		Dietary Phosphorus Requirements	
Common carp	*Cyprinus carpio*	0.6–0.7 FW (0.002 mg P/L)	Ogino & Takeda (1976)
Rainbow trout	*Oncorhynchus mykiss*	0.7–0.8 FW (0.002 mg P/L)	Ogino & Takeda (1978)
		0.54–0.61 FW	Ketola & Richmond (1994)
Chum salmon	*Oncorhynchus keta*	0.5–0.6 FW (0.002 mg P/L)	Watanabe et al. (1980)
Atlantic salmon	*Salmo salar*	0.6 (available P) FW (< 0.5 mg P/L)	Ketola (1975)
		0.83–0.93 FW	Vielma & Lall (1998)
		1.0 (0.9 available P) FW	Asgard & Shearer (1997); Lall & Bishop (1977)
Jian carp	*Cyprinus carpio* var. Jian	0.52 (available P) FW	Xie et al. (2011)
Grass carp	*Ctenopharyngodon idella*	0.85 (available P) FW (0.02–2.27 mg/L)	Liang et al. (2011)
		Dietary Magnesium Requirements	
Rainbow trout	*Oncorhynchus mykiss*	0.06–0.07 FW (3.1 ppm Mg)	Ogino et al. (1978)
		0.05 FW (1.2 mg Mg/L)	Knox et al. (1981)
		0.06 FW (1.3 mg Mg/L)	Shearer (1989)
		0.03 FW (1.4 mg Mg/L)	Shearer & Asgard (1992)

TABLE 3.7 *(Continued)*

Common Name	Scientific Name	Requirement (g/100 g diet)	References
Common carp	*Cyprinus carpio*	0.06 FW (9.4 mg Mg/L)	Dabrowska et al. (1991)
		0.04–0.05 FW (3.5 mg Mg/L)	Ogino & Chiou (1976)
	Dietary Phosphorus Requirements		
Atlantic salmon	*Salmo salar*	0.03 FW (54 mg Mg/L)	El-Mowafi & Maage (1998)
Grass carp	*Ctenopharyngodon idella*	0.05–0.07 FW (5.6 mg Mg/L)	Wang et al. (2011b)
	Dietary Potassium Requirements		
Chinook salmon	*Oncorhynchus tshawytscha*	0.80 FW (<1 mg K/L)	Shearer (1988)
Rainbow trout	*O. mykiss*	1.10 FW	Kalantarian et al. (2013)
Grass carp	*Ctenopharyngodon idellus*	0.47–0.73 FW (1.86–3.10 mg K/L)	Zhu et al. (2014)
		0.95–1.00 FW (1.71–8.18 mg K/L)	Liang et al. (2014)
		0.74 FW (6.86–9.10 mg K/L)	Chen et al. (2016)
	Dietary Sodium/Chloride and Sodium Requirements		
	Sodium Chloride		
Rainbow trout	*Salmo gairdneri*	Dispensable FW; 11.6%	Salman & Eddy (1988)
Common carp	*Cyprinus carpio*	Required FW; Level up to 1.5% for better growth and FE	Nasir & Hamed (2016)
	Copper Requirement		
Common Name	**Scientific Name**	**Requirement (mg/kg)**	**References**
Juvenile Atlantic salmon	*Salmo salar*	5–10	Lorentzen et al. (1998)
Juvenile Atlantic salmon		37	Berntssen et al. (1999)

TABLE 3.7 *(Continued)*

Common Name	Scientific Name	Requirement (g/100 g diet)	References
Juvenile common carp	*Cyprinus carpio*	3	Ogino & Yang (1980)
Juvenile rainbow trout	*Oncorhynchus mykiss*	3	Ogino & Yang (1980)
Young grass carp	*Ctenopharyngodon idella*	4.70–4.95	Tang et al. (2013)
		Zinc Requirement	
Juvenile Atlantic salmon	*Salmo salar*	37–67	Maage & Julshamn (2001)
Common carp	*Cyprinus carpio*	15	Ogino & Yang (1979)
		15–30	Satoh et al. (1992)
Jian carp	*Cyprinus carpio* (*var. Jian*)	43.2–48.7	Tan et al. (2011)
Rainbow trout	*Oncorhynchus mykiss*	15	Ogino & Yang (1978)
		20	Satoh et al. (1987)
		40	Satoh et al. (1987)
		80	Satoh et al. (1987)
		Selenium Requirement	
Atlantic salmon	*Salmo salar*	<1.2	Lorentzen et al. (1994)
Atlantic salmon	*Salmo salar*	0.14	Poston et al. (1976)
Coho Salmon	Oncorhynchus kisutch	8.6	Felton et al. (1996)
Juvenile golden Mahaseer	*Tor putitora*	0.68	Khan et al. (2016)
Iron Requirement			
Atlantic salmon	*Salmo salar*	60–100	Andersen et al. (1998)
		60	Naser (2000)
Juvenile common carp	*Cyprinus carpio*	147.4	Ling et al. (2010)
Grass carp	*Ctenopharyngodon idella*	300	Su et al. (2007)
		Manganese Requirement	
Atlantic salmon	*Salmo salar*	7.5–10.5	Maage et al. (2001)
Juvenile common carp	*Cyprinus carpio*	12.0–13.0	Satoh et al. (1987)

TABLE 3.7 *(Continued)*

Common Name	Scientific Name	Requirement (g/100 g diet)	References
Juvenile grass carp	*Ctenopharyngodon idella*	20.6	Liang et al. (2015)
Rainbow trout	*Oncorhynchus mykiss*	19	Satoh et al. (1991)
Young grass carp	*Ctenopharyngodon idella*	16.91 and 18.21	Tang et al. (2016)

KEYWORDS

- **amino acid**
- **coldwater fishes**
- **essential oils**
- **minerals**
- **nutrient**
- **protein**
- **vitamin**

REFERENCES

Ahmed, I. I., & Maqbool, A., (2017). Effects of dietary protein levels on the growth, feed utilization and haemato-biochemical parameters of freshwater fish, *Cyprinus carpio* var. specularis. *Fish. Aquacult. J., 8*, 1–12. doi: 10.4172/2150-3508.1000187.

Akiyama, T., & Arai, S., (1993). In: Collie, M. R., & McVey, J. P., (eds.), *Proceedings of the Twentieth U.S.–Japan Symposium on Aquaculture Nutrition* (p. 35). Hatfield Marine Science Center, *Newport*, OR.

Akiyama, T., Arai, S., Murai, T., & Nose, T., (1985). *Bull. Japan. Soc. Sci. Fish., 51*, 1005.

Akiyama, T., Arai, S., Murai, T., & Nose, T., (1985). Threonine, histidine and lysine requirements of chum salmon fry. *B. Jpn. Soc. Sci. Fish., 51*, 635–639.

Akiyama, T., Murai, T., & Mori, K., (1986). Role of tryptophan metabolites in inhibition of spinal deformity of chum salmon fry caused by tryptophan deficiency. *B. Jpn. Soc. Sci. Fish., 52*, 1255–1259.

Akram, L., & Swapna, L., (2014). Effect of dietary protein levels on growth, feed utilization, protein retention efficiency of *Tor tor* (Hamilton, 1822). *J. Aquacult. F. Sci. and Nutri., 6*(1), 1–5.

Amezaga, M. R., & Knox, D., (1990). Riboflavin requirements in on-growing rainbow trout, *Oncorhynchus rnykiss*. *Aquacult., 88*, 87–98.

Andersen, F., Lygren, B., & Maage, A., (1998). Interaction between two dietary levels of iron and two forms of ascorbic acid and the effect on growth, antioxidant status and some non-specific immune parameters in Atlantic salmon (*Salmo salar*) smolts. *Aquacult., 161*, 437–451.

Anderson, J. S., Lall, S. P., Anderson, D. M., & McNiven, M., (1993). Quantitative dietary lysine requirement of Atlantic salmon (*Salmo salar*) fingerlings. *Can. J. Fish. Aquat. Sci., 50*, 316–322.

Aoe, H., & Masuda, I., (1967). Water-soluble vitamin requirements of carp. 2. Requirements for p-aminobenzoic acid and inositol. *B. Jpn. Soc. Sci. Fish., 33*, 674–680.

Aoe, H., Masud, I., Mimura, T., Saito, T., Komo, A., & Kitamura, S., (1969). Water–soluble vitamin requirements for carp. 6. Requirements for thiamin and effects of antithiamins. *B. Jpn. Soc. Sci. Fish., 35*, 459–465.

Aoe, H., Masuda, I., & Takada, T., (1967b). Water–soluble vitamin requirements of carp. 3. Requirement for niacin. *B. Jpn. Soc. Sci. Fish., 33*, 681–685.

Aoe, H., Masuda, I., Mimura, T., Saito, T., & Komo, A., (1968). Requirement of young carp for vitamin A. *B. Jpn. Soc. Sci. Fish., 34*, 959–964.

Aoe, H., Masuda, I., Saito, T., & Komo, A., (1967a). Water–soluble vitamin requirements of carp. 1. Requirement for vitamin B2. *B. Jpn. Soc. Sci. Fish., 33*, 355–360.

Aoe, H., Masuda, I., Saito, T., & Takada, T., (1967c). Water–soluble vitamin requirements of carp. 5. Requirement for folic acid. *B. Jpn. Soc. Sci. Fish., 33*, 1068–1071.

Arai, S., & Ogata, H., (1993). In: Collie, M. R., & McVeY, J. P., (eds.), *Proceedings of the Twentieth U.S.–Japan Symposium on Aquaculture Nutrition* (p. 19). Hatfield Marine Science Center, Newport, OR.

Asgard, T., & Shearer, K., (1997). Dietary phosphorus requirement of juvenile Atlantic salmon, *Salmo salar* L. *Aquacult. Nutr., 3*, 17–23.

Bae, J. Y., Park, G., Yun, H., Hung, S. S. O., & Bai, S. C., (2012). The dietary valine requirement for rainbow trout, *Oncorhynchus mykiss*, can be estimated by plasma free valine and ammonia concentrations after dorsal aorta cannulation. *J. Appl. Anim. Res., 40*, 73–79.

Barnett, B. J., Cho, C. Y., & Slinger, S. J., (1982). Relative biopotency of ergocalciferol and cholecalciferol and the role of and requirement for vitamin D in rainbow trout (*Salmo gairdneri*). *J. Nutr., 112*, 2011–2019.

Berge, G. E., Lied, E., & Sveier, H., (1997). Nutrition of Atlantic salmon (*Salmo salar*): The requirement and metabolism of arginine. *Comparative Biochemistry and Physiology Part A: Physiol., 117*(4), 501–509.

Berntssen, M. H. G., Lundebye, A. K., & Maage, A., (1999). Effects of elevated dietary copper concentrations on growth, feed utilization and nutritional status of Atlantic salmon (*Salmo salar* L.) fry. *Aquaculture, 174*, 167–181.

Bodin, N., Mambrini, M., Wauters, J. B., Abboudi, T., Ooghe, W., Boulenge, E. L., Larondelle, Y., & Rollin, X., (2008). Threonine requirements for rainbow trout (*Oncorhynchus mykiss*) and Atlantic salmon (*Salmo salar*) at the fry stage are similar. *Aquacult., 274*, 353–365. doi: 10.1016/j.aquaculture.2007.11.031.

Castell, J. D., Sinnhuber, R. O., Wales, J. H., & Lee, J. D., (1972). Essential fatty acids in the diet of rainbow trout: Lipid metabolism and fatty acid composition. *J. Nutr., 102*, 77–86.

Chance, R. E., Mertz, E. T., & Halver, J. E., (1964). Nutrition of salmonoid fishes: Xn. Isoleucine, leucine, valine and phenylalanine requirements of Chinook salmon and interrelations between isoleucine and leucine for growth. *J. Nutr., 83*, 177–185.

Chang, J., Niu, H. X., Ji, Y. D., Li, S. G., & Xu, G. F., (2017). Effects of dietary lipid levels on growth, feed utilization, digestive tract enzyme activity and lipid deposition of juvenile

Manchurian trout, *Brachymystax lenok* (Pallas). *Aquacult Nutr., 8*(3), 1–8. https://doi.org/10.1111/anu.12598.
Chen, C., Zhu, W., Wu, F., Liu, M., Tan, Q., Han, D., Xie, S., & Liang, X., (2016). Quantifying the dietary potassium requirement of subadult grass carp (*Ctenopharyngodon idellus). Aqucult. Nutr., 22*, 541–549. doi: 10.1111/anu.12279.
Cheng, Z. J., Hardy, R. W., & Usry, J. L., (2003). Plant protein ingredients with lysine supplementation reduce dietary protein level in rainbow trout (*Oncorhynchus mykiss*) diets and reduce ammonia nitrogen and soluble phosphorus excretion. *Aquacult., 218*, 553–565.
Chiu, Y. N., Austic, R. E., & Rumsey, G. L., (1986). Urea cycle activity and arginine formation in rainbow trout (*Salmo gairdneri*). *Journal of Nutrition, 116*, 164G1650.
Cho, C. Y., & Cowey, C. B., (1991). Rainbow trout, *Oncorhynchus mykiss*. In: Wilson, R. P., (ed.), *Handbook of Nutrient Requirements of Finfish* (pp. 131–143). Boca Raton, CRC Press.
Cho, C. Y., & Woodward, B., (1990). Dietary pantothenic acid requirements of young rainbow trout (*Oncorhynchus mykiss*) (abstract). *FASEB J., 4*, 3747.
Choo, P. S., Smith, T. K., Cho, C. Y., & Ferguson, H. W., (1991). Dietary excesses of leucine influence growth and body composition of rainbow trout. *J. Nutr., 121*, 1932–1939.
Cowey, C. B., & Woodward, B., (1993). The dietary requirement of young rainbow trout (*Oncorhynchus mykiss*) for folic acid. *J. Nutr., 123*, 1594–1600.
Cowey, C. B., Adron, J. W., & Youngson, A., (1983). The vitamin E requirement of rainbow trout (*Salmo gairdneri*) given diets containing polyunsaturated fatty acids derived from fish oil. *Aquacult., 30*, 85–93.
Cowey, C. B., Cho, C. Y., Vak, J. G., Weerheim, J. A., & Stãoeart, D. D., (1992). Methionine intake in rainbow trout (*Oncorhynchus mykiss*), relationship to cataract formation and the metabolism of methionine. *J. Nutr*., 1154–1162. doi: 10.1093/jn/122.5.1154.
Craig, A. D. B., (2009). How do you feel—Now? The anterior insula and human awareness. *Nature Reviews Neuroscience, 10*(1), 59–70. doi: 10.1038/nrn2555.
Dabrowska, H., Meyer-Burgdorff, K., & Gunther, K. D., (1991). Magnesium status in freshwater fish, common carp (*Cyprinus carpio,* L.) J the dietary protein-magnesium interaction. *Fish Physiol. Biochau., 9*, 165–172.
Dabrowski, K. R., (1981). Tryptophan requirement of common carp (*Cyprinus carpio* L.) fry. *Anim. Physiol. A Anim. Nutr., 46*(1–5), 64–71. doi: 10.1111/j.1439-0396.1981.tb01338.x.
Dabrowski, K., Hinterleitner, S., Sturmbauer, C., El-Fiky, N., & Wieser, W., (1988). Do carp larvae require vitamin C? *Aquacult., 72*, 295–306.
Davis, D. A., & Gatlin, D. M., (1996). Dietary mineral requirements of fish and marine crustaceans. *Reviews in Fisheries Science, 4*(1), 75–99.
Delong, D. C., Halver, J. E., & Mertze, T., (1962). Nutrition of salmonid fishes. X. Quantitative threonine requirements of chinook salmon at two water temperatures. *J. Nutr., 76*, 174 178.
Dong, M., Feng, L., Kuang, S. Y., Liu, Y., Jiang, J., Hu, K., Jiang, W. D., et al., (2012). Growth, body composition, intestinal enzyme activities and microflora of juvenile Jian carp (*Cyprinus carpio* var. Jian) fed graded levels of dietary valine. *Aquacult. Nutr.*, doi: 10.1111/j.1365-2095.2011.00926.x.
Du, Z. Y., Clouet, P., Huang, L. M., Degrace, P., Zheng, W. H., He, J. G., & Liu, Y. J., (2008). Utilization of different dietary lipid sources at high level in herbivorous grass carp (*Ctenopharyngodon idella*): Mechanism related to hepatic fatty acid oxidation. *Aquacult. Nutr., 14*, 77–92.

El Naggar, G. O., & Lovell, R. T., (1991). L-Ascorbyl-2-monophosphsie has equal antiscorbutic activity as L-ascorbic acid but L-ascorbyi-2-sulfate is inferior to L-ascorbic acid for channel catfish. *J. Nusr., 121*, 1622–1626.

El-Mowafi, A. R. A., & Maage, A., (1998). Magnesium requirement of Atlantic salmon (*Salmo salar* L.) parr in seawater-treated freshwater. *Aquacult. Nutr., 4*, 31–38.

Encarnação, P., deLange, C., Rodehutscord, M., Hoehler, D., Bureau, W., & Bureau, D. P., (2004). Diet digestible energy content affects lysine utilization, but not dietary lysine requirements of rainbow trout (*Oncorhynchus mykiss*) for maximum growth. *Aquacult., 235*, 569–586.

Espe, M., Hevrøy, E. M., Liaset, B., Lemme, A., & El-Mowafi, A., (2008). Methionine intakes affect hepatic sulphur metabolism in Atlantic salmon, *Salmo salar*. *Aquacult., 274*, 132–141. doi: 10.1016/j.aquaculture.2007.10.051.

Espe, M., Lemme, A., Petri, A., & El-Mowafi, A., (2007). Assessment of lysine requirement for maximal protein accretion in Atlantic salmon using plant protein diets. *Aquacult., 263*, 168–178. doi: 10.1016/j.aquaculture.2006.10.018.

FAO, (2014). *Fish Trade and Human Nutrition (COFI Sub-Committee on Fish Trade, 14th Session. Bergen, Norway, COFI: FT/XIV/2014/4)*. Retrieved from: http://www.fao.org/cofi/29401-083ff934c3ccfd8576005d8d0c19b04d6.pdf (accessed on 3 February 2023).

Felton, S. P., Landolt, M. L., Grace, R., & Palmisano, A., (1996). Effects of selenium dietary enhancement on hatchery-reared coho salmon, ft. *Oncorhynchus kisutch* (Walbaum), enzymes and seawater adaptation evaluated. *Aqucult. Res., 27*, 135–142.

Fournier, V., Gouillou-Coustans, M. F., Métailler, R., Vachot, C., Moriceau, J., Le Delliou, H., Huelvan, C., et al., (2003). Excess dietary arginine affects urea excretion but does not improve N utilization in rainbow trout *Oncorhynchus mykiss* and turbot *Psetta maxima*. *Aquacult., 217*, 559–576.

Gao, Y. J., Liu, Y. J., Chen, X. Q., Yang, H. J., & Tian, L. X., (2015). Total aromatic amino acid requirement of juvenile grass carp. (*Ctenopharyngodon idella*). *Aquacult. Nutr.*, doi: 10.111/anu.12306.

Gouillou-Coustans, M. F., Bergot, P., & Kaushik, S. J., (1998). Dietary ascorbic acid needs of common carp *Cyprinus carpio* larvae. *Aquacult., 161*, 453–461.

Grahl-Madsen, E., & Lie, O., (1997). Effects of different levels of vitamin K in diets for cod (*Gadus morhua*). *Aquacult., 151*, 269—274.

Halver, J. E., (1957). Nutrition of salmonoid fishes. IV. An amino acid test diet for chinook salmon. *J Nutr., 62*, 225–243.

Halver, J. E., (1965). Tryptophan requirement of chinook, sockeye and silver salmon. *Fdn Proc. Fdn Am. Socs. Exp. Biol., 24*, 169 (abstr. No. 229).

Halver, J. E., (1972). The vitamins. In: Halver, J. E., (ed.), *Fish Nutrition* (pp. 29–103). New York: Academic Press.

Halver, J. E., Ashley, L. M., & Smith, R. R., (1969). Ascorbic acid requirements of coho salmon and rainbow trout. *T. Am. Fish. Soc., 90*, 762–771.

Halver, J. E., Delong, D. C., & Mertz, E. T., (1958). Threonine and lysine requirements of chinook salmon. *Fdn Proc. Fdn Am. Socs. Exp. Biol., 17*, 478 (abstr. No. 1873).

Halver, J. E., Delong, D. C., & Mertz, E. T., (1959). Methionine and cystine requirements of chinook salmon. *Fdn Proc. Fedn Am. Socs. Exp. Biol., 18*, 527 (abstr. No. 2076).

Hart, S. D., Brown, B. J., Gould, N. L., Robar, M. L., Witt, E. M., & Brown, P. B., (2010). Predicting the optimal dietary essential amino acid profile for growth of juvenile yellow perch with whole body amino acid concentrations. *Aquacult. Nutr., 16*, 248–253. https://doi.org/10.1111/j.1365-2095.2009.00659.x.

He, W., Zhou, X. Q., Feng, L., Jiang, J., & Liu, Y., (2009). Dietary pyridoxine requirement of juvenile Jian carp (*Cyprinus carpio* var. Jian). *Aquacult. Nutr., 15*, 402–408.

Helland, S., Barbara, J., & Grisdale, H., (2011). Dietary threonine requirement of Atlantic salmon smolts. *Aquacult., 321*, 230–236. doi: 10.1016/j.aquaculture.2011.09.008.

Henrique, M. M. F., Gomes, E. F., Gouillou-Coustans, M. F., Oliva-Teles, A., & Davies, S. J., (1998). Influence of supplementation of practical diets with vitamin C on growth and response to hypoxic stress of seabream, *Sparus aurata*. *Aquacult., 161*, 415—426.

Hilton, J. W., Cho, C. Y., & Slinger, S. J., (1978). Effect of graded levels of supplemental ascorbic acid in practical diets fed to rainbow trout (*Salmo gairdneri*). *J. Fish. Res. Board Can., 35*, 431–436.

Hossain, M. A., & Furuichi, M., (2000). Essentiality of dietary calcium supplement in fingerling scorpion fish (*Sebastiscus marmoratus*). *Aquacult., 189*, 155–163.

Hossain, M. A., & Yoshimatsu, T., (2014). Dietary calcium requirement in fishes. *Aquacult. Nutr., 20*(1), 1–11. doi: 10.1111/anu.12135.

Hossain, M. A., Hasan, N., Azad, S. A. K. M., & Hussain, M. G., (2002). Optimum dietary protein requirement of mahseer, *Tor putitora* (Hamilton) fingerlings. *Asian Fish. Sci., 15*(2002), 203–214.

Huang, F., Zhao, Z., Wen, H., Wu, F., Liu, W., & Yang, C., (2004). Dietary thiamin requirement of juvenile grass carp, *Ctenopharyngodon idella*. *J. Worl. Aquacult. Soci., 45*(4). doi: 10.1111/jwas.12132.

Huang, H. H., Feng, L., Liu, Y., Jiang, J., Jiang, W. D., Hu, K., Li, S. H., & Zhou, X. Q., (2011). Effects of dietary thiamin supplement on growth, body composition and intestinal enzyme activities of juvenile Jian carp (*Cyprinus carpio* var. Jian). *Aquacult Nutr., 17*, e233–e240.

Hughes, S. G., Rumsey, G. L., & Nesheim, M. C., (1983). Dietary requirements for essential branched-chain amino acids by lake trout. *Transact. Ameri. Fish. Soci., 112*, 812–817.

Hughes, S. G., Rumsey, G. L., & Nichum, J. G., (1981). Riboflavin requirement of fingerling rainbow trout. *Prog. Fish Cult., 43*, 167–172.

Hung, S. S. O., Cho, C. Y., & Slinger, S. J., (1980). Measurement of oxidation in fish oil and its effect on vitamin E nutrition of rainbow trout (*Salmo gairdneri*). *Can. J. Fish. Aquat. Sci., 37*, 1248–1253.

Islam, M. S., & Tanaka, M., (2004). Optimization of dietary protein requirement for pond-reared mahseer *Tor putitora* Hamilton (*Cypriniformes*: *Cyprinidae*). *Aquacult. Res., 35*, 1270–1276 https://doi.org/10.1111/j.1365–2109.2004.01149.x.

Islam, M. S., (2002). Evaluation of supplementary feeds for semi-intensive pond culture of mahseer, *Tor putitora* Hamilton. *Aquacult., 212*, 263–276.

Jiang, M., Wang, W., Wen, H., Wu, F., Zhao, Z., Liu, A., & Liu, W., (2007). Effects of dietary vitamin K3 on growth, carcass composition and blood coagulation time for grass carp fingerling (*Ctenopharyngodon idellus*) [in Chinese]. *Freshwater Fish, 37*, 61–64.

Kalantarian, S. H., Rafiee, G. H., Farhangi, M., & Mojazi, A. B., (2013). Effect of different levels of dietary calcium and potassium on growth indices, biochemical composition and some whole-body minerals in rainbow trout (*Oncorhynchus Mykiss*) fingerlings. *Aquac. Res. Develop., 4*, 3. doi: 10.4172/2155-9546.1000170.

Kanazawa, A., (1985). In: Cowey, C. B., Mackie, A. M., & Bell, J. G., (eds.), *Nutrition and Feeding of Fish* (p. 281). Academic Press, London.

Kashiwada, K., Teshima, S., & Kanazawa, A., (1970). Studies on the production of B vitamins by intestinal bacteria of fish. 5. Evidence of the production of vitamin B12 by microorganisms in the intestinal canal of carp, *Cyprinus carpio*. *B. Jpn. Soc. Sci. Fish., 36*, 421–424.

Kaushik, S. J., & Seiliez, I., (2010). Protein and amino acid nutrition and metabolism in fish: Current knowledge and future needs. *Aquacult. Res., 41*, 322–332. doi: 10.1111/j.1365-2109.2009.02174.x.

Kaushik, S., (2002). Mineral nutrition. In: Guillaume, J., Kaushik, S., Bergot, P., & Metailler, R., (eds.), *Nutrition and Feeding of Fish and Crustaceans* (pp. 169–181). Chichester, UK: Springer-Praxis Publishing Ltd.

Keshavanath, P., Varghese, T. J., Shetty, H. P. C., Krishnamurthy, D., & Gogoi, D., (1986). Impact of diets with various protein sources and 17 á Methyl testosterone on the growth of mahseer, *Tor khudree* (Sykes). Pb. *Fish. Bull., 10*, 72–83.

Ketola, H. G., & Richmond, M. E., (1994). Requirement of rainbow trout for dietary phosphorus and its relationship to the amount discharged in hatchery effluent. *T. Am. Fish. Soc., 123*, 587–594.

Ketola, H. G., (1975). Requirement of Atlantic salmon for dietary phosphorus. *T. Am. Fish. Soc., 104*, 548–551.

Ketola, H. G., (1976). Choline metabolism and nutritional requirement of BM trout (*Salvelinus namaycush*). *J. Anim. Sci., 43*, 474–477.

Ketola, H. G., (1983). Requirement for dietary lysine and arginine by fry of rainbow trout. *J. Anim. Sci., 56*, 101–107.

Khajepour, K., Hosseini, S. A., & Imanpour, M. R., (2012). Dietary crude protein, citric acid and microbial phytase and them interacts to influence growth performance, muscle proximate composition and hematocrite of common carp, *Cyprinus carpio* L, juveniles. *Worl. J. Zool., 7*(2), 118–122.

Khan, K. U., Zuberi, A., Nazir, S., Fernandes, J. B. K., Jamil, Z., & Sarwar, H., (2016). Effects of dietary selenium nanoparticles on physiological and biochemical aspects of juvenile Tor putitora. *Turk J. Zool., 40*, 704–712.

Khan, M. A., & Zehra, S., (2020). Principal nutrients and nutritional requirements of cultivable finfishes. *Fish Nutrition and its Relevance to Human Health.* doi: 10.1201/9781003107583-1.

Kim, K. I., (1993). Requirement for phenylalanine and replacement value of tyrosine for phenylalanine in rainbow trout (*Oncorhynchus mykiss*). *Aquacult., 113*, 243–250.

Kim, K. I., Kayes, T. B., & Amundson, C. H., (1992a). Requirements for sulfur amino acids and utilization of D-methionine by rainbow trout *(Oncorhynchus mykiss). Aquacult., 101*, 95–103.

Kim, K. L., Kayes, T. B., & Amundsun. C. H., (1987). Effects of dietary tryptophan levels on growth. feed/gain, carcass composition and liver glutamate dehydrogenase activity in rainbow trout (*Salmogw-dncri*). *Comp. Biochem. Physiol., 888*, 737–741.

Kim, K., Kayes, T. B., & Amundson, C. H., (1992b). Requirement for lysine and arginine by rainbow trout (*Oncorhynchus mykiss*). *Aquacult., 106*, 333–344.

Kitamura, S., Suwa, T., Ohara, S., & Nakagawa, K., (1967). Studies on vitamin requirements of rainbow trout. 3. Requirement for vitamin A – AAC deficiency symptoms. *B. Jpn. Soc. Sci. Fish., 33*, 1126–1131.

Klein, R. G., & Halver, J. E., (1970). Nutrition of salmonoid fishes: Arginine and histidine requirements of Chinook and coho salmon. *J Nutr., 100*(9), 1105–1109. doi: 10.1093/jn/100.9.1105. PMID: 5456554.

Knox, D., Cowey, C. B., & Adron, J. W., (1981). Studies on the nutrition of salmonid fish. The magnesium requirement of rainbow trout (*Salmo guairdneri*). *Br. J. Nutr., 45*, 137–148.

Köprücü, K., (2012). Effects of dietary protein and lipid levels on growth, feed utilization and body composition of juvenile grass carp (*Ctenopharyngodon idella*). *J. Fish. Sci. Com., 6*(3), 243–251. doi: 10.3153/jfscom.2012028.

Krosspy, C., Waggbp, R., Fjelldal, P. G., Wargelius, A., Lock, E. J., Graff, I. E., & Omsrud, R., (2009). Dietary menadione nicotinamide bisulphite (4 tamin K3) does not affect growth or bone health in first-feeding fry Atlantic salmon (*Salmo salar* L.). *Aquacult. Nutr., 15*, 638–649.

Lall, S. P., & Bishop, F. J., (1977). Studies on mineral and protein ulilizaiic® by Atlantic salmon (*Salmo salar*) grown in sea water. *Fish. Mar. Sax Tech. Rep., 688*, 1–16.

Lall, S. P., & Weerakoon, D. E. M., (1990). Vitamin B6 requirement of Atlantic salmon (*Salmo salar*) (abstract). *FASEB J., 4*, 3749.

Lall, S. P., (2002). The minerals. In: Halver, J. E., & Hardy, R. W., (eds.), *Fish Nutrition* (3rd edn, pp. 259–308). London, UK: Academic Presi.

Lall, S. P., Olivier, G., Hines, J. A., & Ferguson, H. W., (1988). The role of vitamin E in nutrition and immune response of Atlantic salmon (*Sairm salar*). *Bull. Aqua. Assoc. Can., 76*, 88–92.

Larentzen, M., Maage, A., & Julshamn, K., (1994). Effects of dietary selenite B or selenomethionine on tissue selenium levels of Atlantic salmon (*Salmo solar*). *Aquacult., 121*, 359–367.

Lee, S. M., & Chao, S. H., (2009). Influence of dietary fatty acid profile on growth, body composition and blood chemistry in juvenile fat cod (*Hexagrammos otakii* jordan et starks). *Aquacult. Nutr., 15*, 19–28.

Leith, D., Holmes, J., & Kaattari, S., (1990). *Effects of Vitamin Nutrition on the Immune Response of Hatchery-Reared Salmonids.* Final Report, Project 84-45A and 84-45B. Portland, OR, Bonneville Power Administration.

Li, J., Liang, F. X., Tan, Q., Yuan, X., Liu, L., Zhou, Y., & Li, B., (2016). Effects of vitamin E on growth performance and antioxidant status in juvenile grass carp *Ctenopharyngodon idellus*. *Aquacult., 430*, 21–27.

Li, S. Q., Feng, L., Jiang, W. D., Liu, Y., Wu, P., Zhao, J., Kuang, S. Y., et al., (2015). Eficiency of dietary niacin decreases digestion and absorption capacities via declining the digestive and brush border enzyme activities and downregulating those enzyme gene transcription related to TOR pathway of the hepatopancreas and intestine in young grass carp (*Ctenopharyngodon idella*). *Aquacult. Nutr., 22*(6). doi: 10.1111/anu.12333.

Li, W., Zhou, X. Q., Feng, L., Liu, Y., & Jiang, J., (2010). Effect of dietary riboflavin on growth, feed utilization, body composition and intestinal enzyme activities of juvenile Jian carp (*Cyprinus carpio* var. Jian). *Aquacult. Nutr., 16*, 137–143.

Liang, J. J., Liu, Y. J., Tian, L. X., Yang, H. J., & Liang, G. Y., (2011). Dietary available phosphorus requirement of juvenile grass carp (*Ctenopharyngodon idella*). *Aquacult. Nutr., 2011.*

Liang, J. J., Liu, Y. J., Yang, Z. N., Tian, L. X., Yang, H. J., & Liang, G. Y., (2012). Dietary calcium requirement and effects on growth and tissue calcium content of juvenile grass carp (*Ctenopharyngodon idella*). *Aquacult. Nutr., 18*, 544–550.

Liang, J. J., Wang, S., Han, B., Tian, L. X., Yang, H. J., & Liu, Y. J., (2015). Dietary manganese requirement of juvenile grass carp (*Ctenopharyngodon idella* val.) based on growth and tissue manganese concentration. *Aquacult. Res., 46*, 2991–2998. doi: 10.1111/are.12455.

Liang, J. J., Yang, H. J., Liu, Y. J., & Tian, L. X., (2014). Dietary potassium requirement of juvenile grass carp (*Ctenopharyngodon idella* val.) based on growth and tissue potassium content. *Aquacult. Res., 45*(4). doi: 10.1111/are.12008.

Lim, C., & Lovell, R. T., (1978). Pathology of the vitamin C deficiency syndrome in channel catfish (*Ictalurus punctatus*). *J. Nutr. 108*, 1137–1146.

Linag, L., Feng, L., Liu, Y., Jiang, J., Jiang, W. D., Hu, K., & Li, S. H., (2010). Effect of dietary iron levels on growth, body composition intestinal enzyme activities of juvenile Jian carp (*Cyprinus carpio*). *Aquacult. Nutr., 16*, 616–624.

Liu, A. L., Wen, H., Jiang, M., Zhao, Z. Y., Wu, F., & Liu, W., (2007). Dietary pantothenic acid requirement of grass carp (*Ctenopharyngodon idella*) fingerlings. *Fish. Sci., 26*, 263–266.

Lorentzen, M., Maage, A., & Julshamn, K., (1998). Supplementing copper to a fish meal-based diet fed to Atlantic salmon parr affects liver copper and selenium concentrations. *Aquacult. Nutr., 4*, 67–72.

Luzzana, U., Mentasti, T., Opstvedt, J., Nygard, E., Moretti, V. M., & Valfrè, F., (1999). Racemization kinetics of aspartic acid in fish material under different conditions of moisture, pH and oxygen pressure. *J. Agri. Food Chemi., 47*, 2879–2884.

Maage, A., Julshamn, K., & Berge, G. E., (2001). Zinc gluconate and zinc sulphate as dietary zinc sources for Atlantic salmon. *Aquacult. Nutr., 7*, 183–187.

Masagounder, K., Firman, J. D., Hayward, R. S., Sun, S., & Brown, P. B., (2009). Apparent digestibilities of common feedstuffs for bluegill *Lepomis macrochirus* and largemouth bass *Micropterus salmoides* using individual test ingredients. *Aquacult. Nutr., 15*(1), 29–37.

McLaren, B. A., Keller, E., O'Donnell, D. J., & Elvehjem, C. A., (1947). The nutrition of rainbow trout. 1. Studies of vitamin requirements. *Arch. Biochem. Biophys., 15*, 169–178.

Miao, S., Zhang, W., Zhang, W., Xu, W., Xu, W., & Ma, K., (2013). Dietary folic acid requirement of juvenile abalone *Haliotis discus hannai* Ino. *Aquacult., 400, 401*, 73–76. doi: 10.1016/j.aquaculture.2013.03.005.

Misieng, J. D., Kamarudin, M. S., & Musa, M., (2011). Optimum dietary protein requirement of Malaysian Mahseer, (*Tor tambroides*) fingerlings. *Paki. J. Biolo. Sci., 14*, 232–235. https://doi.org/10.3923/pjbs.2011.232.235 PMid:21870647.

Mohan, M., (2002). Nutrition and feed development for cold water fishes. In: Vass, K. K., & Raina, H. S., (eds.), *Highland Fisheries and Aquatic Resource Management* (pp. 284–305).

Morito, C. L. H., Conrad, D. H., & Hilton, J. W., (1986). The thiamin deficiency signs and requirement of rainbow trout (*Salmo gairdneri*, Richardson). *Fish Physiol. Biochem., 1*, 93–104.

Naser, N., (2000). *Role of Iron in Atlantic Salmon (salmo Salar) Nutrition: REQUIREMENT, Bioavailability, Disease Resistance and Immune Response* (p. 282). Dalhousie University, Halifax, Nova Scotia.

Nasir, N. A., & Hamed, Q., (2016). Growth development of young common carp *Cyprinus carpio* through dietary sodium chloride supplementation. *Mesopotamia Envir. J., 2*(22), 12–18.

Ng, W. K., & Andin, V. C., (2011). The Malaysian mahseer, *Tor tambroides* (Bleeker), requires low dietary lipid levels with a preference for lipid sources with high omega-6 and low omega-3 polyunsaturated fatty acids. *Aquacult.*, 322. doi: 10.1016/j.aquaculture.2011.09.021.

Nose, T., (1979). Summary report on the requirements of essential amino acids for carp. In: Tiews, K., & Halver, J. E., (eds.), *Finfish Nutrition and Fish Feed Technology* (145–156). Heenemann, Berlin.

NRC (National Research Council), (2011). *Nutrient Requirement of Fish and Shrimp*. National Academic Press, Washington, DC (376 + XVI). doi: https://doi.org/10.17226/13039.

NRC, (1993). *Nutrient Requirements of Fish*. Washington, DC: National Academy Press.

Ogino, C., & Chiou, J. Y., (1976). Mineral requirements in fish-II: Magnesium requirement of carp. *Bull. Jap. Soc. Sci. Fish., 42*, 71–75.

Ogino, C., & Takeda, H., (1976). Mineral requirements in fish-III calcium and phosphorus requirements in carp. *B. Jpn. Soc. Sci. Fish., 42*, 793–799.
Ogino, C., & Takeda, H., (1978). Requirement of rainbow trout for dietary calcium and phosphorus. *B. Jpn. Soc. Sci. Fish., 44*, 1019–1022.
Ogino, C., & Yang, G. Y., (1978). Requirement of rainbow trout for dietary zinc. *B. Jpn. Soc. Sci. Fish., 44*, 1015–1018.
Ogino, C., & Yang, G. Y., (1979). Requirement of carp for dietary zinc. *B. Jpn. Soc. Sci. Fish., 45*, 967–969.
Ogino, C., & Yang, G. Y., (1980). Requirement of carp and rainbow trout for dietary manganese and copper. *B. Jpn. Soc. Sci. Fish., 46*, 455–458.
Ogino, C., (1967). B vitamin requirements of carp. 2. Requirements for riboflavin and pantothenic acid. *B. Jpn. Soc. Sci. Fish., 33*, 351–354.
Ogino, C., (1980). Requirements of carp and rainbow trout for essential amino acids. *B. Jpn. Soc. Sci. Fish., 46*, 171–174.
Ogino, C., Takashima, F., & Chiou, J. Y., (1978). Requirement of rainbow trout for dietary magnesium. *B. Jpn. Soc. Sci. Fish., 44*, 1105–1108.
Ogino, C., Uki, N., Watanabe, T., Iida, Z., & Ando, K., (1970a). B vitamin requirements of carp. 4. Requirement of choline. *B. Jpn. Soc. Sci. Fish., 36*, 1140–1146.
Ogino, C., Watanabe, T., Kakino, J., Iwanaga, N., & Mizuno, M., (1970b). B vitamin requirements of carp. 3. Requirement for biotin. *B. Jpn. See Sci. Fish., 36*, 734–740.
Pei, Z., Xie, S., Lei, W., Zhu, X., & Yang, Y., (2004). Comparative study on the effect of dietary lipid level on growth and feed utilization for gibel carp (*Carassius auratus gibelio*) and Chinese longsnout catfish (*Leiocassis longirostris* Gunther). *Aquacult. Nutr., 10*, 209–216.
Phillips, A. M., & Brockway, D. R., (1947). The niacin and biotin requirement of trout. *T. Am. Fish. Soc., 76*, 152–159.
Phillips, A. M., Jr, H. A., Podoliak, H. A., Poston, D. L., Livingston, H. E., Booke, E. A., Pyle, A., & Hammer, G. L., (1964). *The Production of Anemia in Brown Trout* (pp. 66–70). In Fisheries *Research Bulletin* No. 27. Albany: State of New York Conservation Department.
Phromkunthong, W., Boonyaratpalin, M., & Storch, V., (1997). Different concentrations of ascoryl-2-monophosphate–magnesium as dietary sources of vitamin C for seabass, Lates calcarifer. *Aquacult., 151*, 225–243.
Poston, H. A., & Rumsey, G. L., (1983). Factors affecting dietary requirement and deficiency signs of L-tryptophan in rainbow trout. *J. Nur., 113*, 256882577.
Poston, H. A., & Wolfe, M. J., (1985). Niacin requirement for optimum growth, feed conversion and protection of rainbow trout, *Salmo gairdneri* Richardson, from ultraviolet-B-irradiation. *J. Fish. Dis., 8*, 451–460.
Poston, H. A., (1976a). Relative effect of two dietary water–soluble analogues of menaquinone on coagulation and packed cell volume of blood of lake trout, *Salvelinus namaycush. J. Fish. Res. Board Can., 33*, 1791–1793.
Poston, H. A., (1976b). Optimum level of dietary biotin for growth, feed utilization, and swimming stamina of fingerling lake trout (*Salvelinus namaycush*). *J. Fish. Res. Board Can., 33*, 1803–1806.
Poston, H. A., Combs, G. F. Jr., & Leibovitz, L., (1976). Vitamin E and selenium interactions in the diet of Atlantic salmon {*Salmo salar*): Gross, histological and biochemical deficiency signs. *J. Nutr., 106*, 892–904.
Prabhu, P. A. S., Johan, W. S., & Kaushik, J. S., (2016). Mineral requirements of fish: A systematic review. *Reviews in Aquacult., 8*, 172–219. doi: 10.1111/raq.12090.

Radunzneto, J., Corraze, G., Bergot, P., & Kaushik, S. J., (1996). Estimation of essential fatty acid requirements of common carp larvae using semi-purified diets. *Arch. Anim. Nutr., 49*, 41–48.

Robinson, E. H., & Lovell, R. T., (1978). Essentiality of biotin for channel catfish *Ictalurus punctatus* fed lipid and lipid-free diets. *J. Nutr., 108*, 1600–1605.

Robinson, E., (1990). Réévaluation of the ascorbic acid (vitamin C) requirement of channel catfish (*Ictalurus punctatus*) (abstract). *FASEB J., 4*, 3745.

Rodehutscord, M., Becker, A., Pack, M., & Pfeffer, E., (1997). Response of rainbow trout, *Oncorhynchu mykiss* to supplements of individual essential amino acids in a semi-purified diet, including an estimate of the maintenance requirement of essential amino acids. *J. Nutr., 126*, 1166–1175.

Rodehutscord, M., Jacobs, S., Pack, M., & Pfeffer, E., (1995). Response of rainbow trout (*Oncorhynchus mykiss*) growing from 50 to 170 g to supplements of either L-arginine or L-threonine in a semipurified diet. *J. Nutr., 125*, 970–975.

Rollin, X., (1999). *Etude Critique des Besoins en Acides Aminés Indispensables de L'alevin de Saumon Atlantique (Salmo Salar L.) (Critical Study of the Indispensable Amino Acids Requirements in Atlantic Salmon (Salmo Salar L.) Fry)*. PhD thesis, Catholic University of Louvain.

Royuela, M. S., Casado, M., Celada, J. D., Carral, J. M., & Rodríguez, A. G., (2015). Effect of dietary lipid level on survival, growth performance and body composition of juvenile tench (*Tinca tinca* L.) fed practical diets. *Aquacult.*, 439. doi: 10.1016/j.aquaculture.2015.01.017.

Rumsey, G. L., (1991). Choline–betaine requirements of rainbow trout (*Oncorhynchus mykiss*). *Aquacult., 95*, 107–116.

Rumsey, G. L., Page, J. W., & Scott, M. L., (1983). Methionine and cystine requirement of rainbow trout. *Prog. Fish-Cult., 45*, 139–143.

Ruyter, B., Rosjo, C., Einen, O., & Thomassen, M. S., (2000). Essential fatty acids in Atlantic salmon: Time course of changes in fatty acid composition of liver, blood and carcass induced by a diet deficient in n-3 and n-6 fatty acids. *Aquacult. Nutr., 6*, 109–118.

Salman, N. A., & Eddy, F. B., (1988). Effect of dietary sodium chloride on growth, food intake and conversion efficiency in rainbow trout (*Salmo gairdneri* Richardson). *Aquacult., 70*, 131–144.

Sandnes, K., Torrissen, O., & Waagbp, R., (1992). The minimum dietary requirement of vitamin C in Atlantic salmon *Salmo salar* fry using Ca ascorbat-2-monophosphate as dietary source. *Fish Physiol. Biochem., 10*, 315–319.

Santiago, C. B., & Lovell, R. T., (1988). Amino acid requirement for growth of Nile tilapia. *J. Nutr., 118*, 1540–1546.

Sato, M., Yoshinaka, R., & Ikeda, S., (1978). Dietary ascorbic acid requirement of rainbow trout for growth and collagen formation. *B. Jpn. Soc. Sci. Fish., 44*, 1029–1035.

Satoh, S., Izume, K., Takeuchi, T., & Watanabe, T., (1992). Effect of supplemental tricalcium phosphate on zinc and manganese availability to common carp. *Nippon Suisan Gakk., 58*, 539–545.

Satoh, S., Tabata, K., Izume, K., Takeuchi, T., & Watanabe, T., (1987). Effect of dietary tricalcium phosphate on availability of zinc to rainbow trout. *Nippon Suisan Gakk., 53*, 1199–1205.

Satoh, S., Takeuchi, T., & Watanabe, T., (1991). Availability of manganese and magnesium contained in white fish meal to rainbow trout *Oncorhynchus mykiss*. *Nippon Suisan Gakk., 57*, 99–104.

Sawhney, S., & Gandotra, R., (2010). Growth response and feed conversion efficiency of *Tor putitora* (Ham.) fry at varying dietary protein levels. *Paki. J. Nutr., 9*(1), 86–90. https://doi.org/10.3923/pjn.2010.86.90.

Schwarz, F. J., (1995). Determination of mineral requirements of fish. *J. Appl. Ichthyol., 11*(3, 4), 164–174.

Schwarz, F. J., Kirchgessner, M., & Deuringer, U., (1998). Studies on the methionine requirement of carp (*Cyprinus carpio* L.). *Aquacult., 161*, 121–129.

Scott, L., Lall, S. P., & Anderson, D. M., (1996). Dietary histidine requirement of Atlantic salmon (*Salmo salar*) reared in sea water. *The CFIA Eastern Nutrition Conference* (Vol. 259, p. 1996). Dartmouth/Halifax, Canada, CFIA, Canada.

Serrini, G., Zhang, Z., & Wilson, R. P., (1996). Dietary riboflavin requirement of fingerling channel catfish (*Ictalurus punctatus*). *Aquacult., 139*, 285–290.

Shang, X. D., Luo, L., Wen, H., Gao, W., Wang, Q., & Xu, H., (2009). Study on isoleucine requirement for juvenile grass carp, *Ctenopharyngodon idellus. J. Fish. Chi., 33*(5), 813–822.

Shankar, T. J., (1988). Effect of dietary protein level on the growth of deccan mahseer fry *Tor khudree* sykes. *J. India. Fish. Asso., 18*, 135–140.

Shao, L., Zhu, X., Yang, Y., Jin, J., Liu, H., Hana, D., & Xi, S., (2016). Effects of dietary vitamin A on growth, hematology, digestion and lipometabolism of on growing gibel carp (*Carassius auratus* gibelio var. CAS III). *Aquacult., 460*(1), 83–89.

Shearer, K. D., & Åsgård, T., (1992). The effect of water-borne magnesium on the dietary magnesium requirement of the rainbow trout (*Oncorhynchus mykiss*). *Fish Physiol. Biochem., 9*, 387–392.

Shearer, K. D., (1988). Dietary potassium requirement of juvenile C' salmon. *Aquacult., 73*, 119–129.

Shearer, K. D., (1989). Whole body magnesium concentration as an in magnesium status in rainbow trout (*Salmo gairdneri*). *Aqucult., 77*, 201–210.

Shyama, S., (1990). *Growth Performance of Carps in Different Diet Treatments and Under Polyculture* (p. 380). Ph.D. Thesis. University of Agricultural Sciences, Bangalore.

Su, C., Luo, L., Wen, H., Sheng, X., & Li, S., (2007). Effects of dietary iron on growth performance, nutritional composition and some blood indices of grass carp (*Ctenopharyngodon idellus*). *Freshwater Fish., 37*, 48–52.

Takeuchi, L., Takeuchi, T., & Ogino, C., (1980). Riboflavin requirements in carp and rainbow trout. *B. Jpn. Soc. Sci. Fish., 46*, 733–737.

Takeuchi, T., Watanabe, K., Satoh, S., & Watanabe, T., (1992). Requirement of grass carp fingerlings for a-tocopherol. *Nippon Suisan Gakk., 58*, 1743–1749.

Takeuchi, T., Watanabe, T., & Nose, T., (1979). Requirement for essential fatty acids of chum salmon (*Oncorhyncus keta*) in freshwater environment. *B. Jpn. Soc. Sci. Fish., 45*, 1319–1323.

Tan, L. N., Feng, L., Liu, Y., Jiang, J., Jiang, W. D., Hu, K., Li, S. H., & Zhou, X. Q., (2011). Growth, body composition and intestinal enzyme activities of juvenile Jian carp (*Cyprinus carpio* var. Jian) fed graded levels of dietary zinc. *Aquacult. Nutr., 17*, 338–345.

Tang, Q. Q., Feng, L., Jiang, W. D., Liu, Y., Jiang, J., & Li, S. H., (2013). Effects of dietary copper on growth, digestive, and brush border enzyme activities and antioxidant defense of hepatopancreas and intestine for young grass carp (*Ctenopharyngodon idella*). *Biol Trace Elem Res., 155*, 370–380.

Teshima, S. I., Kanazawa, A., Koshio, S., & Itoh, S., (1993). L-Ascorbyl-2-phosphate-Mg as vitamin C source for the Japanese flounder *Paralichthys oliſaceus*. In: Kaushik, S. J.,

& Luquet, P., (eds.), *Fish Nutrition in Practice* (pp. 157–166). Biarritz, France, INRA Editions, Les Colloques.

Tian, J. J., Ji, H., Oku, H., & Zhou, J. S., (2014). Effects of dietary arachidonic acid (ARA) on lipid metabolism and health status of juvenile grass carp, *Ctenopharyngodon idellus*. *Aquacult., 430*, 57–65.

Tian, J., Lei, C. X., & Ji, H., (2016). Influence of dietary linoleic acid (18:2n-6) and α-linolenic acid (18:3n-3) ratio on fatty acid composition of different tissues in freshwater fish *Songpu mirror* carp, *Cyprinus carpio*. *Aquacult. Res., 47*, 3811–3825.

Vielma, J., & Lall, S. P., (1998). Phosphorus utilization by Atlantic salmon (*Salmo salar*) reared in freshwater is not influenced by higher dietary calcium intake. *Aquacult., 160*, 117–128.

Wagle, S. U., Shrestha, M. K., Rai, S., & Pandit, N. P., (2016). Effect of different protein levels on feed utilization, growth and body composition on fry of snowtrout *Schizothorax richardsonii*. *Inter. J. Fish. Aqu. Stud., 4*(5), 89–95.

Walton, M. J., Coloso, R. M., Cowey, C. B., Adron, J. W., & Knox, D., (1984a). The effects of dietary tryptophan levels on growth and metabolism of rainbow trout (*Salmo gairdneri*). *Brit. J. Nutr., 51*, 279–287.

Walton, M. J., Cowey, C. B., & Adron, J. W., (1982). Methionine metabolism in rainbow trout fed diets of differing methionine and cystine content. *J. Nutr., 112*, 1525–1555.

Walton, M. J., Cowey, C. B., & Adron, J. W., (1984b). The effect of dietary lysine levels on growth and metabolism of rainbow trout (*Salmo gairdner;*). *Br. J. Nuts., 52*, 1 t S–122.

Walton, M. J., Cowey, C. B., Colaso, R. M., & Adron, J. W., (1986). Dietary requirements of rainbow trout for tryptophan, lysine and arginine determined by growth and biochemical measurements. *Fish Physiol. Biochem., 2*, 161–169.

Wang, D., Zhao, L., & Tan, Y., (1995). Requirement of the fingerling grass carp (*Ctenopharyngodon idella*) for choline [in Chinese]. *J. Fish. China, 19*, 132–139.

Wang, F. B., Luo, L., Lin, S. M., Li, Y., Chen, S., Wang, Y. G., Wen, H., & Hu, C. J., (2011b). Dietary magnesium requirements of juvenile grass carp, *Ctenopharyngodon idella*. *Aquacult. Nutr., 17*, 691–700.

Wang, J., Xie, S., Zhu, X., Lei, W., Han, D., & Yang, Y., (2011a). Dietary vitamin B6 requirement of juvenile gibel carp. *Carassius auratus* gibelio. *Acta Hydrobiol. Sin., 35*, 98–104. (In Chinese with English abstract).

Wang, S., Liu, Y. J., Tian, L. X., Xie, M. Q., Yang, H. J., Wang, Y., & Liang, G. Y., (2005). Quantitative dietary lysine requirement of juvenile grass carp *Ctenopharyngodon idella*. *Aquacult., 249*(6), 419–429.

Wang, T., Gao, Q., Nie, P. J., & Secombes, C. J., (2010). Identification of suppressor of cytokine signaling (SOCS) 6, 7, 9 and CISH in rainbow trout *Oncorhynchus mykiss* and analysis of their expression in relation to other known trout SOCS. *Fish. Shell. Immun., 29*(2010), 656–667. doi: 10.1016/j.fsi.2010.06.015.

Watanabe, T., Kiron, V., & Satoh, S., (1997). Trace minerals in fish nutrition. *Aquacult., 151*, 185–207.

Watanabe, T., Murakami, A., Takeuchi, T., Nose, T., & Ogino, C., (1980). Requirement of chum salmon held in freshwater for dietary phosphorus. *B. Jpn. Soc. Sci. Fish., 46*, 361–367.

Watanabe, T., Takashima, F., Ogino, C., & Hibiya, T., (1970). Requirements of young carp for a-tocopherol. *B. Jpn. Soc. Sci. Fish., 36*, 972–976.

Watanabe, T., Takeuchi, T., Wada, M., & Vehara, R., (1981). The relationship between dietary lipid levels and a-tocopherol requirement of rainbow trout. *B. Jpn. Soc. Sci. Fish., 47*, 1463–1471.

Wen, H., Zhao, Z., Jiang, M., Liu, A., Wu, F., & Liu, W., (2007). Dietary myoinositol requirement for grass carp, *Ctenopharyngodon idella* fingerling [in Chinese]. *J. Fish. China, 14*, 794–800.

Wen, Z. P., Zhou, X. Q., Feng, L., Jiang, J., & Liu, Y., (2009). Effect of dietary pantothenic acid supplement on growth, body composition and intestinal enzyme activities of juvenile Jian carp (*Cyprinus carpio* var. Jian). *Aquacult. Nutr., 15*, 470–476.

Wilson, R. P., Bowser, P. R., & Poe, W. E., (1984). Dietary vitamin E requirement of fingerling channel catfish. *J. Nutr., 114*, 2053–2058.

Woodall, A. N., Ashley, L. M., Halver, J. E., Olcott, H. S., & Van, D. V. J., (1964). Nutrition of salmonid fishes. XIII. The a-tocopherol requirement of Chinook salmon. *J. Nutr., 84*, 125–135.

Woodward, B., & Frigg, M., (1989). Dietary biotin requirements of YR trout (*Salmo gairdneri)* determined by weight gain hepatic biotin is centration and maximal biotin-dependent enzyme activities in liver white muscle. *J. Nutr., 119*, 54–60.

Wu, F., Jiang, M., Zhao, Z., Liu, A., Liu, W., & Wen, H., (2007a). The dr-niacin requirement of juvenile *Ctenopharyngodon idellus* (in Chi). *J. Fish. China, 29*, 695–669.

Wu, F., Wen, H., Jiang, M., Zhao, Z., Liu, A., & Liu, W., (2007b). Effects dietary vitamin B12 on growth, body composition and hemopoiesis juvenile grass carp (*Ctenopharyngodon idellus*) [in Chinese]. *J. Agric. Univ., 29*, 695–669.

Xie, N. B., Feng, L., Liu, Y., Jiang, J., Jiang, W. D., Hu, K., Li, S. H., & Zhou, X. Q., (2011). Growth, body composition, intestinal enzyme activities and microflora of juvenile Jian carp (*Cyprinus carpio* var. Jian) fed graded levels of dietary phosphorus. *Aquacult. Nutr., 17*, 645–656.

Yang, C., Jiang, M., Wen, H., Tian, J., Wu, F., Liu, W., & Yu, L., (2017). Dietary vitamin B6 requirement of juvenile grass carp (*Ctenopharyngodon idella*). *Isra. J. Aquacult. Bamidgeh IJA., 69*.2017.1376.

Yu, T. C., & Sinnhuber, R. O., (1979). Effect of dietary co3 and to6 fatty acids on growth and feed conversion efficiencies of coho salmon (*Oncorhyneus kisutch*). *Aquacul., 16*, 31–38.

Yun, H., Park, G., Imho, O., Katya, K., Heung, S., & Bai, S. C., (2015). Evaluation of optimum dietary threonine requirement by plasma free threonine and ammonia concentrations in surgically modified rainbow trout, *Oncorhynchus mykiss*. *Asia-Australas. J. Anim. Sci., 28*(4), 551–558. doi: 10.5713/ajas.14.0495.

Yun, H., Park, G., Ok, I., Katya, K., Hung, S., & Bai, S. C., (2016). Determination of the dietary lysine requirement by measuring plasma free lysine concentrations in rainbow trout *Oncorhynchus mykiss* after dorsal aorta cannulation. *Fish. Aqua. Sci., 19*, 1–7.

Zehra, S., & Khan, M. A., (2020). Dietary pantothenic acid requirement of fingerling *Channa punctatus* (Bloch) based on growth, feed conversion, liver pantothenic acid concentration and carcass composition. *Aquacult Nutr., 86*(31), 1–8. https://doi.org/10.1111/anu.12680.

Zehra, S., & Khan, M., (2012). Dietary protein requirement for fingerling *Channa punctatus* (Bloch), based on growth, feed conversion, protein retention and biochemical composition. *Aquacult Int., 20*, 383–395. doi: 10.1007/s10499-011-9470-8.

Zeng, Y. Y., Jiang, W. D., Liu, Y., Wu, P., Zhao, J., Jiang, J., Kuang, S. Y., et al., (2016). Optimal dietary alpha-linolenic acid/linoleic acid ratio improved digestive and absorptive capacities and target of rapamycin gene expression of juvenile grass carp (*Ctenopharyngodon idellus*). *Aquacult. Nutr., 22*, 1251–1266.

Zhao, J., Liu, Y., Jiang, J., Wu, P., Chen, G., Jiang, W., Li, S., et al., (2012). Effects of dietary isoleucine on growth, the digestion and absorption capacity and gene expression in hepatopancreas and intestine of juvenile Jian carp (*Cyprinus carpio* var. Jian). *Aquacult.,* 368–369. http://dx.doi.org/10.1016/j.aquaculture.2012.09.019.

Zhao, S., Feng, L., Liu, Y., Kaung, S. Y., Tang, L., Jiang, J., Hu, K., et al., (2012). Effects of dietary biotin supplement on growth, body composition, intestinal enzyme activities and microbiota of juvenile Jian carp (*Cyprinus carpio* var. Jian). *Aquacult. Nutr., 18*(4). doi: 10.1111/j.1365-2095.2011.00905.x.

Zhao, Z., Wen, H., Wu, F., Liu, A., Jiang, M., & Liu, W., (2008). Dietary folic acid requirement for grass carp fingerling, *Ctenopharyngodon idella*. (Chinese). *J. Shanghai Fish. Univ., 17*, 187–192.

Zhu, W., Liu, M., Chen, C., Wu, F., Yang, J., Tan, Q., Xie, S., & Liang, X., (2014). Quantifying the dietary potassium requirement of juvenile grass carp (*Ctenopharyngodon idellus*). *Aquacult., 430*, 218–223.

CHAPTER 4

Bacterial Diseases of Finfish Prevalent in Coldwater Aquaculture

PRIYANKA ASHWATH[1], RAMYA PREMANATH[1], RAJESHWARI VITTAL[1], DEEKSHIT[1], PRARTHANA AITHAL[2], FEROZ A. SHAH[3], and AKHILA DHARNAPPA SANNEJAL[1]

[1]*Nitte University Center for Science Education and Research, Paneer Campus, Deralakatte, Mangalore, Karnataka, India*

[2]*Coldwater Fisheries Research, Nainital, Uttarakhand, India*

[3]*Division of Aquatic Animal Health Management, Faculty of Fisheries, Sher-e-Kashmir University of Agricultural Sciences and Technology (SKUAST) of Kashmir, Rangil, Ganderbal, Jammu and Kashmir, India*

ABSTRACT

Coldwater aquaculture is a rapidly growing industry functioning worldwide. The challenges faced by this sector are sophisticated and multilateral. Among the constraints encountered, infectious diseases take a lion's share, creating a huge loss annually. Bacterial diseases of salmonids are considered to be important in causing severe loss in the cold water fisheries. Therefore, preventive strategies based on scientifically recommended principles are advocated. These strategies should ideally focus on preventing the occurrence of the disease rather than treating the diseased ones. As a single approach is not always successful, a combination of different schedules is recommended. The current chapter highlights some of the important bacterial diseases of coldwater aquaculture and their preventive approaches.

Coldwater Fisheries and Aquaculture Management: Technology for Sustainable Food Production.
Mohd Ashraf Rather, Faisal Rashid Sofi, Adnan Amin, & Kawkabul Saba (Eds.)

4.1 INTRODUCTION

Coldwater aquaculture holds an important place in the aquaculture sector throughout the globe. The term 'cold water' refers to the aquatic ecosystem where the temperature of the water body remains below 20°C which offers the predominance of fish belonging to Salmonidae, particularly the salmons and trout. Coldwater aquaculture encompasses the fisheries activity in water where the temperature does not go beyond 20°C throughout the year. Cold-water aquaculture is prevalent in upland rivers, lakes, and streams found in high altitudes. In India, cold water fisheries are found in the Himalayan and the sub-Himalayan regions (Mahanta & Sarma, 2010).

The major constraint in aquaculture that results in severe loss annually is mainly due to the infectious diseases caused by various bacteria, viruses, fungi, and parasites. Cold water fisheries are also hampered by a serious loss that causes concern to aqua-culturists. Various abiotic and biotic factors create an ideal environment for the disease outbreak. Furthermore, the higher density of a single species of fish also adds to the rapid disease spread. Some of the most devastating and insidious diseases in coldwater aquaculture are due to the bacterial pathogens. To overcome the losses often, antibiotics are excessively used which creates selective pressure on the bacteria to develop resistance. The emergence of antibiotic resistant bacteria possessing the resistance genes is of concern to food safety as these genes could be transferred to human pathogens. Thus, management practices become very important in reducing the use of antibiotics. As vaccines are not available in the market for a few bacterial cold water diseases, it becomes imperative to prevent the disease occurrence. The implementation of scientifically proven and recommended management practices such as the use of genetically resistant stocks, dietary supplements, probiotics, nonspecific immunostimulants, water disinfection and movement restrictions are contemplated to be the best approaches in controlling infectious bacterial cold water diseases (Kumar et al., 2016; Pokhrel et al., 2018). This chapter emphasizes the important bacterial diseases of the cold water fish, the clinical manifestation of these diseases, diagnostic approaches, and management practices.

4.2 ENTERIC RED MOUTH (ERM) DISEASE

Yersinia ruckeri is one of the important fish pathogens that causes enteric red mouth (ERM) disease of yersiniosis in rainbow trout. A Gram-negative rod-shaped bacterium, first isolated from *Onchorynchus mykiss* in USA (Ross et al., 1966). Although infections are frequently reported in other fishes throughout the world, rainbow trout are found to be highly susceptible to ERM (Tobback et al., 2007; Shaowu et al., 2013). It has also been isolated from the tissues of major carps in West Bengal (Manna et a., 2003; Surendraraj et al., 2009; Pajdak-Czaus et al., 2019) and in Andhra Pradesh (Ummey et al., 2021). The bacterium belonging to *the Enterobacteriaceae* family is a facultative anaerobe with a genome size of 3.7 Mb (Navas et al., 2014). Based on the production of serological markers, there are four serotypes of *Yersinia* of which serotype 1 has two subgroups O1a and O1b while serotype 2 has three subgroups O2a, O2b, and O2c. However, majority of the *Yersinia* infection in salmonids is associated with serotype O1a (Romalde & Toranzo, 1993). Based on the biochemical characteristics, they are also categorized as biotype 1 that shows positive for motility and lipase secretion while biotype is negative for both the tests (Tobback et al., 2007; Ormsby et al., 2016).

4.2.1 PATHOLOGY

Yersinia ruckeri causes chronic infections in adult fishes, while acute infections are highly prominent in young fry and fingerlings. The infected fishes usually become lethargic with loss of appetite and swim near the surface. In addition, exophthalmia, darkening of the skin, subcutaneous hemorrhages in and around mouth and throat, petechial hemorrhages on the liver, pancreas, swim bladder and in the lateral muscles were also seen (Horne & Barnes, 1999; Tobback et al., 2007). Histopathological (HP) examinations show the inflammation in most of the organs including kidney, liver, spleen, heart, and gills. In addition, edema, and desquamation of epithelial cells of the secondary lamellae of the gills were also seen in severely infected fish (Tobback et al., 2009). Multiple virulence factors are known to contribute to the clinical signs associated with the hemorrhagic form of the disease (Romalde & Toranzo, 1993). This include several extracellular products such as iron regulated *Serratia* like hemolysin, an azocasein protease and the 47 kDa metalloprotease that usually degrades fibronectin, actin, and myosin (Secades & Guijarro, 1999).

4.2.2 EPIDEMIOLOGY AND TRANSMISSION

Infection caused by *Y. ruckeri* can be transmitted by direct contact of an infected fish with non-infected fishes. As an enteric pathogen, the bacterium also gets excreted through feces and survives at least four months outside the host (Busch & Lingg, 1975). This is an important for the successful transmission of the pathogen to another host. In certain cases, fishes exhibit carrier state of infection (Rodgers, 1992; Tobback et al., 2007). Such asymptomatic carriers when unstressed were unable to transmit the bacterium to a healthy recipient fish. However, when asymptomatic carriers are stressed with elevated temperatures, the bacterium was successfully transmitted from carrier to recipient fish causing recipient fish to undergo carrier state with mild lower intestinal infections without any mortality (Hunter et al., 1980). In addition, birds and mammals can also act as important vectors that contribute to the spread of the disease (Kumar et al., 2015). The bacterium is known to form biofilm by over expressing flagellar proteins, which is important for the initial development of the biofilm. This further enhances the survival of the pathogens on surfaces of aquatic systems (Coquet et al., 2002). Such bacterial biofilms form a frequent source of recurrent infections in rainbow trout farms (Coquet et al., 2002). The progression of the infection usually starts from the gills which is an important portal of entry. Thereafter, it spreads to other organs through blood stream (Tobback et al., 2009). The optical projection tomography and immunohistochemistry confirms that the gut is the major affected organ upon immersion treatment. It has been found to reach the gut within 30 mins of post-infection and kidney at 3 days post-infection while other organs after 7 days post-infection (Mendez & Guijarro, 2013; Ohtani et al., 2014).

4.2.3 DIAGNOSIS

Several diagnostic techniques have been developed for *Y. ruckeri*. However, classical bacteriological assays are mainly based on culturing of *Y. ruckeri* on Columbia blood agar and MacConkey agar. They grow as circular, elevated, shining, and whitish colonies at a temperature range between 20°C and 28°C (Pajdak-Czaus et al., 2019). In addition, commercially available API-20E strips can be used to detect the bacterium based on the biochemical characteristics (Kumar et al., 2015). Serological tests such as ELISA (enzyme linked immunosorbent assay), agglutination, and immunofluorescence tests can be used to detect *Y. ruckeri* antibodies (Smith et al., 1987). Molecular methods include PCR to detect 16 rRNA gene (Tobback et al., 2007), restriction fragment length polymorphism (Gracia et al., 1998) and loop mediated isothermal amplification (LAMP) to amplify yruR gene

(Saleh et al., 2008). Most often specimens include internal organs such as gills, spleen or intestinal swabs. However, for PCR tests blood samples are usually preferred (Altinok, Grizzle, & Liu, 2001). A PCR test developed by Gibello et al. (1999) could detect low levels of *Y. ruckeri*, thus making it possible to detect the pathogen even in asymptomatic carriers. To improve sensitivity, a qPCR assay was also developed by Bastardo et al. (2012), that target recombinase gene (*recA*). Most recently, nanoparticle-based detection methods that detect DNA or RNA hybridized with pathogen specific probes that are attached with gold nanoparticles have been developed (Saleh et al., 2015). This rapidly emerging technology detects hybridization at a specific temperature and time with the aggregation of nanoparticles and concomitant appearance of red or blue color.

4.2.4 PREVENTION AND CONTROL

The treatment of ERM mainly based on antimicrobial therapy. Antibiotics such as oxolinic acid, amoxicillin, oxytetracycline, sulfadiazine in combination with trimethoprim and florfenicol are most commonly used (Michel et al., 2003). However, emergence resistance towards oxytetracyclines, oxolinic acid and sulfadiazine has been identified recently (Austin & Austin, 2016). To overcome the problem of antimicrobial resistance, alternative methods such as probiotics can be developed. For instance, it was found that oral administration of *Bacillus subtilis* and *Bacillus licheniformis* could protect the rainbow trout from subsequent infections with *Y. ruckeri* (Raida et al., 2003). This might be due to the secretion of antimicrobial agents by the probiotic bacteria or due to imuunostimulating effect. These results further encouraged scientists to look for several other bacteria such as *Carnobacterium maltaromaticum* B26, *Carnobacterium divergens* (Kim & Austin, 2006) and *Lactobacillus lacti* (Balcázar et al., 2008) showed antagonistic effect to *Y. ruckeri*.

The economic loss caused by ERM in the rainbow trout farm can be controlled to some extent by effective vaccination. The vaccine contains monovalent inactivated *Y. ruckeri* serotype O1 and biotype 1 whole cell suspension (Soltani et al., 2014; Villumsen et al., 2014). Newer vaccines based on *Y. ruckeri Yrp1* protease, *aroA* gene and lipopolysaccharide are known to provide good protection against ERM (Temprano et al., 2005; Ispir & Dorucu, 2014). However, these monovalent vaccines are not effective in controlling infections caused by biotype 2. Thus, a bivalent vaccine containing formalin inactivated biotype 1 and 2 was found to show good protection against infection caused by biotype 2 (Deshmukh et al., 2012).

4.3 FURUNCULOSIS

The causative agent furunculosis is *Aeromonas salmonicida* which is an obligate fish pathogen belonging to the family *Aeromonadaceae*. The organism was first isolated from a German freshwater brown trout hatchery by Emmerich & Weibel (1894). It is a gram-negative, nonmotile, facultative bacterium that grows well at temperature range between 22°C and 25°C. It has five subspecies, *salmonicida*, *achromogenes*, *smithia*, *masoucida*, and *pectinolytica* (Graf, 2015). Among these, *A. salmonicida* subspecies *salmonicida* is known to cause furunculosis in marine fish species (Valderrama et al., 2017). The bacterium is positive for catalase and cytochrome oxidase and ferments and oxidizes glucose.

4.3.1 PATHOLOGY

A. salmonicida produces several extracellular factors that includes 25 proteins, enzymes, and toxins. These extracellular virulence determinants contribute to the pathogenesis of infection caused by the bacterium that can be further correlated with the stress of the susceptible fish (Austin & Austin, 2016). In addition, surface polysaccharides such as capsule, glucan, and lipopolysaccharide, iron binding systems, secretory systems, fimbriae, and flagella also contributes to the pathogenesis of the disease. However, it is the combination of all the factors that is responsible for clinical signs of the disease during infection (Ali et al., 1996). Further, the type III secretion system that translocates the effector proteins into the host cell is responsible for subsequent immune response and inflammation (Ebanks et al., 2006; Frey & Origgi, 2016). Based on the severity, the infection can be acute, subacute, chronic or latent. The subacute form of the disease usually presents with few clinical signs such as darkening of the skin, exophthalmia, and distended abdomen (Cipriano & Austin, 2011). The acute form of the disease shows the same clinical signs as that of subacute form, in addition to anorexia, petechia, and hemorrhages (Cipriano & Austin, 2011). The pathological signs chronic form include boil like lesions referred to as furuncles that initially seen at the internal organs and then to the external surfaces. This further leads to hemorrhages in the fin bases, mouth, viscera, and reproductive organs (Roberts, 2012) (Figure 4.1).

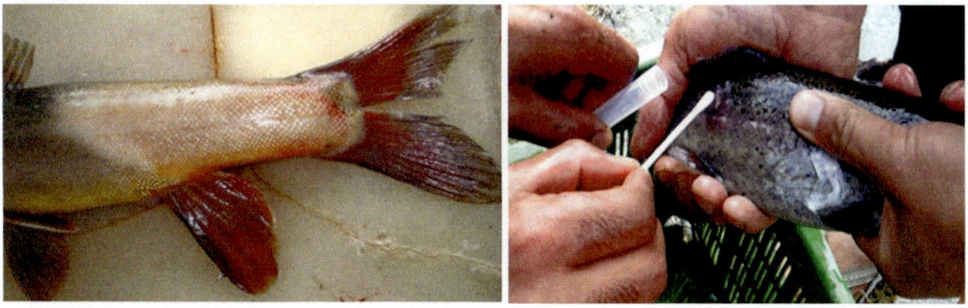

FIGURE 4.1 Fishes showing the caudal lesion.

4.3.2 EPIDEMIOLOGY AND TRANSMISSION

Transmission of furunculosis from fish to fish mainly occur through direct contact or via ingestion. Horizontal transmission of the pathogen from seawater to freshwater or vice versa has been reported in sea reared rainbow trout farms in Denmark (Bartkova et al., 2017). Water plays an important role in transmission of *A. salmonicida* from infected fish to healthy fish (Skrodenytė-Arbačiauskienė et al., 2012). Further, changes in the environmental conditions such as pH, temperature, oxygen content, etc., stimulate the replication and spread of *A. salmonicida* infection through water bodies. However, vertical transmission has not been reported in *A. salmonicida* infection.

4.3.3 DIAGNOSIS

Routine culture-based identification includes growth of the bacterium on blood agar wherein they produce brown diffusible pigmented large colonies within two to four days at 25°C. While on TSA it gives rise to colonies with dark brown water-soluble pigments when incubated at 22°C for three days (Austin & Austin, 2007). Biochemically the strains are positive for oxidase, catalase, lysine decarboxylase, gelatin hydrolysis and methyl red. Molecular confirmation can be done by REP-PCR and 16 rRNA sequencing (Skrodenytė-Arbačiauskienė et al., 2010) and by qPCR (Bartkova et al., 2017). More rapid serological tests such as serum agglutination, ELISA, fluorescent antibody test (FAT) can be performed on tissue or isolated bacterial samples (Austin et al., 1986). A PCR based DNA probe test developed by Mooney et al. (1995) was found to be successful at a rate of 88% in detecting furunculosis in Atlantic Salmon. Further, a reverse transcription-multiplex PCR (RT-MPCR) developed by Rattanachaikunsopon & Phumkhachorn (2012) was able to discriminate between viable and nonviable cells with a sensitivity of detection of 30 CFU (colony forming units) bacterium. MALDI-TOF is another technique which could be successfully used to identify and discriminate between various subspecies of *A. salmonicida* (Benagli et al., 2012; Jansson et al., 2015).

4.3.4 PREVENTION AND CONTROL

The control of furunculosis outbreaks mainly relay on the distribution of antibiotics. However, the extensive use of antibiotics such as oxytetracycline, sulfadiazine/trimethoprim has led to the development of drug resistance (Schmidt et al., 2001; Krikan et al., 2003). To combat the problem of antimicrobial resistance, the aquaculture industries have now widely used alternatives such as bacterial vaccines, which are generally administered as intraperitoneal injections (Midtlyng, 2014). These vaccines result in the nonspecific activation of macrophages in the kidney that can eliminate *A. salmonicida* (Midtlyng, 2014). However, intraperitoneal injection vaccination has also been linked with side effects such as impaired growth and inflammation. To avoid such side effects, vaccination through immersion method has been demonstrated though with short lived protection (Midtlyng, 2014). The ability of recombinant vaccines using iron regulated outer membrane proteins and L-forms of the bacterium have been investigated for its use against *A. salmonicida* infection (Lutwyche et al., 1995; Robertson et al., 2005). The use of disease resistance fish strains through selective breeding has also received considerable attention recently. For instance, higher bactericidal activity was found in sera of rainbow trout (Hollebecq et al., 1995) and salmon families (Zhang et al., 2011) that are bred for their resistance to *A. salmonicida*. Use of probiotic bacteria *viz Carnobacterium* sp. (Robertson et al., 2000), *Lactobacillus plantarum* and *L. fermentum* (Balcazar et al., 2008) were found to be either antagonistic or prevent the adhesion of *A. salmonicida*. The immunostimulant substances such as glucan and chitosan were also found to reduce mortality caused due to *A. salmonicida* infection by improving the nonspecific immunity and thereby survival of the fishes (Anderson & Siwicki, 1994). Finally, specific lytic bacteriophages can also be used as an effective alternative to antibiotics. For instance, Kim et al. (2012) identified 2 lytic bacteriophages (phiAS5 and PAS-1) against *A. salmonicida* was found to reduce mortalities during challenge with *A. salmonicida* (Kim et al., 2015).

4.4 BACTERIAL COLD-WATER DISEASE

Bacterial cold water disease is a disease of freshwater fish, particularly the salmonid fish that results in significant mortality in the salmon and trout populations. It causes considerable economic losses and difficulties are encountered during commercial aquaculture and in conservation hatcheries (Antaya, 2008). The disease is referred by other names including cold water disease (Lumsden et al., 2006), fin rot disease (Post, 1987), peduncle disease (Davis, 1947), saddleback disease (Borg, 1960), fry mortality syndrome (Toranzo & Barja, 1993), and rainbow trout fry mortality syndrome (Barnes & Brown, 2011).

4.4.1 CAUSATIVE AGENT

Flavobacterium psychrophilum, belonging to the family Flavobacteriaceae and previously known as *Cytophaga psychrophila* is the etiological agent of cold water disease is a serious fish pathogen. *F. psychrophilum* is prevalent in the aquatic environment, particularly freshwater (Groff & LaPatra, 2001).

The bacterium was discovered in 1948 from *Oncorhynchus kisutch* from the Pacific Northwest United States. It is a Gram-negative, aerobic, catalase-positive, oxidase-positive, psychrophilic bacilli with growth optimum below between 18°C and 20°C. The size of the bacterium is approximately 1.5 to 7.5 μm in length and 0.75 μm in breadth. It can resist the action of the lysozyme for up to 2 mg/mL and survive at 100 ppm of povidone-iodine for 30 min. By changing its morphology, it can withstand starvation for months (Vatsos et al., 2001). The recovery of *F. psychrophilum* from the brain of a newt, a non-fish host, from algae (Amita et al., 2000), and from benthic diatoms (Izumi et al., 2005) indicates its adaptability (Brown et al., 1997) and perhaps indicates the reservoir of this bacterium. The bacterium exhibits gliding motility which is influenced by the nutrient concentrations. Moreover, the extent of gliding motility varies between strains (Perez-Pascual et al., 2009). It can tolerate a salt concentration of up to 0.5% (Bernardet & Kerouault, 1989).

The chromosome of a virulent *F. psychrophilum* has been mapped. It shows 2,861,988 base pairs which is comparatively smaller than other bacteria within the family (Duchaud et al., 2007). The percentage G+C content has been found to be 33% (Bernardet & Grimont, 1986). The pathogen produces a wide array of virulence factors including fibronectin type adhesins, proteases, cytolysins, and hemolysin like proteins which help the bacterium in attachment and invasion (Duchaud et al., 2007). There are at least three main serotypes of *F. psychrophilum* with distinct genetic variation which has an influence on the virulence of the pathogen (Izumi et al., 2003).

4.4.2 EPIDEMIOLOGY

The disease is prevalent throughout the world. Although the fatal disease outbreaks are linked to water temperatures below 10°C, the commercial trout industry has seen a severe disease at constant temperatures of 15°C. The younger fish (fry and fingerling) are found to be at risk of contracting an infection (Brown et al., 1997).

Horizontal spread of the infection occurs in the presence of predisposing conditions such as damage to the skin or lesion (Madetoja et al., 2000). Invasion into fish is most likely to occur if there is a break in the tegument (Miwa & Nakayasu, 2005). It has been indicated that the bacterium can be transmitted within eggs making the control using povidone-iodine disinfectant difficult (Cumagai et al., 1998). Studies by Brown et al. (1997); and Ekman et al. (1999) has shown the vertical transmission of the pathogen from adult to progeny. It has been recovered from intraovum, ovarian fluids, milt, mucus, egg surfaces, and kidneys from sexually mature trout and salmon (Taylor, 2004; Cipriano, 2005).

Studies by Dalsgaard & Madsen (2000); and Chen et al. (2008) has reported the susceptibility of coho salmon and juvenile rainbow trout to *F. psychrophilum*. The pathogen can survive for months to years in fresh-water outside a fish host (Michel & Garcia, 2003). Overcrowding of fish has been found to be a predisposing factor for disease outbreaks (Davis, 1947). Water hardening in aquaculture ponds free from the pathogen seems to prevent surface contamination (Kumagai et al., 2000). Spleen size has been correlated with the innate immunity of the fish. Hadidi et al. (2008) proved that specimens with a larger spleen index were notably more resistant to *F. psychrophilum*. A greater number of bacteria are released from dead fish (Madetoja et al., 2000). Bacteria have an increased affinity for the fin, lower jaw, and caudal peduncle (Martinez et al., 2004). The presence of ectoparasites has shown an increase in the invasiveness of *F. psychrophilum* (Busch et al., 2003). High nitrite or organic content in water may play a role in the disease incidence (Nematollahi et al., 2003). Human activities are likely the reason for the spread of the clonal complexes of *F. psychrophilum* (Nicolas et al., 2008). The mortality rate is shown to be varied in different countries with an average mortality of 70% in trout from Western Europe (Santos et al., 1992), 35% from Denmark (Jensen et al., 2003), and 10–30% from the UK (Bruno, 1992). The variation in the mortality rates could be attributed to water temperature.

4.4.3 CLINICAL SIGNS

The disease exhibits varied manifestations with the most frequent or the classic form of the disease (CWD: chronic wasting disease) giving rise to typical open lesions on the exterior surface of the fish. Initially, these lesions may be observed as areas of rough appearing skin or with a fraying fin tip along with loss of appetite and lethargy. The regions of bacterial colonization on the fins appear as pale white areas with separation of the fin rays as found in some fish. Necrosis erupts at the location of the bacterial colonization as the disease progress, manifesting in the dorsal region with adipose fin pathology. Systemic infection sets in along with the external symptoms and causes substantial damage. Other signs of the disease may involve abdominal enlargement with a higher volume of ascites, and pale gills. In acute infections, the presence of exterior lesions will be less prevalent and presents with an extensive systemic infection (Starliper, 2011). In chronic infections, the infected fish reveals erratic or spiral swimming, darkened black caudal regions and/or spinal abnormality with spinal compressions in the posterior, mid, or anterior of the fish (Kent et al., 1989; Madsen et al., 2001).

In advanced infection, the damage to the caudal region will be severe that progresses to the caudle vertebra which gets exposed. The bacterial load will be more in the liver, spleen, intestine, peritoneum, heart, and pancreas indicating septicemia (Decostere et al., 2001; Ekman & Norrgren, 2003). Lesions can also be found on the snout jaw region, lateral sides, and muscular tissue between the back of the head and the dorsal fin. Necrosis of liver, spleen, and kidneys, elevated hemosiderin in the kidney and eosinophils, higher vacuolar degeneration, pyknosis, and lymphocyte infiltration in the lateral musculature of the skin lesions and dermis (Nematollahi et al., 2003). The bacteria is also observed in the retina and choroid gland of the eye eventually leading to blindness (Ostland et al., 1997).

The other disease manifestation set forth is known as the rainbow trout fry syndrome. As the name suggests, it attacks the sac fry to the early feeding stages of the fish. It is an acute form of the disease resulting in a higher mortality rate (50% or more). There is a development of bacteremia in addition to ample internal pathology causing pale liver and kidneys along with anemia. Additional characteristic features of the disease are exophthalmia, dark skin pigmentation, pale gills and dullness (Toranzo & Barja, 1993).

The clinical signs of the disease vary with the age of the fish as well as the outbreaks. In general, the disease severity of the disease is greater in younger fish.

4.4.4 HOST RANGE

Cold water disease is believed to affect nearly all species of salmonids. Steelhead and rainbow trout and coho salmon are found to be infected the most. Although the disease is considered primarily to be of the young fish, the old fish can also be infected. The bacterial pathogen has been isolated from a number of cultured and free ranging salmonid fish and also from non-salmonid fish hosts. The broad host range has been represented in Table 4.1.

TABLE 4.1 Host Range of *F. psychrophilum*

Hosts	References
Salmonids	
Atlantic salmon, sockeye salmon, chinook salmon, cutthroat salmon, chum salmon, brook trout, brown trout	Rucker et al. (1953); Bullock et al. (1971); Holt et al. (1993)
Rainbow trout	Rangdale et al. (1997)
Arctic char, coho salmon	Hesami et al. (2008)
Grayling, sea trout, whitefish	Madetoja et al. (2001, 2002)
Amago salmon, Yamame salmon, Iwana salmon	Amita et al. (2000)
Non-Salmonids	
Carp, crucian carp, eel, tench, ayu	Lehmann et al. (1991); Wakabayashi et al. (1994); Lee & Heo (1998)
Japanese eel	Izumi et al. (2003)
Pale chub	Iida & Mizokami (1996)
Perch, roach	Madetoja et al. (2002)
Sea lamprey	Elsayed et al. (2006)
Zacco platypus	Fujiwara-Nagata et al. (2019)
Goby	Amita et al. (2000)

4.4.5 DIAGNOSIS

A presumptive diagnosis of the disease can be made on the basis of case history, rearing condition for the fish, the pattern of mortality, characteristic clinical signs, age, and water temperature. Culture and characterization of the pathogen from the affected tissues, including spleen, brain, reproductive parts, liver, kidney, and skin lesion is considered to be the definitive diagnosis of the disease (Cipriano & Holt, 2005).

Infected tissue examination reveals the presence of thin, rod-shaped *F. psychrophilum* with an approximate size of 0.75–1.0 μm wide by 3–5 μm long. The cells may be found attached end to end, giving the appearance of a lengthy structure. Histological specimens exhibit osteitis, meningitis, periostatis, and gangloneuritis (Bruno, 1992).

The culture involves the use of various infected tissues. However, not all infected fish will have adequate viable cells in the internal tissues for primary culture. The isolation of the pathogen from skin lesions is often challenging as compared to the internal specimen due to the presence of oomycetes fungi and environmental bacteria that will easily grow on the culture media. The recovery of bacteria is enhanced by increasing the number of samples from a greater number of fish and homogenization of tissue samples. Inoculation can be carried out either by using direct streak plating or serial dilution. Poor growth or no growth is found on nutrient rich media such as blood agar, brain heart infusion agar, and tryptic soy agar. As the bacterium is somewhat fastidious, it might require specific nutrients in the culture medium (Alvarez & Guijarro, 2007). Various culture media may be used to grow *F. psychrophilum.* Variations of cytophaga medium are routinely used in diagnostic laboratories (Lorenzen et al., 1997; Daskalov et al., 1999). Pale yellow colonies with a diameter of 2–3 mm develop after a period of 2–3 days of incubation which later transforms into a characteristic fried egg colony with a raised center and irregular margin. The suspected colonies are subjected to various biochemical tests for the confirmation of the bacterium. They are catalase and oxidase-positive, hydrolyze casein and gelatin, break down tyrosine, produce flexirubin like pigments, and lyse killed *Escherichia coli* cells. Variability in the results for nitrate reduction and elastin hydrolysis is found which could be attributed to the difference in the strains. Supplementation of cytophaga medium with 10% calf serum, the addition of glucose, galactose, and skimmed milk, increasing the concentration of tryptone and beef extract resulted in improved growth of *F. psychrophilum* (Obach & Laurencin, 1991; Daskalov et al., 1999; Rangdale et al., 1997) Tryptone-yeast extract-salts agar, tryptone-yeast extract agar, and antibiotic containing media have also been used (Wiklund et al., 2000; Madetoja et al., 2002; Kumagai et al., 2004). A study by Cepeda et al. (2004) found that addition of glucose to tryptone-yeast extract-salts agar was effective at 18°C in isolating *F. psychrophilum*. Tryptone-yeast extract-salts agar in combination with activated charcoal also improved the growth of *F. psychrophilum* (Alvarez & Guijarro, 2007).

Immunological techniques have increased the speed, sensitivity, and specificity of identifying *F. psychrophilum*. Immunohistochemistry and immunofluorescence antibody techniques have been employed to confirm the culture results (Lorenzen & Karas, 1992; Lindstrom et al., 2009). Detection of the pathogen on the cell surface components by using enzyme-linked immunosorbent assay has been developed (Crump et al., 2003). Colony blotting and immunostaining for the detection and quantification have been used (Misaka et al., 2008).

Nucleic acid based techniques are increasingly being used for the detection of *F. psychrophilum* or to confirm the culture based results. Polymerase chain reaction (PCR) using DNA gyrase subunit gene, *gyrB*, and the peptidyl-prolyl cis-trans isomerase C gene, *ppiC* are the preferred primers for the identification of the pathogen (Starliper, 2011). Nested and multiplex PCR techniques have been described for the characterization of *F. psychrophilum* (Amita et al., 2000; Del Cerro et al., 2002).

4.4.6 TREATMENT

An accurate diagnosis is of utmost importance before treating any fish for cold water disease. Oxytetracycline was widely used for the control of the disease (Lumsden et al., 2006). In Europe, amoxicillin and oxolinic acid were extensively used (Bruun et al., 2000). The development of resistance to these antibiotics has made florfenicol the antibiotic of the current use (Gultepe & Tanrikul, 2006). The drug is given orally to affected fish through medicated feed at a concentration of 10 mg/kg of fish for a period of 10 days. To prevent the development of antibiotic resistance in the strains, indiscriminate antimicrobial therapy should be avoided. Before the administration of antibiotics, it is desirable to isolate, confirm, and check the susceptibility of the pathogen to the antimicrobial agent intended to be used.

The use of non-antibiotics and chemicals has also been reported. Gultepe & Tanrikul (2006) detailed the use of hydrogen peroxide bath followed by feeding florfenicol medicated diets. A similar approach was followed with quaternary ammonium baths and potassium permanganate treatment in combination with antibiotics (Schachte, 1983). The treatment should always be used in combination with better environmental conditions that reduce the stressors to fish. Maintaining clean holding tanks and removal of dead fish minimizes the transmission of the pathogen in the water column.

4.4.7 PREVENTION

The ubiquitous nature of the pathogen makes prevention difficult. Good management practices can reduce the transmission or introduction of pathogens, and minimize the severity of the disease outbreaks. The best preventive measure often involves reducing stress and decreasing the damage to the fish skin. Removing the sick, dying, and dead fish from aquaculture ponds is extremely important as it reduces the bacterial load and risk of transmission of disease. Good management practices in conjunction with the available treatments are the only methods for the control of cold water disease.

Iodophor disinfection of eggs just prior to hatching is the most frequently used preventive method (Taylor, 2004). As the pathogen is found in the fluid surrounding the eggs in sexually mature salmonids, surface disinfection of eggs reduces the contamination. Treatment with 100 ppm of the active ingredient for 10 to 30 min is preferred. However, studies have indicated the tolerance of iodine by the pathogen by the virtue of which it can survive even after treatment (Brown et al., 1997; Kumagai, 1998). The relative ineffectiveness of iodine treatment as evident in some studies might be one of the reasons for the spread of virulent strains around the world. Although treatment with povidone-iodine is not 100% effective under all situations, it can reduce the egg associated pathogen transmission. Sodium chloride approved by the FDA (Food and Drug Administration) as a "low regulatory priority" can be effectively used for treating external infection.

To prevent the disease outbreaks, it is very important to reduce physical handling and stress (Cipriano & Holt, 2005). As physical handling increases the chance of skin abrasions that serve as the entry point for the pathogen, it is highly recommended to minimize physical handling. Decreasing the rearing densities is also preferred (Taylor, 2004). Early detection reduces the chance of disease spread. It is very important to take care while the fish is being moved between culture facilities in case of disease suspicion or if the facility has a disease history. As dead fish serve as the reservoirs for *F. psychrophilum* it becomes very important to remove them from rearing areas. The introduction of novel or wild fish into the existing fish stocks should be avoided.

High quality fish diet to a certain extent can help in preventing the disease. Malnutrition has been listed as a probable cause for the disease. Daskalov et al. (2000) demonstrated a link between diet and cold water disease. In their study, high oxidized lipid diet and control diet were fed to rainbow trout. They observed higher mortality in fish fed with a high oxidized lipid diet when challenged with *F. psychrophilum*. Better protection was observed in rainbow trout fed with a diet naturally contaminated with deoxynivalenol compared to controls (Wald et al., 2016). A health enhancing diet could increase the effectiveness of a vaccine. A probiotic *Enterobacter* sp. administered orally protected fry from disease outbreaks with *F. psychrophilum* (Ghosh et al., 2016). Probiotic treatment of water also reduced the populations of *F. psychrophilum* (Boutin et al., 2013).

Increasing the quality of water used in aquaculture can reduce disease outbreaks. Decreasing the nitrite and organic content of water helps in reducing the infectivity of *F. psychrophilum*. Use of filtration, UV treatment, or ozonation for pathogen free water was suggested by Cipriano & Holt (2005). Increasing the water temperature may also reduce *F. psychrophilum* infections.

Broodstock screening may serve as a tool for the prevention of cold water disease. The use of FATs and ELISA has been suggested for broodstock selection (Lindstrom et al., 2009). In the absence of potent vaccines, selective breeding has shown a significant promise in reducing the incidence of the disease (Henryon et al., 2005).

There are currently no available vaccines for cold water disease. Various groups are working on developing an effective vaccine for the disease. Researchers have tried using live attenuated strains and formalin killed strains of *F. psychrophilum* as vaccine candidates. A study by Alvarez et al. (2008) demonstrated a significant disease resistance in fish injected with attenuated strains. Similarly, Kondo et al. (2003) found a notable survival efficiency in fish challenged with virulent strains. Specific antigens of *F. psychrophilum* particularly of the outer membrane identified by LaFrentz et al. (2007) could provide better protection against the disease. Another problem hampering the vaccination strategy is the delivery method by which it can be administered effectively to a large number of fish.

The use of *F. psychrophilum* bacteriophages has shown promising results in laboratory experiments. They have been shown to survive the fish stomach, penetrate the intestine, and enter the circulatory system after oral delivery. This indicates the possibility of delivering phages through coated feed pellets (Christiansen et al., 2016).

Other preventive measures include maintaining a clean fish holding tank with water free from pathogens, regular health inspection on statistically significant numbers of fish, sanitization of equipment, and appropriate storage of quality fish food.

4.5 RAINBOW TROUT GASTROENTERITIS (RTGE)

Rainbow trout gastroenteritis (RTGE) is characterized by the presence of large numbers of "*Candidatus arthromitus*," a segmented filamentous bacterium within the intestine. Several reports have recorded the mortality rate of fish infected with RTGE in the affected sites. Around 10–40% proportion of the productive units remain affected by the RTGE (Branson, 2003; Michel et al., 2002; McCarthy et al., 2016). RTGE is frequent in fishes larger than 800 g (Michel et al., 2002). The presence of outbreaks lasted for 2–4 weeks, with the peak mortality range on the second week after the onset of the infection (Branson, 2003; Michel et al., 2002).

This syndrome was seen preferentially in summer wherein the temperature of water might be a major factor in its development. Several studies suggested that despite the water temperature being an important factor, there is no such constant temperature threshold for RTGE (Branson, 2003; Michel et al., 2002; McCarthy et al., 2016; Del-Pozo et al., 2009; Toranzo et al., 2004). In the disappearance of clinical RTGE, changes in the environment and diet played a major role (Michel et al., 2002). Factors like low-energy diets could be added as an advantage with RTGE related infections (McCarthy et al., 2016).

Filamentous bacteria seen in the distal intestine of the RTGE infected fishes have been recognized as a part of the *Candidatus arthromitus* group of segmented filamentous bacteria (SFB even though culturing of the SFB could not be possible outside the gastrointestinal tract of the host, populations were maintained as a mono-associations with mice (Schubert et al., 2021). The term *Candidatus* is a provisional status given to describe prokaryotes which are incompletely termed (Murray et al., 1995). Arthromitus in Greek meant joint thread. Basically, *C. arthromitus* represent bacterial group with identical morphological and ecological niches. The endospore-forming SFB are Gram-positive and have found to be attached to the intestine of a wide range of vertebrates (Urdaci et al., 2001; Snel et al., 1995; Murray et al., 1995; Smith, 1997; Lowden & Heath, 1995; Angel et al., 1990) and invertebrates (Klaasen et al., 1991; Margulis et al., 1999).

The presence of the pathogen by light microscopy in the intestine of healthy individuals including the carps was observed (Klaasen et al., 1991). This demonstrates the presence of SFB in the animal kingdom is ubiquitous. Despite repeated trials to transform the rat or chicken SBF to mice, gave a clue that the host specificity of the SFB was slightly different depending on the host species (Allen, 1992; Tannock & Archibald, 1984). Subsequent phylogenetic 16S rRNA analysis of *C. arthromitus* from crab eating monkey, mice, rats, and chickens further validated that they were in fact different species (Tannock & Archibald, 1984; Imaoka et al., 1997). These findings stated that different species of SFB colonize different host species specifically, and the fact has been accepted even till date. The same theory holds good as far as the trout SFB is concerned, indicating they represent different species within the *C. arthromitus* group.

Several authors studied the life cycle and multiplication of *C. arthromitus* (Snel et al., 1995; Smith, 1997; Ferguson & Birch-Andersen, 1979; Chase & Erlandsen, 1976).

The endospore formation in the *C. arthromitus* is accompanied with the production of active intracellular offspring for dispersal/reproduction through unfavorable environments like aerobic conditions. The spores have been shown to be actively responsible for the spread of SFB amongst susceptible mice and chickens (Ivanov et al., 2009; Herbert et al., 2016). The same mechanism of spread has been reported in trout *C. arthromitus* (Michel et al., 2002).

Whole growth cycle of the SFB in mice was noted to occur within 20–30 h villous transit time of epithelial cells, except if the attachment is not hindered or there is inhibition in the cell migration (Klaasen et al., 1993; Pamp et al., 2012). This may also be accepted to be the same in the case in *C. arthromitus* infecting the trout.

Arguments arise in the role of *C. arthromitus* as an etiological agent in RTGE in trout as these have never been reported in the digestive system of healthy trout. There are reports suggesting its involvement and recovery from the RTGE-affected trout (Michel et al., 2002; Toranzo, 2004; Urdaci et al., 2001). Yet, there are no reports to prove the association of *C. arthromitus* in causing diseases in other animal species, though it was shown to cause malabsorption, diarrhea, gas, and fluid-filled intestines in poultry (Goodwin et al., 1991).

The same scientists reported that the bacteria might be just a part of normal microflora of the intestine and may not be necessarily a pathogen (Umesaki et al., 1999).

Fragility of trout SFB could be a result of impaired *C. arthromitus* membranes in the intestine of RTGE-infected trout (Michel et al., 2002).

4.5.1 PATHOLOGY

Fishes affected with RTGE shows loss of appetite and lethargy. Affected fishes tend to gather on the surface of the ponds with occasional uncoordinated swimming and reminiscent of neuropathological toxic mechanisms (Branson, 2003; Michel et al., 2002; McCarthy et al., 2016; Del-Pozo et al., 2009). Mucous content of the intestine extruded from the anus and abdominal distension are some of the gross lesions. In few cases the infected fishes presented yellowish mucoid excretions from the vent, dark streaks scattered along the flanks and dyschromia (Branson, 2003; McCarthy et al., 2016; Michel et al., 2002; Toranzo, 2004). Few other cases reported an internal lesion in diarrheic trout including acute hemorrhagic enteritis more significantly in the hind gut accompanied with edematous appearance of the intestinal mucosa and hemorrhage. Straw-colored mucoid material was seen to fill the entire digestive tract, including the pyloric caeca and enlarged stomach causing a dense occluding plug in the terminal portion (Branson, 2003; Del-Pozo et al., 2009; Toranzo, 2004).

The most significant and defining clinical feature of the RTGE syndrome is the large quantity of SFB which could be observed in the smears of the intestinal contents of the infected fishes (Del-Pozo et al., 2010). SFB seen within the digestive tract of the RTGE-affected trout were characterized to be nonbranched and found to be approximately 1 μm (0.6–0.12 μm) of width and up to 60 μm in length. However, the length of the bacteria varied between 1.2 and 2.6 μm in segments (Urdaci et al., 2001).

In the wet mount preparations stained with metachromatic toluidine blue, several forms corresponding to different stages of maturation of the SFB was observed. The bacteria are gram-variable, but usually Gram-positive producing spores that readily stain with malachite green (Del-Pozo et al., 2009, 2010; Ericssonl et al., 2014).

Reports suggest varied locations of the filamentous bacteria harboring the digestive system of the diseased trout. Few reports (Michel et al., 2002) observed their presence throughout the digestive system, while others demonstrated its presence more frequently in the distal intestine (Branson, 2003; Michel et al., 2002; Del-Pozo et al., 2009, 2010).

Affected trout exhibited HP lesions in the digestive system, severe lesions were seen in the caeca and pyloric stomach (Del-Pozo et al., 2010; Urdaci et al., 2001; Ericssonl et al., 2014). Congestion was observed in the mucosa, thickening of the intestinal wall by edema. A most significant change was epithelial necrosis with extensive mucosal detachment (Ronza et al., 2009). There are no reports on the HP changes in the other organs.

4.5.2 DIAGNOSIS

Progression from the culture-dependent techniques like the phenotypic characterization including the morphology, biochemical, and physiology, molecular techniques, in particular the 16S rRNA gene sequencing provides insights into the phylogenetics. The culture-independent approaches have resulted in the identification of new uncultured bacterial groups like *Candidatus.* Sequencing of the 16S rRNA gene is more reliable in allocating names to the bacterial pathogens. It is marked advantageous over the phenotyping identifications (Jolley et al., 2012).

Antibody-based procedures are effective in detecting the exposure of fish to viruses, while the identification of bacterial pathogens becomes complicated due to cross reactivities unless the specific known molecules/antigens are used to coat the ELISA plates rather than whole pathogens (Adams & Thompson, 2006). However, these techniques are specific, sensitive, reliable, and rapid.

Chemotaxonomy comprehends the use of chemical constituents of bacteria for the investigation. In particular, in the study of Gram-positive bacteria. Polar lipids, lipopolysaccharide, fatty acids, ubiquinones, menaquinones, naphthoquinones, mycolic acids, teichuronic, and teichoic acids, peptidoglycan, can be used in the detection of the RTGE related pathogens (Austin, 2011). Molecular methods employing techniques like PCR-based gene sequencing, reverse transcriptase-sequencing and 16S rRNA gene sequencing have been additional advantages to the armamentarium of methods applicable to the bacterial taxonomy (Yin et al., 2013). However, the DNA hybridization introduced into bacterial taxonomy during the 1960s is marked the "gold standard" to determine the absence or presence of new species.

Nucleic acid fingerprinting methods that include pulsed field gel electrophoresis (PFGE), amplified fragment-length polymorphism PCR (AFLP), REP-PCR (repetitive extragenic palindromic-PCR), ERIC-PCR (enterobacterial repetitive intergenic consensus sequences-PCR), rep-PCR (repetitive element primed PCR) and ribotyping may provide some data at/below the subspecies level. Amongst all of the above, ribotyping, and AFLP are considered as standards and are extremely useful (Laing et al., 2011). The methodologies might be culture-independent, those enable the study of uncultured organisms, but there are issues with genomic fluidity. *Candidatus*" reports uncultured prokaryotes for which phylogenetic relationships have been discussed, and the authenticity has been confirmed by methods such as *in situ* probing (Austin, 2011). Above all, the molecular techniques allowed to apprehend unculturable pathogens that belong to new groups of *Candidatus.*

4.5.3 PREVENTION AND CONTROL

Anecdotal observation is the only possible route of possible management option for the RTGE. Changes in the environment, diet, and feeding restrictions have been found effective in reducing RTGE losses (Branson, 2003; Michel et al., 2002). Treatment of the affected sites with RTGE infection is a trial and error-based experiment, due to lack of knowledge and literature on the etiology and pathogenesis. Addition of antibiotic treatments to the feed liquid paraffin or sodium chloride at varying concentrations. Keeping the affected fishes on fast for a week or more has also been used as one of the treatment options in Italy (Rashidian et al., 2020). Tested drugs for the control of RTGE includes bacitracin, clindamycin, amoxycillin, doxycycline, cotrimoxazole, streptomycin, gentamicin, trimethoprim, neomycin, ciprofloxacin, vancomycin, cefotaxime, polymyxin, and metronidazole (Singh et al., 2013). All of these drugs were observed to reduce the number of SFB in the ileum to different degrees, nevertheless suggesting the sensitivity of the SFB to the antimicrobial drugs (Caselli et al., 2010). The cause of RTGE will not be completely eliminated from the population by the treatment, the affected populations will be further susceptible to the disease on continuous exposure.

4.6 BACTERIAL GILL DISEASES (BGD)

Bacterial gill disease (BGD) at first was demonstrated by Davis (1926) as a colonization of the gill by thread-like filamentous bacteria, causing a rapid increase in the epithelium ultimately leading to lamellar fusion. Colonization of the lamellae of the gills by the bacteria results in a decreased rate of exchange of gas. BGD is a bacterial infection of the gills that impairs the oxygen uptake of the fish (Wakabayashi, 1980). It is predominately believed to be a disease of hatchery-bred juvenile salmonids and is not a considerable problem in the wild fish populations (Starliper & Schill, 2011). The disease is prevalent among salmonids, fish of all ages and species including Rohu (*Labeo rohita*), Sheatfish (*Silurus glanis*), Silver carp (*Hypophthalmichthys molitrix*), Walleye (*Sander vitreus*) and Catla (*Catla catla*).

The main causative agent of the BGD is the *Flavobacterium branchiophilum* (Touchon et al., 2011). At first, BGD was thought to be caused by various forms of gliding bacteria, referred to as "myxobacteria" (Bullock, 1972). Subsequently, the agents of BGD, including *Flavobacterium* and *Cytophaga* were grouped into a general category, a yellow-pigmented bacteria, isolated on Anacker and Ordal's medium referred to as Cytophaga Agar (Anacker, 1959). *Flavobacterium branchiophilum* was first isolated from gill lesions of a rainbow trout (*Oncorhynchus mykiss*) and Yamame (*Oncorhynchus masou*) suffering from BGD in hatcheries in Gunma, Japan in 1975 and 1977, respectively (Wakabayashi et al., 1989).

Bacteria belonging to the genus *Flavobacterium* are Gram-negative, exhibit aerobic yellow-pigmented rod. They are capable of gliding motility and can degrade many polysaccharides except for cellulose (Ko & Heo, 1997). Molecular techniques namely DNA-DNA hybridization, whole-cell protein, fatty acid profiles and DNA-rRNA hybridization were the basis of classification of the classified the Flavobacterium species. Distribution of the species are commonly soil environments and the freshwater.

F. branchiophilum, *F. psychrophilum*, and *F. columnare* are three important fish pathogens belongs to the members of the *Flavobacterium* genus. *F. columnare* is said to be the etiological agent of the columnaris disease which generally occurs in water temperatures of 20°C, affecting many fishes of wild and cultured species (Tort, 2000). Rainbow trout fry syndrome is caused by *F. psychrophilum*, a thin rod-shaped bacterium, growing at colder temperatures (10–16°C) (Mohammed, 2015). It also causes diseases in many fishes including tail rot, salmon species, wild, and cultured trout (Tort, 2000).

Both *F. psychrophilum* and *F. columnare* cause systemic and co-infections in both hatchery-raised and wild fish (Skulska, 2014).

4.6.1 PATHOLOGY

Increased brachial movements, decreased interest in feeding lethargy and fatigue leading to fish floating near the water surface are some of the signs associated with BGD (Good et al., 2015). However, the conditions for the onset of BGD are undiscovered, it is speculated that BGD outbreaks are associated with stress factors, such as poor water quality, overcrowding, and increased turbidity in the water supply low oxygen availability (Tort, 2000).

Good et al. (2009) observed that the hatcheries with low water exchange and poor quality of water could be considered a risk factor for the BGD outbreaks. Mortality of the BGD affected fishes could be reduced by the addition of ozone to the water supply, thus by improving its quality (Tort, 2000). Increased amount of feed at the hatcheries could be a possible reason for the infection by *F. branchiophilum*, as observed by Good et al. (2009). This may be because the water quality being affected by the increased feed, by an increase in the amounts of ammonia concentration and suspended solids. While, increased ammonia and decreased dissolved oxygen in the water are thought to be involved with BGD outbreaks (Tort, 2000). It was observed that the effect of fish feeding was a factor for consideration than the food itself, because the fishes unable to feed required feeding through stomach tubes for BGD to manifest, suggesting that a physiological change triggers the disease. Presence of the feed in water directly may not be responsible for the BGD outbreaks but definitely could amplify the effects of the disease (Faisal et al., 2011).

The fishes affected with BGD exhibit lethargy, in appetent, and seem to swim on the surface of the water. Bilaterally flared opercula and gasping are some of the common symptoms. Mortality rate depends on the age, size, stocking density, and environmental conditions of the fishes (Loch & Faisal, 2015). Around 10–15% of the mortality has been recorded within 24–48 h post-infection. Since, *F. branchiophilum* is non-invasive pathology remains limited to the gills. Flared arches or ragged gills and lamellar congestion, excessive debris/mucus are quite evident. It elicits fusion of the primary lamellae, lamellar fusion and epithelial hyperplasia in severe cases. This in turn causes a reduction in the respiratory surface area leading to the inhibition of gaseous, ionic, and osmotic homeostasis (Loch & Faisal, 2015) (Figure 4.2).

Infections of the flavobacteria are diagnosed depending on several factors like the behavioral changes, case history, histopathology, gross pathology with subsequent isolation, phenotypic characterization (Austin & Austin, 2016).

Cytophaga Agar was commonly used for culturing of *Flavobacterium* (Loch & Faisal, 2015). Light microscopy can be carried out for the identification of *F. branchiophilum* colonies on culture. Casitone-yeast agar and Modified Sheih's medium have been used to culture *F. branchiophilum*, of which Casitone-yeast agar has worked best (Kolton et al., 2013).

FIGURE 4.2 Fishes affected with bacterial gill disease caused by *Flavobacterium.*

Though the isolation of *F. branchiophilum* has been proved successful, culturing it is a challenge. For the same reason, as a presumptive diagnosis, examination of gill clippings under light microscopy revealing the presence of long filamentous bacteria on the gill surface has been a better alternative.

In addition to the above-mentioned methods, confirmation is achieved best when preformed the molecular and serological assays. FATs, *in situ* hybridization, Whole-cell agglutination, enzyme-linked immunosorbent assays, PCR, quantitative PCR, loop-mediated isothermal amplification (LAMP), DNA array-based multiplex assay, immunomagnetic separation in conjunction with flow cytometry are some of the molecular bases techniques employed for the identification and detection of the *Flavobacteria* species (Loch & Faisal, 2015; Racheal, 2017; Loch, 2012; Soltani et al., 2019; Noble & Summerfelt, 1996). However, it should be noted that some of the diagnostic reagents specific for the fish associated flavobacteria are lesser known, making their identification hard and time-consuming.

4.6.2 PREVENTIVE AND CONTROL

Optimizing the husbandry conditions by maintaining good water quality, minimizing stress, providing adequate water flow, reduction of rearing densities can help to reduce infections caused by the *Flavobacteria* (Long et al., 2014). Low amounts of dissolved oxygen and elevated amounts of ammonia are the predisposing factors for BGD outbreaks. Suboptimal exchange of water is considered a risk (Wahli & Madsen, 2018).

For the prevention infection by *F. psychrophilum* the eggs of the salmonid must be disinfected with povidone iodine (50 ppm) in the pre-water hardening phase (fertilized eggs before the chorion becomes rigid due to exposure to the water) has proven promising (Oplinger et al., 2015). Disinfecting the sperm of the rainbow trout with streptomycin and penicillin was found successful in the subsequent reduction of the *F. psychrophilum* infection (Pérez-Pascual et al., 2021). However, this needs to be further investigated because there are some drawbacks like lower egg fertilization and survival. Usage of the above-mentioned antibiotics in addition to 0.5% NaCl to the hatchery water prior to the hardening phase, significantly reduced *F. psychrophilum* (Webster & Thompson, 2015). The combination of streptomycin and penicillin with elevations in the water temperature killed the pathogen under laboratory conditions (Gonçalves et al., 2019). However, there might be a contradiction among the *in vitro* results and *in vivo* effectivity as far as florfenicol, enrofloxacin, and doxycycline in treating RTFS-affected trout was concerned. Copper sulfate when used for the control of *F. psychrophilum* infections was not found successful as it was found to be toxic to the eggs. Broodstock screening was considered a potential method for the evaluation of *F. Psychrophilum* infection levels (Perez-Pascual et al., 2021).

Various substances related to the fish feed were tested which were efficient in preventing the infections by Flavobacteria. In this regard, trout by-product hydrolysates, produced by trout pepsin, recorded the highest antibacterial activity against *F. Psychrophilum*. Restrictions on the feed intake served as a protection for the rainbow trout infected with the pathogen (Ghosh et al., 2016). Additionally, rainbow trout fed a diet naturally contaminated with deoxynivalenol also yielded better results. When the fishes were fed a health enhancing" diet, the effect of the vaccine was seen to be enhanced (Jarau et al., 2019). Oral or intraperitoneal administration of the *Enterobacter* species was an effective strategy to combat disease outbreaks with *F. psychrophilum* (Castillo et al., 2014). Prebiotic supplements with 30% soy bean exhibited minimal effect on the *F. psychrophilum* infected cutthroat trout by a cohabitation method. It is noteworthy that the probiotic treatment decreased the *F. psychrophilum* population in the water (Madsen et al., 2013).

Phage based therapy of the *F. psychrophilum* was suggested and also laid promising hopes when experimented *in vitro*. Ability of the phages to survive passage through the stomach of the fish, penetrate the intestinal barrier and gain entry to the circulatory system after oral delivery constituted an efficient method to treat as well as prevent the diseases caused by *F. psychrophilum* (Lahnsteiner, 2021; Farmer et al., 2017). For the prevention of the infections by *F. columnare* various chemicals like the chloramine-T and hydrogen peroxide treatment were effective. They also were able to control the mortality rates associated with the bluegills and external columnaris in bass (Beck et al., 2015). By the addition of extra copper sulfate to the feed, a positive response was observed in regard to the survival of several fish species affected by columnaris disease (Yang et al., 2015).

Chitosan can be a promising candidate and an alternative to antibiotics. Inhibition of the growth of *F. Columnare* could be achieved by *Nigella sativa* (black cumin) seeds when added to the fish feed, its oil has been effective in enhancing the survival rates of zebrafish and catfish (Perez-Pascual et al., 2021). Resistance against *F. Columnare* infections could progress when the Grass carp fed diets supplied with *Ficus carica* polysaccharide (Laanto et al., 2015).

In addition, the ungeremine and its analogs were efficient as bactericides against *F. Columnare* (Perez-Pascual et al., 2021). The activity of Flavone (Wogonin) *in vitro* was effective against *F. columnare*. Optimal dietary levels of vitamin C in the diet improve the barrier functioning while its deficiency was found to depress the gill physical and immune barriers in grass carp infected with *F. columnare* (Tie et al., 2021). Susceptibility towards columnaris infection could be achieved by the use of different prebiotic dietary additives. Increased survival of fishes infected with *F. columnare* disease in walleye was achieved when two strains of *Pseudomonas fluorescens* isolated the from skin and gut of healthy walleye (*Sander vitreus*) was used as a probiotic (Perez-Pascual et al., 2021). As an alternative to chemotherapy, administration of phages against columnaris disease in aquaculture as demonstrated in controlled experiments (Laanto et al., 2015; Pedersen & Lazado, 2020).

4.7 COLD WATER VIBRIOSIS

In the late 1970's there were low levels of mortalities observed in Atlantic salmon smolts, in the Northern part of Norway (Egidius et al., 1984; Kent, 1982). Initially, mortalities were related with environment and nutritional status of fishes. Later observation of abnormal fishes had the necrosis in mucosa of the caeca and intestine, sloughing into the lumen. Hemorrhages in lamina propria, with extensive vasodilatation, congestion of the blood vessels resulting in increase in numbers of eosinophilic granular cells. Regions in the kidney responsible for giving rise to red blood cells were sparse, with large areas of necrosis filled with macrophages with grayish cytoplasm predominated. Spleen had considerable enlargements with signs indicating inflammation densely populated with macrophages (Soleim, 1985). But identification of isolates from internal regions of moribund fishes indicated bacterial infection. Cold water vibriosis or Hitra disease or hemorrhagic syndrome is a bacterial infection caused by motile gram-negative rod *Vibrio salmonicida* mainly affecting Atlantic salmon (*Salmo salar*), rainbow trout (*Oncorhynchus mykiss*) and Atlantic cod (*Gadus morhua*) (Egidius et al., 1981, 1986).

Vibrio salmonicida is gram-negative motile rod. Tryptic soy with added sodium chloride could be used for its growth. For the primary isolation nutrient agar with supplements like, 1 to 1.5% sodium chloride and 5% sheep blood are often used. Optimum growth of the bacteria could be achieved at a medium salinity (1.5%) and temperature of 15°C (Egidius et al., 1986). The biochemical properties of these isolates, however, did not resemble those of *Vibrio anguillarum* or *V. ordalii*. The isolates were sensitive to all of the antimicrobial tested, including oxytetracycline, furazolidone, tribrissen, and oxolinic acid (Bruno et al., 1985).

4.7.1 EPIDEMIOLOGY

Vibrio salmonicida has more frequently reported in the marine waters of Eastern Canada, Scotland, and Norway. Specifically, infecting host Species Atlantic salmon and rainbow trout. The bacteria turn pathogenic at a water temperature of below 10°C. Smolts, juveniles, and adults are said to be affected by the pathogen. Water is the primary root of fish-to-fish transmission of Vibriosis (Bruno et al., 1985).

A temperature range 5–20°C is usually optimum for the growth of Salmonids. At a temperature range of 13–17°C (Wallace, 1993), efficient growth of salmon is achieved. Unfortunately, this range of water temperature also favors the growth of *Vibrio salmonicida*.

The bacteria establish latency period until it proliferates in large numbers to show the signs of the symptoms. The clinical manifestation include hemorrhage to intestines, body cavity, spleen, and muscle, distended mucoid and necrotic intestine and petechiation, erosion, and darkened coloration to the skin and fins. Changes to the eyes include distension and cloudiness and periorbital swelling. White/gray lesions can be found on the intestines and spleen with swollen spleen. Isolation of this bacterial pathogen from ascites and a variety of internal organs could be achieved (Totland et al., 1987).

4.7.2 PREVENTION AND CONTROL

Reduction and control of vibriosis could be achieved by implicating the following methods. Oxolinic acid and florfenicol are commonly used in the treatment of vibriosis Norway (Frans et al., 2011). Flumequine and quinolones are some of the widely used antibiotics in the control cold water vibriosis (Shao, 2001; Defoirdt et al., 2011). Erythromycin, cotrimoxazole, streptomycin, and chloramphenicol are some of the reported antibacterial to combat *V. harveyi* infection in giant tiger prawns. Use of a wide range of chemicals like benzalkonium chloride, copper sulfate, formalin, and malachite green could help eliminate the opportunistic pathogens affecting the fishes. Overdose of these chemicals have adverse effect on fishes including gill damages (Punitha et al., 2008). Hypochlorite's were used by the farmers of Malaysiato control vibrios (Mohamed et al., 2015).

4.8 MYCOBACTERIOSIS

Mycobacteriaceae is a Gram-positive bacterium family that includes a huge number of harmful bacteria. Fish mycobacteriosis is a chronic, progressive condition caused by nontubercolous mycobacteria, which are acid-fast in nature (Chinabut, 1999). Mycobacterium was initially discovered in a diseased marine aquarium fish in 1897. Formerly, the organism was named as *M. piscium*. Even though unproven, the organism originally referred to as *M. piscium* is alleged to be the same thing as the one named as *M. marinum* currently (Noga, 2010). The genus Mycobacterium contains more than 150 species, including the obligate pathogens that cause tuberculosis in mammals and environmental saprophytes that can cause opportunistic infections. Members of the *Mycobacterium tuberculosis* such as *M. tuberculosis*, *M. bovis, M. caprae, M. pinnipedii* cause tuberculosis in mammals, and non-tuberculous mycobacteria are typically separated into two families (Antuofermo et al., 2014). The organisms in the latter group include environmental saprophytes, which can cause opportunistic infections, and other species such as *M. marinum*, which are common pathogens. Nontuberculous mycobacteria can be further divided between fast-growing mycobacteria such as *M. fortuitum*, *M. chelonae*, *M. abscessus* and slow-growing mycobacteria such as *M. marinum*, *M. ulcerans*, *M. haemophilum*, *M. gordonae* which might take weeks or even months to develop visible colonies on solid agar. At least 20 species of nontuberculous mycobacteria, including both fast and slow growers, can cause disease in fish. They include *Mycobacterium marinum*, *M. shottsii, M. pseudoshottsii, M. ulcerans, M. haemophilum, M. gordonae, M. fortuitum M. fortuitum, M. peregrinum, M. saopaulense*, and the *M. chelonae M. abscessus, M. chelonae, M. salmoniphilum, M. chesapeaki, M. montefiorense, M. neoaurum, M. simiae, M. scrofulaceum and M. stephanolepidis*. Some of these species, including *M.*

marinum and *M. haemophilum*, appear to be very harmful to fish, while others are often carried sub clinically (Asakura et al., 2016).

M. marinum, M. ulcerans, M. haemophilum, M. abscessus, M. chelonae, M. fortuitum, M. peregrinum, M. gordonae, M. iranicum, M. neoaurum, M. florentinum, and M. arupense are some of the mycobacteria that have been isolated in fish and are known to affect humans (Bercovier et al., 2001). Mycobacteriosis was produced by *M. senegalese*, a bacterium generally associated with cattle, in cuts from a shattered fish tank in a toddler who had no apparent link to cattle. Other organisms that can cause piscine mycobacteriosis should almost certainly be termed zoonotic. *M. pseudoshottsii* and *M. shottsii*, in particular, are closely related to *M. marinum* and are difficult to distinguish from this organism (Bhatty et al., 2000).

M. marinum has been defined in at least 150 species of freshwater and marine fish, including eels and syngnathid fish namely; seahorses, pipefish, sea dragons. Outbreaks appear to be more common in captive Anabantidae (bettas and gouramis), Characidae (tetras and piranhas), and Cyprinidae (danios and barbs), though this could be related to their lengthy lifetimes rather than unique susceptibility. There is no much information available about the members of the *M marinum* complex. *M. shottsii* has been found in striped bass *Morone saxatilis*, which is the host for *M. pseudoshottsii and M. chesapeaki.* Yellow tail (*Seriola quinqueradiata*), sea bream (*Sparus aurata*), white perch (*Morone americana*), Atlantic menhaden (*Brevoortia tyrannus*), and bay anchovy (*Anchoa mitchilli*) are among the marine species known to be infected with *M. pseudoshottsii.* The majority of cases have been linked to fish farms. *M. ulcerans* is generally associated with stagnant water, and while it has been identified in fish, it does not appear to serve as a reservoir. *M. marinum* is one of the most prevalent *Mycobacterium* species found in sick amphibians. Captive bullfrogs (Rana catesbeiana), Japanese forest green tree frogs (*Rhacophorus arboreus*), leopard frogs (*Rana pipiens*), a red-eyed tree frog (*Agalychnis callidryas*) and Hong Kong warty newts (*Paramesotriton hongkongensis*) it has caused clinical cases. It has also been found in reptiles, including Egyptian spiny-tailed lizards, a Fly River tortoise and a bearded dragon. *M. marinum* has been reported in marine animals, pigs, cattle, camels, a European hedgehog (*Erinaceus europaeus*), a bilby (*Macrotis lagotis*), and an armadillo, as well as a blue-fronted Amazon parrot (*Amazona aestiva*). In mice and bats, experimental infections have been established.

Many organisms found in fish, such as the *M. fortuitum* group and the *M. chelonae*/*M. abscessus* complex, are prevalent environmental organisms in soil, water, and plants. They are not specifically associated with fish, but have been documented in a variety of species, including amphibians, reptiles, mammals, and birds, with or without clinical indications. Few of them may favorably affect certain piscine hosts. For instance, *M. chelonae* and *M. salmoniphilum* have been reported mainly in coldwater fish, specifically salmonids. *M. salmoniphilum* also cause disease in burbot which is a freshwater species, and experimentally infected Atlantic cod (Gadus morhua). *M. haemophilum* has been encountered in zebrafish (Danio rerio) and other aquarium fishes such as Cuchu's blue tetra, *Boehlkea fredcochui*; peacock gudgeon, *Tateurdina ocellicauda*.

4.8.1 PATHOLOGY

Piscine mycobacteriosis is frequently a long-term illness, with infected fish living for months or even years. It is characterized by granulomatous inflammation in internal organs and muscles as well as the integument. Clinical indications vary and are sometimes nonspecific, reflecting those of other fish diseases. Affected fish may appear in appetent, malnourished, or less active than usual; they may swim erratically and/or isolate themselves from other fish, seeking out corners of the holding facility. In the group, there may be a decreased growth rate or increased mortality. Some fish develop shallow, irregular, nonhealing skin ulcers; nodular skin lesions, such as farmed Atlantic salmon infected with *M. chelonae*; cutaneous hemorrhages from muscle lesion rupture; or nonspecific changes in cutaneous pigmentation, such as fading in tropical fish or bright colors in salmonids (Bouricha et al., 2014). Exophthalmos (bulging eyes), pale gills, profuse cutaneous mucus, elevated scales, abdominal distention, skeletal abnormalities such as spine curvature or stunting deficiencies, and fin and tail rot are among of the other apparent symptoms. External indications of disease in fish with internal granulomas are not always visible. Acute cases have only been described following experimental injection, and they appear to be uncommon. Experimentally infected fish perished within 2–3 weeks of infection, with few clinical indications other than edema, fin erosions, and/or fluid accumulations in the coelomic cavity (Antuofermo et al., 2014).

4.8.2 DIAGNOSIS

Mycobacteria can be isolated from internal organs to make a definitive diagnosis. External lesions may also include these organisms; however, these findings should be interpreted with caution because they could be contaminants.

Fast-growing species (e.g., *M. chelonae*, *M. fortuitum*, or *M. abscessus*) can grow on non-selective conditions and appear possible in as short as 5 days. Slow-growing mycobacteriums like *M. marinum*, *M. haemophilum*, and *M. ulcerans* are more difficult to recover and are frequently isolated on mycobacterial selective medium (Middlebrook 7H10 agar or Lowenstein-Jensen slants). *M. haemophilum* is one of the few bacteria that needs iron (heme supplementation) in its media. The optimal growth temperature varies by species, although most organisms that influence fish prefer temperatures between 20 and 30°C. *M. marinum* thrives best at temperatures about 30°C. *M. marinum* can take several weeks to isolate and cultures containing this organism should be preserved for at least 6–8 weeks. *M. haemophilum* and *M. ulcerans* can take even longer to produce visible colonies, thus PCR may be a better way to detect them. Biochemical tests can distinguish mycobacterial species to a limited extent; however, genetic approaches are now favored for identification. Smears or tissue sections of lesions stained with Ziehl-Neelsen (or its variants), Fite's acid fast stain, or fluorochrome may include acid-fast bacilli (auramine). Histopathology can also be beneficial.

PCR-based approaches, such as multiplex PCR assays, are effective for detecting mycobacteria, but they cannot discriminate between some species, such as the *M. marinum* complex's individual members. *M. marinum* complex has also been detected using a loop-mediated isothermal amplification (LAMP) assay. If necessary, DNA sequencing, DNA fingerprinting, MALDI-TOF (mass spectral fingerprinting), and other techniques can be performed to confirm the species (Sanguinetti et al., 1998).

4.8.3 PREVENTION AND CONTROL

Quarantines for new stock (though many mycobacterial illnesses do not manifest clinically within this time), egg disinfection, frequent cleaning and disinfection of tanks and equipment, and water treatment with a germicidal UV light are also preventive measures. Since mycobacteria can be found in water, food, and other fomites, screening eggs and fish is the possible option to prevent the infection (Enzensberger et al., 2002). Sentinel procedures are in place at some facilities to detect agents as soon as they are introduced. If trash fish or fish tissues (salmon viscera) are utilized as a protein source in the feed, they should be pasteurized or heated for 30 minutes at 76°C (169°F). Any ill or dead fish should be removed and destroyed as soon as possible. Reptiles and other animals that are prone to mycobacteriosis should not have fed them. When mycobacteriosis is introduced into a commercial facility, the standard reaction is to cull the fish, thoroughly clean and disinfect the tanks and equipment, and then replenish. In recreational aquaria or valuable fish in displays, depopulation may or may not be maintained. If mycobacteria aren't completely eradicated, husbandry practices, including reducing stress and crowding and ensuring adequate water quality can help lower the likelihood of infection (Delghandi et al., 2020).

Treatment is rarely undertaken because it looks unlikely to completely eradicate *Mycobacterium* from infected fish colonies, and it is also impractical in fish meant for human consumption. Clarithromycin, doxycycline, minocycline, trimethoprim sulfamethoxazole, fluoroquinolones, rifampin, rifabutin, and ethambutol have all been used to treat *M. marinum* in humans. Some of these agents have been used in fish and other vertebrates in various combinations, with variable degrees of success. Isoniazid, pyrazinamide, and streptomycin resistance are common in *M. marinum*.

4.9 BACTERIAL KIDNEY DISEASE (BKD)

Bacterial kidney disease (BKD) of salmonid fishes is a slowly progressive, systemic illness with a long course and an insidious nature. The bacterium that causes BKD is a small Gram-positive diplobacillus known as *Renibacterium salmoninarum*. Initial reports in the 1930s recorded a new complex disease mainly affecting the kidneys from the rivers of Aberdeenshire Dee and Spey-Spey in Scotland (Furunculosis Committee, 1933; Smith, 1964), brown trout (*Salmo trutta*), rainbow trout (*Oncorhynchus mykiss*) in the Western United States (Belding & Merrill, 1935), salmon (*Salmo salar*) from the rivers Aberdeenshire Dee and from hatchery reared brook trout (*Salvelinus fontinalis*).

Lesions describing metabolic disturbances and reactions against the etiological agents were traced. From the smears of the kidney, Gram-positive rods were recovered. Unfortunately, with the available conventional techniques for fish pathogens, propagation could not be studied (Weins, 2011). Later on, the Gram-positive bacteria was cultivated by the injection of an emulsion from internal organs, Koch's postulate could be fulfilled. The disease was termed as white boil disease or Dee disease or Kidney disease (Sanders & Fryer, 1980).

The identical etiology of the disease was confirmed by the comparison of the clinical picture, pathological lesions and its characterization (Ordal & Earp, 1956; Smith, 1964). The disease was finally known to be as *Renibacterium salmoninarum* (Sanders & Fryer, 1980). Bacteria *R. salmoninarum* is an obligate pathogen, which has been proved by its limited survival in water up to a few weeks, either in a diseased or in a latent carrier fish (Evelyn, 1993). Propagation of *R. salmoninarum* has been observed in a few other species of fishes like sablefish *Anoplopoma fimbria* (Bell et al., 1990) and cyprinids (Sakai et al., 1989).

BKD is a major threat to the wild and cultivated salmonid fishes. The disease accounted in several European countries like Germany, Scotland, England, France, Denmark, Norway, Iceland, and Finland. And from South and North America, Japan (Kinkelin, 1974; Fryer & Sanders, 1981; Hoffmann et al., 1984; Kimura & Awakura, 1977; Fryer & Lannan, 1993). Serious implications of the disease were observed in areas where Pacific salmon occur, i.e., the Great Lakes region of the United States and the Pacific Northwest.

R. salmoninarum is a Gram-positive rod, non-motile bacteria of length 0.3–1.0 by 1.0 to 1.5 μm. They often occur in pairs. It is a non-alcohol fast bacterium and does not produce endospores (Goodfellow et al., 1986).

4.9.1 PATHOLOGY

Diseased fish may look normal. The presence of several small, open sores in the epidermis that reveal the underlying musculature can give rainbow and brown trout a "buckshot" appearance. In brook trout and coho salmon fingerlings and yearlings, massive boils filled with a pinkish milky fluid containing vast quantities of Renibacterium salmoninarum bacteria can be discovered on the flanks of the fish.

Exophthalmia (popeye) is a common symptom of BKD caused by osmoregulatory disturbance. However, because other reasons, such as gas supersaturation (gas bubble disease), enteric redmouth (ERM) infections, some viral disorders, or parasites, might induce this symptom, it is not diagnostic. As indicated by its name, BKD affects the kidneys and, to a lesser extent, the spleen and liver. In contrast to the smooth, concave surface of healthy kidneys, the kidneys are frequently enlarged, convex, and have a corrugated or bumpy surface. Massive populations of the pathogenic organism are represented by creamy, soft, off-white cysts. Cysts of this type are prevalent in the posterior kidney, and their size and number might vary. They are not to be confused with normal stannous masses in the mid-kidney or nephrocalcinosis (kidney stones) that can fill the kidney's excretory tubules. In the abdominal cavity and around the heart, a bloody, turbid, or yellow-brown fluid frequently accumulates. The appearance of other internal organs and visceral fat may be normal or exceptionally white. A white or yellow viscous fluid may be present in the intestine.

4.9.1.1 EXTERNAL LESIONS

Loss of balance occurs in the fishes infected with *R. salmoninarum*. Superficial blisters on skin and cavities in the musculature are described in the infected salmonids. Cavitation's and blisters are narrated to contain a white or yellowish hemorrhagic fluid. Hemorrhagic areas around the fins and the lateral line, petechiae, distended abdomen are a few common external indications of BKD (Belding & Merrill, 1935; Ordal & Earp, 1953; Bruno, 1986; Evelyn, 1993).

4.9.1.2 INTERNAL LESIONS

Macroscopic lesions are one of the most common signs observed in connection to the fishes infected with *R. salmoninarum.* A swollen kidney, white-grayish nodules of various sizes on the surface of the kidney are some of the additional characters of the disease. Nodules of similar appearance could be seen in the heart, liver, and spleen. An increase in size of the spleen and bright coloration of the liver could be noted. Accumulation of ascitic fluid in the abdominal cavity and petechial hemorrhages of the muscle lining the peritoneum are described. In complicated cases, necrosis, grayish discoloration, extensive kidney damage is experienced (Belding & Merrill, 1935; Wood & Yasutake, 1956; Bell, 1961; Fryer & Sanders, 1981; Bruno, 1986).

Histologically, BKD is described as a granulomatous, chronic, and systemic inflammation (Snieszko & Griffin, 1955; Wood & Yasutake, 1956). Chronic infections in the infected fishes are distinguished by dispersed melanin (Bruno, 1986) and increased amount of melanomacrophages. Some of the typical observations are focal necrosis edged by granulomatous tissue of epithelioid and lymphoid cells with varied stages of encapsulation. Epithelioid appearance due to the development of granulomas as an activation of macrophages adhering to each other could be observed. Activated macrophages could fuse to a multinucleated giant cell (MGC) (Holland et al., 2003) demonstrated in chronic infections in fishes of parasitic or fungal origin (Ferguson, 1989), and also in BKD infected fish (Snieszko & Griffin, 1955; Wood & Yasutake, 1956). Tissue necrosis occurs as a result of the release of huge amounts of MGCs and activated macrophages.

Salinity (Fryer & Sanders, 1981), stress (Mesa et al., 1998) and temperature (Sanders et al., 1978) are few of the environmental factors responsible for mortality and pathogenesis in connection with BKD. Morphological and macroscopical lesions are due to the dual effect of the bacteria itself and a result of the immunological response of the fish to the infection (Young & Chapman, 1978; Fryer & Lannan, 1993). Complications in the pathogenesis of BKD arise due to the differences in susceptibility among the salmonid species. Species considered most susceptible and associated with higher mortality rates are chinook salmon (*O. tshawytscha*), salmon (*O. nerka*) and Pink salmon (*Oncorhynchus gorbushca*). Rainbow trout (*O. mykiss*), Atlantic salmon (*Salmo salar*) and brown trout are considered fairly resistant to the BKD infection (Sanders et al., 1978; Sakai et al., 1991; Dale et al., 1997; Starliper et al., 1997).

4.9.2 DIAGNOSIS

Advanced cases of BKD are the occurrence of white nodules in internal organs such as heart, liver, kidney, spleen, and enlarged grayish kidney. Gram-staining of smears, immunological techniques could be used to identify the characteristic lesions, since the bacterial growth is time consuming. The pathogen is aerobic and fastidious in its growth requirements. Ever first achieved propagation of the bacteria was on Chicken embryo among the various media tested for growth and sub-cultivation.

Dorset's medium was proven successful. Increased possibility of isolation of the kidney disease bacterium was by the addition of L-cysteine to the Dorset's medium. Though modifications of several different agar media for *R. salmoninarum* like kidney disease medium (KDM) Evelyn, 1981) Charcoal agar (Daly & Stevenson, 1985) and L-cysteine supplemented Muller Hinton medium (Wolf & Dunbar, 1959) were performed, secondary bacterial infections were a serious issue in the diagnostics. These contaminations by other bacteria and fungi could be eliminated, by the use of (SKDM) containing cycloserine, polymyxin B sulfate, oxolinic acid and cycloheximide (Austin et al., 1983). Optimal temperature for the cultivation of *R. salmoninarum* was 15–18°C, no growth was observed at 30°C. Primary isolation and incubation time for the initial growth of the organism was reported to be as long as 19 weeks post-inoculation of the kidney sample (Benediktsdóttir et al., 1991). Serum-free broth, consisting of 0.1% L-cysteine, 1% peptone and 1% yeast extract are suitable for the growth of *R. salmoninarum* in large batches (Daly & Stevenson, 1993). Cultivation of *R. salmoninarum* is possible by cell culture techniques. The bacteria were propagated intracellularly in the fish cells lines (EPC, RTG-2, CHSE) (McIntosh et al., 1996; González et al., 2004).

The API-ZYM test has been found appropriate, as it is an enzymatic test for which propagation of the bacterium is not necessary (Goodfellow et al., 1986; Austin & Rayment, 1985). Proteolytic activity and a strong catalase-positive reaction are described (Ordal & Earp, 1956; Goodfellow et al., 1986). The β-Hemolytic activity of the bacterium declines after prolonged cultivation (Bruno, 1986b).

BKD is defined as chronic, a systemic granulomatous inflammation. Recommended staining methods to distinguish the pathogen are the standard gram-staining, Lillie's allocrome stain and modified Ollet's gram-stain (Midtlyng et al., 2000).

In addition to the observations made on the morphological lesions, immunohistochemical techniques could be used for the identification of the BKD pathogen. Dispersed melanin and melanomacrophages are the characteristic feature of chronic inflammations in fishes (Bruno, 1986). Differentiation among the melanin pigments from the intact bacteria is difficult by the immunohistochemistry techniques due to a similarity in size and in color of the oxidation product of commonly used chromogens. This problem could be avoided by the use of strong hydrogen peroxide solution for bleaching, without any negative effects in the immunoreactivity of the identified epitopes (Jansson et al., 2002).

The utilization of antigen-antibody affinity in the naked eye detection of antigens is one the common immunological technique (de la Pena, 2001). Stability of the antibody-antigen complex (avidity) and strength of the antibody-epitope interaction (affinity) are the two crucial factors in the immunological technique. However, the sensitivity and specificity of the technique depends on the quality of the antibodies employed. Specificity of the monoclonal antibodies, produced by a single B-cell, was considered more reliable, until the polyclonal antibodies, from several B-cell lines, was used for the identification and was completely reliable (Lu et al., 2020).

Indirect fluorescent antibody technique (Laidler, 2006), Real time PCR (Jansson et al., 2008). ELISA (Laurin et al., 2019) was reported to be more sensitive, in comparison to the culture based or the has been FAT for detection of *R. salmoninarum* in the affected fishes.

4.9.3 PREVENTION AND CONTROL

Temporary effect observed upon antibiotic treatment was not recommended as a therapy, due to the possibility of intracellular survival of the bacteria. Frequent screening in order to control the vertical and horizontal transmission of BKD has been the most effective prevention strategy (Elliott et al., 1989; Fryer & Lannan, 1993). Separation of the rearing eggs and juvenile fishes into separate groups have helped in areas of prevalence of BKD (Elliott et al., 1995).

Injection of erythromycin to the brood stock, before spawning was observed in reducing the transmission of *R. salmoninarum* to the eggs. Destruction of gametes by brood stock culling has been reported to reduce the prevalence of the disease (Gudmundsdottir et al., 2017). This stage requires a sensitive procedure in identifying the carriers and a careful separation of the eggs until further confirmation of the diagnosis has been established. Administration of *R. salmoninarum* expressing low levels of cell associated p5 orally have resulted in increased the duration of time from challenge of infection to death (Piganelli et al., 1999). However, this protection was not associated with the humoral antibody level. This is supportive of the previous HP indications which suggested the involvement of the cell-mediated immune response in recovery of the diseased fish.

Advancements in the control and prevention of the BKD depends on the knowledge with regard to the basics of the immune functions in fish and the pathogenesis of the infection.

KEYWORDS

- **coldwater aquaculture**
- **diseases**
- **polymerase chain reaction**
- **preventive**
- **rainbow trout gastroenteritis**
- **segmented filamentous bacteria**

REFERENCES

Adams, A., & Thompson, K. D., (2006). Biotechnology offers revolution to fish health management. *Trends Biotechnol., 24*, 201–205.

Ali, A., Carnahan, A., Altwegg, M., Lütthyhottenstein, J., & Joseph, S., (1996). *Aeromonas bestiarum* sp. Nov. (formerly genomospecies DNA group 2 *A. hydrophila*), a new species isolated from non-human sources. *Med Microbiol Lett., 5*, 156–165.

Allen, P. C., (1992). Comparative study of long, segmented, filamentous organisms in chickens and mice. *Lab. Anim. Sci., 42*, 542–547.

Alvarez, B., & Guijarro, J. A., (2007). Recovery of *Flavobacterium psychrophilum* viable cells using a charcoal-based medium. *Lett. Appl. Microbiol., 44*, 569–572.

Alvarez, B., Alvarez, J., Menendez, A., & Guijarro, J. A., (2008). A mutant in one of two exbD loci of a TonB system in *Flavobacterium psychrophilum* shows attenuated virulence and confers protection against cold water disease. *Microbiology, 154*, 1144–1145.

Amita, K., Hoshino, M., Honma, T., & Wakabayashi, H., (2000). An investigation on the distribution of *Flavobacterium psychrophilum* in the Umikawa river. *Fish Pathol., 35*(4), 193–197.

Anacker, R. L., & Ordal, E. J., (1959). Studies on the myxobacterium *Chondrococcus columnaris*: I. serological typing. *J. Bacteriol., 78*, 25–32.

Anderson, D. P., & Siwicki, A. K., (1994). Duration of protection against *Aeromonas salmonicida* in brook trout immuno-stimulated with glucan or chitosan by injection or immersion. *Prog. Fish-Cult., 56*, 258–261.

Angel, C. R., Sell, J. L., Fagerland, J. A., Reynolds, D. L., & Trampel, D. W., (1990). Long-segmented filamentous organisms observed in poults experimentally infected with stunting syndrome agent. *Avian Dis.*, 994–1001.

Antaya, C., (2008). Current eco-economical impacts of *Flavobacterium psychrophilum*. MMG 445. *Basic Biotechnol. J., 4*, 7–12.

Antuofermo, E., Pais, A., Nuvoli, S., Hetzel, U., Burrai, G. P., Rocca, S., Caffara, M., Giorgi, I., Pedron, C., & Prearo, M., (2014). *Mycobacterium chelonae* associated with tumor-like skin and oral masses in farmed Russian sturgeons (*Acipenser gueldenstaedtii*). *BMC Vet. Res., 10*, 18.

Asakura, T., Ishii, M., Kikuchi, T., Kameyama, K., Namkoong, H., Nakata, N., Sugita, K., et al., (2016). Disseminated *Mycobacterium marinum* infection with a destructive nasal lesion mimicking extranodal NK/T Cell lymphoma: A case report. *Medicine (Baltimore), 95*(11), e3131.

Austin, B., & Austin, D. A., (2007). *Bacterial Fish Pathogens: Disease of Farmed and Wild Fish.* Springer-Verlag, Berlin, Heidelberg, New York.

Austin, B., & Austin, D. A., (2016). *Bacterial Fish Pathogens: Disease of Farmed and Wild Fish* (6th edn.). Springer, Dordrecht.

Austin, B., & Austin, D. A., (2016). Flavobacteria and cytophagas. In: *Bacterial Fish Pathogens* (pp. 397–465). Springer; Cham.

Austin, B., & Rayment, J. N., (1985). Epizootiology of *Renibacterium salmoninarum*, the causal agent of bacterial kidney disease in salmonid fish. *J. Fish Dis., 8*, 505–509.

Austin, B., (2011). Taxonomy of bacterial fish pathogens. *Vet. Res., 42*, 1–3.

Balcázar, J. L., Vendrell, D., De Blas, I., Ruiz-Zarzuela, I., Muzquiz, J. L., & Girones, O., (2008). Characterization of probiotic properties of lactic acid bacteria isolated from intestinal microbiota of fish. *Aquaculture, 278*, 188–191.

Barnes, M. E., & Brown, M. L., (2011). A review of *Flavobacterium psychrophilum* biology, clinical signs, and bacterial cold water disease prevention and treat. *The Open Fish Science Journal, 4*(1), 40–48.

Bartkova, S., Leekitcharoenphon, P., Aarestrup, F. M., & Dalsgaard, I., (2017). Epidemiology of Danish *Aeromonas salmonicida* subsp. salmonicida in fish farms using whole genome sequencing. *Front Microbiol., 8*, 2411.

Bastardo, A., Ravelo, C., & Romalde, J. L., (2012). Highly sensitive detection and quantification of the pathogen *Yersinia ruckeri* in fish tissues by using real time PCR. *Appl. Microbiol. Biotechnol., 96*, 511–520.

Beck, B. H., Barnett, L. M., Farmer, B. D., Peatman, E., & Carter, D., (2015). Kaolinitic clay protects against *Flavobacterium columnare* infection in channel catfish *Ictalurus punctatus* (Rafinesque). *J. Fish Dis., 38*, 241–248.

Belding, D. L., & Merrill, B., (1935). A preliminary report upon a hatchery disease of the *Salmonidae. Trans. Am. Fish. Soc., 65*, 76–84.

Bell, G. R., Higgs, D. A., & Traxler, G. S., (1984). The effect of dietary ascorbate, zinc, and manganese on the development of experimentally induced bacterial kidney disease in sockeye salmon (*Oncorhynchus nerka*). *Aquaculture, 36*, 293–311.

Bell, G. R., Hoffmann, R. W., & Brown, L. L., (1990). Pathology of experimental infections of the sablefish, *Anoplopoma fimbria* (Pallas), with *Renibacterium salmoninarum*, the agent of bacterial kidney disease in salmonids. *J. Fish Dis., 13*, 355–367.

Benagli, C., Demarta, A., Caminada, A., Ziegler, D., Petrini, O., & Tonolla, M., (2012). A rapid MALDI-TOF MS identification database at genospecies level for clinical and environmental *Aeromonas* strains. *PLoS One, 7*, e48441.

Benediktsdóttir, E., Helgason, S., & Gudmundsdóttir, S., (1991). Incubation time for the cultivation of *Renibacterium salmoninarum* from Atlantic salmon, *Salmo salar* L., broodfish. *J. Fish Dis., 14*, 97–102.

Bercovier, H., & Vincent, V., (2001). Mycobacterial infections in domestic and wild animals due to Mycobacterium marinum, M. fortuitum, M. chelonae, M. porcinum, M. farcinogenes, M. smegmatis, M. scrofulaceum, M. xenopi, M. kansasii, M. simiae and M. genavense. *Rev. Sci. Tech., 20*(1), 265–290.

Bernardet, J. F., & Grimont, P. A. D., (1989). Deoxyribonucleic acid relatedness and phenotypic characterization of *Flexibacter columnaris* sp. Nov., nom. rev., *Flexibacter psychrophilus* sp. Nov., nom. rev., and *Flexibacter maritimus* Wakabayashi, Hikida and Masumura, 1986. *Int. J. Syst. Bacteriol., 39*(3), 346–354.

Bernardet, J. F., & Kerouault, B., (1989). Phenotypic and genomic studies of "*Cytophaga psychrophila*" isolated from diseased rainbow trout (*Oncorhynchus mykiss*) in France. *Appl. Environ. Microbiol., 55*, 1796–1800.

Bhatty, M. A., Turner, D. P. J., & Chamberlain, S. T., (2000). *Mycobacterium marinum* hand infection: Cases reports and review of literature. *Br. J. Plast. Surg., 53*(2), 162–165.

Borg, A. F., (1960). Studies on myxobacteria associated with diseases in salmonid fishes. *J. Wildl. Dis., 8,* 1–85.

Bouricha, M., Castan, B., Duchene-Parisi, E., & Drancourt, M., (2014). *Mycobacterium marinum* infection following contact with reptiles: Vivarium granuloma. *Int. J. Infect. Dis., 21*, 17, 18.

Boutin, S., Audet, C., & Derome, N., (2013). Probiotic treatment by indigenous bacteria decreases mortality without disturbing the natural microbiota of *Salvelinus fontinalis*. *Can J. Microbiol., 59*(10), 662–670.

Branson, E., (2003). Rainbow trout gastro-enteritis (RTGE): First diagnosis in the UK. *Fish Vet. J., 7*, 71–76.

Brown, L. L., Cox, W. T., & Levine, R. P., (1997). Evidence that the causal agent of bacterial coldwater disease *Flavobacterium psychrophilum* is transmitted within salmonid eggs. *Dis. Aquat. Organ., 29*, 213–218.

Bruno, D. W., (1986). Histopathology of bacterial kidney disease in laboratory infected rainbow trout, *Salmo gairdneri* Richardson, and Atlantic salmon, *Salmo salar* L., with reference to naturally infected fish. *J. Fish Dis., 9*, 523–537.

Bruno, D. W., (1986). Scottish experience with bacterial kidney disease in farmed salmonids between 1976 and 1985. *Aquac. Res., 17*, 185–190.

Bruno, D. W., (1992). *Cytoghaga psychrophila (='Flexibacter psychrophilus')*(Borg), histopathology associated with mortalities among farmed rainbow trout, *Oncorhynchus mykiss* (Walbaum) in the UK. *Bull. Eur. Assoc. Fish Pathol., 12*, 215, 216.

Bruno, D. W., Hastings, T. S., Ellis, A. E., & Wootten, R., (1985). Outbreak of a cold water vibriosis in Atlantic salmon in Scotland. *Bulletin of the European Association of Fish Pathologists, 5*, 62–63.

Bruun, M. S., Schmidt, A. S., Madsen, I., & Dalsgaard, I., (2000). Antimicrobial resistance patterns in Danish isolates of *Flavobacterium psychrophilum*. *Aquaculture, 187*, 201–212.

Bullock, G. L., & Key, H. M., (1975). Fluorescent antibody identification and detection of the Corynebacterium causing kidney disease of salmonids. *J. Fish. Res. Board Can., 32*, 2224–2227.

Bullock, G. L., (1972). *Studies on Selected Myxobacteria Pathogenic for Fishes and on Bacterial Gill Disease in Hatchery-Reared Salmonids*. US Department of the Interior, Fish and Wildlife Service, Bureau of Sport Fisheries and Wildlife.

Bullock, G. L., Conroy, D. A., & Snieszko, S. F. (1971). Myxobacterioses. In: Snieszko, S. F., & Axelrod, H. R., (eds.), *Diseases of Fishes* (pp. 60–88). T.F.H. Publications, Jersey City.

Bullock, G. L., Stuckey, H. M., & Mulcahy, D., (1978). *Corynebacterial Kidney Disease: Egg Transmission Following Iodophore Disinfection* (Vol. 7, pp. 51, 52). U.S. Fish Wildl. Serv. Keamevsville. WV. Fish Health News.

Busch, R. A., & Lingg, A. J., (1975). Establishment of an asymptomatic carrier state infection of enteric redmouth disease in rainbow trout (*Salmo gairdneri*). *J. Fish Res. Board Can, 32*, 2429–2432.

Busch, S., Dalsgaard, I., & Buchmann, K., (2003). Concomitant exposure of rainbow trout fry to *Gyrodactylus derjavini* and *Flavobacterium psychrophilum*: Effects on infection and mortality of host. *Vet. Parasitol., 117*, 117–122.

Caselli, M., Holton, J., Boldrini, P., Vaira, D., & Calò, G., (2010). Morphology of segmented filamentous bacteria and their patterns of contact with the follicle-associated epithelium of the mouse terminal ileum: Implications for the relationship with the immune system. *Gut Microbes, 1*, 367–372.

Castillo, D., Christiansen, R. H., Espejo, R., & Middelboe, M., (2014). Diversity and geographical distribution of *Flavobacterium psychrophilum* isolates and their phages: Patterns of susceptibility to phage infection and phage host range. *Microb. Ecol., 67*, 748–757.

Cepeda, C., Garcia-Marquez, S., & Santos, Y., (2004). Improved growth of *Flavobacterium psychrophilum* using a new culture medium. *Aquaculture, 238*, 75–82.

Chase, D. G., & Erlandsen, S. L., (1976). Evidence for a complex life cycle and endospore formation in the attached, filamentous, segmented bacterium from murine ileum. *J. Bacteriol., 127*, 572–583.

Chen, Y. C., Davis, M. A., LaPatra, S. E., Cain, K. D., Snekvik, K. R., & Call, D. R., (2008). Genetic diversity of Flavobacterium psychrophilum recovered from commercially raised rainbow trout, *Oncorhynchus mykiss* (Walbaum), and spawning coho salmon, O. kisutch (Walbaum). *J. Fish Dis., 31*, 765–773.

Chinabut, S., (1999). *Fish Disease and Disorders: Viral, Bacterial, and Fungal Infections* (pp. 319–340). CAB International: Wallingford, UK.

Christiansen, R. H., Madsen, L., Dalsgaard, I., Castilo, D., Kalatzis, P. G., & Middelboe, M., (2016). Effect of bacteriophages on the growth of *Flavobacterium psychrophilum* and development of phage resistant strains. *Microb. Ecol., 71*(4), 845–859.

Cipriano, R. C., & Austin, B., (2011). Furunculosis and other aero-monad diseases. In: Woo, P. T., & Bruno, D. W., (eds.), *Fish Diseases and Disorders: Viral, Bacterial and Fungal Infections* (2nd edn., Vol, 3, pp. 424–483). CABI, Wallingford.

Cipriano, R. C., & Holt, R. A., (2005). *Flavobacterium Psychrophilum, Cause of Bacterial Cold-Water Disease and Rainbow Trout Fry Syndrome*. Fish Disease Leaflet No. 86. United States Dept. of the Interior. U.S. Geological Service, National Fish Health Research Laboratory, Kearneysville, WV.

Cipriano, R. C., (2005). Intraovum infection caused by *Flavobacterium psychrophilum* among eggs from captive Atlantic salmon broodfish. *J. Aquat. Anim Health., 17*(3), 275–283.

Coquet, L., Cosette, P., Junter, G. A., Beucher, E., Saiter, J. M., & Jouenne, T., (2002). Adhesion of *Yersinia ruckeri* to fish farm materials: Influence of cell and material surface properties. *Colloids Surf B: Biointerfaces, 26*, 373–378.

Crump, E. M., Perry, M. B., Gale, S., Crawford, E., & Kay, W. W., (2003). Lipopolysaccharide O-antigen antibody-based detection of the fish pathogen *Flavobacterium psychrophilum. J. Mol. Microbiol. Biotechnol., 6*(3, 4), 182–190.

Dale, O. B., (1994). *Bacterial Kidney Disease in Salmonid Fish*. A study on virulence and diagnostic techniques. Norges Veterinaerhoegskole.

Dalsgaard, I., & Madsen, L., (2000). Bacterial pathogens in rainbow trout *Oncorhynchus mykiss* (Walbaum), reared at Danish freshwater farms. *J. Fish Dis., 23*, 199–209.

Daly, J. G., & Stevenson, R. M., (1985). Charcoal agar, a new growth medium for the fish disease bacterium *Renibacterium salmoninarum*. *Appl. Environ. Microbiol., 50*, 868–871.

Daly, J. G., & Stevenson, R. M., (1993). Nutrient requirements of *Renibacterium salmoninarum* on agar and in broth media. *Appl. Environ. Microbiol., 59*, 2178–2183.

Danzeisen, J. L., Clayton, J. B., Huang, H., Knights, D., McComb, B., Hayer, S. S., & Johnson, T. J., (2015). Temporal relationships exist between cecum, ileum, and litter bacterial microbiomes in a commercial turkey flock, and subtherapeutic penicillin treatment impacts ileum bacterial community establishment. *Front. Vet. Sci., 2*, 56.

Daskalov, H., Austin, D. A., & Austin, B., (1999). An improved growth medium for *Flavobacterium psychrophilum*. *Lett. Appl. Microbiol., 28*(4), 297–299.

Daskalov, H., Robertson, P. A. W., & Austin, B., (2000). Influence of oxidized lipids in diets on the development of rainbow trout fry syndrome. *J. Fish Dis., 23*, 7–14.

Davis, C. P., & Savage, D. C., (1976). Effect of penicillin on the succession, attachment, and morphology of segmented, filamentous microbes in the murine small bowel. *Infect. Immun., 13*, 180–188.

Davis, H. S., (1947). *Care and Diseases of Trout*. Rep.no. 12. Fish and Wildlife Service, United States Department of the Interior.

De Kinkelin, P., (1974). Corynebacteriose of the salmonides: Premiere observation in France. *Bull. fr. Peche Piscic.*, 3–7.

De La Pena, L. D., (2001). *Immunological and Molecular Biology Techniques in Disease Diagnosis* (pp. 137–158). In Health management in Aquaculture. Aquaculture Department, Southeast Asian Fisheries Development Center.

Decostere, A., D'Haese, E., Lammens, M., Nells, H., & Haesebrouck, F., (2001). *In vivo* study of phagocytosis, intracellular survival and multiplication of *Flavobacterium psychrophilum* in rainbow trout, *Oncorhynchus mykiss* (Walbaum), spleen phagocytes. *J. Fish Dis., 24*, 481–487.

Defoirdt, T., Sorgeloos, P., & Bossier, P., (2011). Alternatives to antibiotics for the control of bacterial disease in aquaculture. *Current Opinion in Microbiology, 14*, 251.

Del Cerro, A., Marquez, I., & Guijarro, J. A., (2002). Simultaneous detection of *Aeromonas salmonicida, Flavobacterium psychrophilum*, and *Yersinia ruckeri*, three major fish pathogens, by multiplex PCR. *Appl. Environ. Microbiol., 68*(10), 5177–5180.

Delghandi, M. R., El-Matbouli, M., & Menanteau-Ledouble, S., (2020). Mycobacteriosis and infections with non-tuberculous mycobacteria in aquatic organisms: A review. *Microorganisms, 8*, pii: E1368. doi: 10.3390/microorganisms8091368.

Del-Pozo, J., Crumlish, M., Ferguson, H. W., & Turnbull, J. F., (2009). A retrospective cross-sectional study on "*Candidatus arthromitus*" associated rainbow trout gastroenteritis (RTGE) in the UK. *Aquaculture, 290*, 22–27.

Del-Pozo, J., Crumlish, M., Turnbull, J. F., & Ferguson, H. W., (2010). Histopathology and ultrastructure of segmented filamentous bacteria–associated rainbow trout gastroenteritis. *Vet. Pathol., 47*, 220–230.

Deshmukh, S., Raida, M. K., Dalsgaard, I., Chettri, J. K., Kania, P. W., & Buchmann, K., (2012). Comparative protection of two different commercial vaccines against *Yersinia ruckeri* serotype O1 and biotype 2 in rainbow trout (*Oncorhynchus mykiss*). *Vet Immunol Immunopathol., 145*, 379–385.

Duchaud, E., Boussaha, M., Loux, V., Bernardet, J. F., Michel, C., Kerouault, B., et al., (2007). Complete genome sequence of the fish pathogen *Flavobacterium psychrophilum. Nat. Biotechnol., 25*(7), 763–769.

Ebanks, R. O., Knickle, L. C., Goguen, M., Boyd, J. M., Pinto, D. M., Reith, M., et al., (2006). Expression of and secretion through the *Aeromonas salmonicida* type III secretion system. *Microbiology, 152*, 1275–1286. doi: 10.1099/mic.0.28485-0.

Egidius, E., Andersen, K., Clausen, E., & Raa, J., (1981). Cold-water vibriosis or "Hitra disease" in Norwegian salmonid farming. *J. Fish Dis., 4*, 353, 354.

Egidius, E., Soleim, O., & Anderson, K., (1984). Further observations on coldwater vibriosis or Hitra disease. *Bull. Eur. Ass. Fish Pathol., 4*, 50–21.

Egidius, E., Wiik, R., Andersen, K., Hoff, K. A., & Hjeltnes, B., (1986). *Vibrio salmonicida* sp. nov., a new fish pathogen. *International Journal of Systematic Bacteriology, 36*, 518–520.

Ekman, E., & Norrgren, L., (2003). Pathology and immunohistochemistry in three species of salmonids after experimental infection with *Flavobacterium psychrophilum. J. Fish Dis., 26*, 529–538.

Ekman, E., Börjeson, H., & Johansson, N., (1999). *Flavobacterium psychrophilum* in Baltic salmon *Salmo salar* brood fish and their offspring. *Diseases of Aquatic Organisms, 37*, 159–163.

Elliott, D. G., Pascho, R. J., & Palmisano, A. N., (1995). Brood stock segregation for the control of bacterial kidney disease can affect mortality of progeny chinook salmon (*Oncorhynchus tshawytscha*) in seawater. *Aquaculture, 132*, 133–144.

Elsayed, E. E., Eissa, A. E., Faisal, M. Isolation of *Flavobacterium psychrophilum* from sea lamprey, *Petromyzon marinus* L. with skin lesions in Lake Ontario. *J. Fish Dis., 29*(10), 629–632.

Emmerich, R., & Weibel, E., (1894). Über eine durch bakterien erzeugte seuche unter den forellen (About a disease caused by bacteria among trout) (in German). *Arch, für Hyg., 21*, 1–21.

Enzensberger, R., Hunfeld, K. P., Elshorst-Schmidt, T., Boer, A., & Brade, V., (2002). Disseminated cutaneous *Mycobacterium marinum* infection in a patient with non-Hodgkin's lymphoma. *Infection, 30*(6), 393–395.

Ericssonl, A. C., Hagan, C. E., Davis, D. J., & Franklin, C. L., (2014). Segmented filamentous bacteria: Commensal microbes with potential effects on research. *Comp. Med., 64*, 90–98.

Evelyn, T. P., (1993). Bacterial kidney disease-BKD. *Bacterial Disease of Fish*.

Evelyn, T. P., Ketcheson, J. E., & Prosperi-Porta, L., (1981). The clinical significance of immunofluorescence-based diagnoses of the bacterial kidney disease carrier. *Fish Pathol., 15*, 293–300.

Faisal, M., Loch, T. P., Fujimoto, M., Woodiga, S. A., Eissa, A. E., Honeyfield, D. C., Wolgamood, M., et al., (2011). Characterization of novel *flavobacterium* spp. Involved in the mortality of coho salmon (*Oncorhynchus kisutch*) in their early life stages. *J. Aquac. Res. Development S., 2*, 2.

Farmer, B. D., Beck, B. H., Mitchell, A. J., Rawles, S. D., & Straus, D. L., (2017). Dietary copper effects survival of channel catfish challenged with *Flavobacterium columnare. Aquac. Res., 48*, 1751–1758.

Ferguson, D. J., & Birch-Andersen, A., (1979). Electron microscopy of a filamentous, segmented bacterium attached to the small intestine of mice from a laboratory animal colony in Denmark. *Acta Pathologica Microbiologica Scandinavica Section B Microbiology, 87*, 247–252.

Ferguson, H. W., (1989). *Systemic Pathology of Fish*. A text and atlas of comparative tissue responses in diseases of teleosts. Iowa State University Press.

Frans, I., Michiels, C. W., Bossier, P., Willems, K. A., Lievens, B., & Rediers, H., (2011). *Vibrio anguillarum* as a fish pathogen: Virulence factors, diagnosis and prevention. *Journal of Fish Diseases, 34*, 643–661.

Frey, J., & Origgi, F. C., (2016). Type III secretion system of *Aeromonas salmonicida* undermining the host's immune response. *Front Mar. Sci., 3*, 130. doi: 10.3389/fmars.2016.00130.

Fryer, J. L., & Lannan, C. N., (1993). The history and current status of *Renibacterium salmoninarum*, the causative agent of bacterial kidney disease in Pacific salmon. *Fish. Res., 17*, 15–33.

Fryer, J. L., & Sanders, J. E., (1981). Bacterial kidney disease of salmonid fish.*Annu. Rev. Microbiol., 35*, 273–298.

Fuente, S., Egert, M., Jimenez-Valera, M., Monteoliva-Sanchez, M., Ruiz-Bravo, A., & Smidt, H., (2008). A strain of *Lactobacillus plantarum* affects segmented filamentous bacteria in the intestine of immunosuppressed mice. *FEMS Microbiol. Ecol., 63*, 65–72.

Fujiwara-Nagata, E., Shindoh, Y., Yamamoto, M., et al., (2019). Distribution of *Flavobacterium psychrophilum* and its gyrA genotypes in a river. *Fish Sci., 85*, 913–923.

Garcia, J. A., Dominguez, L., Larsen, J. L., & Pedersen, K., (1998). Ribotyping and plasmid profiling of *Yersinia ruckeri*. *J Appl Microbiol, 85*, 949–955.

Ghosh, B., Cain, K. D., Nowak, B. F., & Bridle, A. R., (2016). Microencapsulation of a putative probiotic *Enterobacter* species, C6-6, to protect rainbow trout, *Oncorhynchus mykiss* (Walbaum), against bacterial coldwater disease. *J. Fish Dis., 39*, 1–1.

Gonçalves, R. A., Menanteau-Ledouble, S., Schöller, M., Eder, A., Schmidt-Posthaus, H., Mackenzie, S., & El-Matbouli, M., (2019). Effects of deoxynivalenol exposure time and contamination levels on rainbow trout. *J. World Aquac. Soc., 50*, 137–154.

González, E. A., Sachdeva, A., Oliver, D. A., & Martin, K. J., (2004). Vitamin D insufficiency and deficiency in chronic kidney disease. *Am. J. Nephrol., 24*, 503–510.

Good, C., Davidson, J., Wiens, G. D., Welch, T. J., & Summerfelt, S., (2015). *Flavobacterium branchiophilum* and *F. Succinicans* associated with bacterial gill disease in rainbow trout

Oncorhynchus mykiss (Walbaum) in water recirculation aquaculture systems. *J. Fish Dis., 38*, 409.

Goodfellow, M., & Williams, E., (1986). New strategies for the selective isolation of industrially important bacteria. *Biotechnol. Genet. Eng. Rev., 4*, 213–262.

Goodwin, M. A., Cooper, G. L., Brown, J., Bickford, A. A., Waltman, W. D., & Dickson, T. G., (1991). Clinical, pathological, and epizootiological features of long-segmented filamentous organisms (bacteria, lsfos) in the small intestines of chickens, turkeys, and quails. *Avian Dis*., 872–876.

Graf, J., (2015). *Aeromonas*. Norfolk: Caister Academic Press.

Groff, J. M., & LaPatra, S. E., (2001). In: Lim, C. E., & Webster, C. D., (eds.), *Nutrition and Fish Health* (pp. 11–78). Haworth Press: New York.

Guðmundsdóttir, S., Applegate, L. J., Árnason, I. O., Kristmundsson, Á., Purcell, M. K., & Elliott, D. G., (2017). Detecting *Renibacterium salmoninarum* in wild brown trout by use of multiple organ samples and diagnostic methods. *Bull. Eur. Assoc. Fish Pathol., 37*, 31–40.

Gultepe, N., & Tanrikul, T. T., (2006). Treatment methods of *Flavobacterium psychrophilum*: Cause of rainbow trout fry syndrome (RFTS) and bacterial coldwater disease (BCWD) in Turkey. *J. Fish Int., 1*, 102–105.

Hadidi, S., Glenney, G. W., Welch, T. J., Silverstein, J. T., & Wiens, G. D., (2008). Spleen size predicts resistance of rainbow trout to *Flavobacterium psychrophilum* challenge. *J. Immunol., 180*(6), 4156–4165.

Henryon, M., Berg, P., Olesen, N. J., et al., (2005). Selective breeding provides an approach to increase resistance of rainbow trout (*Oncorhynchus mykiss*) to the diseases, enteric redmouth disease, rainbow trout fry syndrome, and viral haemorrhagic septicaemia. *Aquaculture, 250*, 621–636.

Heo, G. J., Kasai, K., & Wakabayashi, H., (1990). Occurrence of *Flavobacterium branchiophila* associated with bacterial gill disease at a trout hatchery. *Fish Pathol., 25*, 99–105.

Herbert, T., (2016). *Optimization of the Microscopic Detection of Segmented Filamentous Bacteria in Piglets Before, During, and After Weaning*. Doctoral dissertation. North Carolina Agricultural and Technical State University.

Hesami, S., Allen, K. J., Metcalf, D., Ostland, V. E., MacInnes, J. I., & Lumsden, J. S., (2008). Phenotypic and genotypic analysis of *Flavobacterium psychrophilum* isolates from Ontario salmonids with bacterial coldwater disease. *Can. J. Microbiol., 54*(8), 619–629.

Hoffmann, R., & Lommel, R., (1984). Haematological studies in proliferative kidney disease of rainbow trout, *Salmo gairdneri* Richardson. *J. Fish Dis., 7*, 323–326.

Holland, J. W., Gould, C. R., Jones, C. S., Noble, L. R., & Secombes, C. J., (2003). The expression of immune-regulatory genes in rainbow trout, *Oncorhynchus mykiss*, during a natural outbreak of proliferative kidney disease (PKD). *Parasitology, 126*, S95–102.

Hollebecq, M. G., Faivre, B., Bourmaud, C., & Michel, C., (1995). Spontaneous bactericidal and complement activities in serum of rainbow trout (*Oncorhynchus mykiss*) genetically selected for resistance or susceptibility to furunculosis. *Fish Shellfish Immunol., 5*, 407–426.

Holt, R. A., Rohovec, J. S., & Fryer, J. L., (1993). Bacterial coldwater disease. In: Inglis, V., Roberts, R. J., & Brombage, N. R., (eds.), *Bacterial Diseases of Fish* (pp. 3–22). Blackwell Scientific Publications, Oxford.

Horne, M. T., & Barnes, A. C., (1999). Enteric redmouth disease (*Yersinia ruckeri*). In: Woo, P. T. K., & Bruno, D. W., (eds) *Fish Diseases and Disorders: Viral, Bacterial, and Fungal Infections* (pp. 445–477). CABI Publishing, Wallingford.

Hunter, V. A., Knittel, M. D., & Fryer, J. L., (1980). Stress-induced transmission of *Yersinia ruckeri* infection from carriers to recipient steelhead trout *Salmo gairdneri* Richardson. *J. Fish Dis., 3*(6), 467–472.

Iida, Y., & Mizokami, A., (1996). Outbreaks of coldwater disease in wild ayu and pale chub. *Fish Pathol., 31*, 157–164.

Imaoka, A., Okada, Y., Matsumoto, S., Setoyama, H., & Umesaki, Y., (1997). 16S ribosomal DNA sequence divergence of segmented filamentous bacteria with special reference to inter-species and within-species variation of host animals. *Syst. Appl. Microbiol., 20*, 418–422.

Ispir, U., & Dorucu, M., (2014). Efficacy of lipopolysaccharide antigen of *Yersinia ruckeri* in rainbow trout by intraperitoneal and bath immersion administration. *Res. Vet. Sci, 97*, 271–273.

Ivanov, I. I., Atarashi, K., Manel, N., Brodie, E. L., Shima, T., Karaoz, U., Wei, D., et al., (2009). Induction of intestinal Th17 cells by segmented filamentous bacteria. *Cell, 139*, 485–498.

Izumi, S., Fujii, H., & Aranishi, F., (2005). Detection and identification of *Flavobacterium psychrophilum* from gill washings and benthic diatoms by PCR-based sequencing analysis. *J. Fish Dis., 28*(9), 559–564.

Izumi, S., Liu, H., Aranishi, F., & Wakabayashi, H., (2003). A novel serotype of *Favobacterium* psychrophilum detected using antiserum against an isolate from amago, *Oncorhynchus masou* rhodurus Jordan & Gilbert, in Japan. *J. Fish Dis., 26*, 677–680.

Jansson, E., (2002). *Bacterial Kidney Disease in Salmonid Fish*. Doctoral thesis. Swedish University of Agricultural Sciences, Uppsala.

Jansson, E., Eriksson, E., Säker, E., Dalsgaard, I., Nonnemann, B., Roozenburg, I., & Haenen, O., (2015). Matrix-assisted laser desorption/ionization time of flight, MALDI-TOF, mass spectrometry for identification of fish pathogenic bacteria. In: *17th International Conference on Diseases of Fish and Shellfish* (p. 384). European Association of Fish Pathologists, Las Palmas de Gran Canaria.

Jansson, E., Lindberg, L., Säker, E., & Aspan, A., (2008). Diagnosis of bacterial kidney disease by detection of *Renibacterium salmoninarum* by real-time PCR. *J. Fish Dis., 31*, 755–763.

Jarau, M., Macinnes, J. I., & Lumsden, J. S., (2019). Erythromycin and florfenicol treatment of rainbow trout *Oncorhynchus mykiss* (Walbaum) experimentally infected with *Flavobacterium psychrophilum. J. Fish Dis., 42*, 325–334.

Jensen, P. A., Henriksen, N. H., Michelsen, K., Madsen, L., & Dalsgaard, I. (2003). *Prevention of RTFS (Rainbow Trout Fry Syndrome) and Limitation of Medicine Requirements in Egg and Fry Stocks on Danish Fish Farms*. DGU-Rapport No. 124-03; Charlottenlund: Danish Institute for Fisheries Research.

Jolley, K. A., Bliss, C. M., Bennett, J. S., Bratcher, H. B., Brehony, C., Colles, F. M., Wimalarathna, H., et al., (2012). Ribosomal multilocus sequence typing: Universal characterization of bacteria from domain to strain. *Microbiology, 158*, 1005.

Kent, M. L., (1982). Characteristics and identification of *Pasteurella* and *Vibrio* species pathogenic to fishes using API-20E (Analytab products) multitube test strips. *Can. J. Fish. Aquart. Sci., 39*, 1725–1729.

Kent, M. L., Groff, J. M., Morrison, J. K., Yasutake, W. T., & Holt, R. A., (1989). Spiral swimming behavior due to cranial and vertebral lesions associated with *Cytophaga psychrophila* infection in salmonid fishes. *Dis. Aquat. Organ., 6*, 11–16.

Kim, D. H., & Austin, B., (2006). Innate immune responses in rainbow trout (*Oncorhynchus mykiss, Walbaum*) induced by probiotics. *Fish Shellfish Immunol., 21*, 513–524.

Kim, J. H., Choresca, C. H., Shin, S. P., Han, J. E., Jun, J. W., & Park, S. C., (2015). Biological control of *Aeromonas salmonicida* subsp. salmonicida infection in rainbow trout (*Oncorhynchus mykiss*) using *Aeromonas* phage PAS-1. *Transbound Emerg. Dis., 62*, 81–86.

Kim, J. H., Son, J. S., Choi, Y. J., Choresca, C. H., et al., (2012). Complete genome sequence and characterization of a broad-host range T4-like bacteriophage phiAS5 infecting *Aeromonas salmonicida* subsp. salmonicida. *Vet. Microbiol., 157*, 164–171.

Kimura, N., Wakabayashi, H., & Kudo, S., (1978). Studies on bacterial gill disease in salmonids—I selection of bacterium transmitting gill disease. *Fish Pathol., 12*, 233–242.

Kimura, T., & Awakura, T., (1977). *Bacterial Kidney Disease of Salmonid: First Observation in Japan*. Bulletin of the Japanese Society of Scientific Fisheries (Japan).

Kirkan, Göksoy, E. Ö., & Kaya, O., (2003). Isolation and antimicrobial susceptibility of *Aeromonas salmonicida* in rainbow trout (*Oncorhynchus mykiss*) in Turkey hatchery farms. *J. Vet. Med. Ser B, 50*, 339–342.

Klaasen, H. L., Koopman, J. P., Poelma, F. G., & Beynen, A. C., (1992). Intestinal, segmented, filamentous bacteria. *FEMS Microbiol. Rev., 8*, 165–179.

Klaasen, H. L., Koopman, J. P., Van, D. B. M. E., Bakker, M. H., Poelma, F. G., & Beynen, A. C., (1993). Intestinal, segmented, filamentous bacteria in a wide range of vertebrate species. *Lab. Anim., 27*, 141–150.

Klaasen, H. L., Koopman, J. P., Van, D. B., M. E., Van, W. H. P., & Beynen, A. C., (1991). Mono-association of mice with non-cultivable, intestinal, segmented, filamentous bacteria. *Arch. Microbiol., 156*, 148–151.

Kline, K. A., Falker, S., Dahlberg, S., Normark, S., & Henriques-Normark, B., (2009). Bacterial adhesins in host-microbe interactions. *Cell Host. Microbe., 5*, 580–592.

Ko, Y. M., & Heo, G. J., (1997). Characteristics of *Flavobacterium branchiophilum* isolated from rainbow trout in Korea. *Fish Pathol., 32*, 97–102.

Kolton, M., Sela, N., Elad, Y., & Cytryn, E., (2013). Comparative genomic analysis indicates that niche adaptation of terrestrial *Flavobacteria* is strongly linked to plant glycan metabolism. *PloS One, 8*, e76704.

Kondo, M., Kawai, K., Okabe, M., Nakano, N., & Oshima, S., (2003). Efficacy of oral vaccine against bacterial coldwater disease in ayu *Plecoglossus altivelis*. *Dis. Aquat. Organ., 55*, 261–264.

Kumagai, A., Nakayasu, C., & Oseko, N., (2004). Effect of tobramycin supplementation to medium on isolation of *Flavobacterium psychrophilum* from Ayu *Plecoglossus altivelis*. *Fish Pathol., 39*, 75–78.

Kumagai, A., Takahashi, K., Yamaoka, S., & Wakabayashi, H., (1998). Ineffectiveness of iodophore treatment in disinfecting salmonid eggs carrying *Cytophaga psychrophila*. *Fish Pathol., 33*(3), 123–128.

Kumagai, A., Yamaoka, S., Takahashi, K., Fukuda, H., & Wakabayashi, H., (2000). Waterborne transmission of *Flavobacterium psychrophilum* in coho salmon eggs. *Fish Pathol., 35*(1), 25–28.

Kumar, G., Menanteau-Ledouble, S., Saleh, M., & El-Matbouli, M., (2015). *Yersinia ruckeri*, the causative agent of enteric redmouth dis- ease in fish. *Vet. Res., 46*, 10. https://doi.org/10.1186/ s13567-015-0238-4.

Kumar, V., Roy, S., Meena, S. K., & Sarkar, U. K., (2016). Application of probiotics in shrimp aquaculture: Importance, mechanisms of action, and methods of administration. *Rev. Fish. Sci. Aquac., 24*(4), 342–368.

Laanto, E., Bamford, J. K., Ravantti, J. J., & Sundberg, L. R., (2015). The use of phage FCL-2 as an alternative to chemotherapy against columnaris disease in aquaculture. *Frontiers in Microbiology, 6*, 829.

LaFrentz, B. R., Lindstrom, N. M., LaPatra, S. E., Call, D. R., & Cain, K. D., (2007). Electrophoretic and western blot analyses of the lipopolysaccharide and glycocalyx of *Flavobacterium psychrophilum*. *Fish Shellfish Immunol., 23*, 770–780.

Lahnsteiner, F., (2006). Effect of disinfection of non-hardened *Salmo trutta* eggs with Chloramine T®, Wofasteril®, and hydrogen peroxide on embryo and larvae viability, microorganism load, lipid peroxidation, and protein carbonylation. *Aquac. Int.*, 1–4.

Laidler, P., Lityńska, A., Hoja-Łukowicz, D., Łabędz, M., Przybyło, M., Ciołczyk-Wierzbicka, D., Pocheć, E., et al., (2006). Characterization of glycosylation and adherent properties of melanoma cell lines. *Cancer Immunol. Immunother., 55*, 112–118.

Laing, C. R., Zhang, Y., Thomas, J. E., & Gannon, V. P., (2011). Everything at once: Comparative analysis of the genomes of bacterial pathogens. *Vet. Microbiol., 153*, 13–26.

Laurin, E., Morrison, D., Gardner, I. A., Siah, A., Powell, J. F., & Kamaitis, M., (2019). Bayesian latent class analysis of ELISA and RT-rPCR diagnostic accuracy for subclinical *Renibacterium salmoninarum* infection in Atlantic salmon (*Salmo salar*) broodstock. *J. Fish Dis., 42*, 303–313.

Lee, K. B., & Heo, G. J., (1998). First isolation and identification of *Cytophaga psychrophila* from cultured ayu Korea. *Fish Pathol., 33*, 37–38.

Lehmann, J., Mock, D., Sturenberg, F. J., & Bernardet, J. F., (1991). First isolation of *Cytophaga psychrophila* from a systemic disease in eel and cyprinids. *Diseases of Aquatic Organisms., 10*, 217–220.

Lindstrom, N. M., Call, D. R., House, M. L., Moffitt, C. M., & Cain, K. D., (2009). A quantitative enzyme-linked immunosorbent assay and filtration-based fluorescent antibody test as potential tools to screen broodstock for infection with *Flavobacterium psychrophilum*. *J. Aquat. Anim. Health, 21*(1), 43–56.

Loch, T. P., & Faisal, M., (2015). Emerging flavobacterial infections in fish: A review. *J. Adv. Res., 6*, 283–300.

Loch, T. P., (2012). *Identification of Novel Flavobacteria from Michigan and Assessment of Their Impacts on Fish Health*. Doctoral dissertation, Michigan State University.

Long, A., Call, D. R., & Cain, K. D., (2014). Investigation of the link between broodstock infection, vertical transmission, and prevalence of *Flavobacterium psychrophilum* in eggs and progeny of rainbow trout and coho salmon. *J. Aquat. Anim. Health, 26*(2), 66–77.

Lorenzen, E., & Karas, N., (1992). Detection of *Flexibacter psychrophilus* by immunofluorescence in fish suffering from fry mortality syndrome: A rapid diagnostic method. *Dis. Aquat. Org., 13*, 231–234.

Lorenzen, E., Dalsgaard, I., & Bernardet, J. F., (1997). Characterization of isolates of *Flavobacterium psychrophilum* associated with Coldwater disease or rainbow trout fry syndrome I: Phenotypic and genomic studies. *Dis. Aquat. Organ., 31*, 197–208.

Lowden, S., & Heath, T., (1995). Segmented filamentous bacteria associated with lymphoid tissues in the ileum of horses. *Res. Vet. Sci., 59*, 272–274.

Lu, R. M., Hwang, Y. C., Liu, I. J., Lee, C. C., Tsai, H. Z., Li, H. J., & Wu, H. C., (2020). Development of therapeutic antibodies for the treatment of diseases. *J. Biomed. Sci., 27*, 1–30.

Lumsden, J. S., Young, K., Welsh, K., MacInnes, J., Russell, S., & Hesami, S., (2006). Management approaches for coldwater disease caused by *Flavobacterium psychrophilum*.

In: *Proceedings of the Canadian Freshwater Aquaculture Symposium–Aquaculture* (pp. 111–117). Canada 2004. St. Andrews, New Brunswick. Aquaculture Association of Canada Special Publication No. 11.

Lutwyche, P., Exner, M. M., Hancock, R. E. W., & Trust, T. J., (1995). A conserved *Aeromonas salmonicida* porin provides protective immunity to rainbow trout. *Infect Immun., 63*, 3137–3142.

Madetoja, J., Dalsgaard, I., & Wiklund, T., (2002). Occurrence of *Flavobacterium psychrophilum* in fish-farming environments. *Dis. Aquat. Organ., 52*(2), 109–118.

Madetoja, J., Hanninen, M. L., Hirvela-Koski, V., Dalsgaard, I., & Wiklund, T., (2001). Phenotypic and genotypic characterization of *Flavobacterium psychrophilum* from Finnish fish farms. *J. Fish Dis., 24*(8), 469–479.

Madetoja, J., Nyman, P., & Wiklund, T., (2000). *Flavobacterium psychrophilum*, invasion into and shedding by rainbow trout *Oncorhynchus mykiss*. *Dis. Aquat. Organ., 43*, 27–38.

Madsen, L., Arnbjerg, J., & Dalsgaard, I., (2001). Radiological examination of the spinal column in farmed rainbow trout *Oncorhynchus mykiss* (Walbaum): Experiments with *Flavobacterium psychrophilum* and oxytetracycline. *Aquaculture Research, 32*, 235–241.

Madsen, L., Bertelsen, S. K., Dalsgaard, I., & Middelboe, M., (2013). Dispersal and survival of *Flavobacterium psychrophilum* phages *in vivo* in rainbow trout and *in vitro* under laboratory conditions: Implications for their use in phage therapy. *Appl. Environ. Microbiol., 79*, 4853–4861.

Mahanta, P. C., & Debajit, S., (2010). *Coldwater Fisheries Management*. DCFR, ICAR, Nainital (Uttarakhand), India.

Manna, S. K., Samanta, S., Das, M. K., & Mishra, S. S., (2003). An outbreak of *Yersinia ruckeri* septicaemia in Indian major carps. *J. In. Fisher Soc. India, 35*, 28–31.

Manzano, M., Giusto, C., Iacumin, L., Patthey, C., Cecchini, F., Fontanillas, R., & Comi, G., (2010). Nested PCR for the detection of *Candidatus arthromitus* in fish. *FEMS Microbiol. Lett., 308*, 35–39.

Margulis, L., Olendzenski, L., & Afzelius, B. A., (1990). Endospore-forming filamentous bacteria symbiotic in termites: Ultrastructure and growth in culture of *Arthromitus*. *Symbiosis*., 95–116.

Martinez, J. L., Casado, A., & Enriquez, R., (2004). Experimental infection of *Flavobacterium psychrophilum* in fins of Atlantic salmon *Salmo salar* revealed by scanning electron microscopy. *Dis. Aquat. Organ., 59*, 74–84.

Mccarthy, U., Pettinello, R., Feehan, L., Ho, Y. M., & White, P., (2016). Experimental transmission of segmented filamentous bacteria (SFB) in rainbow trout *Oncorhynchus mykiss*. *Dis. Aquat. Org., 119*, 45–57.

McIntosh, D., & Austin, B., (1996). The validity of Western blotting for the diagnosis of bacterial kidney disease based on the detection of the p57 antigen of *Renibacterium salmoninarum*.*J. Microbiol. Methods, 25*, 329–335.

Mendez, J., & Guijarro, J. A., (2013). *In vivo* monitoring of *Yersinia ruckeri* in fish tissues: Progression and virulence gene expression. *Environ. Microbiol. Rep., 5*, 179–185. https://doi. org/10.1111/1758-2229.12030.

Mesa, M. G., Maule, A. G., Poe, T. P., & Schreck, C. B., (1999). Influence of bacterial kidney disease on smoltification in salmonids: Is it a case of double jeopardy? *Aquaculture, 174*, 25–41.

Michel, C., & Garcia, C., (2003). Virulence stability in *Flavobacterium psychrophilum* after storage and reservation according to different procedures. *Vet. Res., 34,* 127–132.

Michel, C., Bernardet, J. F., Daniel, P., Chilmonczyk, S., Urdaci, M., & De Kinkelin, P., (2002). Clinical and aetiological aspects of a summer enteritic syndrome associated with the sporulating segmented filamentous bacterium '*Candidatus arthromitus*' in farmed rainbow trout, *Oncorhynchus mykiss* (Walbaum). *J. Fish Dis., 25*, 533–543.

Michel, C., Kerouault, B., & Martin, C., (2003). Chloramphenicol and florfenicol susceptibility of fish-pathogenic bacteria isolated in France: Comparison of minimum inhibitory concentration, using recommended provisory standards for fish bacteria. *J. Appl. Microbiol., 95*, 1008–1015.

Midtlyng, P. J., (2014). Vaccination against furunculosis. In: Gudding, R., Atle, L. A., & Evensen, Ø., (eds.), *Fish Vaccination* (pp. 185–199). John Wiley & Sons, Chichester.

Midtlyng, P. J., Bleie, H., Helgason, S., Jansson, E., Larsen, J. L., Olesen, N. J., Olsen, A. B., & Vennerstrøm, P., (2000). *Nordic Manual for the Surveillance and Diagnosis of Infectious Diseases in Farmed Salmonids*. Nordic Council of Ministers.

Misaka, N., Nishizawa, T., & Yoshimizu, M., (2008). Quantitative detection of viable *Flavobacterium psychrophilum* in chum salmon *Oncorhynchus keta* by colony blotting and immunostaining. *Fish Pathol., 43*, 117–123.

Miwa, S., & Nakayasu, C., (2005). Pathogenesis of experimentally induced bacterial coldwater disease in ayu *Plecoglossus altivelis*. *Dis. Aquat. Organ., 67*, 93–104.

Mohamed, S., Nagaraj, G., Chua, F. H. C., & Wang, Y. G., (2015). *The Use of Chemicals in Aquaculture in Malaysia and Singapore*. Southeast Asian Fisheries Development Center, Iloilo, Philippines.

Mohammed, H., (2015). *Studies on the Epidemiology, Vaccination, Susceptibility, and Treatment of Columnaris Disease in Fishes.* Doctoral dissertation, Auburn University.

Murray, R. G., & Stackebrandt, E., (1995). Taxonomic note: Implementation of the provisional status candidatus for incompletely described procaryotes. *Int. J. Syst. Evol., 45*, 186–187.

Navas, E., Bohle, H., Henríquez, P., Grothusen, H., Bustamante, F., Bustos, P., & Mancilla, M., (2014). Draft genome sequence of the fish pathogen *Yersinia ruckeri* strain 37551, serotype O1b, isolated from diseased, vaccinated Atlantic salmon (*Salmo salar*) in Chile. *Genome Announc, 2*, e00858–14.

Nematollahi, A., Decostere, A., Pasmans, F., & Haesebrouck, F., (2003). *Flavobacterium psychrophilum* infections in salmonid fish. *J. Fish Dis., 26*(10), 563–574.

Nematollahi, A., Decostere, A., Pasmans, F., Ductaelle, R., & Haesebrouck, F., (2003). Adhesion of high and low virulence *Flavobacterium psychrophilum* strains to isolated gill arches of rainbow trout *Oncorhynchus mykiss*. *Dis. Aquat. Organ., 55*, 101–107.

Nicolas, P., Mondot, S., Achaz, G., Couchenot, C., Bernardet, J., & Duchaud, E., (2008). Population structure of the fish-pathogenic bacterium *Flavobacterium psychrophilum. Appl. Environ. Microbiol., 74*, 3702–3709.

Noble, A. C., & Summerfelt, S. T., (1996). Diseases encountered in rainbow trout cultured in recirculating systems. *Annu. Rev. Fish Dis., 6*, 65–92.

Noga, E. J., (2010). Problem 55: Mycobacteriosis. In: *Fish Disease Diagnosis and Treatment* (2nd edn., pp. 204–208). Ames, IA: Wiley-Blackwell Publishing.

Obach, A., Baudin, & Laurencin, F., (1991). Vaccination of rainbow trout *Oncorhynchus mykiss* against the visceral form of coldwater disease. *Dis. Aquat. Organ., 12*(13), 15.

Ohtani, M., Villumsen, K. R., Strom, H. K., & Raida, M. K., (2014). 3D visualization of the initial *Yersinia ruckeri* infection route in rainbow trout (*Oncorhynchus mykiss*) by optical projection tomography. *PLoS One, 9*, e89672. https://doi.org/10.1371/journal.pone.0089672.

Oplinger, R. W., Wagner, E. J., & Cavender, W., (2015). Use of penicillin and streptomycin to reduce spread of bacterial coldwater disease II: Efficacy of using antibiotics in diluents and during water hardening. *J. Aquat. Anim. Health, 27*, 32–37.

Ordal, E. T., & Earp, B. J., (1956). Cultivation and transmission of etiological agent of kidney disease in salmonid fishes. *Proc. Soc. Exp. Biol. Med., 92*, 85–88.

Orellana, J., (2007). *Identification and Quantification of Suspended Solids and Their Effects in Modern Marine Recirculation Systems*. Doctoral dissertation, Christian-Albrechts Universitat Kiel.

Ormsby, M. J., Caws, T., Burchmore, R., Wallis, T., Verner-Jeffreys, D. W., & Davies, R. L., (2016). *Yersinia ruckeri* isolates recovered from diseased Atlantic salmon (*Salmo salar*) in Scotland are more diverse than those from rainbow trout (*Oncorhynchus mykiss*) and represent distinct subpopulations. *Appl. Environ. Microbiol., 82*, 5785–5794.

Ostland, V. E., Byrne, P. J., Lumsden, J. S., Macphee, D. D., Derksen, J. A., Haulena, M., Skar, K., Myhr, E., & Ferguson, H. W., (1999). Atypical bacterial gill disease: A new form of bacterial gill disease affecting intensively reared salmonids. *J. Fish Dis., 22*, 351–358.

Ostland, V. E., McGrogan, D. G., & Ferguson, H. W., (1997). Cephalic osteochondritis and necrotic scleritis in intensively reared salmonids associated with *Flexibacter psychrophilus*. *Journal of Fish Diseases., 20*, 443–451.

Pajdak-Czaus, J., Platt-Samoraj, A., Szweda, W., Siwicki, A. K., Terech-Majewska, E., (2019). *Yersinia ruckeri*—A threat not only to rainbow trout. *Aquacul. Res., 50*(11), 3083–3096.

Pamp, S. J., Harrington, E. D., Quake, S. R., Relman, D. A., & Blainey, P. C., (2012). Single-cell sequencing provides clues about the host interactions of segmented filamentous bacteria (SFB). *Genome Res., 22*, 1107–1119.

Paterson, W. D., Lall, S. P., & Desantels, D. (1981). "Studies on Bacterial Kidney Disease in Atlantic Salmon (*Salmo salar*) in Canada." *Fish Pathol., 15*, 283–292.

Pedersen, L. F., & Lazado, C. C., (2020). Decay of peracetic acid in seawater and implications for its chemotherapeutic potential in aquaculture. *Aquac. Environ. Interact., 12*, 153.

Perez-Pascual, D., Menendez, A., Fernandez, L., Mendez, J., Reimundo, P., Navais, R., & Guijarro, J., (2009). Spreading versus biomass production by colonies of the fish pathogen *Flavobacterium psychrophilum*: Role of nutrient concentration. *Int. Microbiol., 12*, 207–214.

Pérez-Pascual, D., Vendrell-Fernández, S., Audrain, B., Bernal-Bayard, J., Patiño-Navarrete, R., Petit, V., Rigaudeau, D., & Ghigo, J. M., (2021). Gnotobiotic rainbow trout (*Oncorhynchus mykiss*) model reveals endogenous bacteria that protect against *Flavobacterium columnare* infection. *Plos Pathogens, 17*, e1009302.

Piganelli, J. D., Wiens, G. D., Zhang, J. A., Christensen, J. M., & Kaattari, S. L., (1999). Evaluation of a whole cell, p57-vaccine against *Renibacterium salmoninarum*. *Dis. Aquat. Org., 36*, 37–44.

Pokhrel, H., Baishya, S., Phukan, B., Pillai, D., & Rather, M. A., (2018). Occurrence and distribution of multiple antibiotic resistance bacteria of public health significance in backwaters and aquaculture farm. *Int. J. Curr. Microbiol. App. Sci., 7*(2), 975–987.

Post, G. P., (1987). *Textbook of Fish Health*. New Jersey; T.F.H. Publications.

Proietti, M., Cornacchione, V., Jost, T. R., Romagnani, A., Faliti, C. E., Perruzza, L., Rigoni, R., et al., (2014). ATP-gated ionotropic P2X7 receptor controls follicular T helper cell numbers in Peyer's patches to promote host-microbiota mutualism. *Immunity., 41*, 789–801.

Punitha, S. M. J., Babu, M. M., Sivaram, V., Shankar, V. S., Dhas, S. A., Mahesh, T. C., Immanuel, G., & Citarasu, T., (2008). Immunostimulating influence of herbal biomedicines on nonspecific immunity in grouper *Epinephelus tauvina* juvenile against *Vibrio harveyi* infection. *Aquaculture International, 16*, 511–523.

Racheal, A. M., (2017). *Identification and Characterization of Flavobacteriaceae from Farmed Nile Tilapia (Oreochromis niloticus) and African Catfish (Clarius gariepinus) in Uganda*. Doctoral dissertation, Sokoine University of Agriculture.

Raida, M. K., Larsen, J. L., Nielsen, M. E., & Buchmann, K., (2003). Enhanced resistance of rainbow trout, *Oncorhynchus mykiss* (Walbaum), against *Yersinia ruckeri* challenge following oral administration of *Bacillus subtilis* and *B. licheniformis* (BioPlus2B). *J. Fish Dis., 26*, 495–498.

Rangdale, R. E., Richards, R. H., & Alderman, D. J., (1997). Minimum inhibitory concentrations of selected antimicrobial compounds against *Flavobacterium psychrophilum* the causal agent of rainbow trout fry syndrome (RTFS). *Aquaculture, 158*(3, 4), 193–201.

Rashidian, G., Bahrami, G. S., Farsani, M. N., Prokić, M. D., & Faggio, C., (2020). The oak (*Quercus brantii*) acorn as a growth promotor for rainbow trout (*Oncorhynchus mykiss)*: Growth performance, body composition, liver enzymes activity and blood biochemical parameters. *Nat. Prod. Res., 34*, 2413–2423.

Rattanachaikunsopon, P., & Phumkhachorn, P., (2012). Detection of *Aeromonas salmonicida* by reverse transcription-multi-plex polymerase chain reaction. *Biosci. Biotechnol. Biochem., 76*, 665–670.

Reimschuessel, R., Miller, R. A., & Gieseker, C. M., (2013). Antimicrobial drug use in aquaculture. *Antimicrobial Therapy in Veterinary Medicine*, 6456–6461.

Roberts, R. J., (2012). The bacteriology of teleosts. In: Roberts, R. J., (ed.), *Fish Pathology* (4th edn., pp. 339–382). Wiley-Blackwell, Oxford.

Robertson, P. A. W., Austin, D. A., & Austin, B., (2005). Prevention of ulcer disease in goldfish by means of vaccination. *J. Aquat. Anim. Health, 17*, 203–209.

Rodgers, C. J., (1992). Development of a selective-differential medium for the isolation of *Yersinia ruckeri* and its application in epidemiological studies. *J. Fish Dis., 15*, 243–254.

Rodkhum, C., Hirono, I., Crosa, J. H., & Aoki, T., (2005). Four novel hemolysin genes of *Vibrio anguillarum* and their virulence to rainbow trout. *Microb. Pathog., 39*, 109–119.

Romalde, J. L., & Toranzo, A. E., (1993). Pathological activities of *Yersinia ruckeri*, the enteric redmouth (ERM) bacterium. *FEMS Microbiol. Lett., 112*, 291–300.

Ronza, P., Bermúdez, R., Losada, A. P., Robles, A., & Quiroga, M. I., (2011). Mucosal CD3ε+ cell proliferation and gut epithelial apoptosis: Implications in rainbow trout gastroenteritis (RTGE). *J. Fish Dis., 34*, 433–443.

Ross, A. J., Rucker, R. R., & Ewing, W. H., (1966). Description of a bacterium associated with redmouth disease of rainbow trout (*Salmo gairdneri*). *Can. J. Microbiol., 12*, 763–770.

Rucker, R. R., Earp, B. J., & Ordal, E. J., (1953). *Infectious Diseases of Pacific Salmon* (Vol. 83, pp. 297–312). Transactions of the American Fisheries Society.

Sakai, M., Atsuta, S., & Kobayashi, M., (1989). Comparison of methods used to detect *Renibacterium salmoninarum*, the causative agent of bacterial kidney disease. *J. Aquat. Anim. Health., 1*, 21–24.

Saleh, M., Soliman, H., & El-Matbouli, M., (2008). Loop-mediated isothermal amplification as an emerging technology for detection of *Yersinia ruckeri* the causative agent of enteric red mouth disease in fish. *BMC Vet. Res., 4*, 31.

Saleh, M., Soliman, H., & El-Matbouli, M., (2015). Gold nanoparticles as a potential tool for diagnosis of fish diseases. *Methods Mol. Biol., 1247*, 245–252.

Sanders, J. E., & Fryer, J. L., (1980). *Renibacterium salmoninarum* gen. Nov., sp. Nov., the causative agent of bacterial kidney disease in salmonid fishes. *Int. J. Syst. Evol., 30*, 496–502.

Sanders, J. E., Pilcher, K. S., & Fryer, J. L., (1978). Relation of water temperature to bacterial kidney disease in coho salmon (*Oncorhynchus kisutch*), sockeye salmon (*O. nerka*), and steelhead trout (*Salmo gairdneri*). *J. Fish. Res., 35*, 8–11.
Sanguinetti, M., Posteraro, B., Ardito, F., Zanetti, S., Cingolani, A., Sechi, L., De Luca, A., et al., (1998). Routine use of PCR reverses cross-blot hybridization assay for rapid identification of mycobacterium species growing in liquid media. *J. Clin. Microbiol., 36*, 1530–1533.
Santos, Y., Huntly, P. G., Turnbull, A., & Hastings, T. S., (1992). Isolation of *Cytophaga psychrophila* (*Flexibacter psychorphilus*) in association with rainbow trout mortality in the United Kingdom. *Bull Eur. Assoc. Fish Pathol., 12*, 209–210.
Schachte, J. H., Meyer, F. P., Warren, J. W., & Carey, T. G., (1983). *A Guide to Integrated Fish Health Management in the Great Lakes Basin MICHIGAN*, 193–197.
Schmidt, A. S., Bruun, M. S., Dalsgaard, I., & Larsen, J. L., (2001). Incidence, distribution, and spread of tetracycline resistance determinants and integron-associated antibiotic resistance genes among motile aeromonads from a fish farming environment. *Appl Environ Microbiol., 67*, 5675–5682.
Schubert, M. L., Rohrbach, R., Schmitt, M., & Stein-Thoeringer, C. K., (2021). The potential role of the intestinal micromilieu and individual microbes in the immunobiology of chimeric antigen receptor T-cell therapy. *Front. Immunol., 12*, 1836.
Secades, P., & Guijarro, J. A., (1999). Purification and characterization of an extracellular protease from the fish pathogen *Yersinia ruckeri* and effect of culture conditions on production. *Appl Environ Microbiol., 65*, 3969–3975.
Shao, Z. J., (2001). Aquaculture pharmaceuticals and biologicals: Current perspectives and future possibilities. *Advanced Drug Delivery Review, 50*, 229–243.
Shaowu, L., Di, W., Hongbai, L., & Tongyan, L., (2013). Isolation of *Yersinia ruckeri* strain H01 from farm-raised amur sturgeon *Acipenser schrencki* in China. *J. Aquat. Anim. Health, 25*, 9–14.
Singh, B. R., (2013). Antimicrobial sensitivity assay and antimicrobial chemotherapy in animals: A practical approach. In: Singh, B. R., & Somvanshi, R., (eds.), *Diseases of Animals: Diagnosis and Management*. doi: 10.13140/RG.2.1.2987.4322.
Skrodenytė-Arbačiauskienė, V., Kazlauskienė, N., Vosylienė, M. Z., & Virbickas, T., (2010). Identification of *Aeromonas salmonicida* in European perch from North Lithuanian Rivers during mass mortalities in 2008. *Cent. Eur. J. Biol., 5*, 831–838.
Skrodenytė-Arbačiauskienė, V., Kazlauskienė, N., Vosylienė, M., & Virbickas, T., (2012). *Aeromonas salmonicida* infected fish transfer disease to healthy fish via water. *Open Life Sciences, 7*(5), 878–885.
Skulska, I., (2014). *Culture of the Bacterial Gill Disease Organism, Flavobacterium Branchiophilum and Strain Differences Relevant to Epizoology.* Doctoral dissertation. University of Guelph.
Smith, A. M., Goldring, O. L., & Dear, G., (1987). The production and methods of use of polyclonal antisera to the pathogenic organisms *Aeromonas salmonicida, Yersinia ruckeri* and *Renibacterium salmoninarum*. *J. Fish Biol., 31*, 225–226.
Smith, Jr. P., Ambrose, M. E., & Knobl, Jr. G. N., (1964). Improved rapid method for determining total lipids in fish meal. *Comm. Fish. Rev., 26*, 1–5.
Smith, T. M., (1997). Segmented filamentous bacteria in the bovine small intestine. *J. Comp. Pathol., 117*, 185–190.
Snel, J., Heinen, P. P., Blok, H. J., Carman, R. J., Duncan, A. J., Allen, P. C., & Collins, M. D., (1995). Comparison of 16S rRNA sequences of segmented filamentous bacteria isolated

from mice, rats, and chickens and proposal of "*Candidatus Arthromitus*". *Int. J. Syst. Evol., 45*, 780–782.

Snieszko, S. F., & Griffin, P. J., (1955). Kidney disease in brook trout and its treatment. *Progress Fish Cult., 17*, 3–13.

Soleim, O., (1985). Hitra-disease-a confusing concept. *Bull. Eur. Ass. Fish Pathol. J.,* (4), 77.

Soltani, M., Ghosh, K., Hoseinifar, S. H., Kumar, V., Lymbery, A. J., Roy, S., & Ringo, E., (2019). Genus *Bacillus,* promising probiotics in aquaculture: Aquatic animal origin, bio-active components, bioremediation and efficacy in fish and shellfish. *Rev. Fish. Sci. Aquac., 27*, 331–379.

Soltani, M., Shafiei, S., Yosefi, P., Mosavi, S., & Mokhtari, A., (2014). Effect of Montanide™ IMS 1312 VG adjuvant on efficacy of *Yersinia ruckeri* vaccine in rainbow trout (*Oncorhynchus mykiss*). *Fish Shellfish Immunol., 37*, 60–65.

Starliper, C. E., (2011). Bacterial coldwater disease of fishes caused by *Flavobacterium psychrophilum*. *Journal of Advanced Research, 2*, 97–108.

Starliper, C. E., Smith, D. R., & Shatzer, T., (1997). Virulence of *Renibacterium salmoninarum* to salmonids. *J. Aquat. Anim. Health, 9*, 1–7.

Starliper, C., & Schill, W., (2011). Flavobacterial diseases: Columnaris disease, coldwater disease, and bacterial gill disease. *Fish Diseases and Disorders, 3*, 606–631.

Surendraraj, A., Farvin, K. S., Yathavamoorthi, R., & Thampuran, N., (2009). Enteric bacteria associated with farmed freshwater fish and its culture environment in Kerala, India. *Res. J. Microbiol., 4*, 334–344.

Tannock, G. W., & Archibald, R. D., (1984). The derivation and use of mice which do not harbor *lactobacilli* in the gastrointestinal tract. *Can. J. Microbiol., 30*, 849–853.

Taylor, P. W., (2004). Detection of *Flavobacterium psychrophilum* in eggs and sexual fluids of pacific salmonids by a polymerase chain reaction assay: Implications for vertical transmission of bacterial coldwater disease. *J. Aquat. Anim. Health, 16*, 104–108.

Temprano, A., Riano, J., Yugueros, J., Gonzalez, P., De Castro, L., Villena, A., Luengo, J. M., & Naharro, G., (2005). Potential use of a *Yersinia ruckeri* O1 auxotrophic aroA mutant as a live attenuated vaccine. *J. Fish Dis., 28*, 419–427.

Tie, H. M., Feng, L., Jiang, W. D., Wu, P., Liu, Y., Jiang, J., Kuang, S. Y., et al., (2021). Dietary exogenous supplementation of nucleotides strengthens the disease resistance, antioxidant capacity and immunity in the gill of on-growing grass carp (*Ctenopharyngodon idella*) following a challenge with *Flavobacterium columnare*. *Aquaculture, 540*, 736729.

Tobback, E., Decostere, A., Hermans, K., Haesebrouck, F., & Chiers, K., (2007). *Yersinia ruckeri* infections in salmonid fish. *J. Fish Dis., 30*, 257–268.

Tobback, E., Decostere, A., Hermans, K., Ryckaert, J., Duchateau, L., Haesebrouck, F., & Chiers, K., (2009). Route of entry and tissue distribution of *Yersinia ruckeri* in experimentally infected rainbow trout *Oncorhynchus mykiss*. *Dis Aquat. Org., 84*, 219–228.

Toranzo, A. E., & Barja, J. L., (1993). Fry mortality syndrome (FMS) in Spain. Isolation of the causative bacterium *Flexibacter psychrophilis*. *Bull. Eur. Assoc. Fish Pathol., 13*, 30–32.

Toranzo, A. E., (2004). Report about fish bacterial diseases. In: Alvarez-Pellitero, P., Barja, J. L., Basurco, B., Berthe, F., & Toranzo, A. E., (eds.), *Mediterranean Aquaculture Laboratories* (pp. 49–89).

Tort, M. J., (2000). *Studies of Hydrogen Peroxide for the Treatment of Bacterial Gill Disease in Fish*. Cornell University.

Touchon, M., Barbier, P., Bernardet, J. F., Loux, V., Vacherie, B., Barbe, V., Rocha, E. P., & Duchaud, E., (2011). Complete genome sequence of the fish pathogen *Flavobacterium branchiophilum*. *Appl. Environ. Microbiol., 77*, 7656–7662.

Umesaki, Y., Setoyama, H., Matsumoto, S., Imaoka, A., & Itoh, K., (1999). Differential roles of segmented filamentous bacteria and *clostridia* in development of the intestinal immune system. *Infect. Immun., 67*, 3504–3511.

Ummey, S., Khan, S., Vijayakumar, P. P. N., & Ramya, A., (2021). Enteric redmouth disease and its causative bacterium, *Yersinia ruckeri*, in Indian major carps from culture ponds in Andhra Pradesh, India. *Aqua and Fish, 6*(3), 289–299.

Urdaci, M. C., Regnault, B., & Grimont, P. A., (2001). Identification by *in situ* hybridization of segmented filamentous bacteria in the intestine of diarrheic rainbow trout (*Oncorhynchus mykiss*). *Res. Microbiol., 152*, 67–73.

Valderrama, K., Saravia, M., & Santander, J., (2017). Phenotype of *Aeromonas salmonicida* sp. salmonicida cyclic adenosine 3',5'-monophosphate receptor protein (Crp) mutants and its virulence in rainbow trout (*Oncorhynchus mykiss*). *J Fish Dis., 40*, 1849–1856. doi: 10.1111/jfd.12658.

Vatsos, I. N., Thompson, K. D., & Adams, A., (2001). Adhesion of the fish pathogen *Flavobacterium psychrophilum* to unfertilized eggs of rainbow trout (*Oncorhynchus mykiss*) and n-hexadecane. *Lett. Appl. Microbiol., 33*(3), 178–182.

Villumsen, K. R., Neumann, L., Ohtani, M., Strøm, H. K., & Raida, M. K., (2014). Oral and anal vaccination confers full protection against enteric redmouth disease (ERM) in rainbow trout. *PLoS One,* 9, e93845.

Wahli, T., & Madsen, L., (2018). *Flavobacteria*, a never-ending threat for fish: A review. *Curr. Clin. Microbiol. Rep., 5*, 26–37.

Wakabayashi, H., (1980). Bacterial gill disease of salmonid fish. *Fish Pathol., 14*, 185–189.

Wakabayashi, H., Huh, G. J., & Kimura, N., (1989). *Flavobacterium branchiophila* sp. Nov., a causative agent of bacterial gill disease of freshwater fishes. *Int. J. Syst. Evol. Microbiol., 39*, 213–216.

Wakabayashi, H., Toyama, T., & Iida, T., (1994). A study on serotyping of *Cytophaga psychrophila* isolated from fishes in Japan. *Fish Pathology, 29*, 101–104.

Wald, M., Schwarz, K., Rehbein, H., Bumann, B., & Beermann, C., (2016). Detection of antibacterial activity of an enzymatic hydrolysate generated by processing rainbow trout by-products with trout pepsin. *Food Chem., 205*, 221–228.

Warren, J. W., Abrutyn, E., Hebel, J. R., Johnson, J. R., Schaeffer, A. J., & Stamm, W. E., (1999). Guidelines for antimicrobial treatment of uncomplicated acute bacterial cystitis and acute pyelonephritis in women. *Clinical Infectious Diseases, 29*(4), 745–759.

Webster, C. D., & Thompson, K. R., (2015). Protein, amino acids, and ingredients. *Dietary Nutrients, Additives and Fish Health*, 25–45.

Wiens, G. D., (2011). Bacterial kidney disease (*Renibacterium salmoninarum*). *Fish Diseases and Disorders, 3*, 338–374.

Wiklund, T., Madsen, L., Bruun, M. S., & Dalsgaard, I., (2000). Detection of *Flavobacterium psychrophilum* from fish tissue and water samples by PCR amplification. *J. Appl. Microbiol., 88*, 299–307.

Wolf, K. E., & Dunbar, C. E., (1959). Test of 34 therapeutic agents for control of kidney disease in trout. *Trans. Am. Fish. Sot. 88*, 117.124.

Wood, E. M., & Yasutake, W. T., (1956). Histopathology of fish: III. peduncle (coldwater) disease. *Progress Fish Cult., 18*, 58–61.

Yang, X., Guo, J. L., Ye, J. Y., Zhang, Y. X., & Wang, W., (2015). The effects of *Ficus carica* polysaccharide on immune response and expression of some immune-related genes in grass carp, *Ctenopharyngodon idella*. *Fish Shellfish Immunol., 42*, 132–137.

Yin, Y., Wang, Y., Zhu, L., Liu, W., Liao, N., Jiang, M., Zhu, B., et al., (2013). Comparative analysis of the distribution of segmented filamentous bacteria in humans, mice and chickens. *ISME J., 7*, 615–621.

Young, C. L., & Chapman, G. B., (1978). Ultrastructural aspects of the causative agent and renal histopathology of bacterial kidney disease in brook trout (*Salvelinus fontinalis*). *J. Fish. Res., 35*, 1234–1248.

Zhang, Z., Niu, C., Storset, A., Bøgwald, J., & Dalmo, R. A., (2011). Comparison of *Aeromonas salmonicida* resistant and susceptible salmon families: A high immune response is beneficial for the survival against *Aeromonas salmonicida* challenge. *Fish Shellfish Immunol., 31*, 1–9.

CHAPTER 5

Biotechnological Interventions in Coldwater Aquaculture Health Management

MD. IDRISH RAJA KHAN

Department of Aquatic Health and Environment, College of Fisheries, Central Agricultural University (I), Lembucherra, Tripura, India

ABSTRACT

Over the years, owing to the adoption of scientific practices aquaculture industry has flourished and achieved new horizons of production; however, with progression, fish also became vulnerable to an array of pathogens, including aquatic organisms from cold water environment. Concurrently, the fish health management approaches have also advanced then also the episode of disease outbreaks are still frequent and widespread, and leading to fettered aquaculture production. The biotechnological approach has shown promising responses either to improve the health status of cultured organisms or to strengthen the diagnosis of diseases while safeguarding the aquatic environment. Such biotechnological interventions led to avoidance of the pathogen by strengthening the immune response of fish as well as avoiding any possible use of chemical remedial agents. Recently with the development of anti-microbial resistance and the emergence of superbugs, the employment of biocontrol strategy through biotechnology-based tools have gained momentum. The present chapter is a brief account of different biotechnological approach and their application in aquaculture to augment the immune response with appropriate disease diagnosis to achieve sustainable aquaculture production.

Coldwater Fisheries and Aquaculture Management: Technology for Sustainable Food Production.
Mohd Ashraf Rather, Faisal Rashid Sofi, Adnan Amin, & Kawkabul Saba (Eds.)

5.1 INTRODUCTION

Aquaculture has arrived as one of the most flourishing agri-based commodities with a significant share in the global food basket with a noteworthy contribution to total animal protein supply, i.e., 17%, globally (FAO, 2020). When it comes to global fish production India stands at second place just after China with a significant annual growth rate of over 7%. These attainments are the collective bequest of adaptation of intensive aquaculture practices with scientific advancement. India is bestowed with vast water resources ranging from freshwater, brackish water and marine environment. Moreover, coldwater resources stretched over the Himalayan and peninsular regions also behold as a major resource to capitalize. Indian coldwater resources harboring about 272 fish species belonging to 21 families and 76 genera (Singh & Akhtar, 2015). These species may contribute to India's 2030 dream of self-sufficient for all kinds of fish products and by-products with the adoption of modern and scientific culture practices. However, the low productivity level of the aquatic systems and higher degree of resource unpredictability roots to unique challenges including pathogenic infection caused by an array of pathogens. However, fish production through advanced approaches has led more often than not irresponsible culture practice with indiscriminate application of chemical therapeutics which very often made fish vulnerable to various resistant pathogens. Such detrimental circumstances with negligence to sustainable practices often led to deprived growth, severe mortality and eventually fettered aquaculture production.

Over time, there has been significant development in animal biotechnology and its application in various fields, including aquaculture health management. Biotechnology is the practice in which biological organisms or systems are utilized for enriched production. Increased demand for fish food and declining natural available habitats have encouraged researchers to explore ways that biotechnology can offer to enhance the production and make aquaculture one of the rising avenues in animal research. Biotechnology offers a "give and take management system" where biotechnological tools can be employed for healthy aquaculture practices ranging from seed rearing to table size fish production (Vijayan et al., 2013). Biotechnology is already making important contributions to the world, and it perceives that biotechnological tools should be employed as an adjunct and not as a replacement to the conventional approaches in resolving issues, as well as their application, should be need-driven rather than technology-driven. Aquaculture sector requires novel biotechnological approaches to tackle the challenges emerging in sustainable fish production, particularly disease prevention, diagnosis, and their management in the aquatic animal culture systems. In the recent past, there are several examples of biotechnology-based tools that are employed in aquaculture health management to avoid any probable disease outbreak (Figure 5.1).

FIGURE 5.1 Biotechnology-based tools in aquaculture health management.

5.2 DISEASES–A CONSTANT THREAT FOR AQUACULTURE

Disease outbreak is often documented as a major risk factor for the growth of both fish as well as shellfish industry. Globally, numerous pathogenic micro-organisms affected the aquaculture industry and often led to catastrophic results. It has been observed that about 10% of all cultured fish die because of disease outbreaks (Plumb, 1999). Moreover, in catfish culture, about 75% of newly hatched swim-up fry die before they reach marketable size. And, in the case of rearing freshwater fish, the final survival is about 30% in which majority of loss is contributed by varied etiologies including diseases that have often gone undocumented. The frequent episode of disease outbreak is a multifarious interaction between the host, disease-causing agent and the condition of the aquatic environment. The poor management practice in intensive and semi-intensive farming favored the occurrence of disease outbreak due to varied reasons which includes high stocking density and intensive feeding regime with inadequate water exchange leading to increased stress, and ultimately making them vulnerable to an array of pathogenic agent (Kaoud, 2015). Therefore, there is a need for the employment of standard culture practices with scientific developments to minimize any potential disease outbreak.

5.3 BIOTECHNOLOGICAL TOOLS FOR HEALTH MANAGEMENT IN COLDWATER AQUACULTURE

5.3.1 VACCINES

Vaccination approach in aquaculture is a well-established and cost-effective approach to mitigate infectious diseases. A vaccine is any biological preparation that can establish or improve the immunity of the host upon administration and provide protection against a specific pathogen. The recipient's immune system develops a primary immune response following exposure to the antigen either killed, attenuated, sub-unit, recombinant, etc. (Adams & Thompson, 2006). A vaccine not only ease the severity of disease losses but also reduce the need for chemotherapeutics, leave no residues in the organisms and does not lead to the emergence of resistance in the microorganisms (Subasinghe, Bondad-Reantaso, & McGladdery, 2001). An optimal vaccine must be efficient enough to rouse innate immune response and sufficient antibody titer with specific immune memory in the host for next exposure (Mishra et al., 2020). Vaccine application has arrived as the most effective and significant approach to prevent diseases outbreak in aquaculture. Presently, the policy is more effective against bacterial and viral diseases than against parasitic infections. Eventually, numerous momentous developments have been documented to develop efficient vaccines with superior efficiency and easy to administer over diverse culture environment. However, until now very few vaccines are commercially available in the market against viral and bacterial pathogenic agents.

Bacterial pathogens are the most common and significant threat to aquaculture sector with an approximate contribution of 34% to total disease outbreaks in fish culture systems (Mishra et al., 2020). Owing to the frequency of bacterial disease outbreaks and losses incurred by them, most of the commercialized vaccines are developed against bacterial pathogens. Which includes *Aeromonas salmonicida, Vibrio anguillarum V. salmonicida, Renibacterium salmoninarum, Yersinia ruckeri, Flavobacterium columnare, F. psychrophilum, Edwardsiella ictaluri Pseudimonas salmonis, Lactococcus garvieae, Streptococcus iniae, Photobacterium damselae* subsp. *Piscicida*, etc. (Haenen, 2017). There are several commercial vaccines available against viral agents as well such as infectious pancreatic necrosis virus (IPNV), infectious salmon anemia virus (ISAV), spring viremia of carp virus (SVCV), infectious hematopoietic necrosis virus (IHNV), nervous necrosis virus (NNV), viral hemorrhagic septicemia (VHS), etc.

Most of the commercial vaccines available in the market are inactivated (killed) vaccines. However, few reports suggest the failure of such vaccines, particularly against viruses, which lead the foundation for the development and adoption of live attenuated vaccines (Adams & Thompson, 2006). However, the application of live vaccines also brings several concerns that the attenuated pathogen might back-mutate and revert to the virulent nature and may lead to catastrophic results (Thompson & Adams, 2004). There are several commercially available vaccines which include attenuated (against *A. salmonicida* and *E. ictaluri* infection), killed or inactivated (against ISAV and IHNV infection), recombinant protein (against IPNV and SVCV infection), DNA vaccine (against IHNV, IPNV, VHSV, and NNV infection), synthetic peptide vaccine (VHS), etc. (Table 5.1).

5.3.2 IMMUNOSTIMULANTS

Several alternative immune-enhancing practices to avoid any possible disease outbreak has been adopted globally, among them administration of immunostimulants has gained momentum enormously as it can non-specifically strengthen the immune response of cultured organism. Immunostimulants can be defined as any natural or chemical compound, action or stressor that augments the non-specific immune response of fish and crustaceans by activating various immune cells of the organism (Mishra et al., 2020; Barman, Nen, Mandal, & Kumar, 2013). Immunostimulants protect organisms without any side effects and with no or minimal detrimental impact on the environment; however, the approach is still limited to shrimp culture practices only. Yet various immunostimulants are commercially available in the market which depicts the worth of the approach and practical utility as a substitute to improve survival response against myriads of diseases (Apines-Amar & Amar, 2015). Through studies, it has been revealed that the most established

TABLE 5.1 Biotechnology-based Tools Employed in Aquaculture

Type of Vaccine	Antigen	Commercial Name	Host Organism	Mode of Application	Remarks	References
A. Commercially Available Vaccines						
Attenuated	*Aeromonas salmonicida*	Brivax II	Rainbow trout (*Oncorhynchus mykiss*)	Intraperitoneal (IP) injection	The cellular, humoral, and mucosal immune response of host got elevated	Liu et al. (2015)
	Edwardsiella ictaluri	NA (*in vitro*)	Channel catfish (*Ictalurus punctatus*)	IP injection		Liu et al. (2015)
Killed or inactivated	Infectious pancreatic necrosis virus (IPNV)	Alpha Ject® 1000	Salmon	IP injection	Autogenous	Biering et al. (2004)
	Spring viremia of carp virus (SVCV)	Bioveta	Carps	IP injection	Easy administration.	Salgado-Miranda et al. (2013)
	A. salmonicida and *Vibrio anguillarum*	MULTIVaC, Microtek,	Salmonids	IP injection	Safe for use *V. Salmonicida*	Salgado-Miranda et al. (2013)
Vector based vaccine	Infectious salmon anemia virus (ISAV)	NA (*In vitro*)	Atlantic salmon (*Salmo salar*)	IP injection	Apoptosis was detected in the infected cells; additionally further field level efficacy must need to be performed	Wolf et al. (2012)
	Infectious hematopoietic necrosis virus (IHNV)	NA (*In vitro*)	Atlantic salmon (*S. salar*)	IP injection		Adams & Thompson (2006)
Recombinant protein	IPNV/VP2	Microtekz	Salmon	IP injection	Economically viable and safe to host. Sufficient production of immune proteins. Safe and low-cost method	De Kinkelin (1994)

TABLE 5.1 *(Continued)*

Type of Vaccine	Antigen	Commercial Name	Host Organism	Mode of Application	Remarks	References
	Spring viremia of carp virus (SVCV)	International Inc.	Carp	IP injection		Salgado-Miranda et al. (2013)
	Salmon rickettsiae	Bayovac 3.1, Pharos, S.A.	Salmonids	IP injection		Adams & Thompson (2006)
DNA vaccine	Viral hemorrhagic septicemia virus (VHSV)	NA (*in vitro*)	Salmonids	IM injection	Rouses cellular as well as humoral immunological indices in host.	Meeusen et al. (2007)
	IPNV	Novartis	Rainbow trout (*O. mykiss*)	IM injection		Ballesteros et al. (2014)
	IHNV	Aqua Health Ltd.	Salmon	IM (intramuscular) injection		La Patra et al. (2001)
	Pancreatic disease (PD)	NA (*in vitro*)	Rainbow trout (*O. mykiss*)	IM injection		Kurath (2008)
	Nervous necrosis virus (NNV)	NA (*in vitro*)	European sea bass (*Dicentrarchus labrax*)	Oral feeding		Valero et al. (2016)
Genetically attenuated pathogen	*A. salmonicida*	Brivax II	Rainbow trout (*O. mykiss*)	IP injection	Elevation in immunological response of host	Liu et al. (2015)
Synthetic peptide vaccine	Viral hemorrhagic septicemia (VHS)	NA (*in vitro*)	Rainbow trout (*O. mykiss*)	IP injection	Possibility of a vector encoding several antigens to provide multifarious protection	Coeurdacier, Laporte, & Pepin (2003)

TABLE 5.1 *(Continued)*

B. Commercially Available Immunostimulants			
Commercial Name and Company	**Contains**	**Method of Application**	**References**
Nutri-care by Nutricorp Animal Bio Solutions, India	Vitamins, extracts of herbs, enzymes, amino acids and organic minerals, probiotics, etc.	For salinity stress of below 30 ppt–10 g/kg of feed for 5 days, whereas above 30 ppt 15 g kg^{-1} of feed for 5 days	Mishra et al. (2020)
Lysozyme by Hydrochloride by Belovo, Belgium	Hen egg lysozyme	As a dietary supplement	
Selenium Yeast by Alko, Finland	Selenium yeast	As feed ingredient	
Lactoferrin DMV by International, Netherland	Bovine milk lactoferrin	As feed ingredient	
Macroguard Biotec by Mackzymal, Norway	Yeast β (1,6) and β (1,3) glucan	With vaccines through injection; through-feed for 6–8 weeks continuously	
DS 1999 by International Aquaculture Biotechnologies, Ltd.	Bacterin	Direct incorporation in the culture medium	
Levamisole by Janssen Pharmaceutica, Belgium	Tetrahydro-6-phenylimidazolthiazole hydrochloride	As a dietary supplement	
Immustim by Immundyne, USA	β(1,6) branched β(1,3) glucan from yeast	Immersion for larvae/PLs and feed ingredient for grow-out shrimp	

C. Commercially Available Probiotics			
Commercial Name and Composition	**Indications or Result**	**Dose**	**References**
AQUAGEN PRO (probiotic strains)	Induces ammonia-oxidation	2-to-3-liter $acre^{-1}$ in 3 to 5 feet depth	Mishra et al. (2020)
ALGUTPRO (active/inactive forms of *Saccharomyces* spp., *Candida* spp. and several growth enhancers)	Enhanced proliferation of beneficial microorganisms	2 to 3 kg per ton feed	

TABLE 5.1 *(Continued)*

Commercial Name and Composition	Indications or Result	Dose	References
SUPERGUT (combination of multi-strain of probiotics along with seaweed extract and molasses)	Limit the proliferation of pathogenic bacteria and maintain a balance of gut inhabiting microflora	10 to 20 ml kg^{-1} feed	
PROBAC-G (*Cerevisiae, Lactobacillus* spp., *B. subtilis*, fungal diastase, vitamin B12, pepsin, papain, etc.)	Improve feed conversion ratio (FCR) and provide resistance against diseases	500 g per ton of feed	
Pro marine (probiotics mixed with vitamin C and calcium)	Eliminates harmful pathogens and enhances growth	5 g per kg of feed twice daily	
Thiomax (probiotics and micronutrients)	Improve the pond bottom environment	2 to 3 kg per acre	
Spark-PS (probiotic mixture)	Produces various hydrolytic enzymes to digest complex organic substances	2 to 3 liter per hectare	
VIBRION (probiotic strains, lipase, alkaline protease, etc.)	Eliminates harmful pathogens like *Vibrio* spp.	250 to 500 g per acre for 7 to 10 days	
Environ (mixture of probiotics)	Improves aquatic environment for healthy growth	250 to 500 g per acre	
GENTECH P.S. (probiotics)	Facilitates degradation of organic substances	5 liter per hectare	

TABLE 5.1 *(Continued)*

D. Common Prebiotic Used in Aquaculture				
Prebiotic	**Dose and Length of Administration**	**Host**	**Results**	**References**
	Oral feeding with 20 g per kg for one month	Siberian sturgeon (*Acipenser baerii*)	Increased growth rate	Mahious, Van, & Ollevier (2006)
	Oral feeding with 20 g per kg for one month	Turbot (*Psetta maxima*) larvae	Increased growth rate and proliferation of gut microbiota such as *Bacillus* spp.	Mahious et al. (2006)
	5 and 10 g per kg for one week	Gilthead seabream (*Sparus aurata* L.)	Reduces the phagocytosis and respiratory burst activity of leucocytes	Cerezuela et al. (2008)
	Oral feeding with 0.8% for 114 days	Sharpsnout sea bream (*Diplodus puntazzo*)	No significant change in growth and digestibility	Piccolo et al. (2011)
FOS	Oral feeding with 20 g per kg for one month	*P. maxima* larvae	Enhanced growth and proliferation of gut microbiota (*Bacillus* spp.)	Mahious et al. (2006)
	Oral feeding with 10 g kg for 4 months	Atlantic salmon (*Salmo salar*)	No significant effect on feed intake, growth, or digestibility	Grisdale-Helland et al. (2008)
SCFOS	0.8 or 1.2 g kg^{-1} for 8 weeks	Hybrid tilapia (*Oreochromis aureus* x *O. niloticus*)	Enhanced growth rate, FCR along with significant proliferation of beneficial bacteria such as *Lactobacillus* spp.	Hui-Yuan et al. (2007)
	1 g kg^{-1} for 56 days	Hybrid tilapia	Increased proliferation of gut endosymbionts	Zhou et al. (2009)
	Fish meal (FM) based diets enriched with 1% SCFOS for 7 weeks	European sea bass (*Dicentrarchus labrax*)	No significant change in growth and survival	Guerreiro, Oliva-Teles, & Enes (2015)
MOS	Oral feeding of 0.2% Artemia enriched by A1 DHA (docosahexaenoic acid) SelcoTM with MOS addition for 23–43 days	White sea bream (*Diplodus sargus*) larvae	No significant effect on growth, survival	Dimitroglou et al. (2011)

TABLE 5.1 *(Continued)*

D. Common Prebiotic Used in Aquaculture

Prebiotic	Dose and Length of Administration	Host	Results	References
	FM-based diets enriched with 20 g MOS @ 0.16% for 8 weeks	European sea bass (*D. labrax*)	Increased growth	Torrecillas et al. (2015)
	FM-based diets enriched with MOS @ 0.21 g 0.05%, 0.1%, 0.2%, 0.3% and 0.4% till 75 days	European sea bass	Increased growth and disease resistance	Salem et al. (2016)

E. Synbiotics Used in Aquaculture

Synbiotic Combination	Aquatic Organism	Effects	References
Enterococcus faecalis + MOS and polyhydroxybutyrate acid in different combination for 12 weeks	Rainbow trout (*Oncorhynchus mykiss*)	Increased growth and immune response	Rodriguez-Estrada et al. (2009)
B. clausii + MOS, FOS in different combination for 56 days	Japanese flounder (*Paralichthys olivaceus*)	Increased growth performance and digestibility performance	Ye, Wang, & Sun (2011)
Biomin IMBO (*E. faecium* + FOS) in different combinations for 60 days	Rainbow trout (*O. Mykiss*)	Increased growth and disease resistance capacity	Mehrabi, Firouzbakhsh, & Jafarpour (2012)
B. subtilis + FOS in different combination for 10 weeks	Juvenile large yellow croaker (*Larimichthys crocea*)	Increased growth and disease resistance capacity	Ai et al. (2011)
B. subtilis + chitosan in different combination for 8 weeks	Cobia (*Rachycentron canadum*)	Increased growth response and immune response	Geng et al. (2011)
L. plantarum 7–40 + galactooligosaccharide	Pacific white shrimp (*Litopenaeus vannamei*)	Increased growth response and digestibility	Huynh et al. (2018)
B. lincheniformis + MOS	Pacific white shrimp (*L. vannamei*)	Elevate the dietary assimilation, intestinal SCFAs content, and immune response of host to confer healthy growth	Chen et al. (2020)

TABLE 5.1 *(Continued)*

F. Paraprobiotic Application in Aquaculture				
Paraprobiotic	**Organism Under Study**	**Mode of Administration**	**Post-Administration Responses**	**References**
L. plantarum	Giant freshwater prawn (*Macrobrachium rosenbergii*)	*In vivo*	Significant rise in immune response and diseases resistance	Dash et al. (2015)
B. pumilus SE5	Orange-spotted grouper (*Epinephelus coioides*)	*In vivo*	Significant rise in immune response	Yan et al. (2016)
Pseudomonas aeruginosa VSG2	*Labeo rohita*	*In vitro*	Significant expression of immune genes and immune cell activity	Giri et al. (2016)
L. plantarum	Pacific white shrimp (*L. vannamei*)	*In vivo*	Elevation in growth, immune response, and tolerance towards salinity stress	Zheng et al. (2017)
B. amyloliquefaciens FPTB16	*Catla catla*	*In vivo*	Significant rise in immune responses	Singh et al. (2017)
Lactobacillus plantarum	Whiteleg shrimp (*Penaeus vannamei*)	*In vivo*	Positively modulated the gastrointestinal microbiota	Zheng et al. (2020)

G. Effect of Quorum Quenching on the Virulence Capacity of Pathogens						
Pathogens	**Quorum Quenching Method**	**Virulence Capacity**	**Host**	**Challenge Method**	**Bacterial Density**	**References**
V. harveyi	luxS, luxO, luxP gene mutation	Abolished	Brine shrimp artemia	Bath challenge	10^4 CFU ml^{-1}	Defoirdt et al. (2004)
A. hydrophila	ahyR gene mutation	Reduced	*Epithelioma papillosum* of carp (*Cyprinus carpio*) cells	Coculture 30 min	MOI 1	Bi et al. (2007)
			Swordtail (*Xiphophophorus helleri*)	Intraperitoneal injection (0.1 ml)	10^{4-10} CFU ml^{-1}	

TABLE 5.1 *(Continued)*

Pathogens	Quorum Quenching Method	Virulence Capacity	Host	Challenge Method	Bacterial Density	References
	ahyI gene mutation	Reduced	*Epithelioma papillosum cyprini* (EPC) cells	Coculture 30 min	MOI 1	Chu et al. (2011)
			Goldfish (*Carassius auratus gibelio*)	Intraperitoneal injection (0.1 ml)	10^{4-10} CFU ml^{-1}	
V. parahaemolyticus	opaR gene mutation	Increased	Chinese hamster ovary cells	Coculture 5 h	1.5×10^6 CFU ml^{-1}, MOI 15	Gode-Potratz & McCarter (2011)
V. alginolyticus	luxT gene mutation	Slightly reduced	Zebrafish	Intramuscular injection	10^4–10^7 CFU per fish	Liu et al. (2012)
V. harveyi HAH	Probiotic application, *Cobetia* sp.	Reduced	*P. vannamei*	*In vitro* and *in vivo* tests	Immersion @ 10^4–10^6 CFU ml^{-1}	Yu et al. (2019)
A. hydrophila	Probiotic application, *B. licheniformis* T-1	Reduced	Zebrafish	Injected intraperitoneally	2.6×10^8 CFU ml^{-1}	Chen et al. (2020)

H. Commercially Available Phage-based Products in Aquaculture

Name of the Company/Institute	Bacterial Host	Product Description	References
Intralytix	*Escherichia coli, V. tubiashii* and *V. coralliitycis*	Phage cocktail to control infections in oyster	Intralytix (2018)
Fixed Phage Ltd.	*V. parahaemolyticus*	aquaPHIX™ is a phage-based solution that binds with feed pallets applied directly in the aquaculture system	Mattey (2016)
ICAR-CIBA	*Vibrio* spp.	LUMI^{PHAGE} for biocontrol of luminescent bacterial disease (LBD) in shrimp larvae	ICAR-CIBA (2017)
Phage Biotech Ltd.	*V. harveyi*	Phage therapy to treat vibrio infections in shrimp	Phage Biotech (2017)

TABLE 5.1 *(Continued)*

Name of the Company/Institute	Bacterial Host	Product Description	References
ACD Pharma	*Yersinia ruckeri*	CUSTUS®$_{YRS}$ is phage-based solutions against yersiniosis, or enteric red mouth disease (ERM) of Atlantic salmon	ACD Pharma (2017)
Proteon pharmaceutical	*Pseudomonas* spp. and *Aeromonas* spp.	BAFADOR® is a phage-cocktail applied *via* immersion	Grzelak (2017)
Mangalore Biotech Laboratory	*Vibrio* spp.	LUMI-NIL MBL is a phage formulation to control luminous vibriosis in shrimp	Mangalore Biotech Laboratory (2019)

I. Applications of RNAi against Viral Infection in Shrimps

Target Species	Target Gene	Delivery Method	Pathogen	RNAi Inducer	References
Giant tiger prawn (*P. monodon*)	YRP65	Transfection	Yellow head virus (YHV)	*In vitro* transcribed dsRNA	Ongvarrasopone et al. (2011)
	vp15 and vp28	Injection	White spot syndrome virus (WSSV)	siRNA	Ho et al. (2011)
	ns1 and vp	Injection	Densovirus (DNV)	Bacterially expressed dsRNA	Sellars et al. (2011)
	PmRab7	Injection	Laem-Singh virus (LSNV)	Bacterially expressed dsRNA	Sudhakaran et al. (2011)
	GAV, b-actin	Oral	Gill-associated virus (GAV)	Bacterially expressed dsRNA	Posiri, Ongvarrasopone, & Panyim (2011)
Pacific white shrimp (*L. vannamei*)	Rab7	Injection	Taura syndrome virus (TSV)	Bacterially expressed dsRNA	Mejía-Ruíz et al. (2011)
	rr2 and DNApol, ORF	Injection	WSSV-TSV	*In vitro* transcribed dsRNA	Hirono et al. (2011)

TABLE 5.1 *(Continued)*

J. Antimicrobial Peptide Application in Aquaculture					
AMPs	**Administration Method**	**Pathogen**	**Infected Organism**	**Results**	**References**
CEME, cecropin-melittin	Intraperitoneal injection	*V. anguillarum*	Coho salmon (*O. kisutch*)	Increased survival percentage	Jia et al. (2000)
Synthetic hepcidin	Intraperitoneal injection	*V. anguillarum*	European bass (*D. labrax*)	Significant reduction in the mortality	Alvarez et al. (2016)
FSB-AMP	Oral	*V. parahaemolyticus*	Pacific white shrimp (*L. vannamei*)	Significant reduction in the mortality	Cheng et al. (2017)
SCY-hepc	Oral	*A. hydrophila*	Sea bream (*Sparus macrocephalus*) and hybrid grouper	Increased survival percentage	He et al. (2018)
Synthetic hepcidin	Intraperitoneal injection	*Edwarsiella tarda*	Mudskipper (*Boleophthalmus pectinirostris*)	Significant increase in survival	Chen, Nie, & Chen (2018)
Hepcidin	Oral and intraperitoneal injection	*Flavobacterium columnare*	Grass carp (*Ctenopharyngodon idellus*)	Significantly increased survival	Wei et al. (2018)

mode of action of immunostimulants is to through the proper functioning of phagocytic cells and to enhance their anti-microbial response (Barman, Nen, Mandal, & Kumar, 2013). A wide variety of compounds has been found to have immunostimulatory nature, including synthetic chemicals (FK-565, Levamisole, etc.), bacterial derivatives (β-glucan, lipopolysaccharides or LPS, EF203, etc.), polysaccharides (chitin, chitosan, schizophyllan, etc.), nutritional factors (vitamin C, vitamin E, etc.), cytokines, hormones along with several plant and animal-derived products/by-products have been extensively studied in aquaculture (Barman, Nen, Mandal, & Kumar, 2013) (Table 5.1).

5.3.3 PROBIOTICS

Probiotics are live microbial feed supplements that confer health benefits in the host organism through improved microbial balance. Colonization of healthy gut flora helps the host to resist invasion of a broad range of virulent microorganisms (Kesarcodi-Watson et al., 2008). Numerous probiotics or probiotic-derived products from different potential bacterial species are available in the market for use in aquaculture, globally. Throughout the year probiotics have established themselves as an alternative to chemical therapeutics in preventing infectious diseases outbreak in aquaculture throughout the world. There are several well established beneficial effects that have been observed with the application of probiotics in aquaculture such as overall health status of the host, increased survival percentage, bioremediation of nitrogen and phosphorus related metabolites in pond which confer improved water quality and ultimately increased yields with minimal application of chemical remedial agents (Wang, Xu, & Xia, 2005; Shefat, 2018; Khan, Choudhury, & Kamilya, 2020).

Application of probiotics in intensive and semi-intensive aquaculture practices has surfaced as one of the most appropriate and promising alternatives to meet the nutritional needs by producing a premium quality animal protein with minimal application of chemicals. In the last two decades, the exploratory research and probiotic application in aquaculture have expanded a great deal. According to recent studies majority of potential probiotic candidates belongs to 20 bacterial genera where most of the isolates were identified as members of either *Bacillus* or *Lactobacillus* genera (Khan, Choudhury, & Kamilya, 2020). Other commonly administered probiotic candidates in the aquaculture system include *Bifidobacterium* spp., *Enterococcus* spp., *Streptomyces* spp., *Carnobacterium* spp. and yeast (Van et al., 2019) (Table 5.1).

5.3.4 PREBIOTICS AND SYNBIOTIC

The concept of prebiotics is quite recent, which serves as a source of nutrient/food for probiotic microorganisms (Panigrahi & Azad, 2007). Prebiotics can be defined as non-digestible food ingredients that are metabolized by specific bacteria which selectively stimulate the proliferation of gut inhabiting symbiotic bacteria (Ringø et al., 2010; Ganguly, Dora, Sarkar, & Chowdhury, 2012). They are believed to be resistant to endogenous digestive enzymes and hence can reach the site of proliferation of microflora in the gastro-intestinal site and promote their proliferation. Primarily, prebiotics is fragments of carbohydrate or complex oligosaccharides (3–10 sugar moieties) such as galactose, fructose or mannose. Additionally, short-chain fatty acids (SCFA) were also reported as prebiotic later as they also have a positive influence on colonic health (Ringø et al., 2010; Ganguly, Dora, Sarkar, & Chowdhury, 2012). Some of the common prebiotic compounds that are presently used in feed are fructo-oligosaccharide (FOS), mannan-oligosaccharides (MOS), mixed oligo-dextran, etc. (Panigrahi & Azad, 2007). Bacterial cell surface have certain lectins (glycoproteins) that recognize specific sugars to allow attachment of cells with sugar moieties and utilize the same for multiplication and further proliferation (Panigrahi & Azad, 2007). Over the years, despite the potential benefits of prebiotic compounds in health and growth performance of aquatic animals, the application of prebiotics in aquaculture is still under-explored and requires further investigation (Table 5.1).

Gibson & Roberfroid predicted that a combination approach of probiotic with prebiotic can be adopted to achieve elevated proliferation of potential probiotic, which later termed "synbiotic" (De Vrese & Schrezenmeir, 2008). The supplementation synbiotic lead to a positive influence on feed utilization, immunomodulation, survival against varied infection with superior growth performance (Cerezuela, Meseguer, & Esteban, 2011). Intestinal probiotic microorganisms selectively ferment dietary prebiotic compounds to produce short-chain compounds which eventually enhance the bioavailability of nutrients (Breves, Sztkuti, & Schrder, 2001; Bongers, & Van, 2003). Additionally, the synbiotics application has also shown increased fermentation and production of numerous SCFA, vitamins, and peptides leading to reduced intestinal pH thereby eliminating the chances of any probable infection of certain pathogenic bacteria while triggering the augmentation of various endosymbionts (Breves, Sztkuti, & Schrder, 2001; Bongers, & Van, 2003). Furthermore, there is evidence that prebiotic augmentation can directly enhance the health status of fish by inducing the immune response (Ringø et al., 2010) (Table 5.1).

5.3.5 PARAPROBIOTICS AND POSTBIOTICS

The term paraprobiotics was proposed by Taverniti & Guglielmetti (2011) and defined as non-viable beneficial microbial cells or cell fragments or cell extracts, which upon administration confer health benefit to the host organism. The prefix '*para*' is adopted from the Greek word, which means 'atypical' or 'alongside of' (Choudhury & Kamilya, 2019). Because of their crude nature, they are also called "ghost probiotic" (Choudhury & Kamilya, 2019) whereas; Villamil et al. (2002) were the first who observed the immune augmentation potential of paraprobiotics, *Lactococcus lactis* in turbot. This study leads the foundation for future research, which revealed that paraprobiotic preparations may play a vital role in mediating growth and immune responses, resistance from various diseases, etc. Several other studies have revealed the capability of paraprobiotic in uplifting the immune response. The studies advocates for the potency of paraprobiotic to augment immunity levels significantly through enhanced myeloperoxidase content, nitric oxide (NO) production, respiratory burst, phagocytic response, proliferation of leukocytes in fish head-kidney, etc. (Villamil et al., 2002; Salinas et al., 2006; Román et al., 2012; Kamilya et al., 2015) (Table 5.1).

The term 'postbiotics' evolved recently and refers to "soluble factors such as metabolic products or by-products secreted/released by bacterial lysis" (Aguilar-Toalá et al., 2018). These compounds include different enzymes, peptides, vitamins, SCFA, and various cell surface proteins (Tsilingiri & Rescigno, 2012; Konstantinov, Kuipers, & Peppelenbosch, 2013). There are reports suggesting immune-enhancement through postbiotic, however exact mode of action is yet to be known (Kareem et al., 2016; Nawaz et al., 2018) (Table 5.1). A recent study has shed some light on the mode of action of postbiotics led immune response on host, where study has shown that dietary prebiotics and their gut microflora led fermentation radically increase the levels of SCFA including propionate, acetate, and butyrate (Hoseinifar et al., 2015). Reports also suggest that the SCFA are capable of rousing non-specific immune as well as inflammatory response in the host organism (Maslowski & Mackay, 2011).

5.3.6 QUORUM QUENCHING OR QUORUM SENSING INHIBITION

Quorum sensing is a mechanism of bacterial cell-to-cell communication, in which they produce, release, and detect small signaling molecules to coordinate the production or expression of certain proteins or genes (Defoirdt,

Boon, Bossier, & Verstraete, 2004). The mechanism was first reported in a bacterium, *V. fischeri* (Nealson, Platt, & Hastings, 1970). Later on, similar biological communication systems were also reported in many other bacteria, where they produce *Cholerae quorum-sensing* autoinducer-1 (CAI-1), autoinducer-2 (AI-2), acylated homoserine lactones (AHLs), etc., as communication molecules (Defoirdt, Boon, Bossier, & Verstraete, 2004; Nealson, Platt, & Hastings, 1970). Bacterial phenotypic characteristics that are coordinated by the quorum sensing mechanism include nodulation, antibiotic production, biocorrosion, swarming, conjugation, bioluminescence, sporulation, and most importantly expression of virulence proteins (toxins, lytic enzymes, adhesion molecules, siderophores, etc.), and biofilm formation (Defoirdt, Boon, Bossier, & Verstraete, 2004; De Kievit, & Iglewski, 2000; De Windt et al., 2003). Quorum sensing probably increases the probability of pathogen infection to their host successfully by expression and co-ordination of these virulent factors to overcome the host's immune system (Donabedian, 2003). Over the years, it has been reported that the disruption of bacterial communication through probiotic, chemical administration, gene mutation, etc., can significantly reduce the expression of virulence factors and may provide novel avenues for disease control in aquaculture (Swift et al., 1999) (Table 5.1).

5.3.7 BACTERIOPHAGE

Bacteriophages or phages are the viruses that lyse the specific bacterial host cell and release new progenies (Al-Sum & Al-Dhabi, 2014). Harnessing this precise killing capability of phages to control virulent bacterial pathogens is called 'phage therapy or phagotherapy.' Phages employ the bacterial host bio-machinery for all kinds of metabolic support such as different enzymes and proteins to survive (Al-Sum & Al-Dhabi, 2014). As the natural environment is replete with loads of bacterial hosts, the occurrence of bacteriolytic phages is natural. It can grow up to 10^7–10^8 virions g^{-1} in soil and approximately 10^7 virions ml^{-1} in water either fresh or saline environment (Park et al., 2020; Ninawe et al., 2020). The life cycle of phages can be categorized into two stages, the first is lytic (virulent) and the second, temperate (dormant). For phagotherapy, lytic phages are preferred because of their bacteriolytic activity. Although bacteriophages were discovered way back at the beginning of the 19th century, the focus on their therapeutic potential against bacterial diseases was limited because of the poor understanding of phage life cycle and phage-bacteria interactions (Almeida et al., 2009).

However, the emergence of multi-drug resistant bacteria has substantially encouraged researchers to explore the potential of phage therapy. The strain-specific lytic phages can be employed as bioagents against a wide range of bacterial pathogens. Owing to the specificity of phages, the probability of disturbing the natural microbiota of aquatic environment or host inhabiting beneficial bacteria will be null, which is very unlikely with the administration of chemical therapeutics such as common broad-spectrum antibiotics (Fortuna et al., 2008). Nowadays, work associated with phage therapy against bacterial pathogens in aquaculture has been accepted worldwide and encourages researchers to explore the isolation, application, and efficacy of phage therapy in different circumstances under various cultural conditions (Table 5.1).

5.3.8 RNA INTERFERENCE (RNAI)

RNAi is a post-transcriptional gene silencing method that is evolutionally conserved, where dsRNAs are introduced into cell to cause sequence-specific degradation of desired homologous mRNAs (Almeida & Allshire, 2005). The method is a natural defense strategy adopted by fungus, plants, and invertebrates to counter the introduction of unwanted nucleic acids (viruses, etc.). However, the technique has been also reported in several other eukaryotic organisms revealing that the mechanism corresponds to control the endogenous gene expression (Voinnet, 2002). In the RNAi machinery, the post-transcriptional activity to degrade the cytoplasmic RNAs can be adopted as key to advance towards antiviral strategy in aquaculture (Molnár et al., 2005). Recent advancements in the application of RNAi-based technology promise as a potential molecular tool for silencing particular genes involved in the expression of virulence factor or validation of potential drug targets in various organisms (Golding et al., 2006; Hong-Geller & Micheva-Viteva, 2010) (Table 5.1). In this perspective, RNAi-based silencing or molecular scissoring could help in protection from viral infection.

5.3.9 GENE THERAPY

Gene therapy is a genome editing technology for precise genetic modifications, enabling knocking-out or introduction of a specific gene by engineered nucleases (Li et al., 2019). There are four major ways of gene editing by programmed nucleases: transcription activator-like effector nucleases (TALENs), zinc finger nucleases (ZFNs), mega-nucleases, and clustered regularly interspaced short palindromic repeat-associated nuclease Cas9 (CRISPR-Cas9) (Li et al., 2019). These nucleases precisely recognize specific sites in the genome and generate DNA double-strand breaks (DSBs). Production of DSBs will lead to activation of natural DNA repair mechanism, non-homologous end joining (NHEJ) which will cause direct rejoining and ultimately lead to insertions or deletions of gene (Cox, Platt, & Zanhg, 2015). As a result of this mutation, the gene may lose undesired function or gain desired property, this particular approach offers a novel perspective in disease management in aquaculture as we can design the specific nucleases and precisely achieve predictable substitution, insertion, or deletion of desired genes (Osakabe & Osakabe, 2014) (Table 5.1).

5.3.10 ANTIMICROBIAL PEPTIDES (AMPS)

AMPs are a group of conserved oligopeptides (>5–100 amino acids (AAs)) that play a key role in innate immunity and are known for exhibiting broad-spectrum antimicrobial responses. They are cationic, which can precisely target bacterial cell membranes because of their negative charge and cause the destruction of the organism by disruption of the cellular membrane and impaired cellular functioning (Reddy, Yedery, & Aranha, 2004). Owing to their broad-spectrum anti-microbial capacity, these molecules are serving and may provide advanced alternative biodegradable anti-microbial agents in aquaculture without any concerns towards the development of anti-microbial resistance (Lai & Gallo, 2009) (Table 5.1).

5.3.11 BIOREMEDIATION

Bioremediation techniques are well established as environment-friendly, efficient, and cost-effective tools in improving the quality of the aquatic environment by reducing the waste and associated environmental damage caused by the accumulation of various organic and inorganic wastes. Unlike, terrestrial organisms, aquatic animals are comparatively more susceptible to change in their surrounding environment. The aquatic environment serves as a carrier of excretory products of aquatic animals and pollutants from nearby sources, which makes aquatic organisms more prone to diseases caused by different pathological agents under the influence of deteriorated environmental conditions. So, biotechnology-based bio-remediation approaches can resolve this circumstance to sustain the aquaculture activities. In bioremediation, the environmental pollutants are breakdown by friendly microorganisms particularly bacteria to produce a cleaner environment (Ghosh & Pal, n.d.).

There are several biotechnological approaches that can be employed for bioremediation purposes such as biosorption, biofilters, bioaugmentation, etc. Biosorption is a rapid process of elimination of toxic metals ions from wastewater by living or non-living (dried) biomass (Gavrilescu, 2009). The most common example of biosorption is through the algal mass as it is economically feasible and well effective in waste clean-up from polluted sources (Brinza, Dring, & Gavrilescu, 2007). Biological or biological filters allow swift diffusion of oxygen and provide proliferating ground for the aerobic bacteria responsible for nitrification and denitrification of various excretory products (Rijn, 1996). This type of system is commonly used in recirculatory or zero water exchange system to remediate generated waste. The captured waste is drained as thick effluents which later decomposed (up to 90%) by the bacterial microflora present in the biological filter (Rijn, 2013). Bioaugmentation deals with the incorporation of naturally isolated bacterial strains or certain enzymes or both in combination to enhance the degradation process carried out by the different indigenous bacterial flora. The most common example of augmenting agents belongs to genus *Bacillus* as they are known for the production of different enzymes to facilitate mineralization and breakdown of biomolecules (Singh, Khati, & Chauhan, 2020).

5.3.12 DIAGNOSTICS

A modern and scientific biotechnology-based diagnostic test enables early and precise identification of pathogens. The advanced epidemiology based on recent nucleotide sequencing or whole genome sequencing enabled tracing or pinpointing the origin of any infection. Over the years, numerous modern technologies are being employed for aquaculture health management in cold water practices including prevention as well as diagnostics in aquaculture. These modern approaches are based on interactions of particular biomolecules of the pathogenic agent such as nucleic acids, virulent protein, etc. Such biotechnological methods such as enzyme-linked immunosorbent assay (ELISA) and polymerase chain reaction (PCR) have become a gold standard for diagnosis and monitoring of diseases, globally. Molecular means of diagnosis have significantly transformed disease diagnostic programs. The technologies are much more reliable than conventional methods as they are fast, specific, precise, and preceding cultivation of pathogenic microorganisms especially slow-growing pathogens is not required (Subasinghe et al., 2001). A number of antibody-based approaches are now being employed in fish and shellfish diagnosis. Serology based methods, used for screening of particular antibodies, have been a valuable indicator of previous exposure of a specific pathogen and are being regularly used in clinical science (Campbell & Landry, 2006). The major advantage of serology is the rapid and precise detection which is not possible in culture-based or any other method (Kaoud, 2015). Over the years, the advancement in molecular methods has also led to the creation of global databases or repositories that allow the verification and exchange of molecular data throughout the world which has further strengthened the diagnosis as well as providing a global platform for the identification of pathogens.

5.4 CONCLUSION

Aquaculture sector including cold water fisheries is suffering from severe loss caused by diseases of diverse origins. Several strategies either prophylactic or therapeutic were adopted to resolve the issue; however, the problem remains the same. On the other side, biotechnology is growing rapidly and has endowed us with several new tools and technology that availed capability to create new horizons in aquaculture, including health management. Biotechnological research are rising at a very fast rate and have gained vast importance in disease management in recent time. Some of the advanced biotechnological approaches such as vaccination, anti-microbial peptides, gene editing, etc., have shown promising results in managing the health of cultured organisms over different agro-climatic conditions, globally. Biotechnology has enabled us with such tools and power which we can precisely manipulate the genes and can create novel genotypes as per requirement. However, there is a lot of scope in the optimization and commercialization of these methodologies at different culture conditions so that these technologies can be developed into farmer-friendly tool.

KEYWORDS

- **aquaculture**
- **biotechnology**
- **diagnosis**
- **immune response**
- **infectious pancreatic necrosis virus**
- **nervous necrosis virus**
- **short-chain fatty acids**

REFERENCES

ACD Pharma, (2017). *Bacteriophage Therapy.* ACD Pharma. https://acdpharma.com/ (accessed 08 February 2023).

Adams, A., & Thompson, K. D., (2006). Biotechnology offers revolution to fish health management. *Trends Biotechnol., 24*, 201–205.

Aguilar-Toalá, J. E., Garcia-Varela, R., Garcia, H. S., Mata-Haro, V., González-Córdova, A. F., Vallejo-Cordoba, B., & Hernández-Mendoza, A., (2018). Postbiotics: An evolving term within the functional foods field. *Trends Food Sci. Technol., 75*, 105–114.

Ai, Q., Xu, H., Mai, K., Xu, W., Wang, J., & Zhang, W., (2011). Effects of dietary supplementation of *Bacillus subtilis* and fructo oligosaccharide on growth performance, survival, non-specific immune response and disease resistance of juvenile large yellow croaker, *Larimichthys crocea. Aquaculture, 317*, 155–161.

Almeida, A., Cunha, A., Gomes, N. C. M., Alves, E., Costa, L., & Faustino, M. A. F., (2009). Phage therapy and photodynamic therapy: Low environmental impact approaches to inactivate microorganisms in fish farming plants. *Mar. Drugs, 7*, 268–313.

Almeida, R., & Allshire, R. C., (2005). RNA silencing and genome regulation. *Trends Cell Biol., 15,* 251–258.

Al-Sum, A. B., & Al-Dhabi, N. A., (2014). Isolation of bacteriophage from Mentha species in Riyadh, Saudi Arabia. *J. Pure Appl. Micro., 8*, 945–949.

Alvarez, C., Acosta, F., Montero, D., Guzmán, F., Torres, E., Vega, B., & Mercado, L., (2016). Synthetic hepcidin from fish: Uptake and protection against *Vibrio anguillarum* in sea bass (*Dicentrarchus labrax*). *Fish Shellfish Immunol., 55*, 662–670.

Apines-Amar, M. J. S., & Amar, E. C., (2015). Use of immunostimulants in shrimp culture: An update. *Biotechnological Advances in Shrimp Health Management in the Philippines*, 45–71.

Ballesteros, N. A., Rodriguez Saint-Jean, S., & Perez-Prieto, S. I., (2014). Food pellets as an effective delivery method for a DNA vaccine against infectious pancreatic necrosis virus in rainbow trout (*Oncorhynchus mykiss*, Walbaum). *Fish Shellfish Immunol., 37*, 220–228.

Barman, D., Nen, P., Mandal, S. C., & Kumar, V., (2013). Immunostimulants for aquaculture health management. *J. Marine Sci. Res. Dev., 3*, 1–11.

Bi, Z. X., Liu, Y. J., & Lu, C. P., (2007). Contribution of AhyR to virulence of *Aeromonas hydrophila* J-1. *Res. Vet. Sci., 83*, 150–156.

Biering, E., Villoing, S., Sommerset, I., & Christie, K., (2004). Update on viral vaccines for fish. *Dev. Biol., 121*, 97–113.

Bongers, A., & Van, D. H. E., (2003). Prebiotics and the bioavailability of mineral and trace elements. *Food Rev. Int., 19*, 397–422.

Breves, G., Sztkuti, L., & Schrder, B., (2001). Effects on oligosaccharides on functional parameters of the intestinal tract of growing pigs. *Deutsche Tierarztliche Wochenschrifte, 108*, 246–248.

Brinza, L., Dring, M. J., & Gavrilescu, M., (2007). Marine micro and macro algal species as biosorbents for heavy metals. *Environ. Eng. Manag. J., 6*, 237–251.

Campbell, S., & Landry, M. L., (2006). Rapid antigen tests. In: Tang, Y. W., & Stratton, C. W., (eds.), *Advanced Techniques in Diagnostic Microbiology*. Springer Science1Business Media, LLC, New York, NY.

Cerezuela, R., Cuesta, A., Meseguer, J., & Esteban, M. A., (2008). Effects of inulin on gilthead seabream (*Sparus aurata* L.) innate immune parameters. *Fish Shellfish Immunol., 24*, 663–668.

Cerezuela, R., Meseguer, J., & Esteban, M., (2011). Current knowledge in synbiotic use for fish aquaculture: A review. *J. Aquac. Res. Development, 1*, 1–7.

Chen, B., Peng, M., Tong, W., Zhang, Q., & Song, Z., (2020). The quorum quenching bacterium *Bacillus licheniformis* T-1 protects zebrafish against *Aeromonas hydrophila* infection. *Probiotics Antimicrob. Proteins, 12*, 160–171.

Chen, J., Nie, L., & Chen, J., (2018). Mudskipper (*Boleophthalmus pectinirostris*) hepcidin-1 and hepcidin-2 present different gene expression profile and antibacterial activity and possess distinct protective effect against *Edwardsiella tarda* infection. *Probiotics Antimicrob. Proteins, 10*, 176–185.

Chen, M., Chen, X. Q., Tian, L. X., Liu, Y. J., & Niu, J., (2020). Beneficial impacts on growth, intestinal health, immune responses and ammonia resistance of pacific white shrimp (*Litopenaeus vannamei*) fed dietary synbiotic (mannan oligosaccharide and *Bacillus licheniformis*), *Aquacult. Rep., 17*, 100408.

Cheng, A. C., Lin, H. L., Shiu, Y. L., Tyan, Y. C., & Liu, C. H., (2017). Isolation and characterization of antimicrobial peptides derived from *Bacillus subtilis* E20-fermented soybean meal and its use for preventing *Vibrio* infection in shrimp aquaculture. *Fish Shellfish Immunol., 67*, 270–279.

Choudhury, T. G., & Kamilya, D., (2019). Paraprobiotics: An aquaculture perspective. *Rev. Aquac., 11*, 1258–1270.

Chu, W., Yan, J., Liu, Y., & Wei, Z., (2011). Role of the quorum-sensing system in biofilm formation and virulence of *Aeromonas hydrophila*. *Afr. J. Microbiol. Res., 5*, 5819–5825.

Coeurdacier, J. L., Laporte, F., & Pepin, J. F., (2003). Preliminary approach to find synthetic peptides from nodavirus capsid potentially protective against sea bass viral encephalopathy and retinopathy. *Fish. Shellfish. Immunol., 14*, 435–447.

Cox, D. B. T., Platt, R. J., & Zanhg, F., (2015). Therapeutic genome editing: Prospects and challenges. *Nat. Med., 21*, 121–131.

Dash, G., Raman, R. P., Prasad, K. P., Makesh, M., Pradeep, M. A., & Sen, S., (2015). Evaluation of paraprobiotic applicability of *Lactobacillus plantarum* in improving the immune response and disease protection in giant freshwater prawn, *Macrobrachium rosenbergii* (de Man, 1879). *Fish Shellfish Immunol., 43*, 167–174.

De Kievit, T. R., & Iglewski, B. H., (2000). Bacterial quorum sensing in pathogenic relationships. *Infect, Immun., 68*, 4839–4849.

De Kinkelin, P., (1994). A recombinant viral haemorrhagic septicaemia virus glycoprotein expressed in insect cells induces protective immunity in rainbow trout. *J. Gen. Virol., 75*, 1579–1587.

De Vrese, M., & Schrezenmeir, J., (2008). Probiotics, prebiotics, and synbiotics. In: Stahl, U., Donalies, U. E. B., & Nevoigt, E., (eds.), *Food Biotechnology: Advances in Biochemical Engineering/Biotechnology* (pp. 1–66). Springer Berlin Heidelberg, Berlin, Heidelberg.

De Windt, W., Boon, N., Siciliano, S. D., & Verstraete, W., (2003). Cell density related H-2 consumption in relation to anoxic Fe(0) corrosion and precipitation of corrosion products by *Shewanella oneidensis* MR-1. *Environ. Microbiol., 5*, 1192–1202.

Defoirdt, T., Boon, N., Bossier, P., & Verstraete, W., (2004). Disruption of bacterial quorum sensing: An unexplored strategy to fight infections in aquaculture. *Aquaculture, 240*, 69–88.

Defoirdt, T., Boon, N., Bossier, P., & Verstraete, W., (2004). Disruption of bacterial quorum sensing: An unexplored strategy to fight infections in aquaculture. *Aquaculture, 240*, 69–88.

Dimitroglou, A., Merrifield, D. L., Carnevali, O., Picchietti, S., Avella, M., Daniels, C., C., Güroy, D., & Davies, S. J., (2011). Microbial manipulations to improve fish health and production – a Mediterranean perspective. *Fish & Shellfish Immunology, 30*, 1–16.

Donabedian, H., (2003). Quorum sensing and its relevance to infectious disease. *J. Infect., 46*, 207–214.

FAO (Food and Agriculture Organization of the United Nations), (2020). *The State of World Fisheries and Aquaculture 2020—Sustainability in Action*. FAO: Rome, Italy.

Fortuna, W., Miedzybrodzki, R., Weber-Dabrowska, B., & Gorski, A., (2008). Bacteriophage therapy in children: Facts and prospects. *Med. Sci. Monit., 14*, RA126–RA132.

Ganguly, S., Dora, K. C., Sarkar, S., & Chowdhury, S., (2012). Supplementation of prebiotics in fish feed: A review. *Rev. Fish Biol. Fish., 23*, 195–199.

Gavrilescu, M., (2009). Environmental biotechnology: Achievements, opportunities and challenges. *Dyn Biochem Process Biotechnol. Mol. Biol., 4*, 1–36.

Geng, X., Dong, X. H., Tan, B. P., Yang, Q. H., Chi, S. Y., Liu, H. Y., & Liu, X. Q., (2011). Effects of dietary chitosan and *Bacillus subtilis* on the growth performance, non-specific immunity and disease resistance of cobia, *Rachycentron canadum*. *Fish Shellfish Immunol. 31*, 400–406.

Ghosh, S., & Pal, P. (2019). *Application of Biotechnology in Aquaculture*. Editorial Board, 9.

Giri, S. S., Sen, S. S., Jun, J. W., Park, S. C., & Sukumaran, V., (2016). Heat killed whole-cell products of the probiotic *Pseudomonas aeruginosa* VSG2 strain affect *in vitro* cytokine expression in head kidney macrophages of *Labeo rohita*, *Fish Shellfish Immunol., 50*, 310–316.

Gode-Potratz, C. J., & McCarter, L. L., (2011). Quorum sensing and silencing in *Vibrio parahaemolyticus*. *J. Bacteriol. Res., 193*, 4224–4237.

Golding, M. C., Long, C. R., Carmell, M. A., Hannon, G. J., & Westhusin, M. E., (2006). Suppression of prion protein in livestock by RNA interference. *Proc. Natl. Acad. Sci. U S A, 103*, 5285–5290.

Grisdale-Helland, B., Helland, S. J., & Gatlin, D. M. III., (2008). The effects of dietary supplementation with mannanoligosaccharide, fructooligosaccharide or galsctooligosaccharide on the growth and feed utilization of Atlantic salmon (*Salmo salar* L.). *Aquaculture, 283*, 163–167.

Grzelak, J., (2017). *BAFADOR® Presented at the International Bacteriophage Conference in Tbilisi.*

Guerreiro, I., Oliva-Teles, A., & Enes, P., (2015). Improved glucose and lipid metabolism in European sea bass (*Dicentrarchus labrax*) fed short-chain fructooligosaccharides and xylooligosaccharides. *Aquaculture, 441*, 57–63.

Haenen, O., (2017). *Major Bacterial Diseases Affecting Aquaculture* (Vol. 1, pp. 10, 11). In Aquatic AMR Workshop, Mangalore, India.

He, Y., Peng, H., Liu, J., Chen, F., Zhou, Y., Ma, X., Chen, H., & Wang, K., (2018). *Chlorella* sp. transgenic with Scy-hepc enhancing the survival of *Sparus macrocephalus* and hybrid grouper challenged with *Aeromonas hydrophila*. *Fish Shellfish Immunol., 73*, 22–29.

Hirono, I., Fagutao, F., Kondo, H., & Aoki, T., (2011). Uncovering the mechanisms of shrimp innate immune response by RNA interference. *Mar. Biotechnol., 13*, 622–628.

Ho, T., Yasri, P., Panyim, S., & Udomkit, A., (2011). Double-stranded RNA confers both preventive and therapeutic effects against *Penaeus stylirostris* Densovirus (PstDNV) in *Litopenaeus vannamei*. *Virus Res., 155*, 131–136.

Hong-Geller, E., & Micheva-Viteva, S., (2010). Functional gene discovery using RNA interference-based genomic screens to combat pathogen infection. *Curr. Drug Discov. Technol., 7*, 86–94.

Hoseinifar, S. H., Esteban, M. A., Cuesta, A., & Sun, Y. Z., (2015). Prebiotics and fish immune response: A review of current knowledge and future perspectives. *Rev. Fish. Sci. Aquacult., 23,* 315–328.

Hui-Yuan, L., Zhigang, Z., Rudeaux, F., & Respondek, F., (2007). Effects of dietary short chain fructo-oligosaccharides on intestinal microflora, mortality and growth performance of *Oreochromis aureus* x *O. niloticus*. *Chinese J. Anim. Nutr., 19*, 1–6.

Huynh, T. G., Chi, C. C., Nguyen, T. P., Tran, T. T. T. H., Cheng, A. C., & Liu, C. H., (2018). Effects of synbiotic containing *Lactobacillus plantarum* 7–40 and galactooligosaccharide on the growth performance of white shrimp, *Litopenaeus vannamei*. *Aquacult. Res., 49*, 2416–2428.

ICAR-CIBA, (2017). http://www.ciba.res.in/images/aquaticdiv/adv/EHP,WSSV,%20 Lumiphage%202.pdf (accessed on 3 February 2023).

Intralytix, I., (2018). *Intralytix*, Inc. http://www.intralytix.com/ (accessed on 3 February 2023).

Jia, X., Patrzykat, A., Devlin, R. H., Ackerman, P. A., Iwama, G. K., & Hancock, R. E. W., (2000). Antimicrobial peptides protect coho salmon from *Vibrio anguillarum* infections. *Appl. Environ. Microbiol., 66*, 1928–1932.

Kamilya, D., Baruah, A., Sangma, T., Chowdhury, S., & Pal, P., (2015). Inactivated probiotic bacteria stimulate cellular immune responses of catla, *Catla catla* (Hamilton) *in vitro*. *Probiotics Antimicrob. Proteins, 7*, 101–106.

Kaoud, H. A., (2015). Advanced technology for the diagnosis of fish diseases. *Eur. J. Acad. Essays, 2*, 27–36.

Kareem, K., Loh, T., Foo, H., Asmara, S., & Akit, H., (2016). Influence of postbiotic RG14 and inulin combination on cecal microbiota, organic acid concentration, and cytokine expression in broiler chickens. *Poultry Science, 96*, 966–975.

Kesarcodi-Watson, A., Kaspar, H., Lategan, M. J., & Gibson, L., (2008). Probiotics in aquaculture: The need, principles and mechanisms of action and screening processes. *Aquaculture, 274*, 1–14.

Khan, M. I. R., Choudhury, T. G., & Kamilya, D., (2020). Probiotics in finfish aquaculture: An Indian perspective. *Indian J. Anim. Hlth., 59*, 13–26.

Konstantinov, S. R., Kuipers, E. J., & Peppelenbosch, M. P., (2013). Functional genomic analyses of the gut microbiota for CRC screening. *Nat. Rev. Gastroenterol. Hepatol., 10*, 741.

Kurath, G., (2008). Biotechnology and DNA vaccines for aquatic animals. *Rev. Sci. Tech. Off. Int. Epizoot., 27,* 175.

La Patra, S. E., Corbeil, S., Jones, G. R., Shewmaker, W. D., Lorenzen, N., Anderson, E. D., & Kurath, G., (2001). Protection of rainbow trout against infectious hematopoietic necrosis virus four days after specific or semi-specific DNA vaccination, *Vaccine, 19*, 4011–4019.

Lai, Y., & Gallo, R. L., (2009). AMPed up immunity: How antimicrobial peptides have multiple roles in immune defense. *Trends Immunol., 30*, 131–141.

Lawrence, M. L., & Banes, M. M., (2005). Tissue persistence and vaccine efficacy of an O polysaccharide mutant strain of *Edwardsiella ictaluri*. *J. Aquat. Anim. Health, 17*, 228–232.

Li, Q., Qin, Z., Wang, Q., Xu, T., Yang, Y., & He, Z., (2019). Applications of genome editing technology in animal disease modeling and gene therapy. *Comput. Struct. Biotechnol. J., 17*, 689–698.

Liu, H., Gu, D., Cao, X., Liu, Q., Wang, Q., & Zhang, Y., (2012). Characterization of a new quorum sensing regulator luxT and its roles in the extracellular protease production, motility, and virulence in fish pathogen *Vibrio alginolyticus*. *Arch. Microbiol., 194*, 439–452.

Liu, X., Wu, H., Liu, Q., Wang, Q., Xiao, J., Chang, X., & Zhang, Y., (2015). Profiling immune response in zebrafish intestine, skin, spleen and kidney bath-vaccinated with a live attenuated *Vibrio anguillarum* vaccine. *Fish Shellfish Immunol., 18*, 235–242.

Mahious, A. S., Gatesoupe, F. J., Hervi, M., Metailler, R., & Ollevier, F., (2006). Effect of dietary inulin and oligosaccharides as prebiotics for weaning turbot, *Psetta maxima* (Linnaeus, C. 1758). *Aquacult. Int., 14*, 219–229.

Mahious, A. S., Van, L. J., & Ollevier, F., (2006). Impact of the prebiotics, inulin and oligofructose on microbial fermentation in the spiral valve of Siberian sturgeon (*Acipenser baerii*). *J. World Aquac. Soc.*, 564–565.

Mangalore Biotech Laboratory, (2019). *Mangalore Biotech Lab: Products*. http://mangalore-biotech.com/products.html (accessed on 3 February 2023).

Maslowski, K. M., & Mackay, C. R., (2011). Diet, gut microbiota and immune responses. *Nat. Immunol., 12*, 5–9.

Mattey, M, (2016). *Treatment of Bacterial Infections in Aquaculture.* United States Patent Mattey. Patent No.: US 10,849,942 B2.

Meeusen, E. N., Walker, J., Peters, A., Pastoret, P. P., & Jungersen, G., (2007). Current status of veterinary vaccines, *Clin. Microbiol. Rev., 20*, 489–510.

Mehrabi, Z., Firouzbakhsh, F., & Jafarpour, A., (2012). Effects of dietary supplementation of synbiotic on growth performance, serum biochemical parameters and carcass composition in rainbow trout (*Oncorhynchus mykiss*) fingerlings. *J. Anim. Physiol. Anim. Nutr., (Berl), 96*, 474–481.

Mejía-Ruíz, C. H., Vega-Peña, S., Alvarez-Ruiz, P., & Escobedo-Bonilla, C. M., (2011). Double stranded RNA against white spot syndrome virus (WSSV) vp28 or vp26 reduced susceptibility of *Litopenaeus vannamei* to WSSV, and survivors exhibited decreased susceptibility in subsequent re-infections. *J. Invertebr. Pathol., 107*, 65–68.

Mishra, S. S., Das, R., Sahoo, S. N., & Swain, P., (2020). Biotechnological tools in diagnosis and control of emerging fish and shellfish diseases. In: *Genomics and Biotechnological Advances in Veterinary, Poultry, and Fisheries* (pp. 311–360). Academic Press.

Molnár, A., Csorba, T., Lakatos, L., Várallyay, É., Lacomme, C., & Burgyán, J., (2005). Plant virus-derived small interfering RNAs originate predominantly from highly structured single-stranded viral RNAs. *J. Virol., 79*, 7812–7818.

Nawaz, A., Bakhsh, J. A., Irshad, S., Hoseinifar, S. H., & Xiong, H., (2018). The functionality of prebiotics as immunostimulant: Evidence from trials on terrestrial and aquatic animals. *Fish Shellfish Immunol., 76*, 272–278.

Nealson, K. H., Platt, T., & Hastings, J. W., (1970). Cellular control of the synthesis and activity of the bacterial luminescence system. *J. Bacteriol., 104*, 313–322.

Ninawe, A. S., Sivasankari, S., Ramasamy, P., Kiran, G. S., & Selvin, J., (2020). Bacteriophages for aquaculture disease control. *Aquac. Int., 28*, 1925–1938.

Ongvarrasopone, C., Saejia, P., Chanasakulniyom, M., & Panyim, S., (2011). Inhibition of Taura syndrome virus replication in *Litopenaeus vannamei* through silencing the LvRab7 gene using double-stranded RNA. *Arch. Virol., 156*, 1117–1123.

Osakabe, Y., & Osakabe, K., (2014). Genome editing with engineered nucleases in plants. *Plant Cell Physiol. 56*, 389–400.

Panigrahi, A., & Azad, I. S., (2007). Microbial intervention for better fish health in aquaculture: The Indian scenario. *Fish Physiol. Biochem., 33*, 429–440.

Park, S. Y., Han, J. E., Kwon, H., Park, S. C., & Kim, J. H., (2020). Recent insights into *Aeromonas salmonicida* and its bacteriophages in aquaculture: A comprehensive review. *J. Microbiol. Biotechnol., 30*, 1443–1457.

Phage Biotech, (2017). *Developments*. Phage Biotech Ltd.

Piccolo, G., Centoducati, G., Marono, S., Bovera, F., Tudisco, R., & Nizza, A., (2011). Effects of the partial substitution of fish meal by soy bean meal with or without mannanoligosaccharide and fructooligosaccharide on the growth and feed utilization of sharpsnout seabream, *Diplodus puntazzo* (Cetti, 1777): Preliminary results. *Ital. J. Anim. Sci. 10*, 195–199.

Plumb, J. A., (1999). *Health Maintenance and Principal Microbial Diseases of Cultured fishes*. Iowa State University Press. Ames, IO, USA.

Posiri, P., Ongvarrasopone, C., & Panyim, S., (2011). Improved preventive and curative effects of YHV infection in *Penaeus monodon* by a combination of two double stranded RNAs. *Aquaculture, 314*, 34–38.

Reddy, K. V. R., Yedery, R. D., & Aranha, C., (2004). Antimicrobial peptides: Premises and promises. *Int. J. Antimicrob. Agents 24*, 536–547.

Rijn, J. V., (1996). The potential for integrated biological treatment system in recirculating fish culture—A review. *Aquaculture, 139*, 181–201.

Rijn, J. V., (2013). Waste treatment in recirculating aquaculture systems. *Aquac. Engi., 53*, 49–56.

Ringø, E., Olsen, R. E., Gifstad, T. Ø., Dalmo, R. A., Amlund, H., Hemre, G. I., & Bakke, A. M., (2010). Prebiotics in aquaculture: A review. *Aquac. Nutr., 16*, 117–136.

Rodriguez-Estrada, U., Satoh, S., Haga, Y., Fushimi, H., & Sweetman, J., (2009). Effects of single and combined supplementation of *Enterococcus faecalis*, mannanoligosaccharide and polyhydrobutyric acid on growth performance and immune response of rainbow trout *Oncorhynchus mykiss*. *Aquacult. Sci., 57*, 609–617.

Román, L., Real, F., Sorroza, L., Padilla, D., Acosta, B., Grasso, V., Bravo, J., & Acosta, F., (2012). The *in vitro* effect of probiotic *Vagococcus fluvialis* on the innate immune parameters of *Sparus aurata* and *Dicentrarchus labrax*. *Fish Shellfish Immunol., 33*, 1071–1075.

Salem, M., Gaber, M. M., Zaki, M., & Nour, A. A., (2016). Effects of dietary mannan oligosaccharides on growth, body composition and intestine of the sea bass (*Dicentrarchus labrax* L.). *Aquacult. Res., 47*, 3516–3525.

Salgado-Miranda, C., Loza-Rubio, E., Rojas-Anaya, E., & Garcıa-Espinosa, G., (2013). Viral vaccines for bony fish: Past, present and future. *Expert. Rev. Vaccine, 12*, 567–578.

Salinas, I., Díaz-Rosales, P., Cuesta, A., Meseguer, J., Chabrillón, M., & Morinigo, M. A., (2006). Esteban, Effect of heat-inactivated fish and non-fish derived probiotics on the innate immune parameters of a teleost fish (*Sparus aurata* L.). *Vet. Immunol. Immunopathol., 111*, 279–286.

Sellars, M. J., Rao, M., Arnold, S. J., Wade, N. M., & Cowley, J. A., (2011). *Penaeus monodon* is protected against gill-associated virus by muscle injection but not oral delivery of bacterially expressed dsRNAs. *Dis. Aquat. Org., 95*, 19–30.

Shefat, S. H. T., (2018). Probiotic strains used in aquaculture. *International Research Journal of Microbiology, 7*, 43–55.

Singh, A. K., & Akhtar, M. S., (2015). *Coldwater Fish Diversity of India and its Sustainable Development* (pp. 97–105). Biodiversity for Sustainable development Editor-in-chief Pratibha Singh Published by UP Biodiversity Board, Lucknow. International Day for Biological Diversity.

Singh, G., Khati, A., & Chauhan, R. S., (2020). Applications of environment biotechnology in aquaculture–review. *Int. J. Agric. Environ. Biotechnol., 13*, 239–247.

Singh, S. T., Kamilya, D., Kheti, B., Bordoloi, B., & Parhi, J., (2017). Paraprobiotic preparation from *Bacillus amyloliquefaciens* FPTB16 modulates immune response and immune relevant gene expression in *Catla catla* (Hamilton, 1822). *Fish Shellfish Immunol. 66*, 35–42.

Subasinghe, R. P., Bondad-Reantaso, M. G., & McGladdery, S. E., (2001). Aquaculture development, health and wealth. In: Subasinghe, R. P., Bueno, P., Phillips, M. J., Hough, C., McGladdery, S. E., & Arthur, J. R., (eds.), *Aquaculture in the Third Millennium* (pp. 167–191). Technical Proceedings of the Conference on Aquaculture in the Third Millennium, Bangkok, Thailand, 20–25 February 2000. NACA/FAO, Bangkok/Rome.

Sudhakaran, R., Mekata, T., Kono, T., Inada, M., Okugawa, S., Yoshimine, M., Yoshida, T., et al., (2011). Double-stranded RNA-mediated silencing of the white spot syndrome virus VP28 gene in kuruma shrimp, *Marsupenaeus japonicas*. *Aquac. Res., 42*, 1153–1162.

Swift, S., Lynch, M. J., Fish, L., Kirke, D. F., Tomas, J. M., Stewart, G. S. A. B., & Williams, P., (1999). Quorum sensing-dependent regulation and blockade of exoprotease production in *Aeromonas hydrophila*. *Infect. Immun., 67*, 5192–5199.

Taverniti, V., & Guglielmetti, S., (2011). The immunomodulatory properties of probiotic microorganisms beyond their viability (ghost probiotics: Proposal of paraprobiotic concept). *Genes Nutr., 6*, 261–274.

Thompson, K. D., & Adams, A., (2004). Current trends in immunotherapy and vaccine development for bacterial diseases of fish. In: Leung, K. Y., (ed.), *Molecular Aspects of Fish and Marine Biology: Current Trends in the Study of Bacterial and Viral Fish and Shrimp Diseases* (Vol. 3, pp. 313–362). World Scientific Publishing Co.

Torrecillas, S., Montero, D., Caballero, M. J., Robaina, L., Zamorano, M. J., Sweetman, J., & Izquierdo, M., (2015). Effects of dietary concentrated mannan oligosaccharides supplementation on growth, gut mucosal immune system and liver lipid metabolism of European sea bass (*Dicentrarchus labrax*) juveniles. *Fish Shellfish Immunol., 42*, 508–516.

Tsilingiri, K., & Rescigno, M., (2012). Postbiotics: What else? *Beneficial Microbes, 4*, 101–107.

Valero, Y., Awad, E., Buonocore, F., Arizcun, M., Esteban, M. A., Meseguer, J., Chaves-Pozo, E., & Cuesta, A., (2016). An oral chitosan DNA vaccine against Noda virus improves transcription of cell-mediated cytotoxicity and interferon genes in the European sea bass juveniles' gut and survival upon infection. *Dev. Comp. Immunol., 65*, 64–72.

Van, D. H., Hoseinifar, S. H., Ringø, E., Ángeles, E. M., Dadar, M., Dawood, M. A., & Faggio, C., (2019). Host-associated probiotics: A key factor in sustainable aquaculture. *Rev. Fisheries Sci. Aquacult.*, 16–42.

Vijayan, K. K., Sukumaran, S., Pradeep, M. A., Sanil, N. K., & Vijayagopal, P., (2013). *Biotechnological Approaches in Fishery Management*, 119–125.

Villamil, L. A., Tafalla, C., Figueras, A., & Novoa, B., (2002). Evaluation of immunomodulatory effects of lactic acid bacteria in turbot (*Scophthalmus maximus*). *Clin. Diagn. Lab. Immunol., 9*, 1318–1323.

Voinnet, O., (2002). RNA silencing: Small RNAs as ubiquitous regulators of gene expression. *Curr. Opin. Plant Biol., 5*, 444–451.

Wang, Y. B., Xu, Z. R., & Xia, M. S., (2005). The effectiveness of commercial probiotics in northern white shrimp *Penaeus vannamei* ponds. *Fish. Sci., 7*, 1036–1041.

Wei, X., Sarath, B. V., Lin, L., Hu, Y., Zhang, Y., Liu, X., Su, J., et al., (2018). Hepcidin protects grass carp (*Ctenopharyngodon idellus*) against *Flavobacterium columnare* infection via regulating iron distribution and immune gene expression. *Fish Shellfish Immunol., 75*, 274–283.

Wolf, A., Hodneland, K., Frost, P., Braaen, S., & Rimstad, E., (2012). A hemagglutinin esterase – expressing salmonid alphavirus replicon protects Atlantic salmon (*Salmo salar*) against infectious salmon anemia (ISA). *Vaccine, 31*, 4073–4081.

Yan, Y. Y., Xia, H. Q., Yang, H. L., Hoseinifar, S. H., & Sun, Y. Z., (2016). Effects of dietary live or heat-inactivated autochthonous *Bacillus pumilus* SE5 on growth performance, immune responses and immune gene expression in grouper *Epinephelus coioides*. *Aquacu. Nutr., 22,* 698–707.

Ye, J. D., Wang, K., Li, F. D., & Sun, Y. Z., (2011). Single or combined effects of fructoand mannan oligosaccharide supplements and *Bacillus clausii* on the growth, feed utilization, body composition, digestive enzyme activity, innate immune response and lipid metabolism of the Japanese flounder *Paralichthys olivaceus*. *Aquac. Nutr., 17*, 902–911.

Yu, P., Ye, H., Ma, S., & Shan, H., (2019). Study of quorum quenching bacteria *Cobetia* sp. reducing the mortality of *Penaeus vannamei* infected with *Vibrio harveyi*: *In vitro* and *in vivo* tests. *Aquaculture, 534*, 736244.

Zheng, X., Duan, Y., Dong, H., & Zhang, J., (2017). Effects of dietary *Lactobacillus plantarum* in different treatments on growth performance and immune gene expression of white shrimp *Litopenaeus vannamei* under normal condition and stress of acute low salinity. *Fish Shellfish Immunol., 62*, 195–201.

Zheng, X., Duan, Y., Dong, H., & Zhang, J., (2020). The effect of *Lactobacillus plantarum* administration on the intestinal microbiota of whiteleg shrimp *Penaeus vannamei*. *Aquaculture, 526*, 735331.

Zhou, Z., He, S., Liu, Y., Shi, P., Huang, G., & Yao, B., (2009). The effects of dietary yeast culture or short-chain fructo-oligosaccharides on the intestinal autochthonous bacterial communities in juvenile hybrid tilapia *Oreochromis niloticus* x *O. Aureus*. J. *World Aquac. Soc., 40*, 450–459.

CHAPTER 6

Coldwater Fish Diversity: Issues and Sustainable Management in India

UDAI RAM GURJAR[1], SUMAN TAKAR[2], KHEMRAJ BUNKAR[1], DINESH MOHAN[3], and N. N. PANDEY[3]

[1]*ICAR–Central Institute of Fisheries Education, Mumbai, Maharashtra, India*

[2]*TNJFU–Fisheries College and Research Institute, Thoothukudi, Tamil Nadu, India*

[3]*ICAR–Directorate of Coldwater Fisheries Research, Bhimtal, Uttarakhand, India*

ABSTRACT

Coldwater fish diversity is necessary for stabilizing aquatic ecosystems, protecting general environmental water quality, and understanding the inherent value of every species. Coldwater fish diversity holds a lot of promise for sport fishing and ecotourism in hilly areas. The distribution of fish species in Himalayan streams is influenced by water temperature, food availability, flow rate, and substratum composition. From the northwestern to the north-eastern Himalayan region and some portions of the Western Ghats, India offers abundant and diverse coldwater fish resources for food, sport, and ornamental value, spanning around 12 states. Upland rivers, streams, high and low-altitude natural lakes, and reservoirs provide coldwater fisheries resources throughout the Himalayan and peninsular regions. As a result of natural processes and manmade actions, there has been a significant change in climate, as measured by temperature and rainfall patterns, over a long period of time. However, increasing anthropogenic activities, climate

Coldwater Fisheries and Aquaculture Management: Technology for Sustainable Food Production.
Mohd Ashraf Rather, Faisal Rashid Sofi, Adnan Amin, & Kawkabul Saba (Eds.)

change, and changing streamflow regimes are wreaking havoc on coldwater resources and fisheries, resulting in lower overall productivity. The extensive biomonitoring of all coldwater resources in time and space is required to assess threats perspectives in relation to fish diversity for utilizing the resources sustainable manner.

6.1 INTRODUCTION

Mountains are the highest repositories of biological diversity throughout the world. The availability of cold waters considers mountain regions, many of which are home to fish and are primarily supported by subsistence fishing. India is also one of the world's mega biodiversity hotspots, ranking ninth in terms of freshwater mega biodiversity (Mittermeier et al., 1997). Biodiversity is critical for ecosystem stability and general environmental quality and for recognizing the intrinsic wealth of all species (Jena & Gopalakrishnan, 2012). The coldwater commonly refers to the aquatic ecosystem, which keeps thermal and oxygen levels for well-being of mahseers, trout, snow trout, and other species. The temperature mostly falls in the tolerance limits of trout and salmons, which belongs to the family Salmonidae (<20°C) (Sarma et al., 2018). Remarkably, the mean temperature of water in all Kumaon lakes during the summer season remains above the 20°C thresholds; the maximum temperature rises well above 25°C (Johri et al., 1989). The country is blessed with an abundance of diverse coldwater fisheries resources dispersed across the Himalayan and peninsular areas in the form of streams, upland rivers, reservoirs, and natural lakes (Table 6.1). The potential coldwater areas in India include the Himalayas' stretched of about 2,500 km from east to west and 200–400 km from north to south, covering an area of 5,94,400 km^2. In the high altitude, there are about 8,243 km rivers and streams, 50,000 ha of reservoirs, both natural and manmade, 20,500 ha natural lakes and 2,500 ha brackish water lakes (Mahanta & Sarma, 2010). Even though there are some of the limitations of coldwater fisheries such as accessibility, tough terrain, and lack of a viable market, the current achievement of growing aquaculture in mountainous regions of India displays that fishermen in the rural villages can become the direct beneficiaries of the implementation of low-cost aquaculture techniques. As a result, they accomplish substantial development in their living standard. As fisheries sector plays a significant role in supplying food and a source of income to the mountain people. The Coldwater streams and rivers are recognized for their high-velocity waterfall, deep pools, rapid flows, and substratum comprising bedrock, boulder, and sand. These enormous and diverse water resources in the uplands support a diverse ichthyofauna, with large populations of native and exotic, culturable, and non-cultivable fishes (Sehgal, 1999; Mahanta & Sarma, 2010; Raveendar et al., 2018). The low productivity and increased degree of resource seasonality and volatility result in a unique aquatic diversity, which is generally susceptible to a wide range of perturbations (Bhatt et al., 2012; Jena & Gopalakrishnan, 2012; Singh et al., 2014; Raveendar et al., 2021).

TABLE 6.1 The State-Wise Aquatic Resources in the Himalayan Region of India

Length of Rivers/Canals (1,000 km)	Water Bodies (hectare)	State
27.78	30,000	Jammu and Kashmir
3.0	43,000	Himachal Pradesh
2.69	20,000	Uttaranchal
2.0	4,000	Arunachala Pradesh
5.60	10,000	Meghalaya
0.90	3,000	Sikkim
3.36	46,000	Manipur
1.20	17,000	Tripura
1.40	2,000	Mizoram
1.60	67,000	Nagaland

6.2 COLDWATER FISH DIVERSITY OF INDIA

The water resources of the Himalayan region support diverse ichthyofauna. In the Indian context, 17% of fishes were reported from the mountain region, establishing the main origin and development of biotic resources (Ghosh, 1997). The coldwater fisheries support 258 species which belong to 21 families and 76 genera. Out of these, a total of 255 species have been reported from North-East Himalaya, followed by the west and central Himalaya with 203 and the Deccan plateau with 91 species. The commercially important coldwater fishes of India are shown in Table 6.2. The eastern Himalayan region has a higher coldwater fish diversity as compared to the western Himalayan region. Few of these species contribute to the capture fishery, while others are cultured on the farm at various altitudes based on their temperature tolerance limits (Table 6.3). The most common and important fish species inhabiting the mountain waters of India are mahseer, snow trout, and Indian hill trout. Mahseer is a group of sport fish that predominantly

TABLE 6.2 Common Cold-Water Fishes of India

***Neolissochilus hexagonolepis* (McClelland, 1839)**	***Tor putitora* (Hamilton, 1822)**
***Onchorhynchus mykiss* (Walbaum, 1792)**	***Schizothorax richardsonii* (Gray, 1832)**
***Barilius bendelisis* (Hamilton, 1807)**	***Raiamas bola* (Hamilton, 1822)**
***Labeo dyocheilus* (McClelland, 1839)**	***Labeo gonius* (Hamilton, 1822)**
***Labeo pangusia* (Hamilton, 1822)**	***Glyptothorax* spp. (Blyth, 1860)**
***Chagunius chagunio* (Hamilton, 1822)**	***Bangana dero* (Hamilton, 1822)**

TABLE 6.3 Common Indigenous and Exotic Cold-Water Fishes of India

Groups	Species
	Indigenous
Mahaseer	*Tor tor*, *T. putitora*, *T. khudree*, *T. mosal*, *T. mosal mahanadicus*, *T. mussulah*, *T. nelli*, *T. progeneius*, *Neolissochelius hexagonolepis*, *N. wynaadensis*, and *Naziritor cheylinoides*
Snow trout	*Schizothorax richardsonii*, *S. kumaonensis*, *S. progastus*, *S. plagistomus*, *S. molesworthi*, *Schizothoraichthys esocinus*, *S. curvifrons*, *S. niger*, *S. micropogon*, *S. labiatus*, *Lepidopygopsis typus*, *Ptychobarbus conirostris*, and *Diptychus maculatus.*
Indian hill trout/ Barils	*Raiamas bola*, *Barilius bendelisis*, *B. bakeri*, *B. vagra*, *B. barila* and *B. gatensis*.
Minor carps	*Labeo dero*, *L. daycheilus*, *L. gonius*, *L. pangusia*, *Osteobrama belangeri*, *Garra lamta*, *G. gotyla*, *G. hughi*, *Chagunius chagunio*, *Crossochelius latius*, *Semiplotus semiplotus and Puntius ophicephalus*
Catfishes	*Glyptothorax pectiniopterus*, *Pterocryptis wynaadensis*, *Pseudechenesis sulcatus*, *Sperata* sp., *Heteropneutes* sp., *Mastacembelus* sp.
Loaches	*Botia birdil*, *B. almorhae*, *Nemacheilus rupecola*, *N. arafi*, *N. montanus*, *N. fascimaculatus*, *N. multifasciatus*, *Triplophysa choprai*, *T. gracilis*, *T. griffthi*, *T. ladacensis*, *T. microps* and *T. tenuicauda*
	Exotic
Trout	*Salmo gairdneri gairdneri*, *Salmo tutta fario*, *S. lavensis*, *S. salar*, *Onchorhynchus mykiss*, and *O. nerka*
Tench	*Tinca tinca*
Carps	*Ctenopharyngodon idella*, *Hypophthalmichthys molitrix*, *Carassius carassius*, *Cyprinus carpio*, *Cyprinus carpio communis* (scale carp), *Cyprinus carpio nudus* (leather carp) and *Cyprinus carpio specularis* (mirror carp).

belongs to the genus *Tor*, also conjointly consisting of some species from the genus *Neolissochilus* (*Neolissochilus hexagonolepis,* chocolate mahseer). Schizothoracines (family Cyprinidae and sub-family schizothoracinae) are an economically important group of coldwater fishes throughout the Himalayan zone, particularly in Kashmir. Out of concerning 258 cold water fish species (both native and exotic) described from Indian uplands, various authors have been identified 17 members of the snow-trout (locally called Sela, Rasella or Asela in Uttarakhand, Koushargad in Kashmir, and Gulgali in Himanchal) (Sehgal, 1999, 2012; Sunder et al., 1999). Among these, 10 species belong to genus *Schizothoraichthys* (species being *S. esocinus, S. niger, S. longipinnias, S. curvifrons, S. micropogon, S. hugelli, S. planifrons, S. labiatus, S. progastus, S. nasus*), two species in the genus *Schizothorax* (*S. kumaonensis* and *S. richardsonii*) and one each in the genera *Gymnocypris* (*biswasi*), *Diptychus* (*maculates*), *Schizopygopsis* (*stoliczkae*), *Lepidopygopsis* (*typus*), *Ptychobarbus* (*conirostris*). *Lapidopygopisis typus* is exclusively found in Periyar lake (Western ghats), Kerala. All other species are found in the Jammu and Kashmir region. The most widespread and commonly dispersed species in the Himalayan region is *Schizothorax richardsonii* (Sehgal, 1999).

6.3 MAJOR ISSUES AND THREATS TO COLD-WATER FISH DIVERSITY

The aquatic habitats are facing major threats to biodiversity and environment constancy, necessitating a variety of tactics and priorities to tackle this crisis (Jena & Gopalakrishnan, 2012; Singh et al., 2014). Regardless of the critical role of fish biodiversity in an ecosystem, all fish species that collectively contain fish biodiversity suffer unique risks and threats. The risks could be natural or anthropogenic, or both with interconnected impacts. Such threats are diverse, resulting in habitat changes, reduction of resources, shrinkage of habitat area, loss of germplasm, and so on as a result of reduced water discharge in rivers, overfishing, siltation, pollution of water bodies, the introduction of exotic species and climatic variability, among other things. The following serious threats to the diversity of coldwater fishes are:

1. **Overfishing/Overexploitation:** Overexploitation of any species leads to loss of genetic diversity as well as a decrease in the relative abundance of species as individuals and/or groups of interrelated species. The size of the population can be decreased due to the disturbances in sex composition and age structure (Sarkar et al., 2008).

Overfishing also has an effect on heritable life-history traits such as sexual maturity, age, and growth. Proficient fishing gear removes fast-growing larger aquatic individuals. Subsequently, the proportion of slow-growth individuals increases, while the mean size of individuals in a population drops. In its report on fisheries, the National Commission on Agriculture (NCA, 1976; Nautiyal, 2013) specified that there was an overall decrease in the mahseer fishery because of indiscriminate fishing of juvenile and brooder fishes. The overexploitation of resources is not only a single pressure to decline the mahseer population but also caused by an increase in development activities, particularly the rising number of hydroelectric-cum-irrigation projects, which leads to fragments disturb its natural habitats (Nautiyal & Singh, 1989). Therefore, the mahseer population has dropped rapidly in several locations (Kashmir, Nainital); over the last few decades, it has been designated as 'threatened' in India (Khan & Sinha, 2000; Mahanta & Sarma, 2010; Akhtar et al., 2014). In spite of this, the use of destructive fishing methods such as bleaching powder, dynamite, damming, and pesticides which result in mass mortality of brooders and juveniles has continued unabated, particularly in small streams, which are the spawning grounds of mahseer (Nautiyal, 1994), hence adversely affecting the rates of recruitment.

2. **Introduction of Exotic Species:** The introduction of non-native aquatic species, mainly fishes, led to a global alarm because it caused a slew of issues, including the extinction of native species (Singh & Lakra, 2011; Singh et al., 2013). The exotics compete with native species for environs and food (Singh et al., 2013). They may also predate on indigenous fishes, transmit new parasites and diseases, produce hybrids, cause genetic 'erosion' of native species, and degrade the physicochemical characteristics of aquatic environments (Singh et al., 2013). All of this could lead to a loss of biodiversity (Singh & Lakra, 2011). Numerous studies have been conducted on the disastrous effects of alien imports on ecosystems and species (Singh & Lakra, 2011; Singh et al., 2013). The possible hazards influence the standard or amount of biodiversity and the economic aspects of the local people, which rely upon aquatic environments for nourishment (Singh & Lakra, 2011). Over 300 exotic fish species have been introduced into our country at various times for trial aquaculture, mosquito control, aquarium keeping, and sport fishing (Kumar, 2000; Singh & Lakra, 2011). Numerous exotic fishes are

now well recognized in India's natural water resources. The introduction of *Cyprinus carpio* var. *specularis* into Loktak lake and Dal lakes have influenced the native population of *Osteobrama belangeri* and *Schizothorax* sp., respectively (Singh & Lakra, 2011). After the introduction of silver carp significantly reduced the population of indigenous mahseer and catla in Govind Sagar reservoir (Singh & Lakra, 2011).

3. **Habitat Destruction/Modification:** Anthropogenic activities are considered to be the primary cause of habitat alteration, which is directed to the depletion of aquatic resources (Takar & Gurjar, 2020). Natural disasters such as floods and landslides, biological invasion, changes in environmental flow, and other factors have significantly impacted the Himalayan coldwater ecosystem, posing a serious threat to coldwater fish diversity. Reduced ichthyofaunal diversity in Asian countries is attributed to pollution, increased flow alteration, sedimentation, water diversion, and new species (Nguyen & De Silva, 2006; Singh & Lakra, 2011; Singh et al., 2013, 2014). Because of recurrent floods and the resulting ecological degradation, the original wild stock of valuable fish species such as snow trout, mahseer, minor carps, and others in coldwater rivers and lakes has been greatly affected in recent years.
4. **Climate Change:** Climate variability is an important aspect in regulating the distribution and quantity of aquatic animals as well as the structure of the ecosystem (Jena & Gopalakrishnan, 2012). The impact of climate change on fish is determined by the amount of the shift as well as the sensitivity of specific species or ecosystems. The influence of temperature changes caused by climate change will impact the biological capabilities of aquatic organisms, as most of them are cold-blooded in nature. Coldwater fishes are key ecological indicators for climate change in temperate zones because they are very sensitive to variations in water temperature and other environmental parameters. Many Himalayan coldwater fish species are endangered; therefore, they may be the most vulnerable to climate change or warming. They currently exist at the upper limit of their thermal tolerance range in many regions, and ecological models predict that there will be significant losses of temperate fish species as climate change reduces fish habitat and variety (Mohseni et al., 2003).

6.4 METHODS FOR CONSERVATION AND SUSTAINABLE MANAGEMENT OF COLD-WATER RESOURCES

Coldwater fish diversity is under threat due to various factors or anthropogenic activities, such as resources overexploitation, habitat degradation, exotic species introductions, and others. Therefore, quick action and strategies for protecting aquatic biodiversity are necessary to conserve preserve these threatened ecosystems and species for coming generations (Jena & Gopalakrishnan, 2012). All aquatic ecosystems face grave risks to ecosystem stability and biodiversity. Generally, aquatic conservation techniques should encourage sustainable development and management of biological resources in ways to conserve ecosystems and habitats. Several strategies and priorities have been presented for coldwater fish diversity conservation in order to make it effective are discussed in subsections.

6.4.1 IN-SITU CONSERVATION

The definition of *in-situ* conservation is "the conservation of ecosystems and natural habitats and the maintenance and recovery of the feasible population of species in their natural habitat and, in the case of domesticated or cultivated species, in the nature or habitat where they have developed their distinguishing properties." This conservation approach integrates research on fish and habitat diversity, life history features, habitat utilization, and human intervention. The fish breeding grounds need to special safeguard by asserting them as 'no-fishing zone' or 'protected area (PA). *In-situ* conservation is accomplished by designating certain places as PAs and designating them as Wildlife Sanctuaries, Biosphere Reserves and National Parks. PAs conserve threatened, rare, depleted, or endangered species and populations and protect their environments.

6.4.2 EX-SITU CONSERVATION

In this approach, species are protected outside of their natural environments by keeping populations in genetic resource and breeding centers or preserving in gene banks, germplasm banks, and gamete storage. As in the case of fishes and other animals, fast freezing of gametes to ultralow temperatures was discovered to be fruitful. The storage of fish sperm (milt), eggs, and embryos without loss of viability is significant for conservation as well as nourishing of aquaculture.

6.4.3 AQUA RANCHING

This method is designed to enrich resources by seed stocking in open waters with desired aquatic organisms and providing suitable manmade shelters for the animals to protect themselves from natural threats, allowing them to grow to a size where predation and juvenile mortality are significantly affected decreased. This approach is used to replenish the depleted fish stocks of lakes, reservoirs, rivers, streams, and estuaries as a result of captive breeding and hatchery programs (Rout et al., 2007). Juvenile fish are frequently removed from their original habitat and allowed to mature sexually and reproduce within the safe limits of an aquaculture or lab setting, with the young ones raised in captivity being returned or ranched to natural resources (Lakra et al., 2007). The DCFR has stocked the fingerlings of golden mahseer in various rivers and lakes of different hill states of Arunachal Pradesh, Uttarakhand, Sikkim, Meghalaya, and Assam. The Directorate has also given thousands of golden mahseer seeds to the Madhya Pradesh and West Bengal Departments of Fisheries for their reservoir/wetlands stocking augmentation program. The DCFR stocked golden mahseer at Shyamlatal Lake in Kumaon, India, in 2001, and they have thrived, growing to mature sizes, and becoming a tourist attraction. On 4th November 2020, DCFR organized a seed ranching program of endangered golden mahseer in three different lakes Bhimtal, Naukuchiatal, and Sattal lakes of the Kumaon region (Figure 6.1). On this occasion, 10,000 mahseer fingerlings were released into each lake for the sustainability of these resources. However, as a step forward for other diminishing coldwater fish species, stock enhancement programs for other declining species should be prioritized.

6.4.4 CONSERVATION AQUACULTURE AND CAPTIVE BREEDING

Conservation aquaculture is becoming more important in the restoration of threatened/endangered fishes (Moyer et al., 2009). The use of captive multiplication to sustain endangered species or populations and conserve local traits in the face of severe loss is known as conservation aquaculture (Schreier et al., 2012). It entails aquaculture in endangered fish population restoration programs by boosting the threatened species' effective population size (Ne).

Captive breeding programs have become a popular method for compensating for falling fish populations while also supplementing and increasing the yields of wild fisheries. In this regard, the Directorate of Coldwater Fisheries Research (DCFR) has developed breeding and seed growing techniques for golden mahseer and a regular producing golden mahseer seeds for aquaculture and conserving the resources.

FIGURE 6.1 Ranching of mahseer fingerlings into the Bhimtal lake (ICAR-DCFR, Bhimtal).

DCFR has further developed flow through mahseer hatcheries at Iduli fish farm in Roing, Arunachal Pradesh; Eco-camp, ABACA, Nameri National Park, Tezpur, Assam, and Bagua fish farm in Sikkim to meet the entire need for mahseer seeds. The NBFGR developed techniques for captive breeding and larval rearing of various fishes, which have been extremely successful (Jena et al., 2011). White sturgeon populations have been restored using conservation aquaculture tools (Schreier et al., 2012), and the same technology has also been used to conserve Pacific salmon.

Similarly, in India, conservation aquaculture could be an efficient technique for the conservation of coldwater fish species whose populations are declining.

6.4.5 SUSTAINABLE FISH HARVEST

Harvesting fisheries resources from conventional fishing grounds is requisite to regulate and manage appropriately. To the sustainable management of stock, brooders, and juveniles should be protected. The strict mesh size regulations must be imposed to conserve or protect juvenile resources. The use of explosives or poison to kill fish needs to be strictly prohibited (Rout et al., 2007).

6.4.6 MASS AWARENESS

Increasing awareness among the public is one of the most imperative ways to sustain and manage aquatic resources and biodiversity. It is commonly understood that the success of biodiversity protection is ultimately dependent on public understanding of the various ecological, cultural, socio-economical, nutritional, recreational, esthetic, pharmaceutical, and other services available to people. Hence, protection of diminishing diversity is the accountability of each citizen. This can be achieved by implementing incentive schemes, educational initiatives, and volunteer monitoring programs. The DCFR, Bhimtal, undertook various awareness programs to protect the golden mahseer, which raised public knowledge and demonstrated the concept of "feeling of belongingness" and made local people understand the need for conservation. The respective state governments should also step forward to organize awareness programs and activities with local residents' participation to reach out to a larger public in the coldwater states.

6.4.7 HABITAT RESTORATION/MITIGATION MEASURES

Habitat restoration is a powerful strategy for reversing ecological damage. Situation-specific habitat restoration programs want to be carried out based on the type of degradation. Depending on the type of degradation, situation-specific habitat restoration programs should be implemented that can be restored by habitat restoration management (Ali, 2010). For example, in the event of siltation/sedimentation, deforestation efforts must be immediately halted, along with a massive afforestation campaign in erosion-prone catchment areas. Polluted water bodies require immediate attention, which can be achieved by maintaining strict adherence to pollution control legislation. Due to recurring floods and resulting ecological degradation, the original wild stock of valuable fish species such as mahseer, trout, snow trout, minor carps, and others in cold water lakes and rivers has considerably diminished in recent years. For coldwater fish diversity management, several management measures such as the development of riparian buffer zones and the restoration of natural flow patterns and discharge commands will be used.

6.4.8 REGULATORY/LEGAL MEASURES

To protect biodiversity and avoid the exploitation of ecologically sensitive areas under the Wildlife Protection Act of 1972 and the Environmental Protection Act of 1986, the Indian government established a network of PAs. Since 1st February 1982, India has been a signatory member of the Ramsar Convention for the conservation of wetland resources. In order to conserve and maximize the use of its bioresources, India adopted the Biological Diversity Act in 2002. The Indian Fisheries Act, 1897 (as amended in 1956) is a revolutionary moment for fish protection. Aside from requiring that mesh sizes, gears, and fishing or closed seasons be observed, the legislation also forbids the use of poisons or explosives to kill fish in any water. However, the Act's enforcement is minimal, particularly in coldwater states. As a result, state governments in the coldwater region should ensure that the act is effectively enforced to ensure the long-term management of coldwater fishes. Furthermore, the act is quite old and needs to be updated to include all necessary legal issues in order to safeguard the states' fish germplasm resources, with special provisions for coldwater fisheries.

6.5 CONCLUSION

Coldwater fishery resource conservation and management require concerted efforts that integrate cultural, capture fisheries, and environmental monitoring. The expanding population and economic expansion will continue to put pressure on natural resources, resulting in biodiversity loss. Advanced environmental legislation with stringent enforcement measures must be adopted to conserve fish biodiversity and habitat. However, greater biodiversity monitoring, risk assessment, and a more structured approach to biodiversity protection are all critical parts of effective risk management (Jena & Gopalakrishnan, 2012). Extensive biomonitoring of all coldwater resources in time and space is required to assess danger perspectives concerning fish diversity. Because the country's upland regions are vulnerable in nature, they must be conserved and used sustainably.

KEYWORDS

- **biomonitoring**
- **climate change**
- **coldwater**
- **fish diversity**
- **Himalayan region**
- ***in-situ* conservation**
- **upland rivers**

REFERENCES

Akhtar, M. S., Pal, A. K., Sahu, N. P., Ciji, A., & Mahanta, P. C., (2014). Higher acclimation temperature modulates the composition of muscle fatty acid of *Tor putitora* juveniles. *Weather. Clim. Extremes., 4*, 19–21.

Ali, S., (2010). In: Mahanta, P. C., & Debajit. S., (eds.), *Mountain Fisheries in India: Exploration, Exploitation and Sustainable Utilization* (pp. 187–198). Coldwater fisheries management. DCFR: Bhimtal.

Bhatt, J. P., Kumar, M., & Pandit, M. K., (2012). Elevational gradients in fish diversity in the Himalaya: Water discharge is the key driver of distribution patterns. *PLoS One, 7*, e46237.

Ghosh, A. K., (1997). *Himalayan Fauna with Special Reference to Endangered and Endemic Species; Himalayan Biodiversity: Action Plan* (pp. 53–59). GB Pant Institute of Himalayan Environment and Development: Kosi-Katarmal, Almora.

Jena, J. K., & Gopalakrishnan, G., (2012). Aquatic biodiversity management in India. *Proc. Natl. Acad. Sci., India, Sect. B Biol. Sci., 82*, 363–379.

Jena, J. K., Gopalakrishnan, A., & Lal, K. K., (2011). Towards conservation and sustainable utilization of aquatic genetic resources. *Fish. Chimes., 31*, 15–18.

Johri, V. K., Awasthi, S. K., Sharma, S. R., Tandon, N. K., & Joshi, L. M., (1989). Observations on some limnological aspects of four important lakes of Kumaon hills of U.P. and suggestions for their proper exploitation. *Indian J. Fish., 36*, 19–27.

Khan, M. A., & Sinha, M., (2000). Status of mahseer fisheries in north and north eastern India with a note on their conservation. *J. Inland Fish. Soc. India, 32*, 28–36.

Kumar, A. B., (2000). Exotic fishes and freshwater fish diversity. *Zool. Res., 15*, 363–367.

Lakra, W. S., Mohindra, V., & Lal, K. K., (2007). Fish genetics and conservation research in India: Status and perspectives. *Fish Physiol. Biochem., 33*, 475–487.

Mahanta, P. C., & Sarma, D., (2010). *Coldwater Fisheries Management* (p. 451). ICAR-DCFR, Bhimtal, India.

Mittermeier, R. A., Gil, P. R., & Mittermeier, C. G., (1997). *Megadiversity: Earth's Biologically Wealthiest Nation* (p. 501). Cemex.

Mohseni, O., Stefan, H. G., & Eaton, J. G., (2003). Global warming and potential changes in fish habitat in U.S. streams. *Clim. Change, 59*, 289–409.

Moyer, G. R., Rousey, J. D., & Cantrell, M. A., (2009). Genetic evaluation of a conservation hatchery program for reintroduction of the sicklefin redhorse *Moxostoma* sp. in the Tuckasegee River, North Carolina. *N. Am. J. Fish. Manag., 29*, 1438–1443.

Nautiyal, P., & Singh, H. R., (1989). River projects and endangered Mahseer, *Tor putitora* in Indian uplands of Garhwal region: A critical evaluation. *Him. J. Env., 3*, 129–134.

Nautiyal, P., (1994). In: Nautiyal, P., (ed.), *The Himalayan or Putitor Mahseer Tor putitora (Hamilton)* (pp. 5–12). Mahseer the Game Fish. Jagdamba Prakashan, Dehradun.

Nautiyal, P., Babu, S., & Bahera, S., (2013). *Mahseer Conservation in India – Status, Challenges and Way Forward.* Proceedings of the workshop on Mahseer conservation in India: Status, challenges and the way forward. WWF-India, supported by Water Program HSBC.

NCA, (1976). *National Commission on Agriculture. Reports on the Fisheries*. Ministry of Agriculture, New Delhi.

Nguyen, T. T. T., & De Silva, S. S., (2006). Freshwater finfish biodiversity and conservation: An Asian perspective. *Biodivers. Conserv., 15*, 3543–3568.

Raveendar, B., Gurjar, U. R., Takar, S., Gugulothu, R., Mamidala, S., & Guguloth, B., (2021). Soil and water characteristics of Nanak Sagar reservoir, Tarai region of Uttarakhand, India. *Int. J. Chem. Stud., 9*, 942–948.

Raveendar, B., Sharma, A. P., Gurjar, U. R., Gugulothu, R., & Mishra, A., (2018). Assessment the present status of fish diversity in relation to physicochemical characteristics of Nanak Sagar reservoir of Uttarakhand. *J. Entomol. Zool. Stud., 6*, 477–484.

Rout, S. K., Malla, S., Das, B. K., Trivedi, R. K., & Sundaray, J. K., (2007). Conservation of Indian threatened ichthyofaunal-Immediate implications. *Fish. Chimes., 27*, 40–44.

Sarkar, U. K., Pathak, A. K., & Lakra, W. S., (2008). Conservation of freshwater fish resources of India: New approaches, assessment and challenges. *Biodivers. Conserv., 17*, 2495–2511.

Sarma, D., Akhtar, M. S., & Singh, A. K., (2018). In: Tripathi, S. D., Lakra, W. S., & Chadha, N. K., (eds.), *Coldwater Fisheries Research and Development in India* (pp. 93–133). Aquaculture in India, Narendra Publishing House, Delhi, India.

Schreier, A. D., Rodzen, J., Ireland, S., & May, B., (2012). Genetic techniques inform conservation aquaculture of the endangered Kootenai River white sturgeon *Acipenser transmontanus. Endanger. Species Res., 16*, 65–75.

Sehgal, K. L., (1999). Coldwater fish and fisheries in the Indian Himalayas: Rivers and streams. In: Petr, T., (ed.), *Fish and Fisheries at Higher Altitudes: Asia* (pp. 41–63). FAO Fisheries Technical, Rome.

Sehgal, K. L., (2012). In: Debajit, S., Amit, P., Suresh, C., & Gupta, S. K., (eds.), *History of Coldwater Fisheries in India and Some Neighboring Countries* (pp. 13–23). DCFR Silver Jubilee Compendium on Coldwater Fisheries.

Singh, A. K., & Lakra, W. S., (2011). Risk and benefit assessment of alien fish species of the aquaculture and aquarium trade into India. *Rev. Aqua., 3*, 3–18.

Singh, A. K., Dinesh, K., Srivastava, S. C., Ansari, A., Jena, J. K., & Sarkar, U. K., (2013). Invasion and impacts of alien fish species in the Ganga River, India. *Aquat. Ecosyst. Health Manag., 16*, 408–414.

Singh, A. K., Kumar, P., & Ali, S., (2014). Ichthyofaunal diversity of the Ganges River system in central Himalayas, India: Conservation status and priorities. In: Sinha, R. K., & Ahmed, B., (eds.), *Rivers for Life – Proceedings of the International Symposium on River Biodiversity: Ganges-Brahmaputra-Meghna River System, Ecosystems for Life, A Bangladesh–India Initiative, International Union for Conservation of Nature* (pp. 208–214).

Sunder, S., Raina, H. S., & Joshi, C. B., (1999). *Fishes of Indian Uplands* (p. 64*). Bull. No.2,* NRC on Coldwater Fisheries, Bhimtal.

Takar, S., & Gurjar, U. R., (2020). Review on present status, issues and management of Indian marine fisheries. *Innov. Farming., 5*, 34–41.

CHAPTER 7

The Zebrafish: A Trending Model for T-Cell and Thymic Development Along with Prevailing Challenges

SNEHA SABU[1], SOFIA KHANAM[2], and A. JOTHILIN SUBITSHA[1]

[1]*MSc, Department of Microbiology, CMST, Manonmaniam Sundharanar University, Rajakkamangalam, Kanyakumari, Tamil Nadu, India*

[2]*MPharm, Department of Pharmacology, Calcutta Institute of Pharmaceutical Technology and AHS, Uluberia, Howrah, West Bengal, India*

ABSTRACT

Recently, the *Danio rerio* (zebrafish) has become a trending genetic model for the evolution of vertebrates, offering new research opportunities that are not readily available in other model organisms. The research with *D. rerio* has elucidated the molecular and cellular mechanisms underlying these biological processes. T-cell and thymic growth are strongly maintained mechanisms that have been throughout the evolution of vertebrates. *D. rerio*, like mammals, have an adaptive immune system composed of B lymphocytes, as well as the αβ and γδ lineages of T cells that form in the thymus. Zebrafish are a highly recommended research model since it has plenty of advantages over others such as easy and convenient manipulation of genome, rapid embryonic development, less generation time, external fertilization, etc. While using zebrafish as a prominent research tool, it is significant to consider the prevailing challenges. This review chapter points out identifying the similarities and differences between the immunobiology of *D. rerio* and mammalian T-cells, accentuating the benefits of using *D. rerio*

Coldwater Fisheries and Aquaculture Management: Technology for Sustainable Food Production.
Mohd Ashraf Rather, Faisal Rashid Sofi, Adnan Amin, & Kawkabul Saba (Eds.)

as a genetic model to discover T-cell and thymic development, application of zebrafish as a promising research tool in the biomedical sector and prevailing challenges as well.

7.1 INTRODUCTION

The zebrafish is a rapidly growing and very important model organism for understanding vertebrate developmental biology. The brain, digestive tract, muscle, vasculature, and innate immune system of zebrafish are all quite similar to those of humans (Gore et al., 2012; Kanungo et al., 2014; Kalueff et al., 2014; Guyon et al., 2007; Weinstein, 2002; Lieschke et al., 2001; Zhao et al., 2015). Furthermore, 70% of human disease genes share functional similarities with zebrafish genes (Santoriello et al., 2021). Mammals and zebrafish have similar blood cells and use similar molecular pathways to regulate the production of blood cells such as thymocytes (Burns et al., 2005; Rhodes et al., 2005; Bajoghli et al., 2009). T-cell development always takes place in the thymus' epithelial microenvironment (Boehm, 2011).

The zebrafish (*Danio rerio*) has lately emerged as a significant genetic model of vertebrate development, opening up new areas of investigation not available in other model species. Although invertebrate genetic models have provided enormous insight into the mechanisms that control developmental pathways, Caenorhabditis elegans, and Drosophila melanogaster do not have the same blood-cell lineages as vertebrates and lack an adaptive immune system, making dissection of lymphocyte development impossible in these organisms (Chitwood, 1974; Schulenburg et al., 2004; Evans et al., 2003). Furthermore, forward genetic screenings in zebrafish are substantially less time-consuming than those in mice (Appleby & Ramsdell, 2003), and zebrafish have conserved genetic programs that underlie vertebrate blood formation (Thisse & Zon, 2002). Zebrafish have both T and B cells, just like mammals, birds, and reptiles (Hansen & Zapata, 1998; Trede et al., 2001; Trede & Zon, 1998), which enables the investigation of these lymphoid-cell populations in this model. Like mammals, zebrafish have an adaptive immune system that includes B lymphocytes as well as T-cells from both the lineages that mature in the thymus. Furthermore, the molecular pathways underpinning T-cell development in zebrafish appear to be strikingly similar to those in humans. As a result, discoveries from the zebrafish model will be very relevant to similar mechanisms in humans.

Several aspects of T-cell and thymic development have been found to be retained from teleosts to mammals in investigations conducted over the last two decades. Work with zebrafish (*Danio rerio*) in particular has given information on the cellular and molecular mechanisms that underlie these biological processes. The simplicity with which these species may be imaged non-invasively *in vivo* allows for direct viewing of all events connected with these processes, which are technically challenging in mice. In this study, we will define the role of zebrafish in T-cell development and thymus organogenesis, as well as their applications and challenges.

7.2 ROLE OF ZEBRAFISH AS A RESEARCH MODEL

The norms to choose animal models for biomedical research are directly connected to the final goal line of the research. The usage of zebrafish as a research model was recommended by George Streisinger and fellow workers at the University of Oregon, who launched the contemporary era for zebrafish in the field of biomedical research (Clark & Ekker, 2015). Zebrafish are prevalent animal models since they have abundant benefits than other species. The utmost valuable characteristics of zebrafish are an entirely sequenced genome, trouble-free manipulation of its genome, elevated fertility, quick generation period (about 3 months), quick embryonic growth (24 hr.), and exterior fertilization. The luminous zebrafish embryo allows investigations of the various developmental phases initiated from the initial phase of embryogenesis. Furthermore, zebrafish embryos form whole organ systems, together with the heart, intestine, and blood vessels in 48 hr. after reproduction. More than 10,000 mutants in protein-coding genes were created (Howe et al., 2013) also several transgenic lines of zebrafish have been accomplished to study human diseases. The obtainability of numerous species of zebrafish is an additional significant benefit of this species. Besides, it is also very inexpensive to maintain a huge number of zebrafish in a comparatively small capacity of laboratory space. Though zebrafish need comparatively easy management, distinct care must be paid to guaranteeing a healthy diet and suitable water quality to enhance fish health and growth. Several laboratories use zebrafish to investigate human diseases, together with neural disorders, tumors, infectious diseases, cardiac diseases, kidney diseases, diabetes, loss of sight, deafness, intestinal diseases, hematopoiesis, and muscle disorders. Zebrafish are too significant for emerging new treatments or screening novel drugs to treat or avoid human diseases. Although zebrafish are a vital research model, they have some boundaries, together with the variation of some organs like the respiratory system and the reproductive system. Therefore, it is hard to practice zebrafish for disorders in respiration or reproduction in humans. Furthermore, since zebrafish live in an aquatic habitat, screening of some water-solvable medicines in zebrafish is another restriction (Figure 7.1).

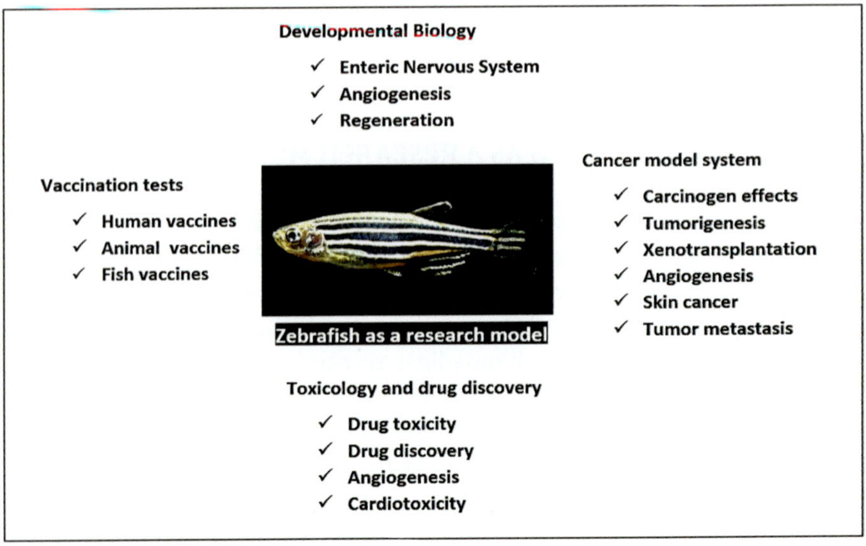

FIGURE 7.1 Zebrafish as a research model.

7.3 T-CELL DEVELOPMENT IN ZEBRAFISH

For *in vivo* genetic studies of human development Zebrafish is becoming an influential craniate model (Amatruda & Zon, 1999; Lieschke & Currie, 2007; Trede et al., 2004). Zebrafish and warm-blooded creatures share comparative blood cells and utilize common molecular pathways to control the blood cell and thymocytes (Burns et al., 2005; Rhodes et al., 2005; Bajoghli et al., 2009). In zebrafish, the primary authoritative HSCs originate from the ventral wall of the dorsal aorta at around 28–30 hours post-fertilization (hpf), at that point move to the caudal hematopoietic tissue, and finally domestic to the thymus and the kidney, where T-cell growth and mature hematopoiesis occur, correspondingly (Murayama et al., 2006). T-cell progenitors marked by *ikaros* expression show up in the thymus by 3 days post-fertilization (dpf) (Murayama et al., 2006). As in warm-blooded creatures, zebrafish T-cell development gives rise to two T lineages, γδ, and αβ and is escorted by the reordering of four T-cell receptors (TCR) loci (*TCRα, TCRβ, TCRγ, and TCRδ*) (Schorpp et al., 2009). As in warm-blooded creatures, TCR genes in zebrafish are accumulated by a V(D)J recombination technique that is reliant on the recombination activating genes, rag1 and rag2 (Willett et al., 1997). The T lymphocyte-specific nonreceptor tyrosine kinase *lck* is similarly found within the lymphoid precursors and developing T lymphocytes within the bilateral thymic lobes of zebrafish (Trede et al., 2004; Langenau & Zon, 2005). A current investigation summarizes extra genes that play a vital role in zebrafish and serve as markers of T lymphocyte development (*c-myb*, *ikaros*, *rag1*, *rag2*, *lck*, *TCRα*, *TCRβ*, *TCRγ*, *TCRδ*, *IL7R*, *jak3*, *ccr9a*, *ccr9b*, *zap70*, *gata3*, *runx1*, *foxn1*, etc.) (Ma et al., 2013).

7.3.1 STAGES IN T-CELL DEVELOPMENT

In this segment, we parted the development of T-lymphocytes into four wide stages (Figure 7.2).

- The entry of lymphoid progenitors into the thymus;
- Commitment of thymocytes;
- Choice of functional also non-self-reactive thymocyte; and
- Egress of thymocytes.

In the initial step, lymphoid progenitors inhabit the thymus; the second spans T-cell commitment and the divergence of αβ and γδ T-cells lineages. The third step is the choice of functional also non-self-reactive T-cells. Finally, developed T-cells exit the thymus into the periphery (Bajoghli et al., 2019).

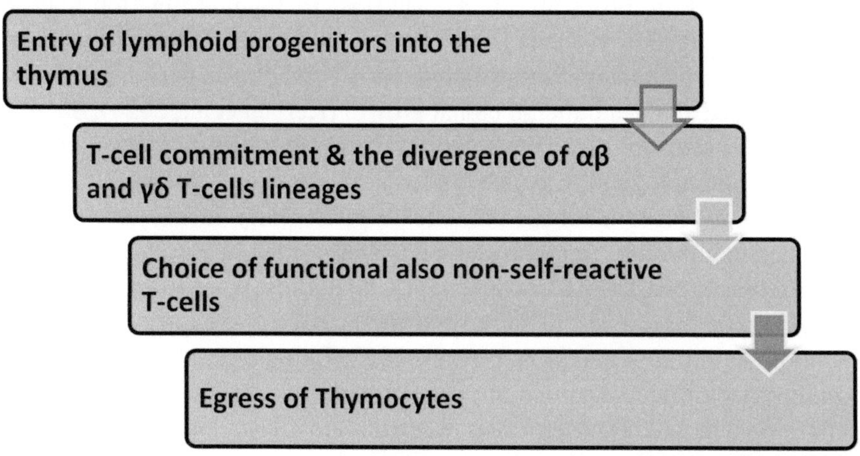

FIGURE 7.2 The development of T-lymphocytes.

7.4 THYMIC DEVELOPMENT IN ZEBRAFISH

The thymus is one of the lymphoid organs where thymocytes differentiate and mature, which is present in all vertebrates (except for jawless fish like the lamprey). It plays an essential role in the development of T-cells repertoire, which is required for immunocompetency (Boehm & Bleul, 2007; Boehm et al., 2012). The formation of the thymic primordium, its colonization by lymphocyte precursors, the expansion of thymocytes, and the histologic regionalization of this organ have all been proposed as four successive processes in the thymus histogenesis (Lam et al., 2002). The structure and function of the thymus in mammals change over the lifespan (Peel & Belov, 2017). Thymus development in humans begins during embryogenesis and achieves its maximal size and output of naive T-cells throughout early life (Palmer, 2013). Following that, the thymus starts a period of slow decline marked by remodeling of the histoarchitecture, shrinkage, and the production of naive T-cells (Manley et al., 2011; Rezzani et al., 2014). The origination of thymus involution was formerly thought to occur during puberty, in conjunction with increased levels of sex steroid hormones and decreased growth hormone synthesis. Human thymus regression is generally widely considered to begin during childhood; however, it may become more visible throughout puberty (Haynes et al., 2000; Manley et al., 2011).

Thymus function comes at a high price in terms of physiological expenses. This is due to the thymus's stringent selection processes because the output of T-cells to the periphery accounts for <10% of all produced thymocytes (Quaglino et al., 2014). The loss of thymus function may be bearable for aged organisms after the development of peripheral T-cell populations, which happens throughout early life because peripheral T-cells are long-lived and can undergo clonal growth (Kernen et al., 2020). Shanley et al. (2009) also addressed the potential life cycle tradeoffs associated with a strong immunological response. To assist successful reproduction, natural selection may favor a robust immune response early in life and a finely regulated inhibition of immune activities later in life.

In teleosts, the thymus matures as a bilaterally paired organ near the operculum on the dorsal side of the branchial cavity. The expression of *foxn1* (Schorpp et al., 2002; Hess & Boehm, 2012; Bajoghli et al., 2009), a transcription factor necessary for the development of TECs (thymic epithelial cells) (Boehm, 2008), determines when thymopoiesis begins in zebrafish (Figure 7.3). *Foxn1*'s role in defining the thymic anlage has been preserved throughout evolution in jawed vertebrates (Swann, 2014). In zebrafish, *foxn1* regulates the expression of two significant factors for T-cell development, *viz*. the delta like 4b (*dll4b*) and the c-c chemokine ligand 25a (*ccl25a*). The former gene is required for lymphoid progenitor attraction into the thymus, while the latter gene is required for lymphoid progenitor specification into T-cells (Koch et al., 2008; Hozumi et al., 2008; Ferrero et al., 2013). Teleosts, unlike mammals, have TECs that express the *foxn4* paralog gene from the larval stage forward (Swann, 2014). On the other hand, *foxn1*-deficient in mice and do not develop thymus (Schlake et al., 1997). Because the ortholog gene does not express in the murine thymus, it has been suggested that *foxn4* thymic expression may have been lost during vertebrate evolution. This is further confirmed by the finding that ectopic *foxn4* expression in TECs cannot compensate for *foxn1* deficiency in nude mice (Swann, 2014).

From one-week post-fertilization (wpf) to four weeks post-fertilization (wpf), the thymus grows from a small spherical shape and develops into a more complicated conical shape (Lam et al., 2002). A comparable change in thymus form has been seen in sharp snout seabream (Romano et al., 1999), showing that this process is conserved among teleosts. The adult thymus of teleosts has a cortical and medullary organization similar to that of mammals, however, it can vary from one lobule to numerous lobules (Flajnik, 2018; O'Neill, 1989). Each side of the body has only one thymic lobule in zebrafish. Thymic compartmentalization occurs later in zebrafish juveniles, between 2 and 3 wpf (Lam et al., 2002). During adolescence, the thymus reaches its peak output, following which it begins to decline with age. In mammals, this age-related thymus involution is well-known (Boehm & Swann, 2013; Lynch et al., 2009). This mechanism starts at 15 wpf in zebrafish (Lam et al., 2002). The point at which the thymus begins to shrink in teleosts, on the other hand, varies greatly (Deanesly, 1927). The thymic structure in some teleosts can even last their entire lives (O'Neill, 1989).

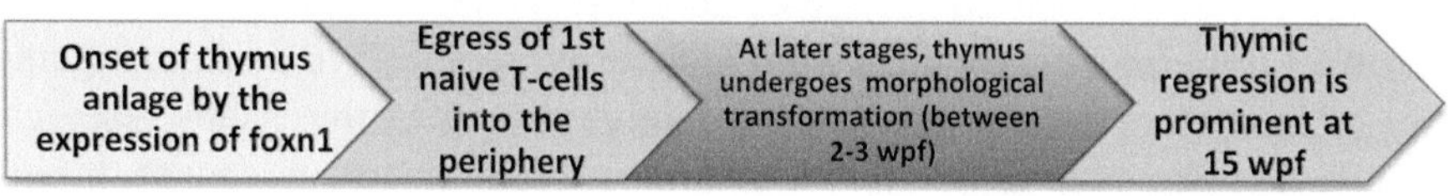

FIGURE 7.3 The timeline of zebrafish T-cell and thymic development.

7.5 APPLICATION OF ZEBRAFISH AS A RESEARCH MODEL

The criteria for selecting animal models for biomedical research are linked to the research's ultimate purpose. George Streisinger and colleagues at the University of Oregon proposed using zebrafish as a biomedical model, which ushered in a new age for zebrafish in biomedical research (Clark & Ekker, 2015). Zebrafish are widely used as animal models due to their multiple benefits over other species. A fully sequenced genome, facile genome manipulation, high fecundity, short generation period (approximately 3 months), rapid embryonic development (24 hours), and external fertilization are the most favorable traits of zebrafish. The transparent zebrafish embryo allows researchers to examine several developmental phases, beginning with the initial stages of embryogenesis. Furthermore, after 48 hours of fertilization, zebrafish embryos create complete organ systems, including the heart, gut, and blood arteries. More than 10,000 protein-coding gene mutants have been generated as well as multiple zebrafish transgenic lines for research into human diseases (Howarth et al., 2013).

In recent years, the Zebrafish model has indeed been frequently used in animal and human health investigations, as well as aquaculture. Regardless of the fact that rodents are the most often used study model in the world, the zebrafish (*Danio rerio*) model has grown in popularity among academics in recent decades. It adheres to the 3Rs (replacement, reduction, and refining) principle, as mandated by a number of national and international regulatory agencies. Furthermore, as compared to the models of more established animals, the usage of the zebrafish model saves time and resources. When compared to *in vitro* outcomes, it also delivers more information and predictability (MacRae & Peterson, 2015). As a result, utilizing the zebrafish model, it is possible to replace and limit the usage of mammals in research while also addressing issues connected to their welfare. Furthermore, zebrafish are utilized as confirmatory models of previously obtained favorable results, allowing researchers to fine-tune their findings (Bailone et al., 2019).

Another significant advantage of zebrafish is the availability of various strains. Furthermore, keeping a high number of zebrafish in a limited quantity of laboratory area is quite cost-effective. Many of these labs research human diseases such as metabolic disorders, brain disorders, cancer, infectious diseases, cardiovascular diseases, kidney diseases, diabetes, blindness, deafness, digestive diseases, hematopoiesis, and muscular problems using zebrafish (Taeme et al., 2019), For aquaculture species, toxicity, and drug discovery (Khan & Alhewairini, 2018). Gene Expression Analysis, Forward Genetic Screens, Transgenesis, Protein Overexpression, Gene Knockout in Zebrafish, Cell Transplantation and Cell-Cell Interactions, and Testing Chemical Compounds in Zebrafish Embryos are all employed as model organisms in developmental pediatric research (Veldman & Lin, 2008).

7.5.1 ZEBRAFISH AS A MODEL FOR HUMAN DISEASES

Except for the lungs and prostate and mammary glands, most tissues and organs in humans and zebrafish are identical. The cloning of mutated genes screened for specific symptoms in zebrafish is analogous to the cloning of mutated genes screened for specific phenotypes in humans, therefore it can be used as a model for human disease and to explore underlying mechanisms. The first human disease discovered using zebrafish was a blood condition caused by a mutation in the ALAS2 gene (Chitramuthu, 2013), which caused a specific failure in hemoglobin formation. A large number of additional mutants with phenotypic parallels to human disease have been found. Neurological problems (Sosa et al., 2012), hematopathological disorders (Berman et al., 2012; Brownlie et al., 1998), cardiovascular diseases (Sehnert et al., 2002), muscular disease (Pappalardo et al., 2013), malignancies (Liu & Leach, 2011; Patton et al., 2005), Parkinson's disease (Sarath et al., 2016), anxiety-posttraumatic stress disorders (Chakravarty et al., 2013) are only a few examples.

7.5.2 ZEBRAFISH AS A RESEARCH MODEL FOR DIFFERENT METABOLIC DISEASES

Duchenne muscular dystrophy, human melanoma, acute lymphoblastic leukemia, polycystic kidney disease, nephronophthisis, acute kidney injury, Parkinson's disease, Huntington's disease, Alzheimer's disease, myocardial infarction, and some metabolic diseases are examples of human diseases that have been successfully modeled in zebrafish. Zebrafish are a good model for studying metabolic disorders because they have all of the organs involved in energy balance and metabolism, such as hunger and insulin control, as well as a lipid storage system that is similar to that of humans (Nishio et al., 2012). Cardiovascular disease is the most common cause of death among metabolism-related human illnesses, as per the World Health Organization (WHO) study (Lozano et al., 2012). Obesity (Ng et al., 2014), type 2 diabetes mellitus (LaBrecque et al., 2014), and non-alcoholic fatty liver disease (Ng et al., 2014) all raise the risk of cardiovascular disease. Zebrafish is a prominent model for studying metabolic problems because they have similar metabolic organs to humans (such as the digestive organs, adipose tissue, and muscle). In particular, various new tools and methods, including as TALENS, CRISPR/Cas9 (Wu et al., 2018), compound treatment

(Poureetezadi et al., 2016), mass spectrometry-based polar metabolomics and lipidomics (Zhang et al., 2018), and *in vivo* imaging of fluorescent dyes (Minchin et al., 2018), make it feasible to explore the molecular pathways. Researchers have also utilized zebrafish as a model organism to investigate a variety of metabolic problems, including congenital metabolic abnormalities, hyper-, and hypothyroidism, disorders of the hypothalamus-pituitary-adrenal axis, circadian clock dysregulation, and cancer metabolism (Gut et al., 2017).

7.5.3 ZEBRAFISH AS A SYSTEMATIC CANCER RESEARCH MODEL

Vertebrate models are essential for better understanding the development, growth, and dissemination of malignant tumors thus zebrafish are a good model for cancer studies. Although humans and fish have a common ancestor, the biology of cancer is the same in both groups of creatures (Amatruda et al., 2002). The zebrafish is adroitly exploited for carcinogenic treatment, transplantation of mammalian tumor cells, and transgenic regulations due to the tremendous benefits of using zebrafish as a research model (Mizgirev & Revskoy, 2010) and malignancies (Liu & Leach, 2011; Patton et al., 2005), Parkinson's disease (Sarath et al., 2016), anxiety, and PTSD (Chakravarty et al., 2013).

Cancer is a disorder in fishes, just like it is in people, as evidenced by melanomas that occur in Xiphophorus hybrids (Walter & Kazianis, 2001; Patton et al., 2005). Using zebrafish as a model organism in cancer studies serves numerous advantages. With zebrafish, highly conserved cancer pathways can be genetically tested. Cancer is primarily an adult illness, although cell cycle phenotype in fast-developing transparent zebrafish embryos could be investigated using mutagenesis screens.

To examine tumors caused by environmental carcinogens, a variety of fish have been employed as models. The zebrafish has proven to be the finest model for studying embryogenesis, organogenesis, and the impact of environmental carcinogens on cancer formation (Bailey et al., 1996). Histopathologically (HP), chemically generated tumors are comparable in both humans and zebrafish (Amatruda et al., 2002), and orthologous oncogenes and tumor suppressor genes (TSGs) have been discovered in both fish and people (Lam et al., 2006). Hepatic transcriptomic profiles in humans and zebrafish at various stages of tumor aggressiveness were shown to be similar in these two phylogenetically distant species (Mirbahai et al.,

2011). Pancreatic cancer, lung cancer, ovarian carcinoma, breast cancer, prostate cancer, retinoblastoma, leukemia, and other tumors have already been implanted into zebrafish (Zhao et al., 2015).

7.5.4 ZEBRAFISH IN TOXICOLOGY AS WELL AS DRUG DISCOVERY RESEARCH

The zebrafish model has been utilized to determine organ function assays and drug effect assessments (Kari et al., 2007). The effects of medicines on specific organs have also been investigated, with organ toxicity being found to be similar in both zebrafish and mammals. Gentamicin, cisplatin, vinblastine, quinine, neomycin, doxorubicin, dexamethasone, cyclosporin A, caffeine, camptothecin, MPA, fluorouracil, and other medications were employed to assess organ damage (Zhang et al., 2003; Daroczi et al., 2006; Langheinrich et al., 2002; McAleer et al., 2005; Ton et al., 2005; Wu et al., 2006). The zebrafish is quickly establishing itself as a promising model animal for pharmacological and chemical toxicology research (Spitsbergen & Kent, 2003; Rubinstein, 2006). The toxicity of some important medications has been investigated using the zebrafish model. For instance, using susceptible zebrafish, Amanuma et al. (2000) developed a test to detect small molecule-induced mutagenesis. The developmental damage caused by ethanol or acetaldehyde exposure was compared using zebrafish embryos (Reimers et al., 2004). The toxicity of antirheumatic medications like diclofenac was studied using zebrafish. The zebrafish now is widely acknowledged as a promising animal model for drug toxicity and environmental toxicology research (Chakraborty et al., 2009).

The zebrafish model has been used to investigate the effects of neurotoxic, ototoxic, and neuroprotective drugs, as well as drug discovery. Lead compound screening, target selection, target validation, and assay development are the four main components of the drug discovery process (Handen, 2002). Target identification is the process of discovering a gene or protein that, when altered by treatment, can have a beneficial effect on disease progression. Following the discovery of a possible target, the target's validation approach begins with the determination of protein function and a draggability assessment (Eckstein, http://www.alzforum.org/drg/tut/tutorial.asp; Lindsay, 2003; Frank, 2005). For zebrafish, each of these drug discovery domains is extremely important.

7.6 OUTLINING SOME CHALLENGES

In biomedical research, the zebrafish (*Danio rerio*) is currently an extensively utilized model organism. The zebrafish has long been used to research heart formation. Exploring the operational inputs and their accompanying physical stimuli, such as mechanical or calcium cues, in greater depth *in vivo* will soon follow, thanks to the expanding number of genetic tools and technological improvements.

Since some genes with a single copy in humans may have two copies in zebrafish, genome duplication is a concern for several zebrafish models (Lu et al., 2012; Semon & Wolfe, 2007; Alsop & Vijayan, 2009). On just one hand, this significantly raises the overall complexity of the genetic analysis required for proper data analysis. Other investigators, on the other hand, see possibilities in this genetic uniqueness, because zebrafish models, for example, might be used to investigate the impact of genes where null mutations induce developmental death when expressed in a single copy (Stewart et al., 2015). The lack of a large amount of possible and characterized zebrafish genetic strains is related to genetic assessments and stems not only from the relative novelty of zebrafish in biomedical and neuroscience research but also from quick fertility loss due to inbreeding in teleosts (Stewart et al., 2014; Nasiadka & Clark, 2012). There are presently no definite inbred lines whereby all animals are similar and homozygous (Nasiadka & Clark, 2012), and the level of genetic, physiological, and behavioral characterization of such strains does not match that of laboratory rodent strains (Stewart et al., 2014). Near-homozygous zebrafish lines were created utilizing heat shock and early pressure, but they were too hard to maintain (Johnson & Zon, 1999), while ongoing research may soon help alleviate this problem (Hou et al., 2015).

In vertebrate animal and human studies, gender is widely recognized as an important biological characteristic that should be considered in research designs, analysis, and reporting. However, because zebrafish lack sex chromosomes and instead rely on a polygenic sex determination, its status is still "complex" for zebrafish models (Nguyen et al., 2013; Lawrence et al., 2008; Liew et al., 2014). As a reason, sex ratios in colonies vary widely, and fish models often display negligible or abnormal sexual dimorphism when contrasted to mammals (Nguyen et al., 2013; Lawrence et al., 2008; Liew et al., 2014). As a result, the ability to obtain sex-dependent variations, which are typically crucial for numerous human diseases, may be hampered. Furthermore, in zebrafish investigations, under-reporting of sex ratio determination lowers research repeatability, which should be addressed by increasing data recording standards in the field. Other physiological variances could also be a problem that needs to be addressed.

The blood-brain barrier, metabolism, thermoregulation, the formation, localization, and function of neurotransmitter signaling pathways, and brain anatomy (e.g., cortex and hippocampus) are all examples of species differences (Stewart et al., 2015; Kalueff et al., 2014). Additionally, certain methodological inconsistencies in zebrafish behavioral testing may have an impact on data repeatability (Facciol et al., 2017). Furthermore, despite significant improvements in mechanization and recording equipment, population-level analysis and movement pattern identification remain technological challenges, particularly in larvae studies (Rennekamp et al., 2015; Ahmad et al., 2012) 29. Finally, individual and contextual factors can influence CNS (central nervous system) models and drug screens in zebrafish (Nichols et al., 2004). As a result, more research is needed to better understand the role of sex, individual, strain, and age disparity in zebrafish models, as well as housing circumstances, nutrition, testing time, and social standing (Hart et al., 2016).

Arrhythmias and ventricular hypertrophy, both of which are underlying illnesses that contribute to sudden cardiac arrest, can be caused by ion channel dysregulation, such as LTCC. Future research that focuses on functionality input modulation will aid in the development of new therapies for common cardiac illnesses such as cardiac hypertrophy, heart failure, and arrhythmias. Many questions about zebrafish vascular and lymphatic research remain unresolved. In parallel to VEGF and Wnt, other signaling pathways such as the Hippo system are being investigated in vascular development (Nagasawa-Masuda & Terai, 2017; Nakajima et al., 2017; Astone et al., 2018). Similarly, hemodynamic factors appear to have a role in the cardiovascular system's growth and differentiation (Chen et al., 2012, 2017; Goetz et al., 2014).

The species is also being used more regularly to examine specifications and repair, as well as to gain a better understanding of human skeletal diseases. When it comes to observing and assessing the skeleton, the model organism's small size is both a plus and a burden. Using the unique methods of zebrafish skeletal tissue imaging and evaluation that we have described in this work can overcome a significant restriction of the zebrafish model for skeletal research: the extreme microscopic size of skeletal structures, particularly in early developmental stages. Technological innovations for imaging live-image-labeled skeletal cells and visualizing small skeletal elements in three dimensions will aid our understanding of the cellular and molecular mechanisms that underlie skeletal development and remodeling (Bruneel & Witten, 2015).

The embryonic zebrafish is a vertebrate model that has proved useful in uncovering fundamental neurodevelopmental pathways and may be used for outstanding chemical screening. In zebrafish, gene editing can also be employed to add genetic vulnerabilities. Despite the fact that the zebrafish model has become more widely used in toxicology studies in recent years, the various tools for imaging structural differences in the developing zebrafish brain have yet to be widely used in studies of the impact of gene-environment interplay on neuronal interconnection in the developing zebrafish brain (Figure 7.4) (Miller et al., 2018).

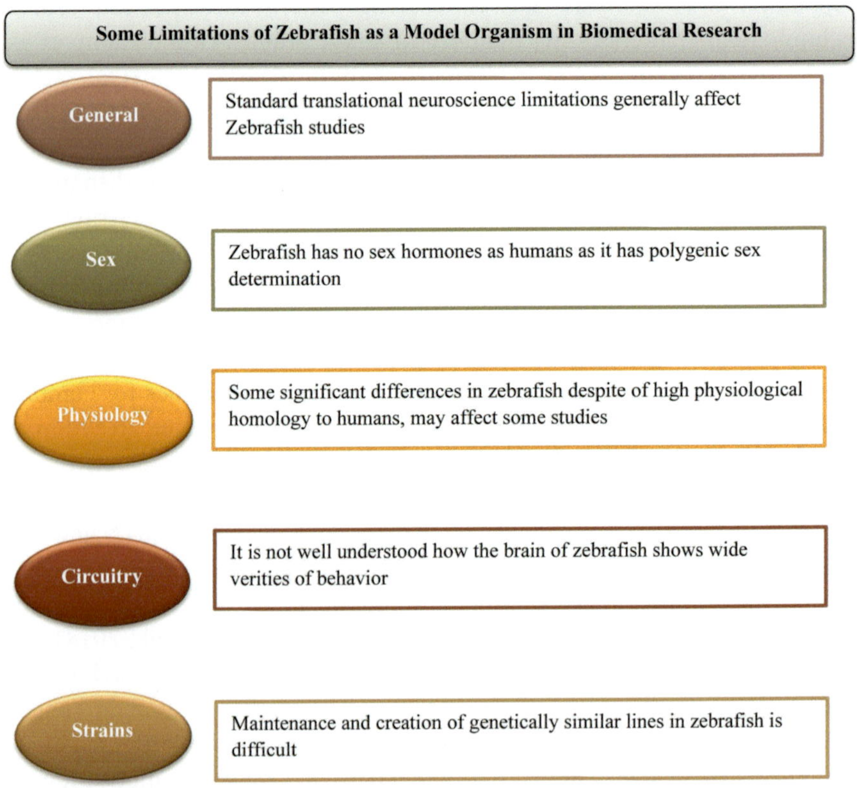

FIGURE 7.4 Some limitations of zebrafish as a model organism in biomedical research.

7.7 CONCLUSION

Zebrafish have significant advantages compared to other vertebrate models used in modeling human diseases, particularly for large-scale genetic mutant and therapeutic compound screenings, as well as other biomedical research applications. The zebrafish provides a unique developmental and genetic model for immune system development, and the thymus has been the focus of study because it is the first primary lymphoid organ to form throughout zebrafish ontogeny. Future studies should look at the role of sex steroids in teleost species' in thymus involution. We know that sex steroids affect both thymus differentiation and thymus involution in mammals. In recent years, various methodologies and technologies have been used to shed light on the mechanisms controlling thymus homing, intrathymic cell trafficking, and the identification of hitherto identified T-cell gene regulators. Although the zebrafish model has some advantages over current T-cell and thymic development models, it also has certain drawbacks. Antibodies specific for blood-cell lineages have yet to be shaped, making it hard to distinguish between diverse T-cell populations at present. The gene-knockout technique for zebrafish has yet to be developed. The zebrafish is fast developing as a new model system for studying T-cell development and vertebrate immunobiology, adding to the elegant research in mice, chickens, and fetal thymic organ culture that have characterized key elements of thymic and lymphoid-cell development.

CONFLICTS OF INTEREST

The authors declare no conflict of interest.

KEYWORDS

- **biomedical applications**
- **chronic unpredictable stress**
- ***Danio rerio***
- **hours post-fertilization**
- **T-cell receptors**
- **T-cells development**
- **thymic development**

REFERENCES

Ahmad, F., Noldus, L. P. J. J., Tegelenbosch, R. A. J., & Richardson, M. K., (2012). Zebrafish embryos and larvae in behavioral assays. *Behavior., 149*, 1241–1281.

Alsop, D., & Vijayan, M., (2009). The zebrafish stress axis: Molecular fallout from the teleost-specific genome duplication event. *Gen. Comp. Endocrinol., 161*, 62–66.

Amanuma, K., Takeda, H., Amanuma, H., & Aoki, Y., (2000). Transgenic zebrafish for detecting mutations caused by compounds in aquatic environments. *Nature Biotechnology, 18*(1), 62–65.

Amatruda, J. F., & Zon, L. I., (1999). Dissecting hematopoiesis and disease using the zebrafish. *Dev. Biol., 216*(1), 1–15.

Amatruda, J. F., Shepard, J. L., Stern, H. M., & Zon, L. I., (2002). Zebrafish as a cancer model system. *Cancer Cell, 1*(3), 229–231.

Appleby, M. W., & Ramsdell, F., (2003). A forward-genetic approach for analysis of the immune system. *Nature Rev. Immunol., 3*, 463–471.

Astone, M., Lai, J. K. H., Dupont, S., Stainier, D. Y. R., Argenton, F., & Vettori, A., (2018). Zebrafish mutants and TEAD reporters reveal essential functions for Yap and Taz in posterior cardinal vein development. *Sci. Rep., 8*, 10189.

Bailey, G. S., Williams, D. E., & Hendricks, J. D., (1996). Fish models for environmental carcinogenesis: The rainbow trout. *Environ. Health Perspect., 104*(suppl 1), 5–21.

Bailone, R. L., Aguiar, L. K De., Roca, R. De. O., Borra, R. C., Corrêa, T., Janke, H., et al., (2019). Zebrafish as an animal model for food safety research: Trends in the animal research. *Food Biotechnol., 33*(4), 283–302.

Bajoghli, B., Aghaallaei, N., Hess, I., Rode, I., Netuschil, N., Tay, B. H., Venkatesh, B., et al., (2009). Evolution of genetic networks underlying the emergence of thymopoiesis in vertebrates. *Cell, 138*(1), 186–197.

Bajoghli, B., Dick, A. M., Claasen, A., Doll, L., & Aghaallaei, N., (2019). Zebrafish and medaka: Two teleost models of T-cell and thymic development. *International Journal of Molecular Sciences, 20*(17), 4179.

Berman, J., Payne, E., & Hall, C., (2012). *The Zebrafish as a Tool to Study Hematopoiesis, Human Blood Diseases, and Immune Function*. Hindawi.

Boehm, T., & Bleul, C. C., (2007). The evolutionary history of lymphoid organs. *Nat. Immunol., 8*, 131–135.

Boehm, T., & Swann, J. B., (2013). Thymus involution and regeneration: Two sides of the same coin? *Nat. Rev. Immunol., 13*, 831–838.

Boehm, T., (2008). Thymus development and function. *Curr. Opin. Immunol., 20*, 178–184.

Boehm, T., (2011). Design principles of adaptive immune systems. *Nat. Rev. Immunol., 11*, 307–317.

Boehm, T., Hess, I., & Swann, J. B., (2012). Evolution of lymphoid tissues. *Trends in Immunology, 33*, 315–321.

Brownlie, A., Donovan, A., Pratt, S. J., Paw, B. H., Oates, A. C., Brugnara, C., et al., (1998). Positional cloning of the zebrafish sauternes gene: A model for congenital sideroblastic anaemia. *Nat. Genet., 20*(3), 244–250.

Bruneel, B., & Witten, P. E., (2015). Power and challenges of using zebrafish as a model for skeletal tissue imaging. *Connective Tissue Research, 56*(2), 161–173.

Burns, C. E., Traver, D., Mayhall, E., Shepard, J. L., & Zon, L. I., (2005). Hematopoietic stem cell fate is established by the notch-runx pathway. *Genes Dev., 19*(19), 2331–2342.

Chakraborty, C., Hsu, C. H., Wen, Z. H., Lin, C. S., & Agoramoorthy, G., (2009). Zebrafish: A complete animal model for *in vivo* drug discovery and development. *Current Drug Metabolism, 10*, 116–124.

Chakravarty, S., Reddy, B. R., Sudhakar, S. R., Saxena, S., Das, T., Meghah, V., et al., (2013). Chronic unpredictable stress (CUS)-induced anxiety and related mood disorders in a zebrafish model: Altered brain proteome profile implicates mitochondrial dysfunction. *PLoS One, 8*(5), e63302.

Chen, C. Y., Bertozzi, C., Zou, Z., Yuan, L., Lee, J. S., Lu, M., et al., (2012). Blood flow reprograms lymphatic vessels to blood vessels. *J. Clin. Invest., 122*, 2006–2017.

Chitramuthu, B. P., (2013). Modeling human disease and development in zebrafish. *Hum. Genet. Embryol., 3*(1), 1000e108.

Chitwood, B. G., & Chitwood, M. B., (1974). *Introduction to Nematology.* University Park, Baltimore.

Clark, K. J., & Ekker, S. C., (2015). How zebrafish genetics informs human biology. *Nat. Educ., 8*(4), 3.

Daroczi, B., Kari, G., McAleer, M. F., Wolf, J. C., Rodeck, U., & Dicker, A. P., (2006). *In vivo* radioprotection by the fullerene nanoparticle DF-1 as assessed in a zebrafish model. *Clinical Cancer Research, 12*, 7086–7091.

Deanesly, R., (1927). Memoirs: The structure and development of the thymus in fish, with special reference to Salmo Fario. *J. Cell Sci., 2*, 113–145.

Eckstein, J., (2022). ISOA/ARF Drug development tutorial, Alzheimer Research Forum, 1–18.

Evans, C. J., Hartenstein, V., & Banerjee, U., (2003). Thicker than blood: Conserved mechanisms in *Drosophila* and vertebrate hematopoiesis. *Dev. Cell, 5*, 673–690.

Facciol, A., Tran, S., & Gerlai, R., (2017). Re-examining the factors affecting choice in the light-dark preference test in zebrafish. *Behav. Brain Res., 327*, 21–28.

Ferrero, I., Koch, U., Claudinot, S., Favre, S., Radtke, F., Luther, S. A., & MacDonald, H. R., (2013). Dl4-mediated notch signaling is required for the development of fetal alpha beta and gamma delta t cells. *Eur. J. Immunol., 43*, 2845–2853.

Flajnik, M. F., (2018). A cold-blooded view of adaptive immunity. *Nat. Rev. Immunol., 18*, 438–453.

Frank, S. D., (2005). Target-based drug discovery: Is something wrong? *Drug Discovery Today, 10*(2), 139–147.

Goetz, J. G., Steed, E., Ferreira, R. R., Roth, S., Ramspacher, C., Boselli, F., et al., (2014). Endothelial cilia mediate low flow sensing during zebrafish vascular development. *Cell Rep., 6*, 799–808.

Gore, A. V., Monzo, K., Cha, Y. R., Pan, W., & Weinstein, B. M., (2012). Vascular development in the zebrafish. *Cold Spring Harbor Perspectives in Medicine, 2*(5), a006684.

Gut, P., Reischauer, S., Stainier, D. Y. R., & Arnaout, R., (2017). Little fish, big data: Zebrafish as a model for cardiovascular and metabolic disease. *Physiol. Rev., 97*(3), 889–938.

Guyon, J. R., Steffen, L. S., Howell, M. H., Pusack, T. J., Lawrence, C., & Kunkel, L. M., (2007). Modeling human muscle disease in zebrafish. *Biochimica et Biophysica Acta, 1772*(2), 205–215.

Handen, J. S., (2002). The industrialization of drug discovery. *Drug Discovery Today, 7*(2), 83–85.

Hansen, J. D., & Zapata, A. G., (1998). Lymphocyte development in fish and amphibians. *Immunol. Rev., 166*, 199–220.

Hart, P. C., et al., (2016). Experimental models of anxiety for drug discovery and brain research. *Methods Mol. Biol., 1438*, 271–291.

Haynes, B. F., Sempowski, G. D., Wells, A. F., & Hale, L. P., (2000). The human thymus during aging. *Immunologic Research, 22*, 253–261.

Hess, I., & Boehm, T., (2012). Intravital imaging of thymopoiesis reveals dynamic lympho-epithelial interactions. *Immunity, 36*, 298–309.

Hou, J., Fujimoto, T., Saito, T., Yamaha, E., & Arai, K., (2015). Generation of clonal zebrafish line by androgenesis without egg irradiation. *Sci. Rep., 5*, 13346.

Howarth, D. L., Yin, C., Yeh, K., & Sadler, K. C., (2013). Defining hepatic dysfunction parameters in two models of fatty liver disease in zebrafish larvae. *Zebrafish, 10*(2), 199–210.

Howe, K., Clark, M. D., Torroja, C. F., Torrance, J., Berthelot, C., Muffato, M., Collins, J. E., Humphray, S., McLaren, K., Matthews, L., et al., (2013). The zebrafish reference genome sequence and its relationship to the human genome. *Nature, 496*, 498–503.

Hozumi, K., Mailhos, C., Negishi, N., Hirano, K., Yahata, T., Ando, K., Zuklys, S., et al., (2008). Delta-like 4 is indispensable in thymic environment specific for t cell development. *J. Exp. Med., 205*, 2507–2513.

https://www.slideserve.com/yered/zebrafish-as-a-model-system (accessed on 3 February 2023).

Johnson, S. L., & Zon, L. I., (1999). Genetic backgrounds and some standard stocks and strains used in zebrafish developmental biology and genetics. *Methods Cell Biol., 60*, 357–359.

Kalueff, A. V., Echevarria, D. J., & Stewart, A. M., (2014). Gaining translational momentum: More zebrafish models for neuroscience research. *Prog. Neuropsychopharmacol. Biol. Psychiatry, 55*, 1–6.

Kalueff, A. V., Stewart, A. M., & Gerlai, R., (2014). Zebrafish as an emerging model for studying complex brain disorders. *Trends in Pharmacological Sciences, 35*(2), 63–75.

Kanungo, J., Cuevas, E., Ali, S. F., & Paule, M. G., (2014). Zebrafish model in drug safety assessment. *Current Pharmaceutical Design, 20*(34), 5416–5429.

Kari, G., Rodeck, U., & Dicker, A. P., (2007). Zebrafish: An emerging model system for human disease and drug discovery. *Clinical Pharmacology & Therapeutics, 82*(1), 70–80.

Kernen, L., Rieder, J., Duus, A., Holbech, H., Segner, H., & Bailey, C., (2020). Thymus development in the zebrafish (Danio rerio) from an ecoimmunology perspective. *J. Exp. Zool.*, 1–15.

Khan, F. R., & Alhewairini, S. S., (2018). Zebrafish (Danio rerio) as a model organism. *Curr. Trends Cancer Manag.*, 3–18.

Koch, U., Fiorini, E., Benedito, R., Besseyrias, V., Schuster-Gossler, K., Pierres, M., Manley, N. R., et al., (2008). Delta-like 4 is the essential, nonredundant ligand for Notch1 during thymic T cell lineage commitment. *J. Exp. Med., 205*, 2515–2523.

LaBrecque, D. R., Abbas, Z., Anania, F., Ferenci, P., Khan, A. G., Goh, K. L., et al., (2014). World gastroenterology organization global guidelines: Nonalcoholic fatty liver disease and nonalcoholic steatohepatitis. *J. Clin. Gastroenterol., 48*(6), 467–473.

Lam, S. H., Chua, H. L., Gong, Z., Wen, Z., Lam, T. J., & Sin, Y. M., (2002). Morphologic transformation of the thymus in developing zebrafish. *Developmental Dynamics, 225*, 87–94.

Lam, S. H., Wu, Y. L., Vega, V. B., Miller, L. D., Spitsbergen, J., Tong, Y., Zhan, H., et al., (2006). Conservation of gene expression signatures between zebrafish and human liver tumors and tumor progression. *Nature Biotechnology, 24*(1), 73–75.

Langenau, D. M., & Zon, L. I., (2005). The zebrafish: A new model of T-cell and thymic development. *Nat. Rev. Immunol., 5*(4), 307–317.

Langheinrich, U., Hennen, E., Stott, G., & Vacun, G., (2002). Zebrafish as a model organism for the identification and characterization of drugs and genes affecting p53 signaling. *Current Biology, 12,* 2023–2028.

Lawrence, C., Ebersole, J. P., & Kesseli, R. V., (2008). Rapid growth and out-crossing promote female development in zebrafish (Danio rerio). *Environ. Biol. Fishes, 81*, 239–246.

Lieschke, G. J., & Currie, P. D., (2007). Animal models of human disease: Zebrafish swim into view. *Nat. Rev. Genet., 8*(5), 353–367.

Lieschke, G. J., Oates, A. C., Crowhurst, M. O., Ward, A. C., & Layton, J. E., (2001). Morphologic and functional characterization of granulocytes and macrophages in embryonic and adult zebrafish. *Blood, 98*(10), 3087–3096.

Liew, W. C., & Orban, L., (2014). Zebrafish sex: A complicated affair. *Brief. Funct. Genomics., 13*, 172–187.

Lindsay, M. A., (2003). Target discovery. *Nature Reviews Drug Discovery, 2*(10), 831–838.

Liu, S., & Leach, S. D., (2011). Zebrafish models for cancer. *Annu. Rev. Pathol. Mech. Dis., 6*, 71–93.

Lozano, R., Naghavi, M., Foreman, K., Lim, S., Shibuya, K., Aboyans, V., et al., (2012). Global and regional mortality from 235 causes of death for 20 age groups in 1990 and 2010: A systematic analysis for the global burden of disease study 2010. *Lancet, 380*(9859), 2095–2128.

Lu, J., Peatman, E., Tang, H., Lewis, J., & Liu, Z., (2012). Profiling of gene duplication patterns of sequenced teleost genomes: Evidence for rapid lineage-specific genome expansion mediated by recent tandem duplications. *BMC Genomics, 13*, 246.

Lynch, H. E., Goldberg, G. L., Chidgey, A., Van, D. B. M. R., Boyd, R., & Sempowski, G. D., (2009). Thymic involution and immune reconstitution. *Trends Immunol., 30*, 366–373.

Ma, D., Wei, Y., & Liu, F., (2013). Regulatory mechanisms of thymus and T cell development. *Dev. Comp. Immunol., 39*(1, 2), 91–102.

MacRae, C. A., & Peterson, R. T., (2015). Zebrafish as tools for drug discovery. *Nat. Rev. Drug Discov., 14*(10), 721–731.

Manley, N. R., Richie, E. R., Blackburn, C. C., Condie, B. G., & Sage, J., (2011). Structure and function of the thymic microenvironment. *Frontiers in Bioscience (Landmark Ed)., 16*, 2461–2477.

McAleer, M. F., et al., (2005). Novel use of zebrafish as a vertebrate model to screen radiation protectors and sensitizers. *International Journal of Radiation Oncology, Biology, Physics., 61,* 10–13.

Miller, G. W., Chandrasekaran, V., Yaghoobi, B., & Lein, P. J., (2018). Opportunities and challenges for using the zebrafish to study neuronal connectivity as an endpoint of developmental neurotoxicity. *Neurotoxicology, 67*, 102–111.

Minchin, J. N., Scahill, C. M., Staudt, N., Busch-Nentwich, E. M., & Rawls, J. F., (2018). Deep phenotyping in zebrafish reveals genetic and diet-induced adiposity changes that may inform disease risk. *J. Lipid Res., 59*(8), 1536–1545.

Mirbahai, L., Williams, T. D., Zhan, H., Gong, Z., & Chipman, J. K., (2011). Comprehensive profiling of zebrafish hepatic proximal promoter CpG island methylation and its modification during chemical carcinogenesis. *BMC Genomics, 12*(1), 1–6.

Mizgirev, I., & Revskoy, S., (2010). Generation of clonal zebrafish lines and transplantable hepatic tumors. *Nat. Protoc., 5*(3), 383–394.

Murayama, E., Kissa, K., Zapata, A., Mordelet, E., Briolat, V., Lin, H. F., Handin, R. I., & Herbomel, P., (2006). Tracing hematopoietic precursor migration to successive hematopoietic organs during zebrafish development. *Immunity, 25*(6), 963–975.

Nagasawa-Masuda, A., & Terai, K., (2017). Yap/Taz transcriptional activity is essential for vascular regression via Ctgf expression and actin polymerization. *PLoS One, 12*, e0174633.

Nakajima, H., Yamamoto, K., Agarwala, S., Terai, K., Fukui, H., Fukuhara, S., et al., (2017). Flow-dependent endothelial YAP regulation contributes to vessel maintenance. *Dev. Cell, 40*, e6.

Nasiadka, A., & Clark, M. D., (2012). Zebrafish breeding in the laboratory environment. *ILAR J., 53,* 161–168.

Ng, M., Fleming, T., Robinson, M., Thomson, B., Graetz, N., Margono, C., et al., (2014). Global, regional, and national prevalence of overweight and obesity in children and adults during 1980–2013: A systematic analysis for the global burden of disease study 2013. *Lancet, 384*(9945), 766–781.

Nguyen, M., et al., (2013). Developing 'integrative' zebrafish models of behavioral and metabolic disorders. *Behav. Brain Res., 256*, 172–187.

Nichols, D. E., (2004). Hallucinogens. *Pharmacol. Ther., 101*, 131–181.

Nishio, S. I., Gibert, Y., Berekelya, L., Bernard, L., Brunet, F., Guillot, E., et al., (2012). Fasting induces CART down-regulation in the zebrafish nervous system in a cannabinoid receptor 1-dependent manner. *Mol. Endocrinol., 26*(8), 1316–1326.

O'Neill, J. G., (1989). Ontogeny of the lymphoid organs in an Antarctic teleost, *Harpagifer antarcticus* (*Notothenioidei*: *Perciformes*). *Dev. Comp. Immunol., 13*, 25–33.

O'Neill, J. G., (1989). Thymic development in two species of marine teleost; an Antarctic silverfish, *Pleuragramma antarcticum* boulenger, and a warmer-water sea bass, *Dicentrarchus labrax* (linnaeus). *Proc. NIPR Symp. Polar Biol., 2*, 83–93.

Palmer, D. B., (2013). The effect of age on thymic function. *Frontiers in Immunology, 4*, 316.

Pappalardo, A., Pitto, L., Fiorillo, C., Donati, M. A., Bruno, C., & Santorelli, F. M., (2013). Neuromuscular disorders in zebrafish: State of the art and future perspectives. *Neuromolecular Med., 15*(2), 405–419.

Patton, E. E., Widlund, H. R., Kutok, J. L., Kopani, K. R., Amatruda, J. F., Murphey, R. D., et al., (2005). BRAF mutations are sufficient to promote nevi formation and cooperate with p53 in the genesis of melanoma. *Curr. Biol., 15*(3), 249–254.

Peel, E., & Belov, K., (2017). Immune-endocrine interactions in marsupials and monotremes. *General and Comparative Endocrinology, 244*, 178–185.

Poureetezadi, S. J., Cheng, C. N., Chambers, J. M., Drummond, B. E., & Wingert, R. A., (2016). Prostaglandin signaling regulates nephron segment patterning of renal progenitors during zebrafish kidney development. *Elife., 5*, e17551.

Quaglino, D., Accorsi, A., Boraldi, F., & Ottaviani, E., (2014). Thymic maturation and programmed cell death. In: Malagoli, D., & Ottaviani, E., (eds.), *Eco-immunology*. Springer.

Reimers, M. J., Flockton, A. R., & Tanguay, R. L., (2004). Ethanol- and acetaldehyde-mediated developmental toxicity in zebrafish. *Neurotoxicology and Teratology, 26*(6), 769–781.

Rennekamp, A. J., & Peterson, R. T., (2015). 15 years of zebrafish chemical screening. *Curr. Opin. Chem. Biol., 24*, 58–70.

Rezzani, R., Nardo, L., Favero, G., Peroni, M., & Rodella, L. F., (2014). Thymus and aging: Morphological, radiological, and functional overview. *Age (Dordr), 36*, 313–351.

Rhodes, J., Hagen, A., Hsu, K., Deng, M., Liu, T. X., Look, A. T., & Kanki, J. P., (2005). Interplay of pu.1 and gata1 determines myelo-erythroid progenitor cell fate in zebrafish. *Dev. Cell., 8*(1), 97–108.

Romano, N., Fanelli, M., Maria, D. P. G., Scapigliati, G., & Mastrolia, L., (1999). Histological and cytological studies on the developing thymus of sharpsnout seabream, *Diplodus puntazzo*. *J. Anat., 194,* 39–50.

Rubinstein, A. L., (2006). Zebrafish assays for drug toxicity screening. *Expert Opinion on Drug Metabolism & Toxicology, 2*(2), 231–240.

Santoriello, C., & Zon, L. I., (2012). Hooked! modeling human disease in zebrafish. *The Journal of Clinical Investigation, 122*(7), 2337–2343.

Sarath, B. N., Murthy, C. L. N., Kakara, S., Sharma, R., Brahmendra, S. C. V., & Idris, M. M., (2016). 1-methyl-4-phenyl-1, 2, 3, 6-tetrahydropyridine induced Parkinson's disease in zebrafish. *Proteomics, 16*(9), 1407–1420.

Schlake, T., Schorpp, M., Nehls, M., & Boehm, T., (1997). The nude gene encodes a sequence-specific DNA binding protein with homologs in organisms that lack an anticipatory immune system. *Proc. Natl. Acad. Sci. USA., 94*, 3842–3847.

Schorpp, M., Bialecki, M., Diekhoff, D., Walderich, B., Odenthal, J., Maischein, H. M., Zapata, A. G., & Boehm, T., (2006). Conserved functions of Ikaros in vertebrate lymphocyte development: Genetic evidence for distinct larval and adult phases of T cell development and two lineages of B cells in zebrafish. *J. Immunol., 177*(4), 2463–2476.

Schorpp, M., Leicht, M., Nold, E., Hammerschmidt, M., Haas-Assenbaum, A., Wiest, W., & Boehm, T., (2002). A zebrafish orthologue (whnb) of the mouse nude gene is expressed in the epithelial compartment of the embryonic thymic rudiment. *Mech. Dev., 118*, 179–185.

Schulenburg, H., Kurz, C. L., & Ewbank, J. J., (2004). Evolution of the innate immune system: The worm perspective. *Immunol. Rev., 198*, 36–58.

Sehnert, A. J., Huq, A., Weinstein, B. M., Walker, C., Fishman, M., & Stainier, D. Y. R., (2002). Cardiac troponin T is essential in sarcomere assembly and cardiac contractility. *Nat. Genet., 31*(1), 106–110.

Semon, M., & Wolfe, K. H., (2007). Rearrangement rate following the whole-genome duplication in teleosts. *Mol. Biol. Evol., 24*, 860–867.

Sosa, M. A. G., De Gasperi, R., & Elder, G. A., (2012). Modeling human neurodegenerative diseases in transgenic systems. *Hum. Genet., 131*(4), 535–563.

Spitsbergen, J. M., & Kent, M. L., (2003). The state of the art of the zebrafish model for toxicology and toxicologic pathology research—Advantages and current limitations. *Toxicologic Pathology, 31*, 62–87.

Stewart, A. M., Braubach, O., Spitsbergen, J., Gerlai, R., & Kalueff, A. V., (2014). Zebrafish models for translational neuroscience research: From tank to bedside. *Trends Neurosci., 37*, 264–270.

Stewart, A. M., et al., (2015). Molecular psychiatry of zebrafish. *Mol. Psychiatry., 20*, 2–17.

Swann, J. B., Weyn, A., Nagakubo, D., Bleul, C. C., Toyoda, A., Happe, C., Netuschil, N., Hess, I., Haas-Assenbaum, A., Taniguchi, Y., et al., (2014). Conversion of the thymus into a bipotent lymphoid organ by replacement of foxn1 with its paralog, foxn4. *Cell Rep., 8*, 1184–1197.

Teame, T., Zhang, Z., Ran, C., Zhang, H., Yang, Y., Ding, Q., et al., (2019). The use of zebrafish (Danio rerio) as biomedical models. *Anim. Front., 9*(3), 68–77.

Thisse, C., & Zon, L. I., (2002). Organogenesis: Heart and blood formation from the zebrafish point of view. *Science, 295*, 457–462.

Ton, C., & Parng, C., (2005). The use of zebrafish for assessing ototoxic and otoprotective agents. *Hearing Research, 208*, 79–88.

Trede, N. S., & Zon, L. I., (1998). Development of T-cells during fish embryogenesis. *Dev. Comp. Immunol., 22*, 253–263.

Trede, N. S., Langenau, D. M., Traver, D., Look, A. T., & Zon, L. I., (2004). The use of zebrafish to understand immunity. *Immunity, 20*(4), 367–379.

Trede, N. S., Zapata, A., & Zon, L. I., (2001). Fishing for lymphoid genes. *Trends Immunol., 22*, 302–307.

Veldman, M. B., & Lin, S., (2008). Zebrafish as a developmental model organism for pediatric research. *Pediatr. Res., 64*(5), 470–476.

Walter, R. B., & Kazianis, S., (2001). Xiphophorus interspecies hybrids as genetic models of induced neoplasia. *ILAR J., 42*(4), 299–321.

Weinstein, B., (2002). Vascular cell biology *in vivo*: A new piscine paradigm? *Trends in Cell Biology, 12*(9), 439–445.

Willett, C. E., Cherry, J. J., & Steiner, L. A., (1997). Characterization and expression of the recombination activating genes (rag1 and rag2) of zebrafish. *Immunogenetics, 45*(6), 394–404.

Wu, R. S., Lam, I. I., Clay, H., Duong, D. N., Deo, R. C., & Coughlin, S. R., (2018). A rapid method for directed gene knockout for screening in G0 zebrafish. *Dev. Cell, 46*(1), 112–125.

Wu, X., Zhong, H., Song, J., Damoiseaux, R., Yang, Z., & Lin, S., (2006). Mycophenolic acid is a potent inhibitor of angiogenesis. *Arteriosclerosis, Thrombosis, and Vascular Biology, 26*, 2414–2416.

Zhang, C., Willett, C., & Fremgen, T., (2003). Zebrafish: An animal model for toxicological studies. *Current Protocols in Toxicology.*

Zhang, M., Di Martino, J. S., Bowman, R. L., Campbell, N. R., Baksh, S. C., Simon-Vermot, T., et al., (2018). Adipocyte-derived lipids mediate melanoma progression via FATP proteins. *Cancer Discov., 8*(8), 1006–1025.

Zhao, S., Huang, J., & Ye, J., (2015). A fresh look at zebrafish from the perspective of cancer research. *Journal of Experimental & Clinical Cancer Research, 34*(1), 80.

CHAPTER 8

Plant-Based Proteins in Fish Diets for Sustainable Coldwater Fisheries

K. T. HAFEEF ROSHAN, MOHAMMED MEHAROOF, and K. A. SAJINA

PhD Scholars, ICAR–Central Institute of Fisheries Education, Mumbai, Maharashtra, India

ABSTRACT

Coldwater fisheries are a crucial component of aquaculture that has been progressing over the years meeting the supply needs and contributing towards the economy. The increase in human population will generate a huge demand for fish and coldwater aquaculture has the potential to gracefully sustain this demand. However, for a sustainable culture, the selection and use of a feed with the required nutrient profile and cost-effective nature is one of the key factors that decide the success and growth. Fish meal (FM) and fish oil are the excellent sources of protein, lipid, and other essential nutrients which are mainly used in aqua feed formulation, but the limited availability of these resources necessitates to look for alternative protein and nutrient sources. While using feed ingredients to supplement fishmeal, it is imperative to consider the key elements such as nutritional value, acceptance, palatability, efficacy, availability, cost, and growth performance of animal. The conventional plant sources like terrestrial plant-based proteins, non-conventional plant sources are being investigated for potential feed contributions to substitute fishmeal but the anti-nutritional factors (ANFs), amino acid (AA) profiles, fatty acid profiles, mineral profiles and palatability are some of the challenges being faced. Several studies demonstrated that the use of feed additives reduces the negative effects of plant-based proteins and

Coldwater Fisheries and Aquaculture Management: Technology for Sustainable Food Production.
Mohd Ashraf Rather, Faisal Rashid Sofi, Adnan Amin, & Kawkabul Saba (Eds.)

increases the feed performance. Fish oil requirements also can be replaced using oils from genetically engineered plants that are rich in omega-3 fatty acids. Future research must be streamlined towards developing alternative plant-based proteins that maximize sustainable yield and triggers mariculture sector growth. The chapter touches on some important principles, practical implications, and potential challenges for the use of plant-based proteins in fish feed.

8.1 INTRODUCTION

Fisheries and aquaculture remain a significant sector and plays a vital role as a source of food, nutrition, income, and livelihood for people around the world. Aquaculture sector now accounts for around half of all food fish eaten and provides substantial food security and economic development and FAO (2014) revealed that the prospect of feeding human population to grow to 9.6 billion by 2050 which further ascertains the importance of the sector. To deal with the rising population and food security, coldwater fisheries could aptly fit as the next feasible alternative for expanding fish production and thus meeting the global demand and providing livelihood security. The word 'coldwater' refers to an aquatic ecosystem that maintains a temperature of less than 20°C and enough oxygen levels for fishes like salmon, trout, mahseers, snow trout, and other minor species (Sarma et al., 2018). Coldwater aquaculture is a specialized branch of aquaculture involving the culture of organisms that inhabits in fairly low water temperature in an enclosed section of the river (cages/pens), or in tanks, ponds or raceways for food and alternative products. Globally, the culture produces many high-value fishes of which the main finfish classes encompass the salmons, trout, Mahseer, and common carps. Within the coldwater body, particularly in exposed areas around the world, there is a huge potential for culture. The experts estimate that, if coldwater aquaculture is developed and promoted responsibly which can expand the production and greatly contribute to global nutritional and livelihood security (Sarma et al., 2018; Singh, 2018). This allows for a roadmap that may enable aquaculture to turn out to be a serious provider of food for humans in the 21st century and in the near future, in line with agriculture.

Despite the growth in aquaculture production, the economic gains of this sector have been systematically affected by increased competition for production factors and higher feed costs (Meharoof et al., 2020). In aquaculture, the cost of feed typically accounts for 60–80% of the operational expenses (Hasan, 2010) which varies with the type of feed (e.g., live, wet, moist, and dry feeds), quality of individual feed ingredients, feed production and the methodology of feed delivery. This will not only influence production, but also will have a huge effect on the composition of the species, where fish that need less and cheaper feed will produce more than a good candidate species. According to FAO (2020), the increase in feed cost coupled with increase in demand of fish will escalate the price to 22% by 2030. In order to ensure the development of the industry and to meet consumer requirements, it is therefore crucial to address these challenges in this sector with environment friendly and economical aquaculture feed ingredients. The use of plant-based resources for aquaculture production can strengthen the sector by reducing the cost of feed, thus meeting the increasing demand of fish without a rise in cost.

Significant progress has been made over the last decade to make a better use of plant-based feed resources for fish and crustaceans (Gatlin et al., 2007; Naylor et al., 2009; Aklakur et al., 2016). The limited accessibility of FM supplies, as well as lower production prices obtained from plant resources, motivated this transition. For aquaculture, the strategy to accelerate the fraction of plant product in existing pellet feeds was successful; a feed supply crisis in cold water fisheries has likely been met and continued to be supported with increased use of plant products as feed ingredients must be correlated with the key means of bringing the cultured species to a lower biological process level, as well as the biological process level of the many cultured species in nature (Kaushik & Troell, 2010; Tacon et al., 2010). FM is one of the most important ingredients required for aqua feed due to its rich content of extremely digested and well-balanced proteins, but it is a limited resource. In parallel, coldwater fisheries production is likely to increase soon and requires more protein derived from FM to be incorporated in feed. As a result, with a high demand for FM and fish oil with limited availability poses a challenge to the development of the rapidly growing and sustainable coldwater fisheries sector. This situation has prompted the exploration of sustainable aquaculture feed alternatives. Therefore, one of the driving forces behind the hunt for cheaper alternatives is the dramatic increase in the price of fishmeal since the beginning of the millennium. This review highlights the short-and long-term alternative feed resources that can be offered and takes a stand on how the feed requirements for coldwater fisheries are often met by and on the far side of 2050.

8.2 REPLACEMENT OF FISH MEAL (FM) IN COLD-WATER FISHERIES WITH PLANT DERIVATIVES: HURDLES AND PROSPECTS

Aquaculture has become the primary means of producing many of the candidate fish species and the use of compounded (formulated) feed to intensively produce farmed fish has increased. For the culture in both marine and inland, fishmeal is mainly used to be the source of protein in compounded feed. Until 2006, several developments had been made in substituting alternative protein sources for fishmeal proportions in aquafeeds and especially in salmon, trout, sea bream and sea bass feeds decreased by 25–50%, depending on species and phase of life history. Similarly, the percentage of fishmeal in feed for omnivorous fish species, especially in grow-out feeds, also declined. However, as aquaculture production and thus aquafeed production continued to increase, the use of fishmeal in the aquafeed sector grew gradually, accounting for 10% of annual fishmeal production in the early 1980s, compared with approximately 29% in 1995 as well as 50% of annual fishmeal production in 2005. Despite significant breakthroughs in farmed fish feed formulations, the substantial increase in FM prices, as well as the ongoing expansion of the spectrum of business coupled with increased feed prices and production costs, has been noticed. While prices have declined, feed costs remain the most urgent concern of the aquaculture industries are under enormous compulsion to develop cost effective formulations that sustain efficient growth at lower price per unit yield. It was relatively easy to substitute plant protein ingredients with fishmeal to provide approximately half the dietary protein, but it is difficult to replace higher percentages of fishmeal. Dependence of aquaculture on agricultural and fishery resources as fertilizer and feed inputs and increasing competition for such resources between aquaculture, human, and livestock sectors. The development of fish feed is mainly dependent on the use of FM in aquafeeds developed as the primary source of dietary protein. A significant number of studies have been undertaken recently to minimize reliance on fishmeal given the limited supply and increasing costs by research institutions and the aquaculture feed industry (Rana et al., 2009; Tacon et al., 2006). Promising candidates to substitute fishmeal as a protein source in aquafeeds are a large variety of plant protein ingredients, and they are easily accessible and moderately priced, facilitating the use of aquafeed plant protein sources. In order to sustain rapid growth rates and feed quality values at higher levels of fishmeal replacement, a range of obstacles need to be addressed. Recognized nutrient deficiencies in plant protein contribute to the central issue as it moves to replace higher levels of FM with plant proteins. A significant alternative source of protein already widespread in aquafeeds is corn gluten meal, but its amino acid (AA) profile and insoluble carbohydrate content counterpart to FM have disadvantages. The crude protein content of corn gluten meal, however, is slightly above its minimum guaranteed level of 60% implies that 40% of the gluten meal of maize is composed of non-protein material, primarily carbohydrates that are not soluble. In fish, non-soluble carbohydrates have no nutritional value (Stone, 2003) and are not added back to the protein fraction during processing, corn gluten meal can be processed to contain higher protein levels, but this method leaves producers with no outlet for the non-soluble carbohydrate fraction.

Nutritional limitations of plant proteins compared to fishmeal is another obstacle faced by the aquafeed industry as it transfers greater replacement of FM with plant proteins. Fishmeal is a complicated product that contains essential nutrients as well as a large number of biologically active compounds. Feed formulators combine plant protein concentrates and AA supplements to ensure that the AA content of feeds with reduced FM levels meets or exceeds the criteria for AAs in farmed fish. They can also supplement feed with mineral supplements such as di calcium phosphate or double the trace mineral premix to increase the levels of calcium, phosphorus, and trace mineral feed when FM is extracted from formulations of fish feed. This may not, however, be sufficient to resolve other deficiencies or imbalances that occur when the amount of FM in feed is lowered. Over time, researchers have detected a variety of dietary constituents that have been added to poultry feed, allowing formulators to reduce FM as a feed ingredient and eventually the removal of it. It's not the same as in aquafeeds because the undiscovered growth factors required for fish are less likely to be trace elements and amines like taurine and possibly steroids. However, like with nutritional deficiencies in all-plant diets, macro, and trace mineral deficiencies cannot be eliminated. Anti-nutritional compounds in plant proteins are the major challenge replacing fishmeal with plant protein concentrates. In general, proteins produced from oilseeds contain more anti-nutrients that are important to fish than proteins produced from grains. Some of the anti-nutrients were inactivated by processes involved in the processing of products or during pelleting by extrusion.

8.2.1 ANTINUTRITIONAL FACTORS

Anti-nutrient factors may be identified as substances which, by themselves or through their residual metabolic products, interfere with the use of feed and have an impact on the health and production of animals (Mohanta, 2012). Antinutritional factors can be classified narrowly into a few categories such as: (i) those that influence the digestion and use of proteins such as protease inhibitors, lectins (hemagglutinins), tannins, etc.; (ii) those that affect the use of minerals such as phytates, gossypol (pigment), oxalates, glucosinolates, etc.; and (iii) miscellaneous substances, such as mycotoxins, mimosin, cyanogens, alkaloids, phytoestrogens, saponins, etc. (Francis et al., 2001a). Inclusion of plant protein at improved degrees or entire substitute of animal proteins by plant proteins have resulted in negative boom and feed performance, which are chiefly approved to the existence of ANFs and beside the point AA balance (Francis et al., 2001). Awareness of the essence of antinutrients, their effects, potential modes of action and ways of reducing the effects or removing potent feed ingredients of antinutrients of plant origin have been studied in detail. The following gives a brief understanding of several important anti-nutritional variables:

1. **Protease Inhibitors:** These are the entities found in the entire plant kingdom, especially among legumes such as soybean with the ability to inhibit the proteolytic activity of certain enzymes (Norton, 1991). There may be several protease inhibitors within a single plant, showing variations in their structure, molecular weight, and mode of action. For example, soybeans show as many as five protease inhibitors that are primarily binds to digestive enzymes such as chymotrypsin or trypsin, rendering them partially or entirely inactive. All protease inhibitors are heat-labile and are easily destroyed by roasting or cooking.
2. **Lectins or Hemagglutinins:** Phytohemagglutinin or lectins are commonly distributed glycoproteins in legumes and certain oil seeds (including soybeans) that have an affinity for particular sugar molecules and are distinguished by their ability to treat antinutritional factors of plant origin. Hemagglutinins, the name came since they activate RBC agglutination (red blood corpuscles) and are shown to be present in several plant seeds. The growth-retardant properties of raw legumes due to protein digestion inhibition and pancreatic hyper-activity resulting in increased development of trypsin and chymotrypsin and consequent cystine loss have been reported to be partly responsible for these inhibitors. Lectins are non-immune proteins or glycoproteins, especially abundant in plants such as maize, beans, quinoa, peas, etc., that are capable of binding carbohydrates or glyco-conjugates without altering them. However, they are heat-labile, and are destroyed by heat processing or roasting.

3. **Tannins:** These are water-soluble phenolic compounds with a molecular weight greater than 500 Daltons, capable of precipitating proteins in the aqueous solution. Tannins that are hydrolysable and condensed (leguminous forage and seeds) are two different groups. They are heat stable and decrease the digestibility of proteins in animals, either by partially making protein inaccessible or by inhibiting digestive enzymes and increasing fecal nitrogen. It is evident that in plant feed, they inhibit the activities of trypsin, chymotrypsin, amylase, and lipase, decrease the content of protein in foods and interfere with the absorption of dietary iron. The activity of microbial enzymes, including cellulose and intestinal digestion, may be inhibited if the concentration of tannin in the food becomes too high. Tannins also form protein-insoluble complexes and these complexes may be responsible for the anti-nutritional effects. Tannins can establish a less digestive complex with dietary proteins and can bind and inhibit endogenous proteins, such as digestive enzymes and have documented intestinal disturbance, interference with iron absorption and the possibility of carcinogenic effects (Butler, 1989).
4. **Phytates:** Phytic acid, often known as phytate, is a type of vitamin that is found in plant seeds and is chemically known as vitamin myo-inositol hexa-phosphates. Soybean oilcake, sesame oil cake, rapeseed oilcake, and other regularly used feed components contain levels ranging from 10–75 g/kg. Matyka et al. (1993) estimated that organically bound phytin phosphorus accounts for 62–73% of total phosphorus (TP) in cereal grains and 46–73% in legume seeds, respectively. Phytic acid functions as a strong chelator, forming complexes of protein and mineral-phytic acid; decreased protein and mineral bioavailability are the net result (Khare & Nakajima, 2000). The phosphorus retained as phytic acid can be released by the digestive enzyme phytase. They form a less digestible complex of phytate protein, thereby limiting dietary protein availability and inhibits the function of gastrointestinal tyrosinase, trypsin, pepsin, lipase, and amylase (Liener, 1980; Hendricks & Bailey, 1989).
5. **Gossypol:** This is a secondary metabolite dependent on terpene, mainly present in cottonseed protein flour (Anderson et al., 2016) and plays an important role in the defense mechanisms of the cotton plant against pests and diseases (Romano & Scheffler, 2008). Gossypol is stored in the roots, seeds, leaves, and green bolls of

the plant in structures called "glands," and is considered to have anti-nutritional effects on warm-blooded animals and fish fed with cottonseed products (Romano & Scheffler, 2008). The average gossypol content of cotton seeds with gland ranges from 0.4–2.4% to less than 0.01% free gossypol in certain low-gossypol cottonseed meals (Liener, 1980; Robinson & Brent, 1989). It is hypothesized that its anti-nutritive effect is due to its binding to lysine and thus rendering this essential AA less accessible (Wilson et al., 1981) and thus adversely affecting the use of protein and the production of nitrogen waste. This cottonseed meal antinutrient may be minimized by genetic selection or alteration for "glandless" seeds with low levels of the lysigenous glands (Alam et al., 2018; Anderson et al., 2016). An alternative approach is the extraction of acidified polar solvents, which eliminates about 90–95% of gossypol (Pelitire et al., 2014).

6. **Oxalates:** Oxalic acid is a metabolic product produced in plants and animals through many pathways. Monovalent oxalates formed with divalent ions, such as calcium, magnesium, and iron, are almost insoluble in water, whereas ions such as sodium, potassium or ammonium are well soluble in water (Libert & Franceschi, 2000). These essential ranges are included in aquaculture diets: saltbush (*Atriplex halimus*), 1.0–3.0% 1.0–2.5% and alfalfa (*Medicago sativa*); 2.4%. In peptic digestion, oxalates impair calcium and magnesium metabolism and combine with proteins to form complexes that have an inhibitory effect.
7. **Glucosinolates:** In the *Brassica nigra* plant species of the family Crusiferae, these compounds are usually present. Glucosinolates are the main antinutritional factors present in rapeseed oilcake and mustard oilcake which are potential sources of protein in fish feed. These compounds have been shown to impair the proper functioning of thyroid glands in fish with problems such as decreased feed intake, feed utilization, metamorphosis, and maturation (NRC, 1993). Thyroid dysfunction is shown by continuous intake of feed-containing glucosinolate ingredients such as mustard oilcake ultimately affecting metabolism and fish development. It is difficult to set a glucosinolate threshold level in different formulated fish feeds because of the lack of data and the quantities of toxic products such as isothiocyanates and/or nitriles which are generated.

8. **Saponins:** There are naturally occurring steroid or triterpenoid glycosides in some feed ingredients derived from plants, such as legumes (18–41 mg/kg in various legume seeds) (Francis et al., 2001). The administration of soybean saponins has reported negative growth performance outcomes, but this appears to depend on either the raw material administered, potential interactions between dietary ingredients or species tolerance (Chen et al., 2011; Couto et al., 2015; Kokou et al., 2012). Saponins were treated as toxic because they appeared to be particularly toxic to fish and cold-blooded animals and in many of them, they had high hemolytic activity. Saponins, in high concentrations, add a bitter taste and astringency to dietary plants which is the key factor that limits its use. In Atlantic salmon, major effects of soy saponin with pea protein concentrate in feed have been observed by reducing feed intake, evident lipid digestibility, AAs and ash (Chikwati et al., 2012). Saponin levels greater than 2 g/kg will decrease feed palatability and result in the accumulation of unconsumed feed pellets.
9. **Non-Starch Polysaccharides (NSPs):** These are the key plant ingredients which contribute towards digestive disability and waste production. Since these ingredients primarily affect fecal physico-chemical properties, their effects should be eliminated by *in vitro* digestion or the addition of degrading NSP enzymes as a result of waste processing as well. Experiments involving cell wall NSPs from potential sources of vegetable protein and NSP-degrading feed enzymes at the same time may help to explain the fundamental mechanism by which their effects are exercised by NSPs and to provide a means to mitigate possible anti-nutritional effects. The use of NSP-degrading enzymes has been studied to reduce their effects on red sea bream (*Pagrus major*), golden sea bream, Japanese sea bass (*Lateolabrax japonicus*), rainbow trout and Atlantic salmon (Castillo & Gatlin, 2015; Ai et al., 2007; Carter et al., 1994; Dalsgaard et al., 2009; Farhangi & Carter, 2007; Ogunkoya et al., 2006). Enhancing the utilization of NSP-degrading enzymes and the carbohydrates by fish is a major nutritional bottleneck for the sustainability of the industry, while a mixture of pure protein concentrates will increase the retention of nutrients and minimize the release of waste to the environment.
10. **Alkaloids:** These are present in many legumes, but only seeds of the genus Lupinus have concentrations high enough to cause

problems (Petterson et al., 1997; Wasilewko and Buraczewska, 1999). They are chemical compounds formed by plants and have a bitter taste which serves as a barrier to food. Experiment on *Lupinus luteus*, which is one of the feed ingredients that have the highest protein and sulfur AA content (van Barneveld, 1999) which is fed to rainbow trout and observed a decrease in feed consumption. It is realized that the reason for the decrease in feed consumption is due to the presence of alkaloids like gramine, lupinine, and sparteine (Glencross et al., 2006). In order to prevent palatability concerns, potential inclusion should concentrate on the alkaloid levels in raw materials (Figure 8.1).

FIGURE 8.1 Anti-nutritional factors.

8.3 MEETING THE CHALLENGES IN A CONSTRUCTIVE MANNER

In the case of plant-based products, such as soy or various types of grain, there is an especially broad and promising range of alternative raw materials.

Generally, plant proteins are available in much greater amounts than fishmeal. Increasing nutritional quality through plant genetic manipulation can help in the enrichment of protein which also reduces the antinutritional factors. For aquafeeds, the nutrient and antinutrient profiles of plant products are not currently optimal. Fortunately, it is possible to change specific traits of grains, such as protein and oil content, by breeding and genetic modification. The molecular pathways associated with the expression of various nutrient traits have been determined through genomic and genetic studies. Genetic manipulations have improved over the last decade, with the ability to further enhance some specific nutrients in plants. The following are examples of improvements in many primary antinutrients and nutrients:

- Genetic materials with high lysine;
- Manipulation of starch-structure;
- Enhancing the use of fish through genetic selection;
- Increased content of micronutrients.

There is a great concern about the abundance of anti-nutritional factors (ANFs) and harmful effects in plants used as animal feed and ways and means should be found to remove or reduce their rate to the minimum (Soetan & Oyewole, 2009). By developing innovative processing techniques and technologies to increase the digestibility/nutritional value of nutrients and consistency may help in this regard. Feed-processing methods have been employed to improve the physical qualities and nutritional content of feed for several years, and the outcomes of different production methods are very well documented (Booth, Allan, & Warner-Smith, 2000; Cheng & Hardy, 2003; Venou et al., 2003). Cooking extrusion, for example, enhances carbohydrate digestibility (Allan & Booth, 2004) and creates a more durable pellet that can be managed to make the pellet sink (Barrows & Hardy, 2001). Until combining the ingredients together into a full feed, pre-processing refers to the treatment of particular ingredients. Pre-processing of ingredients using both biological enhancement and mechanical alteration is being studied to improve the nutritional value of plant derived ingredients for fish. Fermentation by yeasts, bacteria, and fungus has been investigated and found to have a lot of potential for removing antinutrients while also providing key nutrients like protein and AAs (Mukhopadhyay & Ray, 1999; Bairagi et al., 2002). The ultimate purpose of these processes is to increase the concentration of protein and decrease antinutrient levels. The reduction of foreign or fermentable carbohydrates and phytate is well within this approach's capabilities (Ng et al., 2002; Skrede et al., 2002). Microorganisms absorb and transform carbohydrates into cell mass that serves as a digestible source of both protein and energy from unavailable energy. The nutrient and antinutrient profile of the initial product will be determined by the type of microorganisms and plant substrates used in the fermentation. The primary factor influencing their use in future aquafeed formulations would be the costs of processing biologically enhanced plant products. An extension of existing pre-processing methods is the mechanical alteration of plant-derived ingredients. Until grinding and mixing, these techniques may be as easy as de-hulling barley to reduce the fiber content and thereby improve the protein content (Hardy & Barrows, 2000). Processing by extrusion can reduce the antinutritional factors in feed products. Extrusion is a short-term processing of high temperature (HTST) that incorporates many methods, including heat and mass transfer, mixing, shearing, reduction of particle size, melting, texture, caramelization, and forming. As growth agents (e.g., cinnamic acid derivatives) and contaminating microorganisms are more effectively eliminated or reduced, HTST processing is considered an efficient approach in terms of nutrient retention. This method has been used to inactivate or completely eliminate aflatoxins, which normally involve with conditions such as high shear and elevated temperatures (Saalia & Phillips, 2011). The benefits of extrusion include ANF degradation, starch gelatinization, increased soluble dietary fiber, and decreased lipid oxidation (Nikmaram, Kamani, & Ghalavand, 2015). Extrusion is also a very effective method of inactivating the function of a-amylase, trypsin, chymotrypsin, and hemagglutinin inhibitors without altering the amount of protein in food products (Soetan & Oyewole, 2009). Several other methods have been implemented in recent years, such as high-pressure processing (HPP), microwave and extrusion, as alternatives to minimize ANF levels (Zarei & Kafilzadeh, 2013). These manufacturing processes, which included non-thermal, limited heating or high temperature cooking over a short period of time, provided a greater capacity to manufacture finished products with a substantial increase in nutritional quality compared to traditional methods (Linsberger-Martin et al., 2013). In addition, some beneficial improvements in the functional properties of legumes and cereals are anticipated through these processing technologies and, in particular, the elimination or reduction of ANFs could be achieved (Ohlsson & Bengtsson, 2002).

Enhancing the palatability of feedstuffs from plants is imperative for the development of plant-based feeds as fish exhibit distinct taste preferences. Fish tastes and preferences are largely species-specific, and the distinctions between fish species may be seen when comparing the width and content of spectra for both stimulants and deterrents. There is a strong link between the evolution of the ontogenic gustatory system in fish and its ability to differentiate between the taste properties of food products. Bioassay and electrophysiological data comparisons suggest that palatability in the gustatory system is not associated with excitability. However, these studies examining the impact on filet consistency of different plant proteins, including SBM, SPC, soy flour, corn gluten, wheat gluten, peanut meal, canola meal, extruded peas, lupin seed meal and CSM, suggests that in about 40% of the studies, product quality was significantly affected by the source of protein. Color was the characteristic most affected by the source of protein, as calculated by instrumental and sensory methods. In some instances, texture, and flavor were significantly affected by the source of protein, particularly when large substitutions were made. Other research, however, in which 30–60% of the diet consisted of plant protein meals, identified no differences between filets in consumer acceptability (Gatlin et al., 2007). Clearly, further research is needed into the effects of plant protein meals on product quality, particularly those subject to different degrees of concentration and/or isolation.

Apart from this, monitoring the effect on the quality of fish products prepared out of fish fed with plant-based feeds and consumer health is very important. Market research on aquatic foods consistently shows that the single most significant factor influencing fish-buying behavior is product consistency. Consumers are becoming more aware of the health implications linked with eating fish and if we move forward towards increasing the use of plant products in aquafeeds, it is important to assess the impact of these newly developed diets on the quality of fish products and consumer health. Salmonids, namely rainbow trout and Atlantic salmon, were the subject of the vast majority of research, particularly those that used sensory approaches to assess quality. Gulf sturgeon, *Acipenser* sp. comprise non-salmonid species sporadically assessed for dietary effect on sensory quality. In most of the studies taste, color, odor, and/or texture have been significantly influenced by diet (Clydesdale, 1993) Given the physiological, nutritional, environmental, and compositional differences among cultured finfish, it is not possible to automatically apply conclusions drawn on the product quality of one species to another. In order to explain the effect of dietary ingredients on the quality of aquaculture species of interest, further research is needed. Although numerous studies have evaluated the effects on fish growth and feed efficiency of different alternative plant proteins, relatively few have tracked the dietary impact on the quality of fish.

8.4 PLANT-BASED FISH FEED SOURCES

For multiple species for commercial fish culture, the use of plant protein in fish feed has been investigated as these feed components are unique to species specifically based on their requirements. Grasses, veggies, aquatic weeds, plant leaves, roots, seeds, and seed extracts, leaves, fruits of certain plants, grains, oil cakes can be used for feed formulation (Mondal & Payra, 2015). Commonly use fish feed additives includes linseed, safflower, sunflower, soybean, cereals, and cereal products, broken rice, rice polish, sweet potato tubers, wheat bran, maize, cassava, and sorghum.

8.4.1 CANDIDATE PLANT PRODUCTS FOR USE IN AQUAFEEDS

8.4.1.1 SOYBEANS (GLYCINE MAX LINNAEUS)

Soybean is a leading oilseed crop that has an estimated production for 2018–2019 is 358.2 mmt and most of this production is used in the extraction of oil, making a high-quality protein cake. This is processed to yield a large variety of soybean products, such as soy flour, soybean meal (SBM), soybean protein concentrate (SPC) and soy protein isolate (SPI) that have been tested in fish. The predominant type of soybean used was SBM and is available either as dehulled (48% crude protein) or with added hulls (44% crude protein) (NRC, 1993). Soy products are known to be inexpensive and nutritious feed products with a high content of crude protein and a relatively balanced profile of AAs. In general, the concentrations of 10 essential amino acids (EAA) and tyrosine in SBM are lower than in FMs, with the exception of cystine, which is present in SBM at higher concentrations. Lysine, methionine, and threonine are the EAA of concern, which can be restrictive in soy-based diets fed to aquatic animals. With the processing of soy flakes, concentrations of these EAAs increase in SPC and SPI which exceed those found in FM. However, these items are not yet economical for large-scale use in aquafeeds due to the manufacturing costs. Crude concentrations of solvent extracted SBM fat, ash, and other soy products appear to be lower than those in the meal, but concentrations of carbohydrates tend to be higher. Appropriate supplementation with mineral premixes and lipids may resolve lower ash and fat concentrations in soy products, but high carbohydrate concentrations remain an area of concern.

8.4.1.2 CANOLA (BRASSICA RAPA LINNAEUS)

Canola is made from cultivars of rapeseed that have been developed to have low levels of erucic acid and glucosinolates. The crop of canola is an oilseed, and the primary product of its production is canola oil. For salmonids and other carnivorous species of farmed fish, Canola protein concentrate (CPC) has been widely evaluated as a protein source and supports growth rates comparable to those fed with FM-based diets. As long as AA supplements are employed to address restricted AA levels and feeding stimulants, such as betaine, are introduced to the diet to overcome reduced intake, this is the principal source of protein in aquafeeds. The protein concentrate of Canola has a comparable protein content to a high-quality FM, but it is not widely available and no market prices have been identified.

8.4.1.3 CORN (ZEA MAY LINNAEUS)

Corn oil is the primary food product of maize production, and while most of the maize grown is fed as an energy source to livestock. Corn starch, a commonly used sweetener in foods and drinks, is used to make over 400 products, including ethanol, paper coatings and corn syrup. The corn kernel is divided into its main components, bran fiber, germ, gluten, and starch, in the wet-milling process. The oil is extracted from the germ and applied to the corn gluten feed; the remaining corn germ meal is used. To form corn gluten meal, the gluten protein is condensed, filtered, and dried, and a minimum of 60% protein on a 'as-is' basis is guaranteed to include the widely traded corn gluten meal. The crude protein content of processed and purified maize gluten protein can be 70–73% crude protein, but there are limitations in the commercial production process, so that these amounts cannot always be met. Although the price will be higher than commercially traded corn gluten meal, a higher crude protein corn gluten meal would be a more acceptable commodity for use in aquafeeds. To decide whether the development of a high-protein corn gluten meal is warranted, economic, and marketing analysis is required. Corn gluten meal is being utilized in fish feed for salmon and other marine animals like European sea bass, *Dicentrarchus labrax Linnaeus,* and *Sparus aurata Linnaeus.*

8.4.1.4 COTTON SEED (GOSSYPIUM HIRSUTE LINNAEUS)

The third leading legume seed by weight (after soya and rapeseed) used worldwide is cottonseed and cottonseed meal (CSM) have an enormous potential for use in high protein aquafeeds due to its high protein content for human consumption and animals, as well as its low market price relative to other legumes and FM. Multiple studies evaluating CSM in catfish, salmonid, and tilapia diets have found that 40% of solvent retrieved may be utilized in aquaculture diets without causing growth decline. However, the technique of replacing one plant protein source with another was most commonly used in these experiments, while replacing FM or other animal protein sources would be more difficult and profitable. Lee et al. (2002) used a method with exceptionally positive results to fully substitute FM with a combination of animal by-products and both SBM and CSM in the juvenile rainbow trout diet. They also pointed out that the use of CSM in fish diets is affected by the source of CSM and associated processing. Lee et al. (2006) summarized a series of studies in rainbow trout in which CSM completely substituted FM over the three year period without significantly affecting female and male rainbow trout growth rates, while CSM diets had significantly lower assimilation of protein and phosphorus. In one study, *Ictalurus punctatus Rafinesque*, a channel catfish, was raised in earthen ponds at high densities and fed a diet containing 51% CSM with lysine supplementation to satisfy the diet (0.65%). Results showed that the rate of growth, dressing percentage and chemical composition of the filets did not vary significantly from the SBM-containing fish fed diets (42%). Fowler (1980) demonstrated that for two Pacific salmon species, up to 34% of CSM can be used in the diet to replace SBM without growth depression.

8.4.1.5 PEAS/PEAS LUPINS

Pisum sativum Linnaeus and *Lupinus* sp. *Linnaeus*, already used or considered for use in aquafeed for marine and freshwater fish worldwide, are traditional plant ingredients throughout the planet, both goods are manufactured in large amounts (Allan et al., 2000; Glencross, Boujardand, & Kaushik, 2003; Thiessen, Campbell, & Tyler, 2003). The nutrient profile of peas and lupin shows that they are capable of replacing a large proportion of FM protein in aquafeeds. The protein content of peas and lupin is moderate compared to animal meals such as FMs, meat, and bone meals, and poultry meals, while lysine and methionine are small and both contain high levels of carbohydrates (Allan et al., 2000). Starch is the primary carbohydrate energy reserve in peas, and can make up around 55% of the total crop. Lupins, on the other hand, contain negligible starch levels (0.3%), with non-starch polysaccharides (NSPs) such as b-(1,4)-galactan serving as the primary carbohydrate storage energy (Van Barneveld, 1999). For a variety of pea products for rainbow trout, Thiessen, Campbell, & Tyler (2003) recorded protein apparent digestibility coefficient (ADC) values in excess of 90% and also showed that dietary levels of 25% dehulled peas could be used to replace soyabean meal. Glencross et al. (2003a), however, published protein ADC values of 495% for a variety of red sea bream lupin varieties, *Pagrus auratus* Forster. When fed with lupin at up to 60% of the diet, Glencross et al. (2003a) have reported acceptable growth of juvenile red sea bream. Energy ADCs for pea and lupine products were poor in all of the studies described above (51–78%).

8.4.1.6 WHEAT

Wheat (*Triticum aestivum Linnaeus* and *T. dicoccoides, var. durum*) unlike other cereals has at least six broad subdivisions of cultivars that are based on the grain's end-use properties. The presence of large seed storage protein that forms the cross-linking networks of gluten when hydrated four is mixed is the functional characteristic of wheat which distinguishes it from other cereal grains. Gluten gives the dough elastic characteristics and the ability to hold leavening agent-generated gasses. Owing to its strength and restricted water solubility, gluten is also an important binder in aquafeeds. Flaking (roll milling) of whole grains or pelleting of mill feed and grains has been used in conventional wheat preparations for diets. In composition publications, the wheat kernel is frequently believed to be 12% protein by weight at 12% humidity (Bushuk & Wrigley, 1974). Like other cereals, whole grain wheat is mainly an energy source rather than a source of protein based on its high starch composition. In addition to protein and starch, wheat kernels are about 2% crude fiber and about 2% ash (Martin, Leonard, & Stamp, 1976). Potassium (0.40%), phosphorus (0.38%) and sulfur (0.20%) are the main components of the ash fraction, with calcium (0.04–0.05%) as a comparatively minor mineral portion (Guttieri et al., 2004). In the aleurone layer in phytin, phytic acid with chelated metal ions, phosphorus in wheat is mainly sequestered. Phytic acid mutants in wheat will dramatically reduce the concentration of phytic acid and increase the nutritionally more readily available amounts of inorganic phosphorus in the diet (Guttieri et al., 2004). Wheat embryo proteins contain up to 8% lysine and 2% methionine (Pomeranz et al., 1970). Mill feed, however, has many water feed disadvantages. Second, since most of the digestible carbohydrates in the processing of flour are eliminated, the energy quotient is decreased. Secondly, in the milling process, the glutenin and gliadin storage proteins that form gluten are also extracted, with the remaining bran proteins usually of lower value for use as a feed binder. The remaining mill feed for monogastric animals is high in crude fiber (10%) with minimal digestibility and has a more concentrated phytic acid composition that reaches almost 0.5% of total weight (Guttieri et al., 2004).

8.5 LEAF PROTEIN SOURCES

8.5.1 MORINGA LEAF MEAL (MORINGA OLEIFERA)

A highly nutritious supplement to the diet of plant-eating fish such as tilapia, barbs, fancy cars, etc., is the fresh leaves of *Moringa oleifera* or 'drumstick.' They contain rich proteins, lipids, vitamins, and minerals, and hence the leaves, kernels, and pods are also used in the feed industry for aquaculture. Except for saponins and phenols, the leaf is free of other antinutritional variables. Ogbe & Affiku (2021) reported that moringa is nutritionally rich and can be used at low levels in the fish diet. Fresh Moringa leaves provide additional protein, vitamins, and AAs such as methionine, cystine, tryptophan that can promote fish growth and health. Amisah et al. (2009) stated that the optimum amount of inclusion of 20% of moringa leaves can be used for African catfish in the formulation diet. Afuang et al. (2003) also reported that 30% of FM in *Oreochromis niloticus* can be substituted by solvent extracts of moringa leaves. However, in contrast, Tilapia fed with raw moringa leaf meal revealed that no adverse effect on growth output was caused by 10% substitution of FM-based dietary protein (Ritcher et al., 2003). Magouz et al. (2016) concluded that the addition of 12% of *Moringa oleifera* leaves can be recommended as a partial feed replacement in the Nile tilapia diet without any adverse effects on perfect growth. Moringa leaves have been successfully used to partially replace traditional diets without compromising the growth potential of Nile tilapia.

Puycha et al. (2017) recorded that moringa leaf could be included at 10% without a negative impact on growth, feed utilization, digestibility, and serum biochemistry in the Bocourti catfish (*Pangasius bocourti*) diet. Adewumi

(2014) reported that *Moringa oleifera* leaf meal can be used without adverse effects on survival and growth efficiency to substitute up to 10% of the FM in the *C. gariepinus* fry diet. Yuangsoi & Masumoto (2012) recorded that up to 20% soybean protein replacement without adverse effect on growth and digestibility could be used for fancy carp (*Cyprinu carpio*) diets with moringa leaf diet containing ingredients.

8.5.2 SWEET POTATO LEAVES

Sweet potato, *Ipomoea batatas* is widely cultivated around the world and is one of the most important food crops in tropical regions as well. Due to the high protein and fiber content, the leaves were used as forage for the cattle. Adewolu (2008) stated that the content of crude protein in sweet potato leaves ranged from 26.5 to 32.5%. According to the study, the plant's leaf meal contains 26–33% crude protein, as well as significant levels of AAs, minerals, and vitamins. He also mentioned that one of the advantages of utilizing this plant meal in fish feed is that it is accessible multiple times a year, making it easier and less expensive to obtain. Oven or sun-drying, boiling or steaming and grinding will eliminate the presence of ANFs such as invertase and protease inhibitors prior to use in fish feed, thereby improving palatability. Experiments with *Tilapia zilli* showed that, sweet potato leaf meal could be used in diets at up to 15% without sacrificing growth and feed quality (Adewolu, 2008).

8.5.3 MULBERRY LEAF MEAL

Due to the high protein content and mineral elements, *Morus alba* or mulberry leaves grown for the silkworm industry are often used in fish feed formulations. Deficiencies in essential AAs, the presence of ANFs and complex carbohydrates are responsible for limiting its use (NRC, 1993). Bairagi et al. (2002) proposed that fermentation may be a simple and inexpensive way of growing the ANFs in it. Mondal et al. (2012) stated that a possible protein source in the diet of *Labeo bata* may be mulberry leaf meal. For Nile tilapia, Cruz & Laudencia (1978) used mulberry leaf meal, resulting in increased growth and lower FCR at 60% inclusion in combination with 40% rice bran. Bag et al. (2012) found that including leaf meal into the stinging catfish diet improved growth and survival, increased fish acceptance, and improved immunity to common illnesses.

8.5.4 AZOLLA

An underwater free-floating fern that has the capacity to fix and assimilate atmospheric nitrogen is a mosquito plant or Azolla. It grows rapidly and in 2–3 days it has the ability to double its weight. It is also known as a "super plant" because of its high nutrient content, which includes vital AAs, vitamins (vitamin A, vitamin B12, and Beta-carotene), growth boosters, and minerals such as calcium, phosphorus, potassium, ferrous, copper, magnesium, and others (Sheeno & Sahu, 2006). It has a high protein content (19–31%), as well as a low lignin content, making it a popular element in animal feed. Azolla pinnata has been identified as a possible source of protein in the fish feed industry (Micha et al., 1988). When azolla was added to the fish diet, Antoine et al. (1986) observed enhanced feed usage in *Tilapia mossambica* and higher growth in Rohu, Nile tilapia, Common carp, Silver carp, and Mrigal. Azolla powder can be used to substitute up to 25% in tilapia fishmeal (El-Sayeed, 1992), and *Etroplus suratensis* (Joseph et al., 1994), whereas 45% in *Cirrhinus mrigala* fry (Gangadhar et al., 2014), and 42% in *Oreochromis niloticus* fry (Santiago & Lovell, 1988).

8.5.5 WATER HYACINTH

A rapidly growing, herbaceous, free-floating aquatic plant native to the Amazon Basin is the water hyacinth, *Eichhornia crassipes*. Owing to their invasive existence and the culture of it is limited in some parts of the world, this plant is considered to be the world's worst aquatic weed." Lareo & Bressani (1982) stated that for various animals, fresh, and composted water hyacinth is used as feed. They recorded 38% protein content and 17–26% minerals in the dry matter of the leaves. Many researchers did not recommend the use of water hyacinth as feed for large-scale fish diets unless the hyacinths were composted or fermented (Wersal & Madsen, 2012). El-Sayed (2003) stated that up to 25% inclusion level of fermented water hyacinth can be integrated into the diet of Nile tilapia. He recommended fermentation in order to increase the fish's palatability and better consumption. They indicated that composted hyacinth water can be used to substitute up to 75% of the Nile tilapia diet for fishmeal. The inclusion of up to 18.9% inclusion amount of water hyacinth meal (dry and ground) did not compromise the growth and survival of Matrincha fish (*Brycon* sp.), according to Saint-Paul et al. (1981). They did say, however, that high-level inclusion was not recommended since the tannin in the plant interferes with the protein's digestion.

8.5.6 DUCKWEEDS

Duckweeds are a floating aquatic plant that grows luxuriously in tropical and subtropical areas in freshwater bodies. It is made up of four genera, viz. Spirodela, Lemna, Wolfilla, and Wolffiella. Duckweeds contain 15–43% crude protein, 5–30% fibers, 5% lipids, and are frequently used in the animal feed industry (Cheng & Stomp, 2009; Mohapatra & Patra, 2013). Duckweed was suggested for use in fish diets because of its rich protein, trace minerals, K and P, and colors, particularly carotene and xanthophyll. Common carp, Thai sharputi, raj puti, silver carp, mrigal, and tilapia diets have all benefited from the usage of new duckweed (Kabir et al., 2009). When duckweed was included in the diet of common carp, Mohapatra & Patra (2013) discovered a substantial difference in growth when 45% duckweed was supplied together with feed. Fishmeal may not be replaceable with duckweed meal, but can be used to supplement the diet with up to 15% inclusion level to decrease costs without sacrificing growth (Ogello et al., 2014). Das & Ray (1989) reported that without affecting the growth and survival of the fish, dried duckweed (*Lemna polyrhiza*) could be potentially used as a feed ingredient for *Labeo rohita* fingerlings. Yilmaz et al. (2004) suggested lemna leaf meal fermentation for the formulation of fish feed. They also recorded that duckweed in the feed of common carp can replace up to 20% of commercial FM. And Fasakin et al. (1999) outlines that the addition of up to 30% solar-dried duckweed (*Spirodela polyrrhiza*) FM substitute in the Nile tilapia diet improves healthy growth and is cost-effective. Mbagwu et al. (1990) reported improved growth and survival of diet-fed mango tilapia fingerlings containing 10% duckweed (*Lemna paucicostata*).

8.5.7 ALFALFA MEAL

Alfalfa (Medicago sativa) is a pea-family flowering plant (Fabaceae), cultivated as a major forage crop. It is known in the United Kingdom, Australia, and New Zealand as Lucerne and in southern Asia as Lucerne grass. Because of its high protein content, Alfalfa is widely cultivated throughout the world as a forage for cattle. Physicians used young alfalfa leaves in early Chinese medicine to treat digestive and kidney-related disorders. Dried alfalfa leaves in powder form have been used in the preparation of fish feed and tested under laboratory conditions on the diet of a freshwater fish, *Cirrhinus mrigala* (Mrigal Carp) and found better growth nutrient utilization (Maity & Patra, 2008). The sole skill of *Medcago sativa* is not to replace conventional FMs. It is therefore used along with the conventional ingredients as a supplementary fish feed ingredient; however, the level of incorporation of traditional ingredients such as groundnut oilcake, rice bran, etc., was held at a lower level of inclusion.

8.5.8 SUBABUL (LEUCAENA LEUCOCEPHALA)

Wee & Wang (1987) have shown that the inclusion of up to 25% of the total protein with soaked *Leucaena* leaf meal (LLM) has no adverse effects on *O. niloticus* development. In Indian Major Carp, with *Leucaena* as a dietary protein source, improved growth performance (Ghatnekar et al., 1983). Osman et al. (1996) reported that the higher percentage of dried or cooked LLM in tilapia diets significantly improved weight gain, basic growth rate, feed conversion ratio (FCR) and protein consumption parameters. Man et al. (2017) have suggested that in diets for *Macrobrachium rosenbergii*, leucocephala meal could substitute 25% for FM in juveniles. Amisah et al. (2009) concluded that *Leucaeana leucocephala* leaf meal can be up to 30% included in *Clarias gariepinus* at inclusion diets. Tiamiyu et al. (2015) recorded that at 20% levels without adverse impact, *Leucaeana leucocephala* leaf meal can be included in the diets of *Clarias gariepinus.*

8.5.9 CASSAVA WASTE (MANIHOT ESCULENTACRANTZ)

Kana et al. (2012) found that body weight was highest in birds fed diets in which 50% of the maize was replaced by cassava flour meal. It was found that 15% cassava meal can replace coconut meal in broiler diets with no adverse effect on growth performance (Ravindran et al., 1986). In 10% of Nile tilapia fingerling diets, cassava leaf meal produced the best growth, FCR and survival ratio (Nnaji et al., 2010). Lacerd et al. (2005) reported that the 30% inclusion level of cassava root meal can replace up to 100% maize grain with no detrimental effect on growth efficiency in the diet of *Ctenopharyngodon idella* fingerlings.

8.5.10 CHROMOLAENA ODORATA (SIAM WEED/DEVIL WEED/ FRENCH WEED/COMMUNIST WEED/COMMUNIST PACHA)

Chromolaena odorata leaves are of high nutritive value and might have the potential to be used as a protein supplement to ruminants (Apori et al., 2000). Igboh et al. (2009) also suggested that *C. odorata* is a source of high-quality protein which could serve as a potential source of protein supplement. Apori et al. (2000) reported the crude protein content of *C. odorata* leaf is above 194 g kg^{-1} dry matter. The nutrient composition of *C. odorata* leaf meal and AA composition, mineral composition and its availability is comparable with cassava leaf meal (Aro et al., 2009). *C. odorata* leaf meal can effectively include up to 30% in in the diet of *Labeo rohita* fingerlings as an alternate ingredient to improve the growth (Sajina et al., 2021). There are some reports that *C. odorata* leaf meal can be incorporated into the feed of rabbits up to the level of 30% (Bamikole et al., 2004). There are studies which found that dietary supplementation of *C. odorata* leaf extract powder improved feed utilization and growth rate and survival of *Clarias gariepinus* (Dada & Sonibare, 2015).

8.6 CONCLUSION

Slow growth rate of fish and high-cost feeds have been seen as major obstacles in the rapid development and expansion of coldwater fish production. Plant based fish feeds are gaining attraction in aquaculture because it is one of the sustainable and cost-effective methods for future cold water aquaculture practice. On the one hand, it can be cost effective, and moreover, it can lead to the sector's sustainability. Many aspects of the study can be evaluated, and innovative methods and tactics for effective incorporation of plant-based nutrients in fish diets can be developed, paving the way for a prosperous future for the aquaculture sector. A better understanding of the nutritional and physiological effects of possible plant-based feeds on fish would aid in the development of corrective techniques to counteract the negative effects. Further studies on bioactive compounds in plants and its effect in animal body could help in improved acceptance and utilization of plant stuffs by the animal. There should be some strategies for improving the effective use of plant-based diets in coldwater aquaculture. Research has to be streamlined in these aspects, and the improved understanding of this field leads to several applications and development of plant-based feed technologies in the coldwater fisheries sector. Slowly but surely the sector will achieve sustainability in the near future.

KEYWORDS

- **alternative feed**
- **anti-nutritional factors**
- **coldwater fisheries**
- **fish meal**
- **high-pressure processing**
- **non-starch polysaccharide**
- **nutrition**
- **plant protein**
- **sustainability**

REFERENCES

Adewolu, M. A., (2008). Potentials of sweet potato (*Ipomoea batatas*) leaf meal as dietary ingredient for Tilapia zilli fingerlings. *Pakistan Journal of Nutrition, 7*(3), 444–449.

Adewumi, A. A., (2014). *Moringa oleifera* (Lam) as a protein supplement in *Clarias gariepinus* diet. *Advances in Research*, 580–589.

Afuang, W., Siddhuraju, P., & Becker, K., (2003). Comparative nutritional evaluation of raw, methanol extracted residues and methanol extracts of moringa (*Moringa oleifera* Lam.) leaves on growth performance and feed utilization in Nile tilapia (*Oreochromis niloticus* L.). *Aquaculture Research, 34*(13), 1147–1159.

Ai, Q., Mai, K., Zhang, W., Xu, W., Tan, B., Zhang, C., & Li, H., (2007). Effects of exogenous enzymes (phytase, non-starch polysaccharide enzyme) in diets on growth, feed utilization, nitrogen and phosphorus excretion of Japanese seabass, *Lateolabrax japonicus*. *Comparative Biochemistry and Physiology Part A: Molecular & Integrative Physiology, 147*(2), 502–508.

Aklakur, M., Asharf, R. M., & Kumar, N., (2016). Nanodelivery: An emerging avenue for nutraceuticals and drug delivery. *Critical Reviews in Food Science and Nutrition, 56*(14), 2352–2361.

Alam, M. S., Watanabe, W. O., Carroll, P. M., Gabel, J. E., Corum, M. A., Seaton, P., & Dowd, M. K., (2018). Evaluation of genetically-improved (glandless) and genetically-modified low-gossypol cottonseed meal as alternative protein sources in the diet of juvenile southern flounder *Paralichthys lethostigma* reared in a recirculating aquaculture system. *Aquaculture, 489*, 36–45.

Allan, G. L., & Booth, M. A., (2004). Effects of extrusion processing on digestibility of peas, lupins, canola meal and soybean meal in silver perch *Bidyanus bidyanus* (Mitchell) diets. *Aquaculture Research, 35*(10), 981–991.

Allan, G. L., Parkinson, S., Booth, M. A., Stone, D. A., Rowland, S. J., Frances, J., & Warner-Smith, R., (2000). Replacement of fish meal in diets for Australian silver perch, *Bidyanus bidyanus*: I. Digestibility of alternative ingredients. *Aquaculture, 186*(3, 4), 293–310.

Amisah, S., Oteng, M. A., & Ofori, J. K., (2009). *Growth Performance of the African Catfish, Clarias Gariepinus, Fed Varying Inclusion Levels of Leucaena Leucocephala Leaf Meal, 13*(1).

Anderson, A. D., Alam, M. S., Watanabe, W. O., Carroll, P. M., Wedegaertner, T. C., & Dowd, M. K., (2016). Full replacement of menhaden fish meal protein by low-gossypol cottonseed flour protein in the diet of juvenile black sea bass *Centropristis striata*. *Aquaculture, 464*, 618–628.

Antoine, T., Carraro, S., Micha, J. C., & Van, H. C., (1986). Comparative appetency for *Azolla* of *Cichlasoma* and *Oreochromis* (tilapia). *Aquaculture, 53*(2), 95–99.

Apori, S. O., Long, R. J., Castro, F. B., & Orskov, E. R., (2000). Chemical composition and nutritive value of leaves and stems of tropical weed *Chromolaena odorata*. *Grass and Forage Science, 55*, 77–81.

Aro, S. O., Osho, I. B., Aletor, V. A., & Tewe, O. O., (2009). *Chromolaena odorata* in livestock nutrition. *Journal of Medicinal Plants Research, 3*, 1253–1257.

Bag, M. P., Ghorai, M., Mahapatra, S. C., Rao, P. S., & Pal, H., (2012). Evaluation of mulberry (*Morus alba*, Linn.) leaf meal as a complete diet for sting fish (*Heteropneustes fossilis*, Bloch.*). International Journal of Pharmacy & Life Sciences, 3*(9).

Bairagi, A., Ghosh, K. S., Sen, S. K., & Ray, A. K., (2002). Duckweed (*Lemna polyrhiza*) leaf meal as a source of feedstuff in formulated diets for rohu (*Labeo rohita* Ham.) fingerlings after fermentation with a fish intestinal bacterium. *Bioresource Technology, 85*(1), 17–24.

Bamikole, M. A., Ikhatua, U. J., Arigbede, O. M., Babayemi, O. J., & Etela, I., (2004). An evaluation of the acceptability as forage of some nutritive and antinutritive components and of the dry matter degradation profiles of five species of Ficus. *Tropical Animal Health and Production, 36*, 157–167.

Barrows, F. T., & Hardy, R. W., (2001). Nutrition and feeding. In: Wedemeyer, G., (ed.), *Fish Hatchery Management* (2nd edn., pp. 483–558). John Wiley & Sons, New York, NY, USA.

Booth, M. A., Allan, G. L., & Warner-Smith, R., (2000). Effects of grinding, steam conditioning and extrusion of a practical diet on digestibility and weight gain of silver perch, *Bidyanus bidyanus*. *Aquaculture, 182*(3, 4), 287–299.

Bushuk, W., & Wrigley, C. W., (1974). Proteins: Composition, structure and function. In: Inglett, G. E., (ed.), *Wheat: Production and Utilization* (p. 119). AVI, Westport, CT, USA.

Butler, L. G., (1989). Effects of condensed tannin on animal nutrition. In: *Chemistry and Significance of Condensed Tannins* (pp. 391–402). Springer, Boston, MA.

Carter, C. G., Houlihan, D. F., Buchanan, B., & Mitchell, A. I., (1994). Growth and feed utilization efficiencies of seawater Atlantic salmon, *Salmo salar* L., fed a diet containing supplementary enzymes. *Aquaculture Research, 25*(1), 37–46.

Castillo, S., & Gatlin, III. D. M., (2015). Dietary supplementation of exogenous carbohydrase enzymes in fish nutrition: A review. *Aquaculture, 435*, 286–292.

Chen, W., Ai, Q., Mai, K., Xu, W., Liufu, Z., Zhang, W., & Cai, Y., (2011). Effects of dietary soybean saponins on feed intake, growth performance, digestibility and intestinal structure in juvenile Japanese flounder (*Paralichthys olivaceus*). *Aquaculture, 318*(1, 2), 95–100.

Cheng, J. J., & Stomp, A. M., (2009). Growing duckweed to recover nutrients from wastewaters and for production of fuel ethanol and animal feed. *Clean–Soil, Air, Water, 37*(1), 17–26.
Cheng, Z. J., & Hardy, R. W., (2003). Effects of extrusion processing of feed ingredients on apparent digestibility coefficients of nutrients for rainbow trout (*Oncorhynchus mykiss*). *Aquaculture Nutrition, 9*(2), 77–83.
Chikwati, E. M., Venold, F. F., Penn, M. H., Rohloff, J., Refstie, S., Guttvik, A., & Krogdahl, Å., (2012). Interaction of soyasaponins with plant ingredients in diets for Atlantic salmon, *Salmo salar* L. *British Journal of Nutrition, 107*(11), 1570–1590.
Clydesdale, F. M., (1993). Color as a factor in food choice. *Critical Reviews in Food Science and Nutrition, 33*(1), 83–101.
Couto, A., Kortner, T. M., Penn, M., Østby, G., Bakke, A. M., Krogdahl, Å., & Oliva-Teles, A., (2015). Saponins and phytosterols in diets for European sea bass (*Dicentrarchus labrax*) juveniles: Effects on growth, intestinal morphology and physiology. *Aquaculture Nutrition, 21*(2), 180–193.
Cruz, E. M., & Laudencia, I. L., (1978). Screening of feedstuffs as ingredients in the rations of Nile tilapia [freshwater fish, in the Philippines]. *Philippine Journal of Biology*.
Dada, A. A., & Sonibare, O. F., (2015). Effect of dietary administration of the herbal additive siam weed (*Chromolaena odorata*) on growth performance and haematological changes in *Clarias gariepinus* fingerlings. *Journal of Fisheries, 3*, 221–226.
Dalsgaard, J., Ekmann, K. S., Pedersen, P. B., & Verlhac, V., (2009). Effect of supplemented fungal phytase on performance and phosphorus availability by phosphorus-depleted juvenile rainbow trout (*Oncorhynchus mykiss*), and on the magnitude and composition of phosphorus waste output. *Aquaculture, 286*(1, 2), 105–112.
Das, I., & Ray, A. K., (1989). Growth performance of Indian major carps *Labeo rohita* (Ham.) on duckweed incorporated pelleted feed: A preliminary study. *J. Inland Fish, 21*(1), 1–6.
El-Sayed, A. F. M., (2003). Effects of fermentation methods on the nutritive value of water hyacinth for Nile tilapia *Oreochromis niloticus* (L.) fingerlings. *Aquaculture, 218*(1–4), 471–478.
El-Sayed, A. F., (1992). Effects of substituting fish meal with *Azolla pinnata* in practical diets for fingerling and adult Nile tilapia, *Oreochromis niloticus* (L.). *Aquaculture Research, 23*(2), 167–173.
FAO, (2014). *The State of World Fisheries and Aquaculture 2014: Opportunities and Challenges*. Rome.
FAO, (2020). *The State of World Fisheries and Aquaculture 2020*. Sustainability in action. Rome.
Farhangi, M., & Carter, C. G., (2007). Effect of enzyme supplementation to dehulled lupin-based diets on growth, feed efficiency, nutrient digestibility and carcass composition of rainbow trout, *Oncorhynchus mykiss* (Walbaum). *Aquaculture Research, 38*(12), 1274–1282.
Fasakin, E. A., Balogun, A. M., & Fasuru, B. E., (1999). Use of duckweed, *Spirodela polyrrhiza* L. Schleiden, as a protein feedstuff in practical diets for tilapia, *Oreochromis niloticus* L. *Aquaculture Research, 30*(5), 313–318.
Fowler, L. G., (1980). Substitution of soybean and cottonseed products for fish meal in diets fed to chinook and coho salmon. *The Progressive Fish-Culturist, 42*(2), 87–91.
Francis, G., Makkar, H. P. S., & Becker, K., (2001a). Effects of *Quillaja saponins* on growth, metabolism, egg production and muscle cholesterol in individually reared Nile tilapia (*Oreochromis niloticus*). *Comparative Biochemistry and Physiology Part C: Toxicology & Pharmacology, 129*(2), 105–114.

Francis, G., Makkar, H. P., & Becker, K., (2001). Antinutritional factors present in plant-derived alternate fish feed ingredients and their effects in fish. *Aquaculture, 199*(3, 4), 197–227.

Gangadhar, B., Sridhar, N., Saurabh, S., Raghavendra, C. H., Hemaprasanth, K. P., Raghunath, M. R., & Jayasankar, P., (2014). Growth response of *Cirrhinus mrigala* fry to Azolla (*Azolla pinnata*) incorporated diets. *Fishery Technology, 51*(3).

Gatlin, III. D. M., Barrows, F. T., Brown, P., Dabrowski, K., Gaylord, T. G., Hardy, R. W., & Overturf, K., (2007). Expanding the utilization of sustainable plant products in aquafeeds: A review. *Aquaculture Research, 38*(6), 551–579.

Ghatnekar, S. D., Auti, D. G., & Kamat, V. S., (1983). Feeding Leucaena to *Mozambique tilipia* and Indian major carps. In: *Leucaena Research in the Asian Pacific Region: Proceedings of a Workshop Held in Singapore*. IDRC, Ottawa, ON, CA.

Glencross, B. D., Boujard, T., & Kaushik, S. J., (2003). Influence of oligosaccharides on the digestibility of lupin meals when fed to rainbow trout, *Oncorhynchus mykiss*. *Aquaculture, 219*(1–4), 703–713.

Glencross, B., Curnow, J., Hawkins, W., Kissil, G. W. M., & Peterson, D., (2003a). Evaluation of the feed value of a transgenic strain of the narrow-leaf lupin (*Lupinus angustifolius*) in the diet of the marine fish, *Pagrus auratus*. *Aquaculture Nutrition, 9*(3), 197–206.

Glencross, B., Evans, D., Rutherford, N., Hawkins, W., McCafferty, P., Dods, K., & Sipsas, S., (2006). The influence of the dietary inclusion of the alkaloid gramine, on rainbow trout (*Oncorhynchus mykiss*) growth, feed utilization and gastrointestinal histology. *Aquaculture, 253*(1–4), 512–522.

Guttieri, M. J., Bowen, D., Dorsch, J. A., Souza, E., & Raboy, V., (2004). Identification and characterization of a low phyticbacid wheat. *Crop Science, 44*, 418–424.

Hardy, R. W., & Barrows, F. T., (2000). Diet formulation and manufacturing. In: Halver, J. E., & Hardy, R. W., (eds.), *Fish Nutrition* (3rd edn., pp. 505–600). Academic Press, New York, NY, USA.

Hasan, M. R., (2010). On-farm feeding and feed management in aquaculture. *FAO Aquaculture Newsletter*, (45), 48.

Hendricks, J. D., & Bailey, G. S., (1989). Adventitious toxins. In: Halver, J. E., (ed.), *Fish Nutrition* (pp. 605–651). New York, USA: Academic Press Inc.

Igboh, M. N., Ikewuchi, C. J., & Ikewuchi, C. C., (2009). Chemical profile of *Chromolaena odorata* L. (King and Robinson) leaves. *Pakistan Journal of Nutrition, 8*, 52–524.

Joseph, A., Sherief, P. M., & James, T., (1994). Effect of different dietary inclusion levels of *Azolla pinnata* on the growth, food conversion and muscle composition of *Etroplus suratensis* (Bloch). *Journal of Aquaculture in the Tropics, 9*(1), 87–94.

Kabir, A. N. M. A., Hossain, M. A., & Rahman, M. S., (2009). Use of duckweed as feed for fishes in polyculture. *Journal of Agriculture & Rural Development*, 157–160.

Kana, J. R., Fulefack, H. D., Gandjou, H. M., Ngouana, R., Noubissie, M. M., Ninjo, J., & Teguia, A., (2012). Effect of cassava meal supplemented with a combination of palm oil and cocoa husk as alternative energy source on broiler growth. *Archiva Zootechnica, 15*(4), 17.

Kaushik, S. A. C. H. I., & Troell, M., (2010). Taking the fish-in fish-out ratio a step further. *Aquaculture, 35*(1).

Khare, S. K., & Nakajima, M., (2000). Immobilization of *Rhizopus* japonicus lipase on celite and its application for enrichment of docosahexaenoic acid in soybean oil. *Food Chemistry, 68*(2), 153–157.

Kokou, F., Rigos, G., Henry, M., Kentouri, M., & Alexis, M., (2012). Growth performance, feed utilization and non-specific immune response of gilthead sea bream (*Sparus aurata* L.) fed graded levels of a bioprocessed soybean meal. *Aquaculture, 364*, 74–81.

Kyeong-Jun, L., Jacques, R., Konrad, D., Igor, B., Joseph, S. O., & James, E. C., (2006). Long-term effects of dietary cottonseed meal on growth and reproductive performance of rainbow trout: Three-year study. *Animal Feed Science and Technology, 126*(1, 2), 93–106.

Lacerd, C. H. F., Hayashi, C., Soares, C. M., Boscolo, W. R., & Kavata, L. C. B., (2005). Replacement of corn Zea mays L. by cassava *Manihot esculenta* crants meal in grass-carp *Ctenopharyngodon idella* fingerlings diets. *Acta Scientiarum-Animal Sciences, 27*(2), 241–245.

Lareo, L., & Bressani, R., (1982). Possible utilization of the water hyacinth in nutrition and industry. *Food and Nutrition Bulletin, 4*(4), 1–6.

Lee, K. J., Dabrowski, K., Blom, J. H., Bai, S. C., & Stromberg, P. C., (2002). A mixture of cottonseed meal, soybean meal and animal byproduct mixture as a fish meal substitute: Growth and tissue gossypol enantiomer in juvenile rainbow trout (*Oncorhynchus mykiss*). *Journal of Animal Physiology and Animal Nutrition, 86*(7, 8), 201–213.

Liang, J. K., & Lovell, R. T., (1971). Nutritional value of water hyacinth in channel catfish feeds. *Hyacinth Control Journal, 9*(1), 40–41.

Libert, B., & Franceschi, V. R., (1987). Oxalate in crop plants. *Journal of Agricultural and Food Chemistry, 35*(6), 926–938.

Liener, I. E., (1980). Heat labile anti-nutritional factors. In: Summerfield, R. J., & Bunting, A. H., (eds.), *Advances in Legume Science* (pp. 157–170). London, United Kingdom: Royal Botanic Gardens, Kew.

Linsberger-Martin, G., Weiglhofer, K., Phuong, T. P. T., & Berghofer, E., (2013). High hydrostatic pressure influences antinutritional factors and *in vitro* protein digestibility of split peas and whole white beans. *LWT-Food Science and Technology, 51*(1), 331–336.

Magouz, F. I., Eweedah, N. M., Khalafalla, M. M., & Seham, A. A., (2016). Nutritional evaluation of *Moringa oleifera* leaves as unconventional feed stuff in the diets of the Nile tilapia (*Oreochromis niloticus*) fingerlings. *Journal of Agriculture Research. Kafr El-Sheikh Univ., 42*(4), 144–155.

Maity, J., & Patra, B. C., (2008). Effect of replacement of fishmeal by *Azolla* leaf meal on growth, food utilization, pancreatic protease activity and RNA/DNA ratio in the fingerlings of *Labeo rohita* (Ham.). *Can. J. Pure Appl. Sci., 2*(2), 323–333.

Man, M. D., Mamat, N. Z., Affendi, I. S. M., Nih, S., & Nadzri, A., (2017). Partial replacement of fish meal by white leadtree meal in diets for juveniles of Giant River Prawn, *Macrobrachium rosenbergii* (De Man, 1879*). International Journal of Fisheries and Aquatic Studies, 5*(2), 154–157.

Martin, J., Leonard, W., & Stamp, D., (1976). *Principles of Field Crop Production.* New York: Macmillan Publishing Co., inc.

Matyka, S., Bogusz, G., & Korol, W., (1993). Phytate contents in cereal grains, legume and rape seeds. *Biul. Inform. Przem. Pasz., 32*, 37–43.

Mbagwu, I. G., Okoye, F. C., & Adeniji, H. A., (1990). Studies on the Use of Duckweed (*Lemna Paucicostata hegelm.*) as Fish Food. *22*(25). 86–92.

Meharoof, M., Siju, R., Kumar, R. R., & Reshma, K. J., (2020). Manipulation of gut microbiota: An aquaculture perspective. In: Sundaray, J. K., (ed.), *Research Trends in Fisheries and Aquatic Sciences* (1st edn., Vol. 8, pp. 37–53). Akinik Publications. New Delhi.

Micha, J. C., Antoine, T., Wery, P., & Van, H. C., (1988). Growth, ingestion capacity, comparative appetency and biochemical composition of *Oreochromis niloticus* and *Tilapia rendalli* fed with azolla. In: *Second International Symposium on Tilapia in Aquaculture: ICLARM Conference Proceedings* (pp. 347–355). Bangkok: Department of Fisheries.

Mohanta, K. N., (2012). Plant feed resources of India. *Invited Papers on Application of Solid-State Fermentation Technology in Aquaculture*, 100–113.

Mohapatra, S. B., & Patra, A. K., (2013). Utilization of water hyacinth (*Eichhornia crasipes*) meal as partial replacement for fish meal on the growth performance of *Cyprinus carpio* fry. *International Research Journal of Biological Sciences, 2*(12), 85–89.

Mondal, K., & Payra, P., (2015). A review on use of plant protein sources in diets for fish feed formulation. *Journal of International Academic Research for Multidisciplinary, 3*(5), 257–264.

Mondal, K., Kaviraj, A., & Mukhopadhyay, P. K., (2012). Effects of partial replacement of fishmeal in the diet by mulberry leaf meal on growth performance and digestive enzyme activities of Indian minor carp *Labeo bata. Int. J. of Aquatic Science, 3*(1), 72–83.

Mukhopadhyay, B. N., & Ray, A. K., (1999). Utilization of copra meal in the formulation of compound diets for rohu, *Labeo rohita*, fingerlings. *Journal of Applied Ichthyology, 15*(3), 127–131.

Naylor, R. L., Hardy, R. W., Bureau, D. P., Chiu, A., Elliott, M., Farrell, A. P., & Nichols, P. D., (2009). Feeding aquaculture in an era of finite resources. *Proceedings of the National Academy of Sciences, 106*(36), 15103–15110.

Ng, W. K., Lim, H. A., Lim, S. L., & Ibrahim, C. O., (2002). Nutritive value of palm kernel meal pretreated with enzyme or fermented with *Trichoderma koningii* (Oudemans) as a dietary ingredient for red hybrid tilapia (*Oreochromis sp.*). *Aquaculture Research, 33*(15), 1199–1207.

Nikmaram, N., Kamani, M. H., & Ghalavand, R., (2015). The effects of extrusion cooking on antinutritional factors, chemical properties and contaminating microorganisms of food. *International Journal of Farming and Allied Sciences, 4*(4), 352–354.

Nnaji, J. C., Okoye, F. C., & Omeje, V. O., (2010). Screening of leaf meals as feed supplements in the culture of *Oreochromis niloticus*. *African Journal of Food, Agriculture, Nutrition and Development, 10*(2).

Norton, G., (1991). Proteinase inhibitors. In: D' Mello, F. J. P., Duffus, C. M., & Duffus, J. H., (eds.), *Toxic Substances in Crop Plants* (pp. 68–106). The Royal Society of Chemistry, Thomas Graham House, Science Park, Cambridge.

NRC (National Research Council), (1993). *Nutrient Requirements of Fish* (p. 114). National Academy Press, Washington, DC, USA.

Ogbe, A. O., & Affiku, J. P., (2021). Proximate study, mineral and anti-nutrient composition of *Moringa oleifera* leaves harvested from Lafia, Nigeria: Potential benefits in poultry nutrition and health. *Journal of Microbiology, Biotechnology and Food Sciences, 2021*, 296–308.

Ogello, E. O., Munguti, J. M., Sakakura, Y., & Hagiwara, A., (2014). Complete replacement of fish meal in the diet of Nile tilapia (*Oreochromis niloticus* L.) grow-out with alternative protein sources. A review. *IJAR, 2*, 962–978.

Ogunkoya, A. E., Page, G. I., Adewolu, M. A., & Bureau, D. P., (2006). Dietary incorporation of soybean meal and exogenous enzyme cocktail can affect physical characteristics of faecal material egested by rainbow trout (*Oncorhynchus mykiss*). *Aquaculture, 254*(1–4), 466–475.

Ohlsson, T., & Bengtsson, N., (2002). Minimal processing of foods with non-thermal methods. *Minimal Processing Technologies in the Food Industry,* 34–60.

Osman, M. F., Omar, A. E., & Nour, A. M., (1996). The use of *Leucaena* leaf meal in feeding Nile tilapia. *Aquaculture International, 4*(1), 9–18.

Pelitire, S. M., Dowd, M. K., & Cheng, H. N., (2014). Acidic solvent extraction of gossypol from cottonseed meal. *Animal Feed Science and Technology, 195*, 120–128.

Petterson, D. S., Sipsas, S., & Mackintosh, J. B., (1997). *The Chemical Composition and Nutritive Value of Australian Pulses* (2nd edn.). Grains Research and Development Corporation.

Pomeranz, Y., Carvajal, M. I., Hoseney, R. C., & Ward, A. B., (1970). Wheat germ in bread making. I. Composition of germ lipids and germ protein fractions. *Cereal Chemistry, 47*, 373–380.

Pratoomyot, J., Bendiksen, E. Å., Bell, J. G., & Tocher, D. R., (2010). Effects of increasing replacement of dietary fishmeal with plant protein sources on growth performance and body lipid composition of Atlantic salmon (*Salmo salar* L.). *Aquaculture, 305*(1–4), 124–132.

Puycha, K., Yuangsoi, B., Charoenwattanasak, S., Wongmaneeprateep, S., Niamphithak, P., & Wiriyapattanasub, P., (2017). Effect of moringa (*Moringa oleifera*) leaf supplementation on growth performance and feed utilization of Bocourti's catfish (*Pangasius bocourti*). *Agriculture and Natural Resources, 51*(4), 286–291.

Ravindran, V., Kornegay, E. T., Rajaguru, A. S. B., Potter, L. M., & Cherry, J. A., (1986). Cassava leaf meal as a replacement for coconut oil meat in broiler diets. *Poultry Science, 65*(9), 1720–1727.

Richter, N., Siddhuraju, P., & Becker, K., (2003). Evaluation of nutritional quality of moringa (*Moringa oleifera* Lam.) leaves as an alternative protein source for Nile tilapia (*Oreochromis niloticus* L.). *Aquaculture, 217*(1–4), 599–611.

Robinson, E. H., & Brent, J. R., (1989). Use of cottonseed meal in channel catfish feeds. *Journal of the World Aquaculture Society, 20*(4), 250–255.

Romano, G. B., & Scheffler, J. A., (2008). Lowering seed gossypol content in glanded cotton (*Gossypium hirsutum* L.) lines. *Plant Breeding, 127*(6), 619–624.

Ryuh, Y. S., Yang, G. H., Liu, J., & Hu, H., (2015). A school of robotic fish for mariculture monitoring in the sea coast. *Journal of Bionic Engineering, 12*(1), 37–46.

Saalia, F. K., & Phillips, R. D., (2011). Degradation of aflatoxins by extrusion cooking: Effects on nutritional quality of extrudates. *LWT-Food Science and Technology, 44*(6), 1496–1501.

Saint-Paul, U., Werder, U., & Teixeira, A. S., (1981). Use of water hyacinth in feeding trials with Matrinchã (*Brycon sp.*). *Journal of Aquatic Plant Management, 19*, 18–22.

Sajina, K. A., Sahu, N. P., Varghese, T., & Gupta, S., (2021). Siam weed (*Chromolaena odorata*) meal as digestive enhancer and replacer of DORB in the diet of *L. rohita*: Effect on growth, digestive enzymes and amylase gene expression. *Aquaculture Research, 52*(5), 2335–2347.

Santiago, C. B., & Lovell, R. T., (1988). Amino acid requirements for growth of Nile tilapia. *The Journal of Nutrition, 118*(12), 1540–1546.

Sarma, D., Akhtar, M. S., & Singh, A. K., (2018). *Coldwater Fisheries Research and Development in India.*

Sheeno, T. P., & Sahu, N. P., (2006). Use of freshwater aquatic plants as a substitute for fishmeal in the diet of *Labeo rohita* fry. *Journal of Fisheries and Aquatic Science, 1*(2), 126–135.

Singh, P., Paul, B. N., & Giri, S. S., (2018). Potentiality of new feed ingredients for aquaculture: A review. *Agricultural Reviews, 39*(4), 282–291.

Skrede, G., Storebakken, T., Skrede, A., Sahlstrøm, S., Sørensen, M., Shearer, K. D., & Slinde, E., (2002). Lactic acid fermentation of wheat and barley whole meal flours improves digestibility of nutrients and energy in Atlantic salmon (*Salmo salar* L.) diets. *Aquaculture, 210*(1–4), 305–321.

Soetan, K. O., & Oyewole, O. E., (2009). The need for adequate processing to reduce the anti-nutritional factors in plants used as human foods and animal feeds: A review. *African Journal of Food Science, 3*(9), 223–232.

Stone, D. A., (2003). Dietary carbohydrate utilization by fish. *Reviews in Fisheries Science, 11*(4), 337–369.

Tacon, A. G. J., Hasan, M. R., & Subasinghe, R. P., (2006). Use of fishery resources as feed inputs to aquaculture development: Trends and policy implications. In: *FAO Fisheries Circular* (Vol. 1018). Rome: FAO.

Tacon, A. G. J., Hasan, M. R., Allan, G., El-Sayed, A. F. M., Jackson, A., Kaushik, S. J., & Viana, M. T., (2010). Aquaculture feeds: Addressing the long-term sustainability of the sector. In: *Proceedings of the Global Conference on Aquaculture* (pp. 193–232).

Thiessen, D. L., Campbell, G. L., & Tyler, R. T., (2003). Utilization of thin distillers' solubles as a palatability enhancer in rainbow trout (*Oncorhynchus mykiss*) diets containing canola meal or air-classified pea protein. *Aquaculture Nutrition, 9*(1), 1–10.

Tiamiyu, L. O., Okomoda, V. T., & Agbo, A. O., (2015). Nutritional suitability of *Leucaena* leaf meal in the diet of *Clarias gariepinus*. *Journal of Fisheries Sciences.com, 9*(1), 351.

Van, B. R. J., (1999). Understanding the nutritional chemistry of lupin (*Lupinus spp.*) seed to improve livestock production efficiency. *Nutrition Research Reviews, 12*(2), 203–230.

Venou, B., Alexis, M. N., Fountoulaki, E., Nengas, I., Apostolopoulou, M., & Castritsi-Cathariou, I., (2003). Effect of extrusion of wheat and corn on gilthead sea bream (*Sparus aurata*) growth, nutrient utilization efficiency, rates of gastric evacuation and digestive enzyme activities. *Aquaculture, 225*(1–4), 207–223.

Wasilewko, J., & Buraczewska, L., (1999). Chemical composition including content of amino acids, minerals, and alkaloids in seeds of three lupin species cultivated in Poland. *J. Anim. Feed Sci., 8*(1), 1–12.

Wee, K. L., & Wang, S. S., (1987). Nutritive value of *Leucaena* leaf meal in pelleted feed for Nile tilapia. *Aquaculture, 62*(2), 97–108.

Wersal, R. M., & Madsen, J. D., (2012). Aquatic plants their uses and risks. *A Review of the Global Status of Aquatic Plants.* FAO, Rome.

Wilson, R. P., Robinson, E. H., & Poe, W. E., (1981). Apparent and true availability of amino acids from common feed ingredients for channel catfish. *The Journal of Nutrition, 111*(5), 923–929.

Yılmaz, E., Akyurt, İ., & Günal, G., (2004). Use of duckweed, Lemna minor, as a protein feed-stuff in practical diets for common carp, *Cyprinus carpio*, fry. *Turkish Journal of Fisheries and Aquatic Sciences, 4*(2), 105–109.

Yuangsoi, B., & Masumoto, T., (2012). Replacing moringa leaf (*Moringa oleifera*) partially by protein replacement in soybean meal of fancy carp (*Cyprinus carpio*). *Songklanakarin Journal of Science & Technology, 34*(5).

Zarei, M., & Kafilzadeh, F., (2013). Effects of using radiation processing in nutrition science and their restriction: A review. *International Journal of Advanced Biological and Biomedical Research, 1*(3), 222–231.

CHAPTER 9

Small-Scale Fisheries by Indigenous Fishing Methods

SHABIR AHMAD DAR[1], GOHAR B. WANI[1], FAISAL RASHID[2], and KAWKABUL SABA[3]

[1]Division of Fishery Engineering, Faculty of Fisheries, SKUAST-K, Jammu and Kashmir, India

[2]Division of Post-Harvest Technology, Faculty of Fisheries, SKUAST-K, Jammu and Kashmir, India

[3]Division of Fish Nutrition and Biochemistry, Faculty of Fisheries, SKUAST-K, Jammu and Kashmir, India

ABSTRACT

The fishing craft and gears operated in the lake were simple tools used by local fishermen for commercial catches of fish production. Against this background, the study was attempted to explore the different fishing methods operated by fisher folk of Dal Lake, Srinagar, Kashmir. During the present investigation one type craft and six types of gears were observed in the lake. The craft was plank built wooden boat, while the gears were long line, Cast net, Narchoo, Scoop net, Gillnet, and Rod and Line. The fishers use indigenous materials and methods for construction, fabrication, and maintenance of fishing gears and crafts. A total of nine species of fishes were recorded during the present study viz., *Cyprinus carpio, Carassius carassius, Schizothorax niger, S. esocinus, S. curvifrons, Crossochielus diplochilus, Trplophysa* spp., *Puntius conchonius,* and *Ctenppharygodon idella.* The dominant species during the study period was *C. carpio* which contributed 63.7% of the total catch by biomass and 41.64% by number on the whole.

Coldwater Fisheries and Aquaculture Management: Technology for Sustainable Food Production.
Mohd Ashraf Rather, Faisal Rashid Sofi, Adnan Amin, & Kawkabul Saba (Eds.)

9.1 INTRODUCTION

Fishing methods of Kashmir valley include a wide range of fishing gears. The gears mainly used for fishing in Kashmir valley are Cast net, Gill net, long line, Hook, and Line, Scoop net, Trap, and Harpoon displayed in Figure 9.1(a) (Dar et al., 2014). Among various fishing nets, cast net is the main type of gear used in the valley (Sunder et al., 1978; Yousuf, 1996; Shabir, Najar, & Ashraf, 2010) and these existing fishing methods in the Dal Lake may be termed as primitive (Anonymous, Annual Report, 1977). The main fishing gear used in the Dal Lake is the cast net which is made of six pieces each with different mesh sizes (Anonymous, Annual Report, 1977). Cast nets are used by hand force to enclose and catch fish as shown in Figure 9.1(b) and (c). These nets are used from the shore, dam, or boat (Nedelec & Prado, 1989).

Gillnet is one of the oldest types of fishing gear and is widely used to harvest diverse fish species (Sainsbury, 1996). Modern nets are made of synthetic fibers such as polyamide and can be monofilament or multifilament or a combination of both in case of more than one panel as shown in Figure 9.1(d) and (e). But the gillnets are in general, considered as having a high degree of selectivity, in terms of fish species, as well as the size of the fish, which directly depends on the size of the mesh (Anonymous, 2000). Technologically, gillnets are simple, easy to mend, require little in the way of on-board equipment and are relatively cheap to purchase (Hovgard & Lassen, 2000).

The other methods of fishing operated in Kashmir are the Scoop net, Narsoo (multi-headed spear), rod & line and long lines (Anonymous, 1977; Shabir, Najar, & Ashraf, 2010). Sunder et al. (1978); and Yousuf (1996) also reported about the fishing techniques employed in the Kashmir Valley. They reported about the use of nets mostly the traditional ones like Long Line locally called *wael raz*, Hook & line and Harpoon (*Narsoo*). There has been no improvement in such gears in Kashmir and as such efforts are needed towards improvement of the gears and crafts used for making them more economic and sustainable. The main fishing crafts used in the Dal Lake are the wooden boats which are not more than 10 meters in length and one meter in depth. The total fishermen population in the State is presently estimated at 91,984 (Anonymous, 2013). About 15,000 professional fishermen derive their livelihood from natural water resources (Anonymous, 2013).

The work done on gear and craft in Kashmir has been limited. Also, the work on the efficiency of fishing gears including the catch composition is also negligible. The changing ecological condition of the Lake over the years has also affected the fish catch and composition. It was with this background that the present research work on the operation of different types of gears and their fish catch composition were carried out.

FIGURE 9.1 Representation of (a) harpooning; (b) & (c) operation of cast net; (d) and (e) operation of gill net.

9.2 CATCH COMPOSITION

Catch composition of any fishing operation gives an idea about the varieties of fish availability in that region, which in turn helps in better understanding of the biodiversity. In addition to this type, fish quality and quantity landed by any gear play an important role in determining its economic viability.

A total of nine species of fishes were recorded in the catches of cast net and gillnet during the study period in Dal Lake. These included *Cyprinus carpio* var. *communis*, *Cyprinus carpio* var. *specularis*, *Carassius carassius*, *Schizothorax niger*, *Schizothorax esocinus* and *Schizothorax curvifrons*, *Crossochielus diplochilus*, *Tryplophysa* spp., *Ctenopharyngodon idella,* and *Puntius conchonius* presented in Figure 9.2. The commercially important species were *Cyprinus carpio* var. *communis*, *Cyprinus carpio* var. *specularis*, *Carassius carassius*, *Schizothorax niger*, *Schizothorax esocinus,* and

Schizothorax curvifrons. As many as 17 species of fish were identified in the Lake by (Das, & Subla, 1963). A presence of *Cyprinus carpio*, *Schizothorax niger, Schizothorax esocinus,* and *Schizothorax curvifrons* in Dal Lake. Along with these some other fish species reported by Anonymous (1977) were *Barbus conchonius*, *Crossochielus latius,* and *Botia birdi* (Shafi et al., 2005) also reported the presence of seven species in Dal Lake namely *Cyprinus carpio var. communis*, *Cyprinus carpio var. specularis*, *Carassius carassius*, *Schizothorax niger*, *Schizothorax esocinus* and *Schizothorax curvifrons*, *Crossochielus diplochilus* and *Puntius conchonius*. Bhat et al. (2010) reported 10 species from the Lake: *Cyprinus carpio* var. *communis*, *Cyprinus carpio* var. *specularis*, *Carassius carassius*, *Crossocheilus diplochilus, Schizothorax esocinus, Schizothorax curvifrons, Schizothorax niger*, *Schizothorax labiatus*, *Botia birdi, Puntius conchonius,* and *Gambusia affinis*.

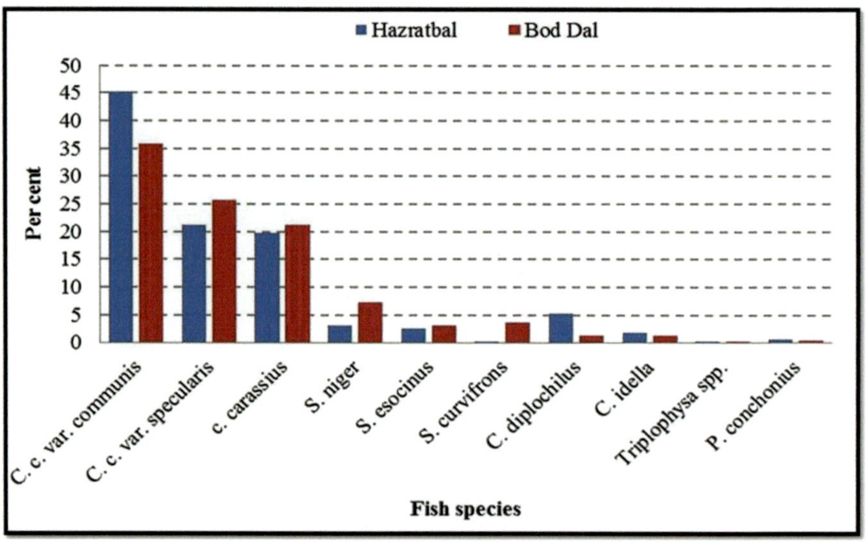

FIGURE 9.2 Image depicts species wise contribution to the total catch.

The analysis of fish catches during the study period has shown that common carp is gaining predominance over the more acceptable indigenous fishes of the sub family *Schizothoracinae*. The catch was dominated by common carp, throughout the study period. By weight it contributed 63.45% at Hazratbal and 56.88% at Bod dal basin. Due to prolong spawning period, plenty of facilities (aquatic vegetation) to spawn and due to excessive rate of reproduction of Common carp, the indigenous fish are losing their ground in the lake (Anonymous, 1977). Common carps have adaptive advantages to utilize Lake resources for its growth and dominate over other species due to its high fecundity (Das & Subla, 1963). The highest catch on the basis of weight was seen in the month of May in both the basins using both the gears. In May its percentage composition by weight was 72.89% (Cast net) and 86.83% (Gill net) at Hazratbal basin while as at Bod dal it was calculated to be 67.65% (Cast net) and 76.14% (Gill net), respectively. This could be attributed to the higher temperature and better food availability in the summer months (Sunder et al., 1978; Yousuf, 1996; Shafi et al., 2005; Bhat et al., 2010; Das, & Subla, 1964) reported similar observation in their study.

Carassius carassius was the second highest catch on the basis of weight with average biomass of 43.04% at Hazratbal and 38.76% at Bod dal during the study period. Highest catch in Hazratbal basin using cast net was recorded in the month of March (28.75%) and in the month of Feb. (34.22) using gill net. While in the Bod dal basin highest catch of 27.1% was recorded in the month of May using cast net and 21.99% in the month of March using gill net, respectively. Shafi et al. (2005) reported the presence of *Carassius carassius* with 12.41% biomass and 20.55% by number. Their percentage in the Lake has increased from the last decade.

Schizothorax niger was seen to contribute 3.86% by biomass during the study period in Hazratbal basin while a higher proportion of 9.56% was observed in Bod dal basin. The highest catch was seen in the month of February in both the basins. The catch of *S. niger* declined towards the summer months. In Hazratbal basin highest catch observed was 8.74% while in Bod dal a slightly higher catch of 13.84% was observed. *S. niger* spawns during March-April. *S. niger* does not show much breeding migration unlike the other Schizothoracids but it becomes quiescent and inactive and thus, is not much seen in the catches from March onwards. Anonymous (1977) observed similar trends.

Schizothorax curvifrons showed a rare presence in Hazratbal basin while it was caught in better proportion in Bod dal basin. Six samples of the fish were seen in the catches of February in Hazratbal basin while no other specimens were recorded for the rest of the study period in the said basin contributing only 0.3% by biomass. However, a slightly higher contribution of 5.64% was observed in Bod dal basin. *Schizothorax esocinus* was seen to contribute 2.88% by biomass in Hazratbal basin and 4.40% at Bod dal basin, respectively. This trend indicates the over exploitation of the fish in Hazratbal basin. Also, the catch of the fish seemed to decline from February onwards. The probable reason could be the breeding migration showed by certain snow trout. Das & Subla (1963) also discussed that breeding migration of the fish starts from March to April towards the incoming streams hence decreasing their catch in the Lake.

In the present study, common carps were seen to dominate the catch by 63.69% by biomass and 41.6% by number. Among these carps *Cyprinus carpio var. communis* contributed 40.51% and *Cyprinus carpio var. specularis* contributed 23.45%. The report by Anonymous (1977) also indicates that Common carp dominated the catches in Dal Lake. Similar trends can be seen in the study conducted by Shafi et al. (2005). According to Shafi et al. (2005) Common carp dominated the catches by 69.13%. The two varieties; *Cyprinus carpio* var. *communis* contributed 59.2% while *Cyprinus carpio* var. *specularis* formed 9.11% of the total catch by weight. Sunder et al. (1978) also found *Cyprinus carpio* to be the dominant contributor to the total catch from this water body. The cyprinid species reported by Shafi et al. (2005) included *Carassius carassius* which contributed about 12.41% of total catch, *C. diplochilus* with 4.35% of the total catch. The three Schizothoracids contributed 14.6% with *S. niger* being the main contributor with 10.37%. In the present study, *Carassius carassius* contributed 20.50% by weight showing increased abundance in the water body. *Crossochielus diplochilus* was found to contribute 3.25% showing a slight decline comparatively while Schizothoracids were seen to contribute 10% of the total catch with *S. niger* being the dominant contributor with 5.25% by weight. The native Schizothoracids show a declining trend. The deterioration of the Lake water has shown a profound effect on the ichthyofauna of the Lake, registering a sharp decline in the fish catches to the extent of 25–35% (Sunder et al., 1978). The introduction of common carp (*Cyprinus carpio*) also has severely affected the indigenous schizothoracid population of the Lake (Saxena & Koul, 1966; Das & Subla, 1967; Qureshi et al., 1971). More than 65% of the present day fish catches from the Lake comprises of common carp, whereas, the endemic schizothoracids contribute only about 20% (Sunder, 1995; Shafi et al., 2005). Their abundance is seen to be diminishing in the water body. Month wise analysis showed that the maximum productive month in the Lake in both the basins was May using both the gears. Thus, it can be concluded that the same varieties of fish were caught all along the two basins by gillnet as well as cast net and the dominance of Common carps persists and our natives species are diminishing in catches.

During the study, wooden plank-built boats were recorded with an overall length (OAL) range between 5 m and 10 m and width range between 0.5 m and 0.75 m. These crafts were made up of locally available wood-Deodar, which is costly but durable. It does not decompose in water and remains sturdy throughout. Locally it is known as *Naav*. They are often navigated by two boatmen. No modifications have been seen as such in the crafts used.

The gears and methods mainly used for fishing in Kashmir valley are long line, Hook and Line, Scoop net, Rod & line, Narsoo, Cast net, Gill net, and Trap. Among various fishing nets, cast net is the main type of gear used in the valley (Sunder et al., 1978; Yousuf, 1996; Shabir, Najar, & Ashraf, 2010). The main fishing gear used in the Dal Lake is the cast net which is made of six pieces each with different mesh sizes (Anonymous, 1977) Similar observations were seen in the current study.

Several types of fishing gears were seen in Dal Lake for fish harvesting, such as, Cast net is a throw net made up of three parts: the upper section (net band), the middle section (a conical-shaped net mesh), and the lower section. The lower section of the net was seen to have pockets with fixed iron weights. One net is operated by one person. No fish attractive device used. The size range used is between 1.0 and 2.0 m in diameter. Length of cast net, when hung is 4.0 m and diameter of the net when spread is 4.8 m. The mesh size varies between 1.2 and 3.0 cm bar to bar. The net is provided with iron or lead sinkers of about 8–10 kg weight around the peripheral cord. They are cylindrical in shape with a length of 17 mm and diameter of 10 mm. This net can only be operated from a boat. The life expectancy of a cast net is 3–4 years. It costs around Rs. 2,500 for its construction. Depending upon the number of meshes the cast net is known by different names such as: *guran zaal* (1,200 meshes), *thapthat zaal* (1,100 meshes), *dal zaal* (1,000 meshes), *Naushath zaal* (900 meshes), *nuchkul zaal* (800 meshes), *pouchkul zaal* (500 meshes) and *Ara zaal* (400 meshes). Among these *noushath zaal*, *guran zaal* and dal zaal are used in Dal Lake. They locally call it *duph* or *zaal*. No detailed work had been done in this regard. Sunder et al. (1978); Yousuf (1996); and Shabir et al. (2010) mentioned cast net as the main gear being used in water bodies of Kashmir (Anonymous, 1977) gave some further details regarding the structure which have not changed much till date. Anonymous (1977) mentioned a six-piece cast net each with different mesh sizes as the main fishing gear being used in Dal Lake.

Gill nets locally called *shaitan zaal* in Dal Lake are 15 to 40 m long and 1.5 to 3 m wide with mesh size ranging between 45 mm and 75 mm. Gill nets used in Dal Lake are made up of thin nylon monofilaments usually sea green in color. They have a short life expectancy of just two to four months, after which they have to be replaced. Locally made sinkers are used. The use of these nets has seriously affected the regenerative capacity of the fish fauna. It has been observed that the communities themselves had imposed restrictions on the use of lower mesh size nets due to decline in fish catch. Promotion of gill nets was also undertaken by the State Government Department at select locations but due to declining fish catch in the Lake, the Department of Fisheries has imposed a ban on gill nets. Despite the fact, gill netting is still being practiced in some parts of the water body, especially in the winter months due to low catch.

Long lining locally known as *'wael razz'* is a primitive method of fishing in which a nylon line measuring about 1,000 meters in length is generally used in the Lakes. In Dal Lake long lines of length 30–100 m is used. In Dal Lake the line fishing is mainly done from June to September months. Rod and line (*Bislai*) is also an important gear used in Dal Lake. It is usually practiced in the warmer months of May, June, July, and August.

In fishing accessories fishers use a long wooden pole called *Shum*, which is made up of wood mainly Daevdor timber (*budhul*), it is used to handle the net to catch the fish; its length is 10–12 ft. Fishers use it while operating cast nets as well as gill nets. It is an essential tool used while fishing. It helps in searching suitable fishing grounds for using cast net and also in reaching back the net once it is thrown in the water body.

9.3 CONCLUSION

For increasing fishing efficiency in reservoirs and lakes, replacement of age old fishing crafts and gears by motorized ones. However, no mechanized fishing methods could be recommended because of scientific reasons. Most of the fishermen of Dal lake are still using traditional fishing craft and gears because of which are obsolete because of their inefficiencies. Fishers complained of a decline in revenue generation during the last few decades. The most important problem faced by the fishers was decline in fish catch and that was the greatest problem to the fishers, because they do not have enough fish to catch. The second vital problem was lack of capital for the purchase of fishing crafts and gears. It can be concluded that cast net is an appropriate fishing gear operating in the Lake throughout the year. Gill net can also be an economic gear provided the appropriate mesh size is regulated. It can prove to be more time and energy saving. It may be summarized that fish catch in the Lake has declined gradually due to pollution, indiscriminate gear use, and lack of management policy.

KEYWORDS

- **biodiversity**
- **catch composition**
- **Dal Lake**
- **fishing gear**
- **indigenous fishing gear**
- **overall length**
- ***Schizothorax niger***

REFERENCES

Anonymous, (1977). Annual report, Published by CIFRI. *Report on Dal Lake for Development, Srinagar, Kashmir with Suggestions for Development.* Bulletin no. 24.

Anonymous, (2000). Fisheries Research, International Council for the Exploration Sea (ICES). In: Walsh. S. J., Engas, A., Ferro, R., Fonteyne, R., & Marlen, B. V., (eds.), *Improvement of Fishing Technology to Catch (or conserve) More Fish: The Evolution of the ICES Fishing Technology and Fish Behavior Working Group During the Past Century.*

Anonymous, (2013). *Published by Economic Survey J&K*, p. 124.

Bhat, F. A., Balkhi, M. H., & Yousuf, A. R., (2010). Fish diversity in Jammu and Kashmir and conservation measures. *Kashmir Speaks*, 104–115.

Dar, S. A., Desai, A. Y., Rather, A. M., Sayani, A. N., Parmar, R., Arjamand, S., & Chesti, A., (2014). Fishing gears operated along the Wular Lake, Jammu and Kashmir, India. *Ecology Environment and Conservation, 2*, 45–47.

Das, S. M., & Subla, B. A., (1963). The ichthyofauna of Kashmir—History, topography, origin, ecology and general distribution. *Ichthyologica, 1*, 68–106.

Das, S. M., & Subla, B. A., (1964). The ichthyofauna of Kashmir, Part II. The speciation of Kashmir fishes. *Ichthyologica, 3*, 57–62.

Hovgard, H., & Lassen, H., (2000). *Manual on Estimation of Selectivity for Gillnet and Long Line Gears in Abundance Surveys* (p. 397). FAO, Fisheries Technical Paper No. 2000.

Nedelec, C., & Prado, J., (1989). *FAO Catalogue of Small-Scale Fishing Gear* (p. 224). Blackwell Science Ltd., Oxford.

Qureshi, M. Y., Singh, J. P., & Das, S. M., (1997). *On the Problem of Depletion of Endemic Fishes of Kashmir by the Introduction of Exotic Carp* (*Cyrprinus carpio*) (pp. 118, 119). All India symposium on icthyology and hydrobiology and fisheries.

Sainsbury, J. C., (1996). *Comemercila Fishing Methods* (3rd edn., p. 310). Oxford: Osney Mead.

Saxena, D. B., & Koul, D. N., (1966). Fish and fisheries of Jammu and Kashmir State, part 1. fisheries resource and problems. *Ichthyologica, 5*, 45–52.

Shabir, A. D., Ferooz, A. B., & Balkhi, M. H., (2015). Study on different fishing methods (gear & craft) used in Manasbal Lake of Kashmir, Himalaya. *J. Himalayan Ecol. Sustain. Dev., 10.*

Shabir, A., Najar, A. M., & Ashraf, M., (2010). *Indigenous Technical Knowledge (ITK) in Fisheries Sector in Central and Northern Region of India* (p. 165). Narendra Publishing House, New Delhi.

Shafi, S., Bhat, F. A., Parveen, M., & Yousuf, A. R., (2005). Catch composition of fish from Dal Lake, Kashmir. *Journal of Research and Development, 5*, 111–114.

Sunder, S., (1995). Some conservation and management strategies for Dal Lake Fisheries. *Punjab Fisheries Bulletin, 19*, 53–63.

Sunder, S., Bhagat, M. J., Joshi, C. B., & Ramakrishna, K. V., (1978). Fishing methods and fish catch composition of Dal Lake, Srinagar (J&K) during 1969–1972. *J. Inland Fish Soc. India, 10*, 9–18.

Yousuf, A. R., (1996). Fishery resources of Kashmir. In: Khan, A. H., & Pandit, A. K., (eds.), *Ecology, Environment and Energy* (pp. 75–120).

CHAPTER 10

Reproductive Physiology and Breeding Biology of Rainbow Trout (*Oncorhynchus mykiss*)

YOUNIS AHMAD HAJAM[1], DISKSHA[1], RAJESH KUMAR[2], and MOHD. SALIM RESHI[3]

[1]*Department of Life Sciences and Allied Health Sciences, Sant Baba Bhag Singh University, Khalia Padhiana, Jalandhar, Punjab, India*

[2]*Department of Biosciences, Himachal Pradesh University, Shimla, Himachal Pradesh, India*

[3]*Toxicology and Pharmacology Laboratory, Department of Zoology, School of Biosciences and Biotechnology, Baba Ghulam Shah Badshah University, Rajouri, Jammu and Kashmir, India*

ABSTRACT

The rainbow trout (*Oncorhynchus mykiss*) is a trout and species of salmonid native to coldwater tributaries of the Pacific Ocean in Asia and North America. Freshwater forms that have been introduced into the Great Lakes and migrate into tributaries to spawn are also called steelhead. Rainbow *Oncorhynchus mykiss* trout likes to live in cold water and are mostly found in lake, reservoirs, rivers, etc., and rainbow trout are also considered as fluvial populations in the whole forest. The population of fluvial of rainbow trout is not figured out clearly because they do not refer to the migration into rivers to spawn lives into larger drainage. Female partner prefers to make their nest where male partner protects it from other interested males and predators. The

Coldwater Fisheries and Aquaculture Management: Technology for Sustainable Food Production.
Mohd Ashraf Rather, Faisal Rashid Sofi, Adnan Amin, & Kawkabul Saba (Eds.)

female digs the nest (a redd) by using her anal fin and then descends sits in a way so that her vent and anal fin goes into the deepest part of the redd. The male partner joins her parallelly so that their vents become opposite to one another. Both the partners open their mouth, arch their backs, and deposit the eggs and milt (fish sperm) at the same time. The eggs are enclosed by a cloud and the fertilized. However, the reproductive processes of the Rainbow trout are affected by various environmental factors and changes the internal endocrine environment of the fish. Therefore, Rainbow trout, has received wide research consideration. This chapter summarized the reproductive patterns of Rainbow trout and impact of environmental and hormonal stress on the reproduction of Rainbow trout.

10.1 INTRODUCTION

Aquaculture is also known as aquafarming. Fish, crustacean algae, etc., are the organism that lives in this, and the high quality of both progeny and gamete shows its success rate (Bromage, 1995). At the time of brood selection, some characters should be analyzed such as eggs spawned, size of an egg and its mortality, size, and mortality percentage of fingerlings (Leitritz & Lewis, 1980). High-quality gametes are required but sometimes it creates a stressful environment for the broodstock (Billard et al., 1981). Transportation, handling, cleaning, crowding, water quality, etc., are some general practices. But sometimes, unfavorable water also affects the reproduction rate by acting as stressors (Billard et al., 1981; Bromage, 1995). Pottinger & Pickering (1990) reported that the fish reproductive system is also influenced due to both acute and chronic environmental stresses. The gamete quality is also affected due to all these stressors (Campbell et al., 1992, 1994). The various researcher reported in previous years that there is significant decrease was observed in the survival of *Oncorhynchus mykiss* and egg size as well constantly (Campbell et al., 1992, 1994). Bromage (1995) studied that the proper management of broodstock also helps in increase in the progeny performance and eggs quality by providing husbandry activities. The period, in which females use huge energy for gamete growth till its maturation and the process is known as vitellogenesis (Tyler et al., 1990; Rather et al., 2016). Therefore, stressor affects the reproductive processes during vitellogenesis because lots of energy is required for this process and stressor divide this energy in two ways.

10.2 BIOLOGY AND ECOLOGY

10.2.1 SYSTEMATICS/TAXONOMY

The detailed study of rainbow trout and other salmonids including taxonomic description was described by Behnke (2002), and for this, these rainbow trout were named *Oncorhynchus mykiss*. On the other hand, they have about three ecological forms (anadromous, stream resident, and lake-dwelling) of rainbow trout, among these anadromous also classified as *Oncorhynchus mykiss irideus* (Behnke, 2002). Behnke (2002) reported that there are some subspecies of rainbow trout, i.e., red-band trout, coastal rainbow trout, and Gulf of California (Mexico) based on their appearance and life history as well (Behnke, 2002).

10.2.2 IDENTIFICATION

Behnke (2002) studied the identification of rainbow trout. It is silver-gray tom greenish brown in color on the back. It has a white color belly on the side in stream-resident form. It has lateral strips of pink or reddish color. But in some cases, lavender and connotations from gill covering (the whole length of fishtail). Behnke (2002) also reported that in caudal fins small dark spots are having on both dorsal and adipose fins as well and the lower fin color is pale pink. In some cases, black spots saw on the head and their sides. The color becomes deep in males with intensive red side strip at the stage of spawning.

10.2.3 MOVEMENTS AND ACTIVITY PATTERNS

Generally, rainbow trout likes cold water, so they are found in lakes, reservoirs, rivers, etc., and rainbow trout are also considered as fluvial populations in the whole forest. The population of fluvial of rainbow trout is not figured out clearly because they do not refer to the migration into rivers to spawn lives into larger drainage such as upper Taylor River, lower Taylor River, Cimarron River, San Miguel River, etc., researchers studied the salmonids life-history adaptation mostly Anadromous fish species. The various researcher reported that the movement of juvenile and adult straying play an important role that allows anadromous populations for catastrophic turbulences continuity (Reeves et al., 1995). In anadromous adaptation, fluvial behavior plays an important role. It is an important mechanism by which a healthy population of rainbow trout forms. Generally, fishes like dynamic conditions and are patchy as well so they are preferably found in that particular forest. Sustainability during the drought periods is successful by proper movement of rainbow trout and its straying when spawning allows "meta-population." It also forms due to change in habitats during maintenance. There are no restriction forms when fishes move in the water bodies. But in some cases, like man-made barriers or sometimes natural barriers trouble them. In 2004, in the forest, about 200 streams were found in crossing sites and about 11% found in elegiacal due to the aquatic organisms' movements, also increase and decrease in velocity and depth, respectively make trouble in fish movement. However, annually about 1–2 stream "crossing improvement projects" were established in the forest, and in 2001–2007 habitat intensive inventories were covered completely up to 62.3 miles from 224 reaches. There are three protocols for stream habitat assessment were designed for the measurement of its parameters. The parameters are stream condition inventory (SCI), R1/R4 method of fish habitat inventory, and Pacfish In fish Biological Opinion (PIBO) (Frazier et al., 2005; Overton et al., 1997; Heitke et al., 2006), and in material and methods they are all slightly different from each other. But a core set of habitat variables equally measured that already identified by hydrologists and fisheries scientists and habitat conditions were also recorded from the entire forest that helps to represent the area of the stream where rainbow trout could not occupy. There are different ways to studied about the rainbow trout habitat conditions and the requirements forms during reproduction, and about 1–7% gradient found in a stream in the forest found in previous data. Also, another data indicates less Bankfull width (BFW), i.e., about 5.5 m (n = 216), in which about 90% samples show the minimum range of BFW, i.e., 1–10 m. Two trout forms, i.e., brook trout and cutthroat trout are well documented found in headwaters. However, the distribution of trout occurs in western mountainous areas. Meanwhile, brown trout and rainbow trout are the forms of trout that are found in mid and lower elevation sections or large rivers (Rahel & Nibbelink, 1999; McHugh & Budy, 2005), and in these rainbow trout are distributed in lower gradient and elevation sections with large width in the whole forest. About 3–100 mm range forms in spawning gravel in rainbow trout (Orcutt et al., 1968; Raleigh et al., 1984), however, substrate size depends upon the spawned size. So, according to that, we can say that rainbow trout have about 50 cm length and the diameter of spawning gravel about 15–100 mm ranges (Orcutt et al., 1968). A data (pebble count) was collected from 210 stream reaches that showed the spawning substrate size ranges, i.e., 3–100 mm that play an important role to constitutes about 56% of substrate because the substrate size is ranging from 15–60 mm particularly that refers the composition of substrate, i.e., 44%. In this study, the biologists reported that there is an availability of spawning gravel in the entire forest that are adapted by trout mainly in large streams. A data that the measurements of fine sediment from pool tails revealed that "percent fines" lower than 2 mm that comprise a huge percentage of typical spawning sites and also 189 reaches showed results in its favor that areas suitable for Spawning comprise (20% fines less than 2 mm). Raleigh et al. (1984) reported that for rainbow trout, the optimum spawning condition may be less than 5% fines because generally, fines have about 30% spawning gravel, decreasing the embryo survival rate found (Raleigh et al., 1984). So, for the survival of rainbow trout, the fine sediments are the limited factor revealed in the data.

About 12–18°C temperature is required for *Oncorhynchus mykiss* (Rainbow trout) reported in different studies. Also, the increase in mortality occurs with an increase in temperature (25°C) (Raleigh et al., 1984). *Oncorhynchus mykiss* (Rainbow trout) requires a particular temperature and water temperature in June–September period on which different data are available. After September, the water temperature gets decreases, i.e., 0°C from November to March. In the winter season, the feeding process in fishes (in adult rainbow trout) decreases with a decrease in metabolic activities. Due to this they could not grow properly or were negligible in extremely cold water studied by various workers (Raleigh et al., 1984). In the winter season for the survival of trout, the proper density of the pool and its depth play a significant role. Mainly, fish survive in the pool in small streams that seem a key factor that affects its diversity and trout abundance as well, the pool depth is about 1–2 m to maintain its good quality (Raleigh et al., 1984). Generally, in the system, 0.02–1.61 depth of pool ranges, an average it is 0.32 m. In large streams, about 1 m depth of the pool is required minimum and if it became less than it (≥1 m) then trout survival rate also become decreased. It particularly occurs when low flow conditions are there in the winter season. For the maintenance of trout, the population pool should have a suitable depth. Different streams on the forest could not have efficacy suitable depth of the pool. It is formed due to water production, basin area, etc.

10.3 BREEDING BIOLOGY

Raleigh et al. (1984) reported that spawning occurs at particular 'natal streams' that form comparatively low straying incidence. The maturity of rainbow trout occurs when they are 2–3 years old and the length of the trout is

100–150 mm revealed from a data. At the stage of maturity, females dig nests commonly known as redds. Females used it for the deposition, fertilization, and covering with gravel of egg. Mostly female forms these redds having proper depth and velocity and bottom configuration that induced flow of water through stream substrate (Young, 1989). Spawning of rainbow trout season, i.e., mid-April to late June having temperatures 7–12°C, respectively (Raleigh et al., 1984). About 28–49 days is the incubation period of rainbow trout that further varies according to the temperature of water (Raleigh et al., 1984). Bjornn & Reiser (1991) reported that after the hatching of larvae they stay up to a weak in redds, and in that period they develop continuously. There are two stages in which rearing is processed in a different way, i.e., summer rearing stage and the winter rearing stage in which various methods of behavior are displayed. The various researcher reported that adult rainbow trout come out from the gravel riffles in late winter and early springtime (temperature get increased). They seek out appropriate spawning sites (Raleigh et al., 1984; Bjornn & Reiser, 1991). There is no feeding activity found in the winter season due to which trout gets aggregated rapidly, mostly at limited pool habitat (Bjornn & Reiser, 1991). The temperature of the water plays an important role in behavior change. It inhibits the energy and its metabolic rate. In summer, due to rearing habitat destruction, the survival rate of fish becomes decreased as compared to another season also the reproduction rate is very low and the fecundity rate varies and at the early stage, predation is severe (Root, 1994). About 3% rates of survivorship were found in trout at post-emergence period reported in a previous study (Pender & Kwak, 2002).

In the Salmonidae family, *Oncorhynchus mykiss* is the species that increase their population rapidly because they adapt themselves according to the system (Bud et al., 2007). Cocan et al. (2013) reported that rainbow trout breed artificially in Romania. So, to evaluate all the systems of artificial breeding, all the technology and activities should be streamlined according to the selection of the broodstock program. The various researcher studied that in the salmonid unit, breeding plays a significant role because it is a method of technological growing flow that effect by environmental conditions and nutrition also (IHUŢ et al., 2015). Age, weight, and structure also affect the formation of broodstock because brood quality is related to the survival rate of offspring, its development, and disease resistance (Kayam, 2004). So, it was also reported by different researchers that for random selection, there is a requirement of breeding stock (Mireşan et al., 2013). For the better improvement in the selection aspect, utilization of correlation occurs in phenotypic characters and breeding indices as well, and in between breeding indices and somatic features (Cocan et al., 2013).

10.4 REPRODUCTIVE ENVIRONMENT IN RAINBOW TROUT

10.4.1 OVARIAN FLUID OF RAINBOW TROUT A STANDARD MEDIUM FOR FERTILIZATION

The various researcher reported salmonids *in-vitro* fertilization that shows a positive effect on ovarian fluid for Caspian brown trout (*S. trutta caspius*) and brown trout (*S. trutta f. fario*) *as* well (Hatef et al., 2009; Lahnsteiner, 2002). But some researchers recommend that the ovarian fluid can be washed out when the fertilization process occurs artificially especially in rainbow trout, because some negative effects are also seen by the researchers in the case of sub-fertile females (Hugunin et al., 2008). They explained as fertilization can be fertilization by impeding the movement of sperm and contact with the egg. Rime et al. (2004) reported that the substance released from it is associated with this released substance from the overripe egg. Different authors also accredited the outcomes of fertilization discrepancies towards the post-copulated effects that designed to select the best sire, i.e., cryptic female choice and especially Butts et al. (2012) reported the spermatozoa performance when ovarian fluids get activated in females. Various researchers also studied that some factors affect ovarian fluids such as pH of maternal fluids and almost 40% variability caused by the factors like iron, carbohydrates, and proteins (Wojtczak et al., 2007). It was also reported by other researchers that in Chinook salmon the performance of protozoa affected in different female ovarian fluids (Rosengrave et al., 2008), and similarly a case was reported by the researchers that the ovarian fluid in the fertilization medium does not form towards "cryptic female choice" in Arctic charr (Kleppe et al., 2018). The experiment that was carried out showed that the fertilization medium prepared by ovarian fluid also helps to increase sperm performance. At the initial stage, this medium was kept in the water as experimental work and that showed about 20% low motility in female albino rats and the number of fertilized eggs was 1,50,000:1 (spermatozoon-egg ratio) which was very low. Rosengrave et al. (2008); and Myers & Cassinelli (2020) also studied this and present observation supports their results that if experiment occurs in water, in which, in salmonids, it emphasized the inadequacy of pre-fertilization motility estimation. In both activation medium and water, the motility rate of sperm was observed and resulted that the rate being low in regular rate as compared to the albino rats and higher in water, respectively. The number of colored eggs embryos increased significantly when ovarian fluids are washed in albino female eggs, but on the other hand, if this fluid was utilized as an activation solution with the use of isosmotic saline, then showed positive results in both males and this showed that the performance of spermatozoa was triggered through non-uniformity of fertilization environment through ovarian fluid confining the batch of eggs. Also, when the fertilization medium is ovarian fluid then it also increases the embryos from these males. Because the ovarian fluid helps to enhance the spermatozoa performance that further wins the competition among the sperm. In the fertilization medium, the ratio of spermatozoa and egg was 1,50,000:1, and the ratio increase 10-fold to 1,500,000 spermatozoa per egg shown in the cases of different experiments. It was also observed in *in-vitro* fertilization conditions, the association of small and dependencies which is due to huge male gametes and the formation of any relationship without proper knowledge about the condition required for it may cause an effect on gametes (the gamete ratio), so the *in-vitro* process shows positive effects and shows maximum fertilization. Some aspects considered below by us on the sperm behavior having potential that understands the mechanisms of the process.

10.4.2 RAINBOW TROUT OVARIAN FLUID ENHANCES THE KINETIC TRAITS OF SPERMATOZOA

Male gamete plays an important role in successful fertilization. It carries genetic material (DNA). It fuses with the female gamete that further changes the features of spermatozoa. Different researchers reported that this change becomes a true velocity in both (single male and the male having no completion among them) and also in case of competition in sperm (Lahnsteiner et al., 1998; Levitan, 2000). The survival rate of spermatozoa is more among them that have huge genetic contributions and vertebrates that are fertilized externally showing sperm competition worldwide, particularly in the fish (Taborsky, 1998). Salmonids are species that have very tough completions in sperm due to variations in the tactics from spawning in pairs. Various researchers reported significant velocity of spermatozoa in a salmonid (Gage et al., 2004; Liljedal et al., 2008). Moreover, Gage et al. (2004) studied that for sperm completion, sperm velocity plays an important role by increasing the winner spermatozoa velocity that penetrated more frequently with the egg micropyle. Different researchers reported that ovarian fluid plays an important role to enhance the performance of spermatozoa in the activation medium with an increase in velocity (Lahnsteiner, 2002; Poli et al., 2019; Butts et al., 2012; Litvak & Trippel, 1998; Turner & Montgomerie, 2002; Rosengrave et al., 2009). Various workers studied that there was an increase in the number of spermatozoa in 20% water solution of ovarian fluid as compared with unrelated male gametes or lake trout *Salvelinus namaycush* (Butts et al., 2012). But in a similar study carried out by Rosengrave et al. (2009) cannot find that any kind of clues represents the absence of correlation in water and ovarian fluid. In the present work, it was confirmed that the ovarian fluid plays a significant role that affects the kinetic features of spermatozoa than the water activation in rainbow trout. Also, Turner & Montgomerie (2002) reported that there was no clue of an increase in curvilinear velocity (10 s post-activation) for salmonid and Arctic charr species, but Poli et al. (2019) reported the same case in Zebrafish (*Danio rerio*). When spermatozoa velocity decreases then the post-activation process also gets reduced in the presence of ovarian fluid. When dilution forms in plain water then only some cases are there that show differences in about 5% aqueous solution of ovarian fluid. Similarly in isotonic saline, the velocity of spermatozoa activated that different from ovarian fluid, in which the initial velocity was higher in saline than that of water or in ovarian fluid as well. However, the spermatozoa can activate in ovarian fluid much faster than that of water solution. Moreover, there was no significant difference in the velocity when ovarian fluid was diluted with the isotonic saline. In the propagation of spermatozoa, motility tracks show its essential role in addition to the velocity, in which ovarian fluid shows an effect on male gametes' path. Also, this was observed by different workers that the activation of spermatozoa in the ovarian fluid found in rainbow trout (Dietrich et al., 2008). There was symmetrical propagation shown with the flagellum and in water, it shows the non-symmetrical flagellar wave of spermatozoa, with a decrease in the ovarian fluid concentration the specific trajectories also get increased as turn and run pattern. These were present in almost even 2%, 5%, and 10% but they do not recognize in "non-diluted ovarian fluid" (up to 50% dilution), so it was observed that it was not the features of sperm when activated in isotonic saline. The ovarian fluid molecular weight cut-off fraction was only 100+kDa that induced a pattern of motility and showed no significant difference. There were circular trajectories in other molecular weight cut-off fractions. In which they show significant results in both (ovarian fluid and 100+ kDa fraction). Alavi et al. (2019) reported that osmolarity plays a significant role as a driver of fish spermatozoa motility that presents in freshwater because both (osmolarity and calcium ion concentration) have a cross effect on motility traits. They also reported that there was a difference between the outer media initializer and inner media initializer (fraction membrane orchestra of channels) thus motility becomes triggered. Alavi & Cosson (2006) also studied that in spermatozoa, calcium ions (an integral part of some physiological processes) are responsible for the initiation of motility and their progress.

In our experiments on the cross-effects of osmolarity and Ca^{2+} concentration, we varied these indices to identify which among these isolated factors or combination of them may be responsible for the chemokinetic effects of spermatozoa when contacting the ovarian fluid. Some examples show that osmolarity and calcium ion concentration both play their role to affect the path linearity (individually as well as in combination form), i.e., when calcium ion is absent then osmolarity increases (0–300 mOsm/L). When the calcium ion concentration was intermediated the path linearity depending upon the medium of osmolarity and both Calcium ion concentration and osmolarity corresponding towards the ovarian fluid, controlled motility traits that further showed the same ovarian fluid effect but osmolarity signally shows more effective results.

10.4.3 THE OVARIAN FLUID CAUSES THE ATTRACTION AND TRAPPING OF SPERMATOZOA

Inactivated spermatozoa media, with the help of fluid-filled micro-injection, play an important role to copy exactly interface found in between water and ovarian fluid as well which mainly surrounds the egg, and when male gamete and maternal fluids interact with each other than male gamete behavior get changed, it showing different behavior as is in a uniform environment. The experiment also shows that spermatozoa can show a positive response for non-uniformity in the medium. It gets trapped and found some conditions such as pH, osmolarity, and ionic content also and all these conditions are caused by ovarian fluid. It is also helpful to detect the factors that affect mostly the response of spermatozoa. The positive results were found in a plain area of trapping and taxis in the experimental trial. And the spermatozoa of rainbow trout are activated in both water and as well as in isosmotic saline and the effects form more effective when the molecular weight becomes decreased and towards thermos-treated ovarian fluid. Somewhere it is caused by active agents' spatial distribution from injected fluids having low viscosity from the ovarian fluid. In some cases, it was noted that in both mediums (isotonic medium or low molecular fraction) in the ovarian fluid was injected in water. After that spermatozoon get separated into different populations, in which some of them showed a "hyper activation-like-motility pattern" (close towards the border of the injected cloud or near to microcapillary tip). About 1 mmol/L Ca^{2+} was observed in isosmotic saline and distilled water at the time when the motility pattern changes. Finally, the positive results were observed of taxis for ovarian fluids from a complex set of reactions of the spermatozoa of rainbow trout for the different factors present in the fluid and also the change in the movement of spermatozoa comes in an optimal environment and non-optimal temperature as well. When ovarian fluid and isotonic saline-injected then due to its trapping behavior, this is also exhibited (where optimal motility conditions have been developed) and it increases straight motion period. On the other hand, negative behavior (negative taxis) found when distilled water injected in the medium (in this, males avoided entry in the medium), so it can be concluded from this behavior that the spermatozoa of rainbow trout are very sensitive to the provided environment that allows to change its direction also and only live in the conditions that provided by ovarian fluid. Different researchers reported that there were similar effects on both the effect of ovarian fluid on spermatozoa and in extreme fertilized species. Various researchers reported that rainbow trout spermatozoa show good performance when in contact with female fluids as compared to the invertebrates found in marine habitats such as sea urchins, ascidians, and squids (Hirohashi et al., 2013). The behavior is considered more spectacular as it is similar which is in rainbow trout form turn and run loops through asymmetric flagellar bending. It also occurs in a segment of flagella that is bent and forms like a burst rise in Ca^{2+} concentration. A study was carried out by different researchers that this process also helps to notice the chemotactic activity of ascidian spermatozoa attached with chemotactic receptors of the membrane (Yoshida et al., 2018). The symmetry of the flagellum of spermatozoon is changed when in rainbow trout spermatozoa the intraflagellar Ca^{2+} increased in all marine invertebrates (sea urchins, ascidians, and squids) and in path curvature. The concentration of Ca^{2+} found constant when path curvature is constant (either when in the ovarian fluid it run straight-forward in water. Kaupp (2012); and Yoshida et al. (2018) reported that all the theories and experiments of "spermatozoon chemotaxis" are mainly based upon certain investigations that were carried out in marine invertebrates. The researchers also studied that in different fishes that are discussed above, the Pacific herring was well studied in sperm-egg encounters only that further works on the model for better activity and sperm entry into the egg (Yanagimachi et al., 2017). The spermatozoa have a unique feature that it survives for an hour to some days till the activation. Mainly, males release their sperm into the water. Further, it is attached to the female and then fuses with the egg. A specific protein "sperm motility-initiation factor"

(SMIF) was used to activate the male gamete. The protein is attached to the ovarian fluid and in the egg membrane (Inagawa et al., 2001). The different researchers reported that interaction occurs in the membrane receptor and activator molecules that are followed by activation of different molecules (adenyl cyclase, potassium, and calcium), and also the attachment of Camp and influx of potassium and calcium ions (Yanagimachi et al., 2017), when it was activated, through micropylar sperm attractant (MISA) protein the spermatozoa bind with micropyle area (Yanagimachi et al., 2013). Also, Yanagimachi et al. (2017) reported that this binding occurs due to the action of Ca ion channels and proton pumps. These help to control the calcium ion concentration. As similar to the pacific herring, the activation of spermatozoa in water is less in rainbow trout. But in an actual way, it occurs in the majority of freshwater species. Female produces eggs with ovarian fluid and the motility of rainbow trout spermatozoa is changed due to ovarian fluid. The "turn and run" process is carried out by the spermatozoa in diluted ovarian fluid and forms positive taxis. Therefore, the cell that includes it gets trapped. When a cell with plenty of water reached the border then they back to its position also. The asymmetric band that is present in the flagella helps to process the activity. In the bent segment, the concentration of calcium ions is increased and presumably formed attachment in spermatozoa and eggs. In Pacific herring and marine vertebrates (sea urchins, etc.), some same similarities were found that are described for them. This also helps to suppose that some basis for some chemotactic response was recognized in the broader spectrum that already formed. The interaction between the egg and sperm in rainbow trout has been attached through particular characteristics of the spawning process.

Some tests were carried out on freshwater spawning fish and rainbow trout to study the gamete features and the results showed that female fluids play an important role in it by increasing male gamete performance showing spawning behavior in trout species. It does not show any agent particularly that affects the motility of spermatozoa. The main aim of this is to identify the natural agents that are present in the male gametes and the different mechanisms formed in the process of reproduction that show a significant role in evolutionally development biology. The processes are well adapted for the particular environmental conditions with modification (tactics and reproductive strategies) due to their different ecological niches. Fishes are good for experimental work in both *in-vivo* and as well as in *in-vitro* conditions especially for the freshwater species because they have limited time for fertilization. Therefore, the future study of gamete encounters helps to understand the reproduction evolutionary aspects, behavior, and vice versa.

10.5 REGULATION OF REPRODUCTION BY ENDOCRINE SYSTEM IN RAINBOW TROUT

Environmental cues are processed through endocrine signal by activating the hypothalamic-pituitary-gonadal (HPG) axis. Hypothalamus releases gonadotropin-releasing hormone (GnRH). GnRH shows their expression in various tissues; however, it is produced in the hypothalamus (typically termed GnRH1) and primary contributes in the activation of the HPG axis by stimulating the synaptic gonadotropin-producing cells (gonadotrophs) of the pituitary (Rather et al., 2020). Further modulation of GnRH synthesis and release occurs through the more recently discovered Figures 10.1(a) and (b).

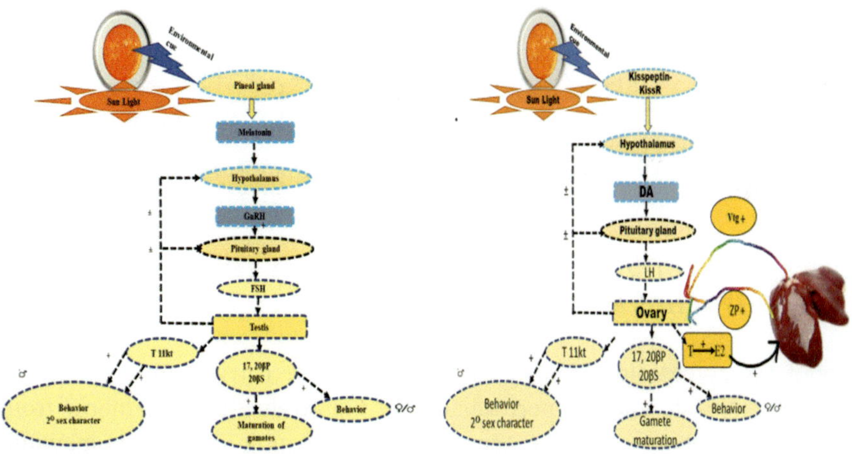

FIGURE 10.1 The role of environmental cues in the activation of endocrine signaling and its contribution in the maturation of gametes and sexual behavior.

10.5.1 EFFECTS OF STRESS ON THE REPRODUCTIVE ENDOCRINE SYSTEM

In female brown, and male rainbow trout, two weeks of confinement cortisol level increase in circulatory plasma, inhibits testosterone level in circulatory plasma, but does not change the level of estradiol and level of plasma VTG decreases in female rainbow trout. On the other hand, plasma E2 level decreases after 2–24 h of imposed confined stress given to rainbow trout. During the post-capture period the level of testosterone and estradiol decreases in migratory wild rainbow trout kept in stream cages for 24 hrs.

after capture, more it has been observed that cortisol level increases. All the changes induced by stress on reproductive performance has been summarized in Figures 10.2(a) and (b).

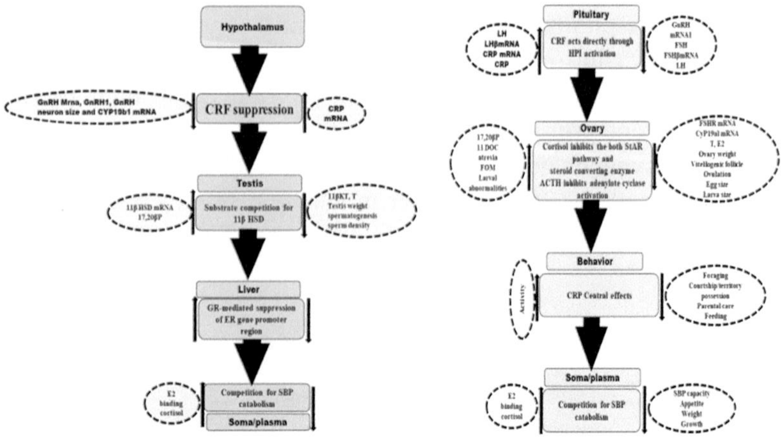

FIGURE 10.2 The effect of stress on the reproductive performance of rainbow trout.

10.5.2 REPRODUCTIVE INDICES OF RAINBOW TROUT (ONCORHYNCHUS MYKISS) FEMALES FROM A TROUT FARM

About three-year-old *Oncorhynchus mykiss* were carried out to study the reproduction rate and the performance of the production of rainbow trout, so based on their weight and samples after spawning and before spawning the relative fecundity (335.88±17.723 g) was determined to have 52.77% variability coefficient. It was the first generation that have high variability coefficient. This activity was performed without any improvement activity or any selection process in the nursery. A criterion in production that is used most of the time is the total number of eggs and egg diameter (TNE & Ed), as the weight of broodstock increases that helps to increase the eggs diameter and their number (Ahmadnia et al., 2014). The value of the total number of eggs was 8127.44±478.532. Also, there was an increase in variability coefficient (58.88) (Kayam, 2004), in the future, the spawn has a good quality of the individual. Different researchers reported that some factors affect the production of rainbow trout eggs (among breeding stock, their weight, quality of water, etc.) (Ozgur & Bayir, 2013). In this, the weight of the egg and its diameter were 0.06±0.001 g and 4.16±0.029 mm, respectively. Similarly, Ozgur & Bayir (2013); and Kocaman et al. (2009) reported the diameter of eggs in four-year-old broodstock, i.e., 5.58±0.18 mm and 5.33±0.10 mm, respectively. The data concluded that both the egg diameter and their number increased with an increase in the broodstock age and some reproduction indices were also calculated (gonadosomatic report (GR); Behning fertilizing coefficient (BFC) and Williams absolute fecundity indices (WAFI). Different studies represent different results. The volume of the egg was 63.63±1.223 mm^3 and the values of the GR were 14.82±0.587. But according to the Coroian et al. (2013); and Bud et al. (2009) the GR values (15.218±0.438 and 22.54±1.13) were decreased in the female who was four years old. So only those females were selected in the future that has higher GR but some researchers reported that the female having 6–8 years old is good because in both the rainbow trout and other females of the species showing good performance with an increase in their age (Cocan et al., 2013). In this, the study was carried out on the females that were three years old at the starting period of reproduction. Various workers reported that 16.82±0.563 was BFC having 33.48% variability. Coroian et al. (2013) reported that the total number of the egg was increased with an increase in Behning coefficient value with age. Coroian et al. (2013) studied that the value of Williams's absolute fecundity is directly proportional to the age of females. The higher value was recorded, i.e., 5.77 ± 0.333 in the study than the previous one (4.278±0.279') in the same age group females. The positive correlation was represented by the GR which was 0.621. It has a higher initial body weight and maximum height (IBW and MH), in which the correlation value decreased, i.e., 0.415 and it was lower from previous results obtained in the same aged female by Cocan et al. (2013). The correlation coefficient value shows that to improve the features, the selection can be done, and it results from an increase in correlation value. In contrast, the GR shows the negative correlation for the total length (TL) that was lower from the value reported by Cocan et al. (2013), i.e., –0.116 and –0.027, respectively. It shows that with an increase in the TL both the coefficient and indices values and breeding get decreased. The positive value has been represented by the BFC for TL, i.e., r = 0.264 which was higher than the value represented in the Cocan et al. (2013), i.e., r = 0.130. On the other hand, it also shows a negative correlation for the initial bodyweight, i.e., r = –0.258. A strong and positive value well was represented by Williams absolute fecundity index for both MH and IBW (r = 0.635 and r = 0.562 which was lesser than previous results (r = 0.562) obtained by Cocan et al. (2013). The negative value of correlation occurs –0.054 for TL but Cocan et al. show a positive correlation value (r = 0.162). It needs further evaluation for better results. There are some strong correlations and positive correlations found in the FBW (final body weight) and IBW, i.e., r = 0.887. Also, in relative fecundity and MH (r = 0.752) and 0.696 in the total number of eggs and MH. It was concluded that for the better selection of female breeders the above correlation showed much better results. It gives the best egg quality and quantity as well at the particular range of age.

10.6 CONCLUSION

Reproduction in rainbow trout are affected by various environmental factors and changes the internal endocrine environment of the fish. Therefore, it can be concluded environmental cues are processed through endocrine signal by activating the hypothalamic-pituitary gonadal (HPG) axis. Hypothalamus releases GnRH and gets expressed in various tissues and contributes to the activation of the HPG axis by stimulating the synaptic gonadotropin-producing cells (gonadotrophs) of the pituitary. It was concluded that for the better selection of female breeders the above correlation showed much better results. It gives the best egg quality and quantity as well at the particular range of age.

KEYWORDS

- **gamete maturation**
- **gonadotropin-releasing hormone**
- **micropylar sperm attractant**
- **rainbow trout**
- **reproduction**
- **stress**

REFERENCES

Ahmadnia, M. H. R., Tohidian, F. Y., Mohamadi, A. M. H., Nafisi, B. M., & ZareKhafri, M., (2014). Examine the fecundity and egg quality connected to biometrical parameters of female broad stocks of rainbow trout (*Oncorhynchus mykiss*) A case study in cold water fish reproduction center (Shahid Motahari in Yasuj). *J. Environ. Sci. Technol., 15*, 127–133.

Alavi, S. M. H., & Cosson, J., (2006). Sperm motility in fishes (II) Effects of ions and osmolality: A review. *Cell Biol. Internat., 30*, 1–14.

Alavi, S. M. H., Cosson, J., Bondarenko, O., & Linhart, O., (2019). Sperm motility in fishes:(III) diversity of regulatory signals from membrane to the axoneme. *Theriogenol., 136*, 143–165.

Behnke, R. J., (2002). Comment: First documented case of anadromy in a population of introduced rainbow trout in Patagonia, Argentina. *T. Am. Fish. Soc., 131*, 582–585.

Billard, R., Breton, B., & Richard, M., (1981). On the inhibitory effect of some steroids on spermatogenesis in adult rainbow trout (*Salmo gairdneri*). *Can. J. Zool., 59*, 1479–1487.

Bjornn, T. C., & Reiser, D. W., (1991). Habitat requirements of salmonids in streams. *Am. Fish. Soc. Speci. Public., 19*, 138.

Bromage, N. R., (1995). Broodstock management and seed quality-general considerations. *Brood-Stock Management and Egg and Larval Quality*, 1–24.

Bud, I., Boaru, M. A., & Petrescu-Mag, I. V., (2009). Influence of food and age on breeding and reproductive performances in a rainbow trout population. *Aquaculture, Aquarium, Conservation & Legislation, 2*, 239–247.

Bud, R., (2007). *Penicillin: Triumph and Tragedy*. Oxford University Press on Demand.

Butts, I. A., Love, O. P., Farwell, M., & Pitcher, T. E., (2012). Primary and secondary sexual characters in alternative reproductive tactics of Chinook salmon: Associations with androgens and the maturation-inducing steroid. *G. Compar. Endocrinol., 175*, 449–456.

Campbell, P. M., Pottinger, T. G., & Sumpter, J. P., (1992). Stress reduces the quality of gametes produced by rainbow trout. *Biol. Reprod., 47*, 1140–1150.

Campbell, P. M., Pottinger, T. G., & Sumpter, J. P., (1994). Preliminary evidence that chronic confinement stress reduces the quality of gametes produced by brown and rainbow trout. *Aquac., 120*, 151–169.

Cocan, D., Mireşan, V., Paşca, A., Răducu, C., & Constantinescu, R., (2013). Body features and growth dynamics of two rainbow trout (*Oncorhynchus mykiss*) varieties, Irideus vs. Palomino. *Aquaculture, Aquarium, Conservation & Legislation, 6*, 137–144.

Coroian, A., Erler, S., Matea, C. T., Mireșan, V., Răducu, C., Bele, C., & Coroian, C. O., (2013). Seasonal changes of buffalo colostrum: Physicochemical parameters, fatty acids and cholesterol variation. *Chemist. Cent. J., 7*, 1–9.

Dietrich, G. J., Wojtczak, M., Słowińska, M., Dobosz, S., Kuźmiński, H., & Ciereszko, A., (2008). Effects of ovarian fluid on motility characteristics of rainbow trout (*Oncorhynchus mykiss* Walbaum) spermatozoa. *J. Appl. Ichthyol., 24*, 503–507.

Frazier, J. A., Chiu, S., Breeze, J. L., Makris, N., Lange, N., Kennedy, D. N., Herbert, M. R., et al., (2005). Structural brain magnetic resonance imaging of limbic and thalamic volumes in pediatric bipolar disorder. *Am. J. Psychi., 162*, 1256–1265.

Gage, W. H., Winter, D. A., Frank, J. S., & Adkin, A. L., (2004). Kinematic and kinetic validity of the inverted pendulum model in quiet standing. *Gait & Posture, 19*, 124–132.

Hatef, A., Niksirat, H., & Alavi, S. M. H., (2009). Composition of ovarian fluid in endangered *Caspian brown trout, Salmo trutta caspius*, and its effects on spermatozoa motility and fertilizing ability compared to freshwater and a saline medium. *Fish Physiol. Biochemis., 35*, 695–700.

Heitke, J. D., Archer, E. J., Dugaw, D. D., Bouwes, B. A., Archer, E. A., Henderson, R. C., & Kershner, J. L., (2006). *Effectiveness Monitoring for Streams and Riparian Areas Within the Upper Columbia River Basin Sampling Protocol for Stream Channel Attributes.* PACFISH/

INFISH Biological Opinion (PIBO) Effectiveness Monitoring Program Staff-Multi-federal Agency Monitoring Program; Logan, UT.

Hirohashi, N., Alvarez, L., Shiba, K., Fujiwara, E., Iwata, Y., Mohri, T., Inaba, K., et al., (2013). Sperm from sneaker male squids exhibit chemotactic swarming to CO_2. *Curr. Biol., 23*, 775–781.

Ihuţ, A., Mireşan, V., Cocan, D. I., Răducu, C., Laţiu, C., & Pop, D., (2015). Reproductive indices of rainbow trout (*Oncorhynchus mykiss*) females from a trout farm. Bulletin of the university of agricultural sciences & veterinary medicine CLUJ-NAPOCA. *ANIM. Sci. Biotechnol., 72*.

Inagawa, H., Kuroda, A., Nishizawa, T., Honda, T., Ototake, M., Yokomizo, Y., Nakanishi, T., & Soma, G. I., (2001). Cloning and characterization of tandem-repeat type galectin in rainbow trout (*Oncorhynchus mykiss*). *Fish shellfish Immunol., 11*, 217–231.

Kaupp, U. B., (2012). 100 years of sperm chemotaxis. *J. Gen. Physiol., 140*, 583–586.

Kayam, S., (2004). The effect of mating different age groups of broodstocks on the reproductive performance, sex ratio, growth, and survival bate of rainbow trout (*Oncorhynchus mykiss*). *J. Freshwate. Ecol., 19*, 695–699.

Kleppe, M., Koche, R., Zou, L., Van, G. P., Hill, C. E., Dong, L., De Groote, S., et al., (2018). Dual targeting of oncogenic activation and inflammatory signaling increases therapeutic efficacy in myeloproliferative neoplasms. *Cancer Cell, 33*, 29–43.

Kocaman, G., Ugur, A., & Dicle, A., (2009). A longitudinal analysis of the self-directed learning readiness level of nursing students enrolled in a problem-based curriculum. *J. Nur. Edu., 48*, 286–290.

Lahnsteiner, F., (2002). The influence of ovarian fluid on the gamete physiology in the *Salmonidae. Fish Physiol. Biochemist., 27*, 49–59.

Lahnsteiner, F., Berger, B., Weismann, T., & Patzner, R. A., (1998). Determination of semen quality of the rainbow trout, *Oncorhynchus mykiss*, by sperm motility, seminal plasma parameters, and spermatozoal metabolism. *Aqua., 163*, 163–181.

Lahnsteiner, F., Mansour, N., & Weismann, T., (2002). A new technique for insemination of large egg batches with cryopreserved semen in the rainbow trout. *Aquacult., 209*, 359–367.

Leitritz, E., & Lewis, R. C., (1980). *Trout and Salmon Culture: Hatchery Methods*. UCANR Publications.

Levitan, D. R., (2000). Sperm velocity and longevity trade off each other and influence fertilization in the sea urchin *Lytechinus variegatus*. Proceedings of the Royal Society of London. Series B: *Biol. Sci., 267*, 531–534.

Liljedal, S., Rudolfsen, G., & Folstad, I., (2008). Factors predicting male fertilization success in an external fertilizer. *Behav. Ecol. Sociobiol., 62*, 1805–1811.

Litvak, M. K., & Trippel, E. A., (1998). Sperm motility patterns of Atlantic cod (*Gadus morhua*) in relation to salinity: Effects of ovarian fluid and egg presence. *Can. J. Fish. Aqua. Sci., 55*(8), 1871–1877.

McHugh, P., & Budy, P., (2005). A comparison of visual and measurement-based techniques for quantifying cobble embeddedness and fine-sediment levels in salmonid-bearing streams. *Nor. Am. J. Fish. Manag., 25*, 1208–1214.

Meyer, K. A., & Cassinelli, J. D., (2020). The effect of raceway grading on return to creel for catchable-sized hatchery rainbow trout. *Nor. Am. J. Aqua., 82*, 224–234.

Mireşan, V., Cocan, D., Răducu, C., Constantinescu, R., & Ihuţ, A., (2013). Female fertility curve of rainbow trout (*Oncorhynchus mykiss*) by different ages. *Bulletin UASVM Anim. Sci. Biotechnol., 70*, 15–18.

Orcutt, D. R., Pulliam, B. R., & Arp, A., (1968). Characteristics of steelhead trout redds in Idaho streams. *T. Am. Fish. Soc., 97*, 42–45.

Overton, J. L., Macintosh, D. J., & Thorpe, R. S., (1997). Multivariate analysis of the mud crab *Scylla serrata* (*Brachyura: Portunidae*) from four locations in Southeast Asia. *Mari. Biol., 128*, 55–62.

Özgür, M. E., & Bayir, İ., (2013). A research on reproduction performance for broodstocks of rainbow trout in a fish farm. *BİBAD, Biyoloji Bilimleri Araştırma Dergisi., 6*, 1–5.

Pender, D. R., & Kwak, T. J., (2002). Factors influencing brown trout reproductive success in Ozark tailwater rivers. *T. Am. Fish. Soc., 131*, 698–717.

Poli, M. A., Legarda, E. C., De Lorenzo, M. A., Pinheiro, I., Martins, M. A., Seiffert, W. Q., & Do Nascimento, V. F., (2019). Integrated multitrophic aquaculture applied to shrimp rearing in a biofloc system. *Aquacul., 511*, 734274.

Pottinger, T. G., & Pickering, A. D., (1990). The effect of cortisol administration on hepatic and plasma estradiol-binding capacity in immature female rainbow trout (*Oncorhynchus mykiss*). *Gen. Comp. Endocrinol., 80*, 264–273.

Rahel, F. J., & Nibbelink, N. P., (1999). Spatial patterns in relations among brown trout (*Salmo trutta*) distribution, summer air temperature, and stream size in Rocky Mountain streams. *Can. J. Fish. Aqua. Sci., 56*, 43–51.

Raleigh, R. F., Zuckerman, L. D., & Nelson, P. C., (1984). *Habitat Suitability Index Models and Instream Flow Suitability Curves: Brown Trout*. Western Energy and Land Use Team, Division of Biological Services, Research and Development, Fish and Wildlife Service, US Department of the Interior.

Rather, M. A., Bhat, I. A., & Sharma, R., (2016). Identification, cDNA cloning, and characterization of luteinizing hormone beta subunit (Lhb) gene in Catla catla. *Animal Biotechnology, 27*(3), 148–156.

Rather, M. A., Dutta, S., Guttula, P. K., Dhandare, B. C., Yusufzai, S. I., & Zafar, M. I., (2020). Structural analysis, molecular docking and molecular dynamics simulations of G-protein-coupled receptor (kisspeptin) in fish. *Journal of Biomolecular Structure and Dynamics, 38*(8), 2422–2439.

Reeves, G. H., Benda, L. E., Burnett, K. M., Bisson, P. A., & Sedell, J. R., (1995). A disturbance-based ecosystem approach to maintaining and restoring freshwater habitats of evolutionarily significant units of anadromous salmonids in the Pacific Northwest. In: *Am. Fish. Soc. Symp., 17*, 334–349.

Rime, H., Guitton, N., Pineau, C., Bonnet, E., Bobe, J., & Jalabert, B., (2004). Post-ovulatory ageing and egg quality: A proteomic analysis of rainbow trout coelomic fluid. *Reprod. Biol. Endocrinol., 2*, 1–10.

Root, H. L., (1994). *The Fountain of Privilege: Political Foundations of Markets in Old Regime France and England* (p. 26). Univ. Californ. Press.

Rosengrave, P., Gemmell, N. J., Metcalf, V., McBride, K., & Montgomerie, R., (2008). A mechanism for cryptic female choice in chinook salmon. *Behavior. Ecol., 19*, 1179–1185.

Rosengrave, P., Taylor, H., Montgomerie, R., Metcalf, V., McBride, K., & Gemmell, N. J., (2009). Chemical composition of seminal and ovarian fluids of chinook salmon (*Oncorhynchus tshawytscha*) and their effects on sperm motility traits. *Comparative Biochemistry and Physiology Part A: Molecule. Integra. Physiol., 152*, 123–129.

Taborsky, M., (1998). Sperm competition in fish: Bourgeois' males and parasitic spawning. *T. Ecol. Evol., 13*, 222–227.

Turner, E., & Montgomerie, R., (2002). Ovarian fluid enhances sperm movement in arctic charr. *J. Fish Biol., 60*, 1570–1579.

Tyler, C. R., Sumpter, J. P., & Witthames, P. R., (1990). The dynamics of oocyte growth during vitellogenesis in the rainbow trout (*Oncorhynchus mykiss*). *Biol. Reprod., 43*, 202–209.

Wojtczak, M., Dietrich, G. J., Irnazarow, I., Jurecka, P., Słowińska, M., & Ciereszko, A., (2007). Polymorphism of transferrin of carp seminal plasma: Relationship to blood transferrin and sperm motility characteristics. *Com. Biochemist. Physiol. Part B: Biochemist. Molecul. Biol., 148*, 426–431.

Yanagimachi, R., Cherr, G., Matsubara, T., Andoh, T., Harumi, T., Vines, C., Pillai, M., et al., (2013). Sperm attractant in the micropyle region of fish and insect eggs. *Biol. Reprod., 88*, 47–41.

Yanagimachi, R., Harumi, T., Matsubara, H., Yan, W., Yuan, S., Hirohashi, N., Iida, T., et al., (2017). Chemical and physical guidance of fish spermatozoa into the egg through the micropyle. *Biol. Reprod., 96*, 780–799.

Yoshida, G. M., Bangera, R., Carvalheiro, R., Correa, K., Figueroa, R., Lhorente, J. P., & Yáñez, J. M., (2018). Genomic prediction accuracy for resistance against *Piscirickettsia salmonis* in farmed rainbow trout. *G3: Genes, Genomes, Genetics, 8*, 719–726.

Young, I. M., (1989). Polity and group difference: A critique of the ideal of universal citizenship. *Ethics, 99*, 250–274.

CHAPTER 11

Feeding Carbohydrates to Fish: Utilization and Looking Beyond Energy Nutrition

GARIMA ANAND[1], PUSPA KUMARI[1], TINCY VARGHESE[1], and SHOWKAT AHMAD DAR[2]

[1]*ICAR–Central Institute of Fisheries Education, Mumbai, Maharashtra, India*

[2]*Department of Aqualife Medicine, College of Fisheries and Ocean Studies, Chonnam National University, South Korea*

ABSTRACT

Carbohydrates are one of the most prevalent organic compounds in the biosphere and are known for the cheapest source of energy. Even though there is an unlimited carbohydrate source present in nature, only a few have nutritional value in animal nutrition. However, in fish nutrition carbohydrates are not indispensable in fish diets, although it constitutes an inexpensive source of energy. In fishes, it is found that there is no dietary requirement for carbohydrates; however, if carbohydrates are not provided in the diet, other nutrients such as protein and lipids are catabolized for energy and metabolic intermediates for the synthesis of other biologically important compounds. The use of carbohydrate-rich ingredients in fish feed is gaining day in day importance because of their ability to spare the more expensive protein. So various methodologies and techniques have been used to utilize and uses of carbohydrates in fish feed range from highly digestible and indigestible hemicellulose and cellulose recovered from various sources, including seaweed, algae, and plankton, to refined grain and other numerous products. The present chapter gives the idea about the importance, methods,

Coldwater Fisheries and Aquaculture Management: Technology for Sustainable Food Production.
Mohd Ashraf Rather, Faisal Rashid Sofi, Adnan Amin, & Kawkabul Saba (Eds.)

mechanism, and ways to utilize more carbohydrate sources in fish nutrition and reduce the cost of fish feed.

11.1 INTRODUCTION

Nowadays, the majority of the developing countries and many developed ones have extended the use of artificial compounded aquafeeds for farmed fishes and other invertebrates. The increased use of aqua-feeds has stimulated the rapid development of the production sector, which became one of the fastest expanding agricultural industries globally. The success of culture-based fishery is partially and or wholly dependent on the quality and cost of feed delivered as the cost of feed depends on the protein source used in the feed. So, it is necessary to use low-cost protein source ingredients for feed formulation, which reduce the overall cost of the feed. Research studies showed that carbohydrate has a protein-sparing effect (Shiau & Peng, 1993; Zhou et al., 2014; Lin et al., 2018) and is the least expensive dietary energy source for humans and animals (Stone, 2003; Knudsen et al., 2012; Ludwig et al., 2018). Carbohydrate-rich ingredients are available in desirable quantities at low prices. Carbohydrates in fish feed range from highly digestible and indigestible hemicellulose and cellulose recovered from various sources, including seaweed, algae, and plankton, to refined grain and soybean products. However, carbohydrates are not indispensable in fish feed, although it constitutes an inexpensive source of energy. In fishes, it is found that there is no dietary requirement for carbohydrates, however, if carbohydrates are not provided in the diet, other nutrients such as protein and lipids are catabolized for energy and metabolic intermediates for the synthesis of other biologically important compounds. Thus, it is necessary to provide the appropriate carbohydrate level in the diet of the fish species being cultured (Wilson, 1994; Kamalam & Panserat, 2016). However, certain fish species, when fed with carbohydrate-free diet, showed in reduced growth rate whereas high levels of dietary carbohydrate resulted in increased glycogen content in body and vital organs and liver size in fishes (Hilton & Atkinson, 1982; Boonanuntanasarn et al., 2018; Singh et al., 2012) optimum levels of dietary incorporation are still unclear.

11.2 SOURCES OF CARBOHYDRATES

Fish are not able to utilize all types of carbohydrates. The relative utilization of dietary carbohydrates by fish varies and appears to be associated with the complexity of carbohydrates. For example, starch is more efficiently used compared to glucose in both marine and freshwater species: Carp, red seabream (Furuichi & Yone, 1982), yellowtail (Furuichi *et al.,* 1986), Tilapia (Shiau & Peng, 1993), white sturgeon (Deng *et al.,* 2005). On the other side, juvenile grass carp and rainbow trout (Hung & Storebakken, 1994) utilize glucose better than starch. However, in grouper (Shiau & Lin, 2001), starch seems to be used efficiently as glucose (Wilson & Poe, 1987). It has been found that channel catfish can efficiently use starch or dextrin but cannot utilize dietary mono and disaccharides as an energy source. The exact mechanism in fish that is responsible for the difference in carbohydrate utilization is not known. Still, Pieper, & Pfeffer (1980) suggested poor performance with simple sugar such as glucose saturation. That might be because simple sugar like glucose requires no digestion and is rapidly absorbed from the gut, whereas complex carbohydrates like starch must undergo enzyme hydrolysis before absorption. Thus, free glucose appears at gut absorption sites more rapidly than glucose from a complex carbohydrate source. Various factors affect the digestible energy provided by complex carbohydrates in fish diets (Bergot, 1993). Some of the factors include their source and treatment of starch.

11.3 CARBOHYDRATE REQUIREMENT OF FISH

The nutritional value of carbohydrates varies among species. In the case of warm water fish, they can utilize much higher levels of dietary carbohydrates than coldwater and marine fish. Grass carp (*Ctenopharyngodon idella*) is herbivorous fish in which the optimum dietary level of digestible carbohydrate is significantly higher than carnivorous categories, or even in other carp (Lin, 1991; Kamalam & Panserat, 2016). It has been reported that larger fish utilize dietary carbohydrates better than fish with small body sizes (Pen-Hsing & Shi-Yen, 1993; Hatlen *et al.,* 2005). No specific requirement of carbohydrates in the fish diet is known. Still, some studies showed that excess carbohydrate levels in the fish diet reduce the growth rate are often accompanied by poor feed utilization (Hemre *et al.,* 2002; Han et al., 2021).

11.4 EFFECT OF CARBOHYDRATE ON GROWTH OF FISH

Carbohydrates are one of the most widespread organic compounds in the biosphere. Although there is an infinite carbohydrate source present in nature, only a few have nutritional value in animal nutrition. Some of them are hexoses (glucose), disaccharides, and some homopolysaccharides, including starch. Fish do not require dietary carbohydrates for their sustenance, they have a higher capability to make glucose through gluconeogenesis for their energy needs than other animals. However, fish nutritionists want to include carbohydrates in aquafeed, as it is the cheapest energy source (Kamalam & Panserat, 2016). Anderson & Coworkers (1984) found that *Oreochromis niloticus* fed diets containing 40% carbohydrate (approximately 30% of gross dietary energy) did not retard growth. An increase in glucose level up to 15% in the rations positively affects weight gain, food efficiency, and PER in juvenile starry flounders (Lee & Lee, 2004). But at the same time, some studies showed the negative effect of carbohydrates on the growth of *C. batrachus* fry when the levels were maintained at more than 15% (Mollah & Alam, 1990). Furuichi & Yone (1980) noted depression in growth and feed efficiency in red sea bream, yellowtail, and common carp fed diets high in carbohydrate content. Tung & Shiau (1991) studied how meal frequency affects carbohydrate utilization in tilapia. Fish were fed with diets containing 44% of different carbohydrates sources (starch, dextrin, and glucose). They found that fish fed starch, dextrin, or glucose diets six times a day gained significantly more body weight than fish fed these diets two times a day.

11.5 PROTEIN-SPARING EFFECT BY CARBOHYDRATE

The use of carbohydrate-rich ingredients in fish feed is gaining day in day importance because of their ability to spare the more expensive protein component of the diet to be used for growth (Shiau, 1997; Zhu et al., 2014; Lin et al., 2018). Although lipids serve as an important source of non-protein energy for fish (Kaushik et al., 1989), they are more expensive than carbohydrates. Protein cost contributes to a large part of the cost of most prepared feeds, so protein fraction must only be utilized for growth, not energy. This knowledge of the optimal protein level and protein-sparing effects of non-protein nutrients such as carbohydrates can be used effectively to help reduce feed costs. Shiau & Peng (1993) conducted a study to evaluate the possible protein-sparing effects of carbohydrates in juvenile tilapia. Three dietary protein levels (32, 28, and 24%) were achieved by substitution with three levels (33, 37, and 41%) and three sources (glucose, dextrin, starch) of dietary carbohydrates. Results indicated that fish fed starch or dextrin had significantly higher weight gain and FER than that fed glucose for all the protein levels, except for weight gain for the 24% protein level, in which the difference was not significant. Fish fed starch and dextrin had similar weight gain. Decreasing the dietary protein level from 28% to 24% by increasing the starch or dextrin content in the diet from 37% to 41% did not reduce weight gain and FER, suggesting that starch or dextrin could spare some protein when the dietary protein level was low. Fynn-Aikins et al. (1992) also did not observe a protein-sparing effect when sturgeon were fed diets with an optimum protein (42%) and different levels of D-glucose (0–35%), which were effectively utilized by sturgeon (Hung et al., 1989). These studies show that carbohydrates can be used in fish feed at an optimum level to fulfill the energy requirement of fish.

11.6 CARBOHYDRATE UTILIZATION IN FISH

Fish, in general, utilize dietary carbohydrates poorly. Apart from being the most economical source of energy in the diet, Carbohydrate-rich ingredients also have good availability throughout the world. But despite all these benefits, fishes have poor ability to make use of dietary carbohydrates because of the scarcity of carbohydrates in the aquatic environment. Two problems are related to carbohydrate nutrition in fish. Firstly, as a general rule, digestion of complex carbohydrates is low in the aquatic animal. Secondly, fishes are poorly adapted to the metabolic utilization of simple sugars (monosaccharides or disaccharides). It is also found that carbohydrate utilizations in fish vary with their feeding habits and habitat.

Warmwater fish can utilize much higher levels of dietary carbohydrates than cold water and marine fish. There are certain reports of limited ability to utilize carbohydrates in carnivorous fish like cod Atlantic salmon (Hemre *et al.,* 1995) and white sturgeon (Deng *et al.,* 2005). However, improved growth performance was observed in different species of fish upon feeding high carbohydrate diets viz., eels (Degani & Viola, 1987), rainbow trout (Grisdale & Helland, 1997), catfish (Hung *et al.,* 2002) and carp (Mohapatra *et al.,* 2003).

11.6.1 IMPROVING CARBOHYDRATE UTILIZATION IN FISH

Several promising strategies have been explored to overcome the challenges in carbohydrate utilization in farmed fishes, especially the carnivorous fish. Therefore, tailoring metabolic pathways to improve carbohydrate stimulus at critical early developmental stages may imprint an adaptive ability to withstand with high carbohydrate diets in progressive life stage (Kamalam & Panserat, 2016). Further dietary modifications of carbohydrates may improve its utilization in fish. These modifications include: (1) pre-gelatinization, which are made by applying starch water slurry to heated drum dryer; (2) chemical modification of starch by treatment with acids or with enzymes; and (3) Starch dextrinization with heat treatment. Starch is a valuable source of energy, and its utilization is enhanced by gelatinization (Bergot & Breque, 1983) due to its increased digestibility (Bergot, 1991; Podoskina et al., 1997; Mohapatra et al., 2003). Gelatinization of starch results in hybridization and swelling of starch granules (Hofer & Sturmbauer, 1985) and thus improves digestibility by increasing the susceptibility of starch granules to the enzyme. Digestibility of starch varies with the source of starch. For example, it has been reported that the digestibility of tuber starch (potato and tapioca) is lower than that of cereal starch in rainbow trout (Bergot, 1991) and Carp (Schwarz & Kirchgessner, 1993). Pigs fed with Corn were found to have a higher feed conversion ratio (FCR) than wheat fed groups (Han et al., 2005). Corn-based rations are also commonly used in poultry for high available energy (Wang et al., 2005). The inclusion level of starch also plays an important role in their digestion; it is found that low digestibility of starch at higher inclusion level has (Hemre et al., 1989). However, a significant increase in digestibility was reported with an increase in gelatinized carbohydrate level in the diet of a carp *L. rohita* fry (Mohapatra et al., 2003).

11.7 IMPACT OF CARBOHYDRATE ON IMMUNE SYSTEM OF FISH

Compared to the immense number of studies in nutritional immunology in humans and livestock, the evaluation of the nutritional impact on fish health is in its infancy. The mechanisms by which the single nutrient modulates immunity are frequently not understood, even if experiments indicate the mode of action. Among all the macronutrients, carbohydrate is considered one of the cheapest source nutrients, but at the same time, it has their limitations. Fish have limited ability for digestion and metabolism of carbohydrates, and hence excessive intake of this nutrient may adversely affect fish health through metabolic disorder and physio-clinical signs such as hyperglycemia, increment of glycogen deposition, liver hypertrophy, and histopathological (HP) development (Lall, 1991; Li et al., 2019; Azaza et al., 2020). Some of the studies show that fish with excess glucose is found to be under constant metabolic stress which may lead to suppressed immune functions (Wiik et al.,

1989; Talukdar et al., 2019; Li et al., 2020; Han et al., 2021). Stress is one of the important factors that increase fish's susceptibility to infectious diseases (Mishra et al., 2017; Sarkodie et al., 2019). Thus high dietary carbohydrates may increase the susceptibility and incidence of infectious. Fewer studies are conducted on the health implications of high dietary carbohydrates on fish. High carbohydrate in the diet of rainbow trout (*Oncorhynchus mykiss*) shows intracellular damage due to deposition of glycogen particles and thus reduce the liver detoxification capacity by influencing the metabolism and cause glucose metabolism disorders (Hilton & Hodson, 1983; Azaza et al., 2020; Wang et al., 2020; Zhou et al., 2021).

11.8 CONCLUSION

Being the cheapest source of energy, carbohydrates are an integral part of aquafeeds. However, carbohydrate utilization in fish depends on many factors such as inclusion level, molecular complexity and physical state of fish. Another factor impacting carbohydrate utilization is species' habitat, as warmwater fishes can more efficiently utilize carbohydrates than cold water and marine fish. They have a role in modulating fish immunity, and they also affect the metabolic fate of other nutrients by sparing action. Thus, the importance of carbohydrates in fish nutrition is beyond that of an energy substitute.

KEYWORDS

- **agricultural industries**
- **carbohydrates**
- **carbohydrates**
- **energy**
- **feed**
- **fish nutrition**
- **macronutrients**
- **metabolic utilisation**
- **protein-sparing effect**

REFERENCES

Anderson, J., Jackson, A. J., Matty, A. J., & Copper, B. S., (1984). Effects of dietary carbohydrate and fiber on the tilapia (*Oreochromis niloticus*). *Aquaculture, 37*, 303–314.

Azaza, M. S., Saidi, S. A., Dhraief, M. N., & El-Feki, A., (2020). Growth performance, nutrient digestibility, hematological parameters, and hepatic oxidative stress response in juvenile Nile tilapia, *Oreochromis niloticus*, fed carbohydrates of different complexities. *Animals, 10*, 1913.

Bergot, F., & Breque, J., (1983). Digestibility of starch by rainbow trout: Effects of the physical state of starch and of the intake level. *Aquaculture, 34*, 203–212.

Bergot, F., (1991). Digestibility of native starches of various botanical origins by rainbow trout (*Oncorhynchus mykiss*). In: Kaushik, S. J., &. Luquet, P., (eds.), *Fish Nutrition in Practice* (pp. 857–865). IV International Symposium on Fish Nutrition and Feeding, INRA, Paris.

Bergot, F., (1993). *Digestibility of Native Starches of Various Botanical Origins by Rainbow Trout (Oncorhynchus Mykiss).* Colloques de l'INRA (France).

Degani, G., & Viola, S., (1987). The protein sparing effect of carbohydrates in the diet of eels (*Anguilla anguilla*). *Aquaculture, 64*, 283–291.

Deng, D. F., Hemre, G. I., Storebakken, T., Shiau, S. Y., & Hung, S. S., (2005). Utilization of diets with hydrolyzed potato starch, or glucose by juvenile white sturgeon (*Acipenser transmontanus*), as affected by Maillard reaction during feed processing. *Aquaculture, 248*(1), 103–109.

Furuichi, M., & Yone, Y., (1980). The utilization of carbohydrate by fishes. 1. Effect of dietary dextrin levels on the growth and feed-efficiency, the chemical composition of liver and dorsal muscle, and the absorption of dietary protein and dextrin in Fishes. *Bulletin of the Japanese Society of Scientific Fisheries, 46*(2), 225–229.

Furuichi, M., & Yone, Y., (1982). Availability of carbohydrate in nutrition of carp and red sea bream. *Bulletin of the Japanese Society of Scientific Fisheries, 48*, 945–948.

Furuichi, M., Taira, H., & Yone, Y., (1986). Availability of carbohydrate [alpha-potato starch and glucose] in nutrition of yellowtail [*Seriola quinqueradiata*]. *Bulletin of the Japanese Society of Scientific Fisheries*.

Fynn-Aikins, K., Hung, S. S., Liu, W., & Li, H., (1992). Growth, lipogenesis and liver composition of juvenile white sturgeon fed different levels of Dglucose. *Aquaculture, 105*(1), 61–72.

Grisdale-Helland, B., & Helland, S. J., (1997). Replacement of protein by fat and carbohydrate in diets for Atlantic salmon (*Salmo salar*) at the end of the freshwater stage. *Aquaculture, 152*(1), 167–180.

Han, H., Wang, Z., Wang, J., Wang, T., Li, Y., Guan, D., & Sun, H., (2021). Impact of high dietary cornstarch level on growth, antioxidant response, and immune status in GIFT tilapia *Oreochromis niloticus*. *Scientific Reports, 11*(1), 1–10.

Han, Y. K., Soita, H. W., & Thacker, P. A., (2005). Performance and carcass composition of growing-finishing pigs fed wheat or corn-based diets. *Asian Australasian Journal of Animal Science, 18*, 704–710.

Hatlen, B., Grisdale-Helland, B., & Helland, S. J., (2005). Growth, feed utilization and body composition in two size groups of Atlantic halibut (*Hippoglossus hippoglossus*) fed diets differing in protein and carbohydrate content. *Aquaculture, 249*(1), 401–408.

Hemre, G. I., Lie, O., Lied, E., & Lambertsen, G., (1989). Starch as an energy source in feed for cod (*Gadus morhua*): Digestibility and retention. *Aquaculture, 80*, 261–270.

Hemre, G. I., Mommsen, T. P., & Krogdahl, A., (2002). Carbohydrates in fish nutrition: Effects on growth, glucose metabolism and hepatic enzymes. *Aquaculture Nutrition, 8*(3), 175–194.

Hemre, G. I., Sandnes, K., & Waagbø, R., (1995). Blood chemistry and organ nutrient composition in Atlantic salmon, *Salmo salar* L., fed graded amounts of wheat starch. *Aquaculture Nutrition, 1*(1), 37–42.

Hilton, J. W., & Atkinson, J. L., (1982). Response of rainbow trout (*Salmo gairdneri*) to increased levels of available carbohydrate in practical trout diets. *British Journal of Nutrition, 47*(03), 597–607.

Hilton, J. W., & Hodson, P. V., (1983). Effect of increased dietary carbohydrate on selenium metabolism and toxicity in rainbow trout (*Salmo gairdneri*). *The Journal of nutrition, 113*(6), 1241–1248.

Hofer, R., & Sturmbauer, C., (1985). Inhibition of trout and carp α-amylase by wheat. *Aquaculture, 48,* 227–283.

Hung, L. T., Liem, P. T., Tu, H. T., & Mariojouls, C., (2002). Comparing growth and protein requirements for fingerlings of three catfish of the Mekong River (*Pangasius Bocourti, Pangagasius Hypothalmus and Pangasius Conchophilus). Journal of Aquaculture in the Tropics, 17*(4), 325–335.

Hung, S. S., & Storebakken, T., (1994). Carbohydrate utilization by rainbow trout is affected by feeding strategy. *The Journal of Nutrition, 124*(2), 223–230.

Hung, S. S., Fynn-Aikins, F. K., Lutes, P. B., & Xu, R., (1989). Ability of juvenile white sturgeon (*Acipenser transmontanus*) to utilize different carbohydrate sources. *The Journal of Nutrition, 119*, 727–733.

Kamalam, B. S., & Panserat, S., (2016). Carbohydrates in fish nutrition. *International Aquafeed,* 20–23.

Kaushik, S. J., Medale, F., Fauconneau, B., & Blanc, D., (1989). Effect of digestible carbohydrates on protein/energy utilization and on glucose metabolism in rainbow trout (*Salmo gairdneri* R.). *Aquaculture, 79*(1), 63–74.

Knudsen, K. E. B., Hedemann, M. S., & Lærke, H. N., (2012). The role of carbohydrates in intestinal health of pigs. *Animal Feed Science and Technology, 173*(1, 2), 41–53.

Lall, S. P., (1991). Salmonid nutrition and feed production. In: Cook, R. H., & Pennel, W., (eds.), *Proceedings of the Special Session on Salmonid Aquaculture. The Journal of World Aquaculture Society*, 107–123.

Lee, S. M., & Lee, J. H., (2004). Effect of dietary glucose, dextrin and starch on growth and body composition of juvenile starry flounder *Platichthys stellatus. Fisheries Science, 70*(1), 53–58.

Li, S., Sang, C., Wang, A., Zhang, J., & Chen, N., (2019). Effects of dietary carbohydrate sources on growth performance, glycogen accumulation, insulin signaling pathway and hepatic glucose metabolism in largemouth bass, *Micropterus salmoides. Aquaculture, 513*, 734391.

Li, S., Wang, A., Li, Z., Zhang, J., Sang, C., & Chen, N., (2020). Antioxidant defenses and non-specific immunity at enzymatic and transcriptional levels in response to dietary carbohydrate in a typical carnivorous fish, hybrid grouper (*Epinephelus fuscoguttatus*♀× E. lanceolatus♂). *Fish & Shellfish Immunology, 100*, 109–116.

Lin, D., (1991). Grass carp, *Ctenopharyngodon idella.* In: Wilson, R. P., (ed.), *Handbook of Nutrient Requirements of Finfish* (pp. 89–96).

Lin, S. M., Shi, C. M., Mu, M. M., Chen, Y. J., & Luo, L., (2018). Effect of high dietary starch levels on growth, hepatic glucose metabolism, oxidative status and immune response of juvenile largemouth bass, *Micropterus salmoides. Fish & shellfish immunology, 78*, 121–126.

Ludwig, D. S., Hu, F. B., Tappy, L., & Brand-Miller, J., (2018). Dietary carbohydrates: Role of quality and quantity in chronic disease. *BMJ*, 361.

Mishra, S. S., Rakesh, D., Dhiman, M., Choudhary, P., Debbarma, J., Sahoo, S. N., & Mishra, C. K., (2017). Present status of fish disease management in freshwater aquaculture in India: State-of-the-art-review. *Journal of Aquaculture & Fisheries, 1*(003).

Mohapatra, M., Sahu, N. P., & Chaudhari, A., (2003). Utilization of gelatinized carbohydrate in diets in *Labeo rohita* fry. *Aquaculture Nutrition, 9*, 189–196.

Mollah, M. F. A., & Alam, M. S., (1990). Effects of different levels of dietary carbohydrate on growth and feed utilization of catfish *Clarius batrachus* L. fry. *Indian Journal of Fisheries, 37*(3), 243–249.

Pen-Hsing, T., & Shi-Yen, S., (1993). Carbohydrate utilization versus body size in tilapia (*Oreochromis niloticus× O. Aureus*). *Comparative Biochemistry and Physiolology, 104*(3), 585–588.

Pieper, A., & Pfeffer, E., (1978). *Carbohydrates as Possible Sources of Dietary Energy for Rainbow Trout (Salmo gairdnerii, Richardson).* In FAO, Rome (Italy). Fisheries Dept. Symposium on Finfish Nutrition and Feed Technology. Hamburg (Germany, FR).

Podoskina, T. A., Podoskin, A. G., & Bekina, E. N., (1997). Efficiency of utilization of some starch modification by rainbow trout (*O. mykiss*). *Aquaculture, 152*, 235–248.

Sarkodie, E. K., Zhou, S., Baidoo, S. A., & Chu, W., (2019). Influences of stress hormones on microbial infections. *Microbial Pathogenesis, 131*, 270–276.

Schwarz, F. J., & Kirchgessner, M., (1993). *Influence of Different Carbohydrates on Digestibility, Growth and Carcass Composition of Carp (Cyprinus carpio).* Colloques de l'INRA (France).

Shiau, S. Y., & Lin, Y. H., (2001). Carbohydrate utilization and its protein-sparing effect in diets for grouper (*Epinephelus malabaricus*). *Animal Science, 73*, 299–304.

Shiau, S. Y., & Peng, C. Y., (1993). Protein-sparing effect by carbohydrates in diets for tilapia *Oreochromis niloticus X 0. aureus. Aquaculture, 117*, 327–334.

Shiau, S. Y., (1997). Utilization of carbohydrates in warmwater fish with particular reference to tilapia (*Oreochromis niloticus X O. aureus*). *Aquaculture, 151*(1), 79–96.

Singh, S. K., Rather, M. A., Mandal, S. C., Das, P., Pawar, N., Singh, Y. J., & Dar, S. A., (2012). *Effects of Dietary fish Oil Substitution with Palm Oil on Growth, Survival, and Muscle Proximate Composition of Cirrhinus Mrigala (Hamilton, 1822), 64*, 1–15.

Stone, D. A., (2003). Dietary carbohydrate utilization by fish. *Reviews in Fisheries Science, 11*(4), 337–369.

Talukdar, A., Kumar, S., Varghese, T., Jain, K. K., Sahu, N. P., & Sahoo, S., (2019). Feeding gelatinized carbohydrate in the diets of magur, *Clarias batrachus* (Linnaeus, 1758): Effects on growth performance, enzyme activities and expression of muscle regulatory factors. *Aquaculture Research, 50*(3), 765–777.

Tung, P. H., & Shiau, S. Y., (1991). Effects of meal frequency on growth performance of hybrid tilapia, *Oreochromis niloticus× O. aureus,* fed different carbohydrate diets. *Aquaculture, 92*, 343–350.

Wang, C. L., Lu, W. Q., Li, D. F., & Xing, J. J., (2005). Effects of alpha-galactosidase supplementation to corn-soybean meal diets on nutrient utilization, performance, serum indices and organ weight in broilers. *Asian and Australaisian Journal of Animal Science, 18*, 761–176.

Wilson, R. P., & Poe, W. E., (1987). Apparent inability of channel catfish to utilize dietary mono and disaccharides as energy sources. *Journal of Nutrition, 117*, 280–285.

Wilson, R. P., (1994). Utilization of dietary carbohydrate by fish. *Aquaculture, 124*(1), 67–80.

Zhou, C., Ge, X., Lin, H., & Niu, J., (2014). Effect of dietary carbohydrate on non-specific immune response, hepatic antioxidative abilities and disease resistance of juvenile golden pompano (*Trachinotus ovatus*). *Fish & Shellfish Immunology, 41*(2), 183–190.

Zhou, W. H., Wu, C. C., Limbu, S. M., Li, R. X., Chen, L. Q., Qiao, F., Luo, Y., et al., (2021). More simple more worse: Simple carbohydrate diets cause alterations in glucose and lipid metabolism in Nile tilapia (*Oreochromis niloticus*). *Aquaculture*, 737–857.

CHAPTER 12

Application of Plant-Derived Natural Preservatives for Shelf-Life Extension of Freshwater Fish

HAFSA MAQBOOL[1], MUDASSIR AZHAR[2], and A. A. ZYNUDHEEN[3]

[1]*Division of Aquatic Animal Health Management, Faculty of Fisheries, Sher-e-Kashmir University of Agricultural Sciences and Technology of Kashmir, Jammu and Kashmir, India*

[2]*Department of Fisheries Science, Doon PG College of Agriculture and Allied Science, Uttarakhand, India*

[3]*Fish Processing Division, Central Institute of Fishing Technology, Kochi, Ernakulam, Kerala, India*

ABSTRACT

Quality loss in fish happens shortly after death, throughout the processing and storage and is linked to enzymatic, microbiological, and chemical reactions. As a result, food additives are required to preserve the shelf life, nutritional content, texture, and flavor of both raw and processed materials. Following the advent of "green consumerism" which resulted in a decrease in consumer demand for synthetic food additives, extensive research is being done on natural additive applications. The natural extracts isolated from various plant sources such as spices and herbs, fruits, and vegetables, seaweed, and so on comprises of bioactive compounds that have demonstrated remarkable free radical inhibitor and antibacterial activities. Plant phenolic compounds are of particular interest due to their antioxidant and antimicrobial properties.

Coldwater Fisheries and Aquaculture Management: Technology for Sustainable Food Production.
Mohd Ashraf Rather, Faisal Rashid Sofi, Adnan Amin, & Kawkabul Saba (Eds.)

Agricultural handling factories produce profuse volume of phenolic-rich by-products, which could be valuable natural sources of antioxidants and anti-bacterial. Some of these by-products have been studied and found to be effective sources of phenolic antioxidants and anti-bacterial. Based on this, plant extract treatments have been effectively practicable to chilled, frozen, and dried fish, fish filets and mince as a natural additive. This chapter will discuss the mode of action and applications of these natural preservatives for maintaining quality and extending the shelf life of freshwater fish and its products.

12.1 INTRODUCTION

Due to the high content of polyunsaturated fatty acids (PUFAs) and protein, fish, and fish products are considered a valuable part of human nutrition (Kykkidou et al., 2009) long-chain polyunsaturated fatty acids (PUFAs) in these products have gained attention because of their prevention of human cardiovascular diseases (Ozogul et al., 2006). These products have a short shelf-life, despite their nutritional significance. This is due to the large amounts of free amino acids (AAs), volatile nitrogen bases and unsaturated fatty acids that are highly susceptible to oxidation (Mexis et al., 2009), which can affect the flavor, texture, aroma, and shelf-life of fish (Banerjee, 2006). In recent years, consumers have demanded fresh, natural, and minimally processed foods, as well as an increase in food safety and quality. This perspective has prompted research into alternative natural antimicrobials and antioxidants, as well as pressure on the food sector to gradually phase out chemical preservatives (Cox et al., 2010). Some plants and vegetables have been found to contain beneficial antioxidants and antibacterial with nutritional and medicinal benefits. Antimicrobial activities of various plant extracts and essential oils (EOs) have been evaluated against a variety of foodborne bacteria (Boziaris et al., 2011).

12.2 SPOILAGE MECHANISM IN FISH

Fresh fish can quickly deteriorate after being harvested and acquire rigor mortis, a condition in which fish lose their elasticity due to muscle stiffening a few hours after death. Enzymatic autolysis, oxidation, and microbial growth are the three mechanisms that cause spoilage.

12.2.1 ENZYMATIC ACTIVITY

Autolysis, the process of destroying proteins by native enzymes, begins immediately after rigor mortis is complete. This procedure provides an environment that is conducive to bacterial growth (Ghaly et al., 2010). The degradation process in fish starts quickly after collection owing to enzymatic deterioration that causes chemical and biological changes and thus the breakdown of basic molecules. Although the autolytic enzymes had a significant impact on textural quality, resulting in spoilage, they did not cause the typical spoilage with off-flavor and off-odor. The presence of gastrointestinal digesting and endogenous muscular enzymes in the fish is found to be responsible for post-mortem changes in fish muscles during processing and storage, resulting in sensory changes. On the other hand, autolysis of fish muscle proteins may lead to the deterioration of fish meat and the generation of biogenic amines. During autolysis, the degree of freshness of the fish must be determined before it begins to decay (Ocanohiguera et al., 2011). It has been suggested that maintaining the low activity of endogenous autolytic enzymes in muscle during storage can be achieved by keeping the temperature and water activity (aw) low (Pachecoaguilar et al., 2010).

12.2.2 MICROBIAL ACTIVITY

Trimethylamine oxide (TMAO) is found in many fish species and promotes bacterial growth over a broad temperature range and has a high content of free AA, a high pH after death, and a high-water content (Chaillou et al., 2015). The growth and metabolism of microorganisms involved in the production of biogenic amines, alcohols, histamine, putrescine, sulfides, organic acids, aldehydes, and ketones are thought to be the primary cause of deterioration of fish quality (Kuley et al., 2017). The majority of the fish microflora from humid sea waters consists of psychotropic Gram-negative bacteria. On the other hand, continuous processing or extended storage of products allows Gram-positive bacteria to take over and ruin the product (Al-Bulushi et al., 2010). Specific spoilage organisms (SSOs) are smaller portions of the fish microbiota that are considered the major causes of seafood spoilage, such as *Shewanella*, *Photobacterium phosphoreum*, and *Pseudomonas* (Gram & Dalgaard, 2002). The SSOs consume the metabolites produced from the breakdown of non-protein nitrogen (NPN) of fish muscle causing the unacceptable off-flavors and thus the rejection by the consumer.

12.2.3 CHEMICAL ACTIVITY

In general, seafood is high in lipids, particularly fats containing long-chain PUFAs (Venugopal & Gopakumar, 2017). Lipids have a key role in the production of off-flavor and off-odor, as well as the nutritional value of seafood (Mariutti & Bragagnolo, 2017). Lipid oxidation also causes a depletion of fat-soluble vitamins and other substances (Souza & Bragagnolo, 2014). Lipid oxidation has a three-stage free radical mechanism: initiation, propagation, and termination, which might lead to spoilage and degradation due to the production of reactive aldehydes such as malonaldehyde (MDA) and hydrononenal (HNE) as depicted in Figure 12.1 (Shahidi & Botta, 1994). Discoloration, protein lipoxidation, cytosolic enzymes, photolytic oxidation, and hydrogen peroxide are some of the processes that initiate and propagate lipid oxidation in fish (Schneider, 2009). There are two types of lipid oxidation: enzymatic and non-enzymatic, together it can result in deterioration of fish quality. The reaction of oxygen with the double bonds of fatty acids is known as oxidation. Fish contains high amount of unsaturated fatty acids and pro-oxidant compounds, both of which are easily oxidized, resulting in rancidity and quality loss (Losada et al., 2007). The smell, color, texture, and nutritional value of these compounds all are affected due to these substances. Protein denaturation, alteration of protein electric field characteristics, nutrient deterioration, and intracellular antioxidant system deficits, and the production of fluorescent substances have all been linked to lipid oxidation products (Lugasi, 2007). Several pro-oxidants (hemoglobin, myoglobin, and cytochrome c) can cause lipid oxidation (Ghaly et al., 2010).

12.3 FOOD ADDITIVES

Codex alimentarius defines additive as any substance that is added in foods for the purposes of restoring esthetic appearance degraded during production and improving aroma or taste, rather than nutritional improvement (WHO/FAO, 2004). Food additives are used in the food sector because of the variety of new technologies and consumer preferences. These can be used to reduce losses, improve quality, extend shelf life, and develop novel formulas. Food preservers, dietary ingredients, food dyes, artificial sweeteners, texturizing agents, and other additives are all types of food additives. The majority of additives have multiple functions (Carocho et al., 2015). Antimicrobials such as nitrites, sulfites, and organic acids, as well as antioxidants such as tertiary butyl hydroxy quinone (TBHQ), butylated hydroxy toluene (BHT) and butylated hydroxy anisole (BHA) were employed in fish for a long time to limit the microbial growth and lipid oxidation (Manju et al., 2007). The moieties of these synthetic additives have been considered as potentially toxicological, in spite of their great activity (Naveena et al., 2008). The practice of functional food ingredients has grabbed a lot of attention, due to the rapid development of the social economy. Because of the negative perception of synthetic preservatives, researchers are focusing on finding safe, effective, and acceptable natural preservatives for suppressing the microbiological and chemical pathways that cause fish to decay.

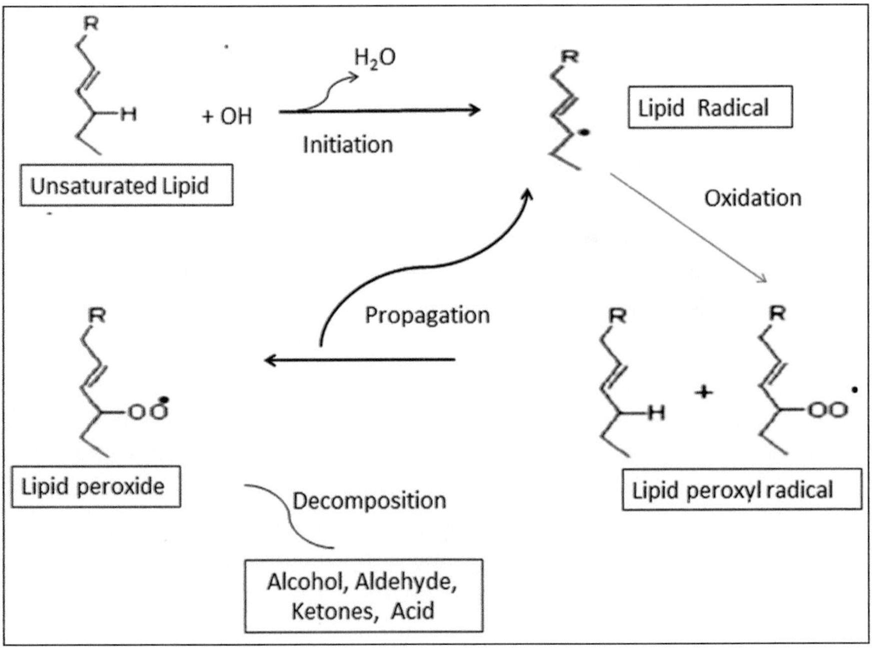

FIGURE 12.1 Mechanism of lipid oxidation in fish.

Source: Shahidi & Botta, 1994.

12.4 PLANT EXTRACT

Plant extracts and EOs derived from plant petals, leaves, fruits, peels, stems, and roots are now considered as natural preservatives or food additives with potent antibacterial, antifungal, and antioxidant properties, and are used in the fish industry to preserve raw and processed fish (Chouliara et al., 2007). Phytochemicals are chemical compounds found in plants that have an impact on the bacteriological and esthetic perception of foods. The efficiency of plant extract varies due to qualitative and quantitative variances in the presence of bioactive phytochemicals. Polyphenolic compounds like flavonoids, terpenoids, isothiocyanates, and polypeptide are illustrations of phytoconstituents. Extracts are dissolved portion of organic materials such as spices, fruits, and vegetables that are produced by solubilizing the components of interest in an aqueous, alcohol, lipid, solvent, or supercritical CO_2 phase and then removing it (Brewer, 2011). Plant extracts are generally regarded as safe (GRAS) due to their long history of usage with no reported adverse from several specialized toxicological investigations (Smid & Gorris, 1999). Plant based natural antioxidant and antimicrobials have been long recognized and are been used as preservatives in a variety of food systems (Kanatt et al., 2007).

12.4.1 EFFECTIVE COMPOUNDS IN PLANT EXTRACTS AND ITS SYSTEM OF ACTIVITY

Plant extracts have bioactivities due to the presence of phytochemicals, particularly polyphenolics. Polyphenols are organic molecules found in plants that have one or more aromatic ring structures with multiple hydroxyl groups (Boudet, 2007). They are largely divided into four groups—flavonoids, phenolic acids, hydroxycinnamic acids and lignans as shown in Figure 12.2 (Maqsood et al., 2013). Flavonoids are categorized as flavanol, flavanones, isoflavones, and anthocyanins based on the degree of oxidation of the heterocyclic ring. Quercetin is a flavonol found in onions and apples; catechin is flavanol mostly exist in tea and some fruits; naringenin is the main flavanone found in grapefruit; and cyanidin-glycoside is an anthocyanin found in berry fruits (black currant and blackberry); and cyanidin-glycoside is an anthocyanin found in berry fruits (black currant and blackberry); and genistein and glycitein are the main isoflavones found in soybean (D'Archivio et al., 2007). There are two categories of phenolic acids: benzoic acid derivatives like gallic acid and cinnamic acid derivatives like coumaric, caffeic, and ferulic acid. The most abundant phenolic acid is caffeic acid, found in many fruits and vegetables, which is often esterified with quinic acid like in chlorogenic acid, the most important phenolic ingredient in coffee. Ferulic acid, found in grains, is another prevalent phenolic acid, while coumaric acid is plentiful in citrus fruits. Polyphenols contain active hydroxyl groups that can interact with free radicals to reduce lipid oxidation (Mitsumoto et al., 2005). Polyphenols can also act as free radicals and metal scavengers, reducing the generation of metal catalyzed free radicals (Rice-Evans et al., 1996). The antioxidant capacity is determined by the presence of "–OH" as the functional group in the structure and its position either on the phenolic or flavonoid molecular ring.

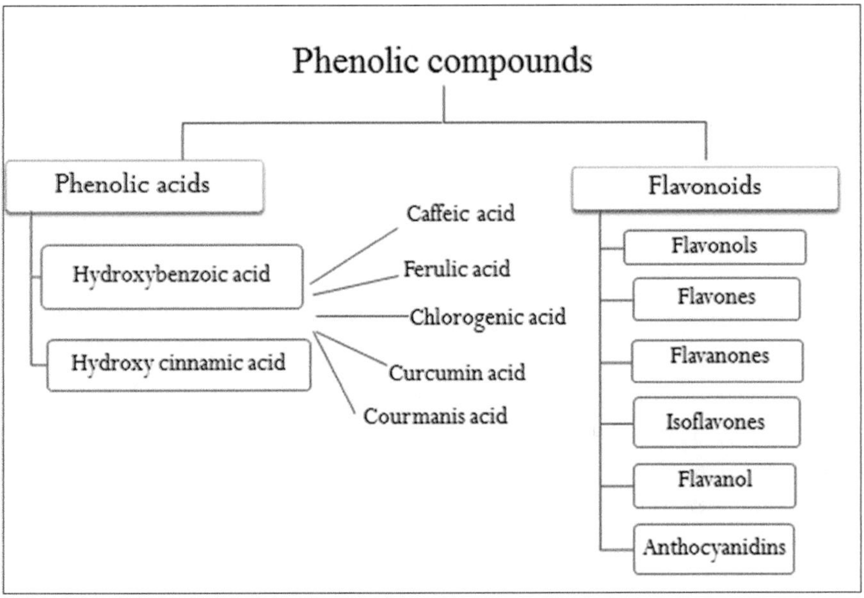

FIGURE 12.2 Classification of phenolic compounds.

12.4.2 PLANT EXTRACT AS ANTIMICROBIAL AND ANTIOXIDANT AGENT

Plant extracts exhibit antibacterial activity due to phytochemical components that may be deadly to microbial cells or impede secondary metabolite formation. The outer cell membrane, also known as the cytoplasmic membrane, is a major site of bacterial contact; damage to this important membrane can cause the bacterium to die in one of several ways: (i) epithelial interruption (Shimamura et al., 2007); this leads in the loss of macromolecules from the inside; (ii) contact with protein, which causes structural and functional deformation; and (iii) disrupting membrane-associated activities such as electron transport and generation of new protein (Bajpai et al., 2009). Antimicrobial activity has been found in a variety of plant extracts against bacteria, yeast, and molds. Phenols, alcohols, aldehydes, ketones, ethers, and hydrocarbons are among the active chemicals that can improve storage stability. Various research have documented antimicrobial properties of plant extracts produced from herbs, spices, vegetables, and fruits (Al-Habib et al., 2010; Hayek et al., 2012). Antioxidants are food additives that delay or prevent food from becoming rancid due to oxidation, hence extending the shelf-life. Natural antioxidant molecules are abundant in plants. Plant-based antioxidants, such as BHT and BHA, have a protective impact. Different vegetables and fruits might be considered intriguing sources of functional additives for food fortification and stability due to their antioxidant characteristics. Antioxidants can slow the oxidation of lipids by preventing the generation of free radicals or delaying their spread by one or more mechanisms: (a) scavenging species that cause peroxidation; (b) chelating metal ions so they do not generate reactive species that cause oxidation; (c) quenching O_2 and so preventing peroxide generation; (d) stopping the auto-oxidative chain reaction; or (e) lowering the O_2 concentration. The ability of antioxidants to disrupt free-radical chain reactions is the most important factor for determining their efficacy. The hydroxyl groups of phenolic compounds are efficient hydrogen donors. The polyphenol becomes a radical after reacting with the original reactive radical species. These radical intermediates are stabilized by the creation of quinone structures and thus cause delocalization of electrons within the aromatic ring. Plant polyphenols can thus inhibit lipid oxidation in seafood (Maqsood et al., 2014).

12.4.3 APPLICATION OF PLANT EXTRACT FOR FISH AND FISH PRODUCTS PRESERVATION

The use of agricultural produce and their peels, seed extract, or powders having antioxidant and antibacterial qualities as organic component utilized to enhance the storage period of fish and their products at chilled temperatures is a potential approach. Hafsa et al. (2019) evaluated the combined effect of beetroot peel extract (BPE) and chilled condition on the biochemical properties of fresh mahseer (*Tor khudree*) steaks and found that BPE dip treatment enhanced the storage stability and extended the shelf life of mahseer by six days compared to control steak during the chilled storage.

Tea polyphenols are gaining popularity as a natural food preservative among researchers. Li et al. (2012) used 0.2% tea polyphenol to treat entire un-gutted crucian carp and found that it effectively inhibited microbiological growth, hindered chemical deterioration, and thus extended the shelf life by 6–8 days as compared to control during refrigerated storage. The white radish (*Raphanus satavis*) extract, a rich source of antioxidant, therapeutic, and antimicrobial properties was found to limit the progress of fish spoilage bacteria and thus extended the shelf life of fish steaks over 15 days of storage (Hafsa et al., 2020). BPE was found to have a significant role in the free radical process and in restricting the bacterial growth in frozen steaks. Control steaks had a shelf-life of about 5 months, according to sensory analysis scores, whereas BPE treated steaks were found to be acceptable for more than 6 months (Hafsa et al., 2021). In vacuum packed sardine filets held at 31°C, Ozogul et al. (2011) found that rosemary and sage tea extracts reduced ammonia and biogenic amine production. Turmeric, a good source of antibacterial substance was found to preserve quality parameters of rainbow trout during vacuum storage of 20 days under refrigerated conditions when used singly or combined with shallot (*Allium cepa*) extract (1.5% each, v/v) (Pezeshk et al., 2011). In comparison to the pure extract treatment and control, Mazandrani et al. (2016) found the least lipid oxidation, TVB-N content, and microbial load during storage of silver carp filets at 4°C treated with liposomal encapsulated fennel extract (FE). Silver carp filets had their shelf life prolonged from 6 to 15 days when treated with 0.5% encapsulated FE and 12 days when treated with 0.5% pure extract. When compared to standard icing, including plant extracts into ice was found to be effective in preventing microbiological and biochemical deterioration of fresh fish. This unique icing approach has been found to be particularly effective in increasing the quality of fish housed on board and transported to interior locations. Chemical, sensory, and microbiological approaches were used by Bensid et al. (2014) investigated the effect of ice extracts containing thyme (0.04% w/v), oregano (0.3% w/v), and clove (0.2% w/v) extracts on anchovy quality parameters (*Engraulis encrasicholus*). When compared to the standard ice batch, these icing techniques resulted in a substantial antioxidant impact as well as a considerably slighter decrease of aerobic and low temperature loving bacteria in anchovy muscle. Oxidation is a major cause of seafood quality decline during frozen storage. To delay oxidation reactions in frozen storage fish and fish products, antioxidant chemicals were applied after freezing, and the fish and fish products were dipped in organic suspension before sub-zero temperature. In recent years, biodegradable protein-based films have become increasingly important in food packaging, acting as a specific hurdle to water movement, aerobic metabolism and the loss of odor components (Maqsood et al., 2013). Most of the edible packages possess barrier properties to gas, organic vapors, and oils when compared to plastic wraps (Jiang et al., 2007). The texture and physical qualities of surimi seafood are mostly determined by the gel strength of fish protein. Exogenous substances of chemical and biological origin can improve the texture of fish protein (Chiou et al., 2006). Plant phenolic compounds have also been demonstrated to have the ability to cross-link proteins, making them potential gel enhancers (Rawel et al., 2002). When phenolic chemicals interact with proteins, changes in protein secondary structure occur, resulting in increased heat stability (Ali, 2002).

12.5 CONCLUSION

Polyphenolic constituents of plant extracts are most active as natural antioxidants and antibacterial in seafood. These ingredients have the potential to replace synthetic additives in food systems to avoid oxidation and quality degradation, particularly in fish and fish products. As a result, plant extracts broaden the range of multifunctional natural food components that can be used in the fish processing industry for a variety of purposes.

ACKNOWLEDGMENT

I am thankful to A. A. Zynudheen for his valuable help, Ashraf Rather and Adnan Amin for providing me with the opportunity to write this chapter.

KEYWORDS

- **antimicrobial properties**
- **antioxidants**
- **malonaldehyde**
- **natural preservative**
- **non-protein nitrogen**
- **polyunsaturated fatty acids**
- **shelf-life**
- **specific spoilage organisms**

REFERENCES

Al-Habib, A., Al-Saleh, E., Safer, A. M., & Afzal, M., (2010). Bactericidal effect of grape seed extract on methicillin-resistant *Staphylococcus aureus* (MRSA). *J. Toxicol. Sci., 35*, 357–364.

Ali, H., (2002). *Protein-Phenolic Interactions in Food.* MSc. Thesis, Department of Food Science Agricultural Chemistry Macdonald campus, McGillb University Monteral, Quebec.

Bajpai, V. K., Al-Reza, S. M., Choi, U. K., Lee, J. H., & Kang, S. C., (2009). Chemical composition, antibacterial and antioxidant activities of leaf essential oil and extracts of *Metasequioa glyptostroboides* Miki ex Hu. *Food Chem.Toxicol., 47*, 1876–1883.

Banerjee, S., (2006). Inhibition of mackerel (*Scomber scombrus*) muscle lipoxygenase by green tea polyphenols. *Food Res. Int., 39*, 486–491.

Bensid, A., Ucar, Y., Bendeddouche, B., & Ozogul, F., (2014). Effect of the icing with thyme, oregano and clove extracts on quality parameters of gutted and beheaded anchovy (*Engraulis encrasicholus*) during chilled storage. *Food Chem., 145*, 681–686.

Boudet, A. M., (2007). Evolution and current status of research in phenolic compounds. *Phytochem., 68*, 2722–2735.

Boziaris, I. S., Proestos, C., Kapsokefalou, M., & Komaitis, M., (2011). Antimicrobial effect of *Filipendula ulmaria* plant extract against selected foodborne pathogenic and spoilage bacteria in laboratory media, fish flesh and fish roe product. *Food Technol. Biotechnol., 49*, 263–270.

Brewer, M. S., (2011). Natural antioxidants: Sources, compounds, mechanisms of action, and potential applications. *Food Sci. Food Safety, 10*, 221–247.

Carocho, M., Morales, P., & Ferreira, I. C. F. R., (2015). Natural food additives: Quovadis? *Trends Food Sci. Technol., 45*, 284–295.

Chaillou, S., Chaulot-Talmon, A., Caekebeke, H., Cardinal, M., Christieans, S., Denis, C., Desmonts, M. H., et al., (2015). Origin and ecological selection of core and food-specific bacterial communities associated with meat and seafood spoilage. *ISME J., 9*, 1105–1118.

Chiou, B. S., Avena-Bustillos, R. J., Shey, J., Yee, E., Bechtel, P. J., & Imam, S. H., (2006). Rheological and mechanical properties of cross-linked fish gelatins. *Polymer, 47*, 6379–6386.

Chouliara, E., Karatapanis, A., Savvaidis, I., & Kontominas, M., (2007). Combined effect of oregano essential oil and modified atmosphere packaging on shelf-life extension of fresh chicken breast meat, stored at 4°C. *Food Microbiol., 24*, 607–617.

Cox, S., Abu-Ghannam, N., & Gupta, S., (2010). An assessment of the antioxidant and antimicrobial activity of six species of edible Irish seaweeds. *Int. Food Res. J., 17*, 205–220.

D'Archivio, M., Filesi, C., Benedetto, D. R., Gargiulo, R., Giovannini, C., & Masella, R., (2007). Polyphenols, dietary sources and bioavailability. *Ann. Ist. Super. Sanita., 43*, 348–361.

Ghaly, A. E., Dave, D., Budge, S., & Brooks, M., (2010). Fish spoilage mechanisms and preservation techniques. *Am. J. Appl. Sci., 7*, 859.

Gram, L., & Dalgaard, P., (2002). Fish spoilage bacteria – problems and solutions. *Curr. Opin. Biotechnol., 13*, 262–266.

Gramza, A., Khokhar, S., Yoko, S., Gliszczynska-Swiglo, A., Hes, M., & Korczak, J., (2006). Antioxidant activity of tea extracts in lipids and correlation with polyphenol content. *Eur. J. Lipid Sci. Technol., 108*, 351–362.

Hafsa, M., Safeena, M. P., Sathish, K. K., Zynudheen, A. A., Ninan, G., & Manoj, B. K., (2019). Beetroot peel extract as a natural preservative for shelf-life extension of Deccan mahseer (*Tor khudree*) steaks during chill storage. *Fish. Technol., 56*, 190–198.

Hafsa, M., Safeena, M. P., Zynudheen, A., Mudassir, A., & Sathish, K., (2021). Effect of beetroot peel dip treatment on the quality preservation of Deccan mahseer (*Tor khudree*) steaks during frozen storage (-18°C). *LWT – Food Sci. Technol., 151*, 112222.

Hafsa, M., Zynudheen, A. A., Safeena, M. P., & Sathish, K. K., (2020). Antioxidant properties and the preservative effect of whole radish extract on quality of Deccan mahseer (*Tor khudree*) steaks during chilled storage. *J. Aquat. Food Prod. Technol.*, 3–15.

Hayek, S. A., & Ibrahim, S. A., (2012). Antimicrobial activity of xoconostle pears (*Opuntia matudae*) against *Escherichia coli* O157:H7 in laboratory medium. *Int. J. Microbiol.*, 368472.

Huss, H. H., (1995). *Quality and Quality Changes in Fresh Fish.* FAO: Rome, Italy.

Jiang, Y., Tang, C. H., Wen, Q. B., Li, L., & Yang, X. Q., (2007). Effect of processing parameters on the properties of transglutaminase-treated soy protein isolate films. *Innov. Food Sci. Emerg. Technol., 8*, 218–225.

Kanatt, S. R., Chander, R., & Sharma, A., (2007). Antioxidant potential of mint (*Mentha spicata* L.) in radiation processed lamb meat. *Food Chem., 100*, 451–458.

Kuley, E., Durmus, M., Balikci, E., Ucar, Y., Regenstein, J. M., & Ozogul, F., (2017). Fish spoilage bacterial growth and their biogenic amine accumulation: Inhibitory effects of olive by-products. *Int. J. Food Prop., 20*, 1029–1043.

Kykkidou, S., Giatrakou, V., Papavergou, A., Kontominas, M. G., & Savvaidis, I. N., (2009). Effect of thyme essential oil and packaging treatments on fresh Mediterranean swordfish fillets during storage at 4°C. *Food Chem., 115*, 169–175.

Li, T. T., Li, J. R., Hu, W. Z., Zhang, X. G., Li, X. P., & Zhao, J., (2012). Shelf-life extension of crucian carp (*Carassius auratus*) using natural preservatives during chilled storage. *Food Chem., 135*, 140–145.

Losada, V., Barros-Velázquez, J., & Aubourg, S. P., (2007). Rancidity development in frozen pelagic fish: Influence of slurry ice as preliminary chilling treatment. LWT-*Food Sci. Technol., 40*, 991–999.

Lugasi, A., Losada, V., Hovari, J., Lebovics, V., Jakoczi, I., & Aubourg, S., (2007). Effect of pre-soaking whole pelagic fish in a plant extract on sensory and biochemical changes during subsequent frozen storage. *LWT-Food Sci. Technol., 40*, 930–936.

Manju, S., Jose, L., Gopal, T. K. S., Ravisankar, C. N., & Lalitha, K. V., (2007). Effect of sodium acetate dip treatment and vacuum-packaging on chemical, microbiological, textural and sensory changes of pearl spot (*Etroplus suratensis*) during chilled storage. *Food Chem., 102*, 27–35.

Maqsood, S., Benjakul, S., & Shahidi, F., (2013). Emerging role of phenolic compounds as natural food additives in fish and fish products. *Crit. Rev. Food Sci. Nutri., 53*, 162–179.

Maqsood, S., Benjakul, S., Abushelaibi, A., & Alam, A., (2014). Phenolic compounds and plant phenolic extracts as natural antioxidants in prevention of lipid oxidation in seafood: A detailed review. *Compr Rev. Food Sci. Food Saf., 13*, 1125–1140.

Mariutti, L. R. B., & Bragagnolo, N., (2017). Influence of salt on lipid oxidation in meat and seafood products: A review. *Food Res. Inter., 94*, 90–100.

Mazandrani, H. A., Javadian, S., & Bahram, S., (2016). The effect of encapsulated fennel extracts on the quality of silver carp fillets during refrigerated storage. *Food Sci. Nutr., 4*, 298–304.

Mexis, S. F., Chouliara, E., & Kontominas, M. G., (2009). Combined effect of an oxygen absorber and oregano essential oil on shelf-life extension of rainbow trout fillets stored at 4°C. *Food Microbiol., 26*, 598–605.

Mitsumoto, M., Grady, O. M. N., Kerry, J. P., & Buckley, D. J., (2005). Addition of tea catechins and vitamin C on sensory evaluation, color and lipid stability during chilled storage in cooked or raw beef and chicken patties. *Meat Sci., 69*, 773–779.

Naveena, B. M., Sen, A. R., Vaithiyanathan, S., Babji, Y., & Kondaiah, N., (2008). Comparative efficacy of pomegranate juice, pomegranate rind powder and BHT in cooked chicken patties. *Meat Sci., 80*, 1304–1308.

Ocanohiguera, V. M., Maedamartínez, A. N., Marquezríos, E., Canizalesrodríguez, D. F., Castilloyáñez, F. J., Ruízbustos, E., Gracianoverdugo, A. Z., & Plascenciajatomea, M., (2011). Freshness assessment of ray fish stored in ice by biochemical, chemical and physical methods. *Food Chem., 125*, 49–54.

Ozogul, F., Kuley, E., & Kenar, M., (2011). Effects of rosemary and sage tea extract on biogenic amines formation of sardine (*Sardina pilchardus*) fillets. *Int. J. Food Sci. Tech., 46*, 761–766.

Ozogul, Y., Ozogul, F., Kuley, E., Ozkutuk, S., Gokbulut, C., & Kose, S., (2006). Biochemical, sensory and microbiological attributes of wild turbot (*Scophthalmus maximus*), from the Black Sea, during chilled storage. *Food Chem., 99*, 752–758.

Pachecoaguilar, R., Lugosanchez, M. E., & Roblesburgueno, M. R., (2010). Post-mortem biochemical and functional characteristic of Monterey sardine muscle stored at 0°C. *J. Food Sci., 65*, 40–47.

Pezeshk, S., Rezaei, M., & Hosseini, H., (2011). Effect of turmeric, shallot extracts and their combinations on quality characteristics of vacuum packaged rainbow trout stored at 4±1°C. *J. Food Sci., 76*, 387–391.

Rawel, H. M., Czajka, D., Rohn, S., & Kroll, J., (2002). Interactions of different phenolic acids and flavonoids with soy proteins. *Int. J. Biol. Macromol., 30*, 137–150.

Rice-Evans, C. A., Miller, J. M., & Paganga, G., (1996). Structure-antioxidant activity relationship of flavonoids and phenolic acids. *Free Radic. Biol. Med., 20*, 933–956.

Schneider, C., (2009). An update on products and mechanisms of lipid peroxidation. *Molecular Nutrition and Food Res., 53*, 315–321.

Shahidi, F., & Botta, J. R., (1994). *Seafoods: Chemistry, Processing Technology and Quality*. Springer: New York, NY, USA.

Shimamura, T., Zhao, W. H., & Hu, Z. Q., (2007). Mechanism of action and potential for use of tea catechin as an anti-infective agent. *Anti-Infect. Agents Med. Chem., 6*, 57–62.

Smid, E. J., & Gorris, L. G. M., (1999). Natural antimicrobials for food preservation. In: Rahman, M. S., (ed.), *Handbook of Food Preservation* (pp. 285–308). Marcel Dekker, New York.

Souza, H. A., & Bragagnolo, N., (2014). New method for the extraction of volatile lipid oxidation products from shrimp by headspace–solid-phase microextraction–gas chromatography–mass spectrometry and evaluation of the effect of salting and drying. *J. Agric. Food Chem., 62*, 590–599.

Venugopal, V., & Gopakumar, K., (2017). Shellfish: Nutritive value, health benefits, and consumer safety. *Compr. Rev. Food Sci. Food Saf., 16*, 1219–1242.

WHO/FAO, (2004). World Health Organization/Food and Agriculture Organization of the United Nations. *Codex Alimentarius Commission Procedural Manual* (14th edn.). WHO/FAO, Rome.

CHAPTER 13

Perspective of eDNA Application to Control Pollution in Freshwater Ecosystems

ADNAN AMIN[1], MONISA M. MALIK[1], D. PAMANNA[2], IMTIYAZ QAYOOM[1], and ADNAN ABUBAKR[1]

[1]*Division of Aquatic Environmental Management, Faculty of Fisheries, Rangil, SKUAST-K, Jammu and Kashmir, India*

[2]*College of Fishery Science, Muthukur, Nellore, Andhra Pradesh, India*

ABSTRACT

The continual loss of biodiversity on earth is a serious concern and task for the 21st century. This mission is affected in large part by a dearth of understanding about the condition and distribution of biodiversity both aquatic and terrestrial. The advancements made in the stream of molecular biology over the last decade have transformed the way we think about biodiversity on the planet. Aquatic ecosystems must be monitored and assessed quickly and accurately. Environmental DNA (eDNA) sequencing has been proved to be a reliable and sensitive technology for detecting and monitoring single or multiple species in a variety of aquatic samples. In this chapter, we look at how eDNA can be used to control and mitigate pollution in aquatic ecosystems, particularly freshwater, as well as how it can be used to monitor aquatic biodiversity, survey species communities, and provide instructions for environmental conservation and when used in conjunction with other sophisticated techniques and conventional ways.

Coldwater Fisheries and Aquaculture Management: Technology for Sustainable Food Production.
Mohd Ashraf Rather, Faisal Rashid Sofi, Adnan Amin, & Kawkabul Saba (Eds.)

13.1 INTRODUCTION

Environmental DNA (eDNA) refers to the genetic material found in environmental materials such as substratum, H_2O, and air, as well as complete cells, extracellular DNA, and maybe whole species (Ficetola et al., 2008). eDNA may be taken from a variety of ecosystems, recognized, improved, sequenced, and categorized depending on the sequence (Deiner et al., 2015). By gaining this information about particular environment, it will be easy to detect and classify various species. eDNA can be of different source such as water, mucous, saliva, sperm, eggs, fishes, benthos, macrophytes, microorganisms' urine, blood, roots, leaves, vegetation, and decomposing remain of larger species, etc. (Bohmann et al., 2014).

Although being a moderately new technique of surveying, eDNA has holds promising potential in biological monitoring of different ecosystems. Traditionally used techniques for evaluating richness and abundance are restricted in terms of taxonomic identification, leading to possible habitat alteration or loss. eDNA can overcome these limitations by targeting various species, sampling larger diversity, and enhancing taxonomic resolution (Deiner et al., 2017). Furthermore, eDNA has ability to identify rare species, however not determining qualitative population traits such as male female ratios and health status, making it suitable replacement for conventional research (Goldberg et al., 2016; Deiner et al., 2017).

In past decades, eDNA investigation has drawn much limelight as a potential and responsive approach for studying biodiversity (Kelly et al., 2016; Pikitch, 2018), which in turn can help in evaluating the changes of different species in various eutrophication levels. Because of the breakthrough and significant growth of high-throughput Sequencing technology, it is now possible to utilize this approach to quickly detect the community structure of organisms (Hamady et al., 2008). Additionally, eDNA metabarcoding was seen to be useful for both qualitative and quantitative monitoring of diverse communities in various habitats (Hanfling et al., 2016). In determining the efficacy of eutrophication, pollution, and HAB control, assessing diversity of species, community architecture, and species change patterns is critical (Figure 13.1). For example, the complete biomass and species compositions of planktonic organisms in aquatic environment are important factors that indicate the water quality of lentic or lotic body (Chen et al., 2017). The analysis of species compositions can aid in the discovery of ecological indicators that can be used to mitigate the problems associated with nutrient enrichment and excessive algal blooms in aquatic environments. The most recent technology of employing eDNA as a molecular marker for species recognition and biodiversity assessment should provide comprehensive and useful data for ecosystem monitoring and conservation (Schwartz et al., 2007).

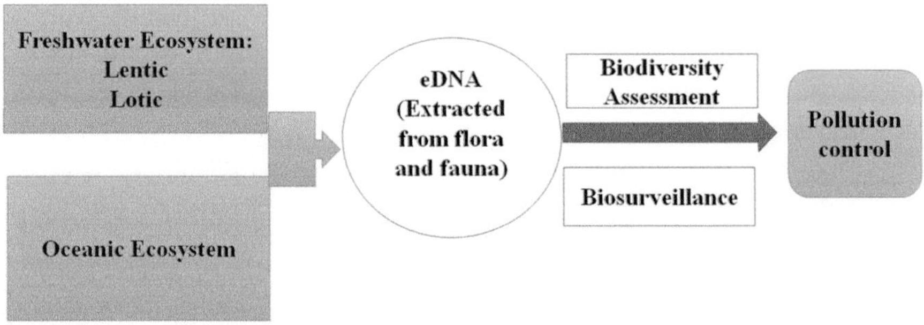

FIGURE 13.1 Application of eDNA for controlling pollution in aquatic ecosystem.

Eutrophication of rivers, lakes, and oceans due to significant nutrient intake has become a pervasive issue in the aquatic environment (Smith, 1998). Increased anthropogenic activities lead to nutrient inputs in aquatic ecosystem which includes population explosion, urban development, agriculture, forestry, and fishery industries (Carpenter et al., 1999). The most important factors which leads in eutrophication of any water body is increasing nitrogen (N) and phosphorus (P) inputs (Conley et al., 2009; Ma et al., 2015) and thus, cutting short supply of N and P concentrations can help in improvement of water quality and ecological status (Sondergaard et al., 2013; Wu et al., 2017). The harmful effects of eutrophic or polluted waters might end in murky water, HABs, anoxia or toxicity of water bodies which can lead to a decrease in species diversity (Smith & Schindler, 2009).

There are various methods which can help in halting cyanobacterial blooms and enhance water quality, such as coagulation and use of copper sulfate (Kenefick et al., 1993; He et al., 2016). These methods, on the other hand, can be time-consuming, costly, and harmful to the environment. Furthermore, due to the strength and residues of chemical reagents, the use of chemical agents may induce heavy metal deposition and secondary pollution of the aquatic environment (Lin et al., 2019). For the protection of aquatic ecosystem, natural approaches are helpful tool in cyanobacterial bloom control (He et al., 2016). It is now widely established that algicidal microorganisms can restrain cyanobacterial blooms efficiently and with specificity (Ahn et al., 2003; Zeng et al., 2015).

eDNA is said to be the most helpful tool for tracking species distributions and classifying different organisms (Deutschmann et al., 2019). eDNA is extracellular DNA released into the environment by organisms by a variety of discharges such as secretions, excrement, sperms, eggs, sloughed cells, or biological remnants (Furlan et al., 2016; Epp et al., 2012). eDNA was introduced in the area of microbiology by Ogram et al. in 1987. eDNA is an important tool, and study of eDNA is more and more being used to identify delicate and nonnative or invasive species, identify low abundance or endangered species and approximate biodiversity (Coble et al., 2019; Robson et al., 2016). In addition to it, earlier studies revealed that eDNA investigation could be more suitable, capable, and complete than conventional techniques used for evaluating diverse taxa across spatial and temporal scales (Jo et al., 2019).

Conventional methods, such as physical identification, are ineffective in aquatic environment for reflecting true population configuration and detecting nonnative species early (Ardura et al., 2015), because living organisms in aquatic ecosystems may be hidden underwater and hard to detect using conventional methods. eDNA is more applicable to a broader spectrum of aquatic bodies, besides other animals and plants (Hanfling et al., 2016). Consequently, eDNA analysis could be a sensitive and effective assay for discovering species that would otherwise go undetected by typical survey methods (Rees et al., 2014; Stoeckle et al., 2017). Furthermore, in aquatic ecosystem, eDNA analysis aids in the discovery of specific groups from multipart samples. Rudko et al. for example, established four species-specific identification methods in aquatic samples from Northern Michigan lakes (Rudko et al., 2019). eDNA is already employed in the numerous studies of both freshwater and marine ecosystems (Table 13.1). DNA metabarcoding, as proposed by Banerji et al. could be used to enhance traditional biological approaches to facilitate to get more thorough profiles of freshwater planktonic organisms commune structure and dispersion (Banerji et al., 2018). The use of eDNA for quick surveillance of biodiversity changes offers considerable potential for monitoring the projected changes as a due to climate change, particularly in vast lake ecosystems (Wagner & Adrian, 2011). eDNA analysis has been employed to investigate and examine microorganisms, fish, and other organisms in marine water samples (Sogin et al., 2006; Thomsen et al., 2012).

13.2 COLDWATER MONITORING

Studies pertaining to assessment of Biodiversity using eDNA are relatively same to those in marine and estuarine ecosystems, among the prime areas of attention is biofilms, bulk specimens, sediment, and water samples (Figure 13.2). In aquatic environment particularly freshwater ecosystem the animal communities are sensitive towards environmental change, making them potential indicators of environmental health, but traditional monitoring is require huge man power. eDNA has the capability to enhance aquatic biodiversity monitoring; however, more research into the characteristics of species and life cycle traits that may affect identification and quantification is required (Evans et al., 2016).

TABLE 13.1 Usage of eDNA for Different Species in Different Habitat

SL. No.	eDNA Usage	Habitat	Evaluating Method	References
1.	Mekong giant catfish (*Pangasianodon gigas*)	Riverine	qPCR	Bellemain et al. (2016)
2.	Crown of thorns (*Acanthaster cf. solaris*)	Coral reef	qPCR+Sanger sequencing	Doyle et al. (2017)
3.	Tilapia	Freshwater Pond	qPCR+Sanger sequencing	Robson et al. (2016)
4.	Mollusks	Ballast water	Metabarcoding	Ardura et al. (2015a)
5.	Eukaryotic plankton community	Freshwater	metabarcoding	Banerji et al. (2018)
6.	Plant DNA	Lake sediments	Metabarcoding	Alsos et al. (2018)
7.	Brown trout (*Salmo trutta*)	Freshwater stream	Metabarcoding	Deutschmann et al. (2019)

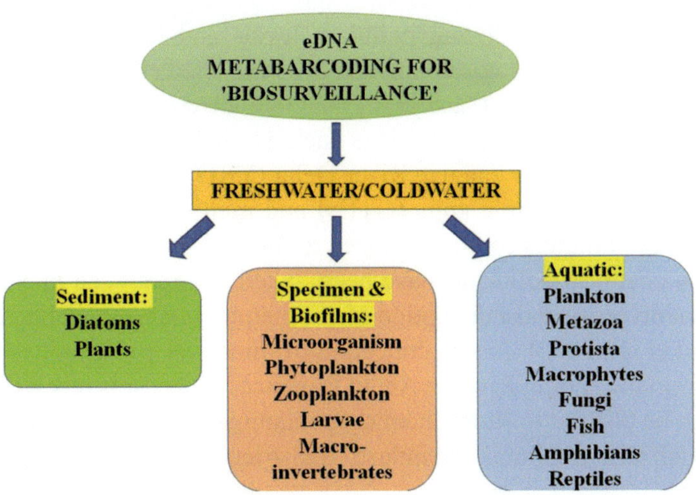

FIGURE 13.2 Schematic diagram of biosurveillance of aquatic ecosystem with eDNA.

13.3 SEDIMENTS

eDNA is being used on sediments from both freshwater and marine water ecosystem to analyze plant communities. Alsos et al. (2018) estimated the biodiversity of vegetation in lake and vascular plant eDNA at 11 different lakes by using both methods eDNA and traditional sampling technique to assess the level to which new taxa are reported in sediment eDNA samples. They revealed 489 taxa by conventional method, of which 17–49% were identified with the help of eDNA, while 47 plant taxa were determined using eDNA, 73% of the samples were found to be in agreement with those documented using traditional methods. They summarized that the taxonomic groups which were available in eDNA were skipped during the sampling and those skipped in eDNA were revealed with increased sampling effort during traditional technique. Additionally, they reported that identification rates differentiated among plant groups, and land based plant DNA because of dependence upon sampling distance (Alsos et al., 2018).

13.4 EDNA AND EUTROPHICATION

The monitoring and management of freshwater habitats, such as lakes, ponds, and rivers, has been the subject of several research (Thomsen et al., 2012; Li et al., 2018). By properly monitoring changes in the microbial, phytoplanktonic, and zooplanktonic population structure in lakes, eDNA could be utilized to identify environmental stressors and analyze the trends of lake ecosystems (Schulz, 2004). It has been suggested that eDNA data could be useful in predicting pollution levels and identifying the major drivers influencing ecological networks. Li et al. used eDNA metabarcoding and principal component analysis (PCA) to determine species distributions and major community stressors such as total nitrogen (TN) and total phosphorus (TP) in the Yangtze River Delta (YRD) (Li et al., 2018). Based on operational taxonomic units (OTUs) and multivariate linear regression models, these findings could be utilized to forecast pollution status (MLRs).

Many chemical pollutants have been linked to a reduction in eukaryotic microbial diversity and a disruption of geochemical element cycling in water (Yergeau et al., 2012). *In situ* collection of microbial communities in sediments from the Nanfei River in Anhui Province, China, was reported by Xie et al. can be utilized to show chemical contamination from various land-use types, such as agricultural and industrial districts (Xie et al., 2017).

13.5 EDNA AND HABS

Cyanobacteria are the dominating species in eutrophicated water and can create cyanobacterial blooms with fast cell multiplication (Vardaka et al., 2005; Otten et al., 2012). Harmful blooms, such as cyanobacterial blooms, may be induced by global warming and eutrophication (Davis et al., 2009). By depleting oxygen and limiting nutrient cycling and light, cyanobacterial blooms can have a severe impact on the environment and aquatic biodiversity (Paerl, 2013). Additionally, cyanobacteria-produced toxins have been reported to be harmful to animals, plants, and humans (Imai et al., 2010). Bell et al. discovered changes in the eukaryotic and bacterial populations in Lake Velence, as well as their correlations with environmental variables (Bell et al., 2018).

In freshwater ecosystems, a large number of microorganisms, such as bacteria, viruses, and fungi, play a key role in managing and mitigating HABs (Yamamoto, 2013). Heterotrophic bacteria have also been identified as a tool for evaluating and reducing cyanobacteria's harmful effects. Around 460 bacterial strains were recovered from lentic, lotic, and drinking water samples by Berg et al. (Berg et al., 2009). They used 16S rRNA DNA sequencing to determine the taxonomic features of these heterotrophic bacteria. Many growth-promoting and growth-inhibiting cyanobacterial strains were discovered in the study's findings, including Flavobacterium and Pedobacter. Other strains capable of degrading poisons or chemical substances, such as *Sphingomonas*, were also discovered. Cyanobacteria's secondary metabolites may have an impact on water quality and be toxic to plants, animals, and humans (Briand et al., 2003; Funari et al., 2008).

Although eDNA is largely valuable tools for detecting diverse species, as well as looking into the link between biodiversity and environmental stress, few studies have looked into using it to find algicidal microbes in freshwater environments. The majority of the research has been on isolated and cultivated bacteria or fungus from freshwater habitats that can directly or indirectly suppress HABs. Growth-inhibiting bacteria (GIBs), which hinder Microcystis aeruginosa growth, have been isolated from the biofilm surface of various aquatic plants, including *Trapa japonica* in Japan (Miyashita et al., 2019). Imai et al. found bacteria from the biofilm surface of aquatic plants (*Egeria densa* and *Ceratophyllum demersum*) in Lake Biwa that induced mortality in Microcystis and GIBs (Imai et al., 2010). Most algicidal microbes are undetectable by conventional procedures, which could create a bottleneck in the termination of HABs. As a result, putting more focus on hiring should be prioritizing the eDNA approach to discover and analyze algicidal microbes.

13.6 SPECIMENS AND BIOFILMS

Sample acquisition for metabarcoding is a good choice for documenting freshwater biodiversity, similar to what is done in marine habitats, with many of these research concentrating on plankton. In a study of the influence of prefiltering on the finding of microeukaryotic planktonic groups, Liu et al. (2017) discovered that about 25% of readings in both categories were from metazoans.

They also discovered no significant variations in richness, diversity, or spatiotemporal trends, implying that prefiltering has little effect on eukaryotic plankton size categories in freshwater ecosystems. Wurzbacher et al. (2017) released a study that year which investigated eDNA metabarcoding for dynamics in total plankton community through time, including eukaryotes and prokaryotes. They found changes in community composition over time that followed the same trends as morphological methods, although morphological methods missed a bloom of a rare dominating bacterium that was affecting the planktonic population. These findings demonstrate the value of microbial and macrobial population metabarcoding research for gaining a better knowledge of whole ecosystems, species interactions, and community dynamics. Banerji et al. (2018) authored a similar study in which they used metabarcoding to identify over 1,300 taxonomic units from samples collected over 4 months from multiple sites in a reservoir, including metazoans, protists, chlorophytes, and fungi. They discovered that estimates of diversity varied between sites and taxonomic groups, implying that distinct plankton groups respond to various environmental factors.

Freshwater biofilms have also been employed for broad microbial metabarcoding. Besemer et al. (2013) tested fluvial biofilms using eDNA metabarcoding to evaluate the variety of ecologically important benthic bacteria's distribution. They discovered that microbial richness dropped downstream, particularly near confluences, based only on metabarcoding data, implying that environmental and biotic factors may influence biofilm connectivity and diversity. Furthermore, they discovered a large degree of diversity in species composition between streams that was not explained by geographic distance, implying that microbial dispersion is constrained by fluvial networks (Besemer et al., 2013).

13.7 CONCLUSION

eDNA has shown to be a highly effective tool for tracking biodiversity and monitoring species populations, different communities, thanks to the rapid advancement of DNA sequencing technologies. For the control and mitigation of Pollution and eutrophication in lentic and lotic water bodies, it is vital to monitor and detect changes in coldwater habitats on a consistent basis. The eDNA procedures when combined with other techniques, such as physical and chemical methods, extensive data can be obtained. When these new technologies are combined with traditional field methods for estimating population structure, abundance, biomass, or individual condition, as well as the extent of pollution, the utility of eDNA as a tool to better understand and predict their impacts on the aquatic environment will be greatly enhanced.

KEYWORDS

- **aquatic biodiversity**
- **environmental DNA**
- **freshwater**
- **growth-inhibiting bacteria**
- **nitrogen**
- **pollution**
- **principal component analysis**

REFERENCES

Ahn, C. Y., Joung, S. H., Jeon, J. W., Kim, H. S., Yoon, B. D., & Oh, H. M., (2003). Selective control of cyanobacteria by surfactin-containing culture broth of *Bacillus subtilis* C1. *Biotechnol. Lett. 25*, 1137–1142.

Alsos, I. G., Lammers, Y., Yoccoz, N. G., Jørgensen, T., Sjeogren, P., Gielly, L., & Edwards, M. E., (2018). Plant DNA metabarcoding of lake sediments: How does it represent the contemporary vegetation. *PLoS One, 13*, e0195403.

Ardura, A., Zaiko, A., Martinez, J. L., Samuiloviene, A., Borrell, Y., & Garcia-Vazquez, E., (2015a). Environmental DNA evidence of transfer of North Sea molluscs across tropical waters through ballast water. *J. Molluscan Stud., 81*, 495–501.

Ardura, A., Zaiko, A., Martinez, J. L., Samulioviene, A., Semenova, A., & Garcia-Vazquez, E., (2015b). eDNA and specific primers for early detection of invasive species—A case study on the bivalve *Rangia cuneata*, currently spreading in Europe. *Mar. Environ. Res., 112*, 48–55.

Banerji, A., Bagley, M., Elk, M., Pilgrim, E., Martinson, J., & Domingo, J. S., (2018). Spatial and temporal dynamics of a freshwater eukaryotic plankton community revealed via 18S rRNA gene metabarcoding. *Hydrobiologia, 818*, 71–86.

Bell, T. A. S., Sen-Kilic, E., Felfoldi, T., Vasas, G., Fields, M. W., & Peyton, B. M., (2018). Microbial community changes during a toxic cyanobacterial bloom in an alkaline Hungarian lake. *Anton Leeuw. Int. J. G., 111*, 2425–2440.

Bellemain, E., Patricio, H., Gray, T., Guegan, F., Valentini, A., Miaud, C., & Dejean, T., (2016). Trails of river monsters: Detecting critically endangered Mekong giant catfish *Pangasianodon gigas* using environmental DNA. *Glob. Ecol. Conserv., 7*, 148–156.

Berg, K. A., Lyra, C., Sivonen, K., Paulin, L., Suomalainen, S., Tuomi, P., & Rapala, J., (2009). High diversity of cultivable heterotrophic bacteria in association with cyanobacterial water blooms. *ISME J., 3*, 314–325.

Besemer, K., Singer, G., Quince, C., Bertuzzo, E., Sloan, W., & Battin, T. J., (2013). Headwaters are critical reservoirs of microbial diversity for fluvial networks. *Proc. R. Soc. B., 280*, 20131760.

Bohmann, K., Evans, A., Gilbert, M. T. P., Carvalho, G. R., Creer, S., Knapp, M., Yu, D. W., & De Bruyn, M., (2014). Environmental DNA for wildlife biology and biodiversity monitoring. *Trends Ecol. Evol., 29*, 358–367.

Briand, J. F., Jacquet, S., Bernard, C., & Humbert, J. F., (2003). Health hazards for terrestrial vertebrates from toxic cyanobacteria in surface water ecosystems. *Vet. Res., 34*, 361–377.

Carpenter, S. R., Ludwig, D., & Brock, W. A., (1999). Management of eutrophication for lakes subject to potentially irreversible change. *Ecol. Appl., 9*, 751–771.

Chen, K., Allen, J., & Lu, J., (2017). Community structures of phytoplankton with emphasis on toxic cyanobacteria in an Ohio inland lake during bloom season. *J. Water Resour. Prot., 9*, 1.

Coble, A. A., Flinders, C. A., Homyack, J. A., Penaluna, B. E., Cronn, R. C., & Weitemier, K., (2019). eDNA as a tool for identifying freshwater species in sustainable forestry: A critical review and potential future applications. *Sci. Total Environ., 649*, 1157–1170.

Conley, D. J., Paerl, H. W., Howarth, R. W., Boesch, D. F., Seitzinger, S. P., Havens, K. E., Lancelot, C., & Likens, G. E., (2009). Ecology controlling eutrophication: Nitrogen and phosphorus. *Science, 323*, 1014–1015.

Davis, T. W., Berry, D. L., Boyer, G. L., & Gobler, C. J., (2009). The effects of temperature and nutrients on the growth and dynamics of toxic and non-toxic strains of *Microcystis* during cyanobacteria blooms. *Harmful Algae, 8*, 715–725.

Deiner, K., Bik, H. M., Machler, E., Seymour, M., Lacoursiere-Roussel, A., Altermatt, F., Creer, S., et al., (2017). Environmental DNA metabarcoding: Transforming how we survey animal and plant communities. *Mol. Ecol., 26*, 5872–5895.

Deiner, K., Walser, J. C., Mächler, E., & Altermatt, F., (2015). Choice of capture and extraction methods affect detection of freshwater biodiversity from environmental DNA. *Biological Conservation, 183*, 53–63.

Deutschmann, B., Muller, A. K., Hollert, H., & Brinkmann, M., (2019). Assessing the fate of brown trout (*Salmo trutta*) environmental DNA in a natural stream using a sensitive and specific dual-labelled probe. *Sci. Total Environ., 672*, 1045.

Doyle, J. R., McKinnon, A. D., & Uthicke, S., (2017). Quantifying larvae of the coralivorous seastar *Acanthaster cf. solaris* on the great barrier reef using qPCR. *Mar. Biol., 164*, 176.

Epp, L. S., Boessenkool, S., Bellemain, E. P., Haile, J., Esposito, A., Riaz, T., Erseus, C., Gusarov, V. I., Edwards, M. E., Johnsen, A., et al., (2012). New environmental metabarcodes for analyzing soil DNA: Potential for studying past and present ecosystems. *Mol. Ecol., 21*, 1821–1833.

Evans, N. T., Olds, B. P., Renshaw, M. A., Turner, C. R., Li, Y., Jerde, C. L., Mahon, A. R., et al., (2016). Quantification of mesocosm fish and amphibian species diversity via environmental DNA metabarcoding. *Mol. Ecol. Res., 16*, 29–41.

Ficetola, G. F., Miaud, C., Pompanon, F., & Taberlet, P., (2008). Species detection using environmental DNA from water samples. *Biology Letters, 4*, 423–425.

Funari, E., & Testai, E., (2008). Human health risk assessment related to cyanotoxins exposure. *Crit. Rev. Toxicol., 5623338*, 97–125.

Furlan, E. M., Gleeson, D., Hardy, C. M., & Duncan, R. P., (2016). A framework for estimating the sensitivity of eDNA surveys. *Mol. Ecol. Resour., 16*, 641–654.

Goldberg, C. S., Turner, C. R., Deiner, K., Klymus, K. E., Thomsen, P. F., Murphy, M. A., Spear, S. F., et al., (2016). Critical considerations for the application of environmental DNA methods to detect aquatic species. *Methods Ecol. Evol., 7*, 1299–1307.

Hamady, M., Walker, J. J., Harris, J. K., Gold, N. J., & Knight, R., (2008). Error-correcting barcoded primers for pyrosequencing hundreds of samples in multiplex. *Nat. Methods, 5*, 235–237.

Hanfling, B., Handley, L. L., Read, D. S., Hahn, C., Li, J. L., Nichols, P., Blackman, R. C., et al., (2016). Environmental DNA metabarcoding of lake fish communities reflects long-term data from established survey methods. *Mol. Ecol., 25*, 3101–3119.

He, X. X., Liu, Y. L., Conklin, A., Westrick, J., Weavers, L. K., Dionysiou, D. D., Lenhart, J. J., et al., (2016). Toxic cyanobacteria and drinking water: Impacts, detection, and treatment. *Harmful Algae, 54*, 174–193.

Imai, I., Kido, T., Yoshinaga, I., Ohgi, K., & Nagai, S., (2010). *Isolation of Microcystis-Killer Bacterium Agrobacterium Vitis from the Biofilm on the Surface of the Water Plant Egeria Densa* (Vol. 150, pp. 150–152). Kalliopi A. Pagou.

Jo, H., Kim, D. K., Park, K., & Kwak, I. S., (2019). Discrimination of spatial distribution of aquatic organisms in a coastal ecosystem using eDNA. *Appl. Sci., 9*, 3450.

Kelly, R. P., Port, J. A., Yamahara, K. M., Martone, R. G., Lowell, N., Thomsen, P. F., Mach, M. E., Bennett, M., Prahler, E., Caldwell, M. R., et al., (2014). Harnessing DNA to improve environmental management. *Science, 344*, 1455, 1456.

Kenefick, S. L., Hrudey, S. E., Peterson, H. G., & Prepas, E. E., (1993). Toxin release from microcystis-aeruginosa after chemical treatment. *Water Sci. Technol., 27*, 433–440.

Li, F. L., Peng, Y., Fang, W. D., Altermatt, F., Xie, Y. W., Yang, J. H., & Zhang, X. W., (2018). Application of environmental DNA metabarcoding for predicting anthropogenic pollution in rivers. *Environ. Sci. Technol., 52*, 11708–11719.

Lin, Y., Chen, A., Luo, S., Kuang, X., Li, R., Lepo, J. E., Gu, J. D., et al., (2019). Cyanobacterial bloom mitigation by sanguinarine and its effects on aquatic microbial community structure. *Environ. Pollut., 253*, 497–506.

Liu, L., Liu, M., Wilkinson, D. M., Chen, H., Yu, X., & Yang, J., (2017). DNA metabarcoding reveals that 200-mm-size-fractionated filtering is unable to discriminate between planktonic microbial and large eukaryotes. *Mol. Ecol. Res., 17*, 991–1002.

Ma, J. R., Qin, B. Q., Wu, P., Zhou, J., Niu, C., Deng, J. M., & Niu, H. L., (2015). Controlling cyanobacterial blooms by managing nutrient ratio and limitation in a large hyper-eutrophic lake: Lake Taihu, China. *J. Environ. Sci., 27*, 80–86.

Miyashita, Y., Hagiwara, T., & Imai, I., (2019). The existence of cyanobactericidal bacteria and growth-inhibiting bacteria on water plants in Lake Ohnuma, Japan. *Limnology, 20*, 39–53.

Ogram, A., Sayler, G. S., & Barkay, T., (1987). The extraction and purification of microbial DNA from sediments. *J. Microbiol. Meth., 7*, 57–66.

Otten, T. G., Xu, H., Qin, B., Zhu, G., & Paerl, H. W., (2012). Spatiotemporal patterns and ecophysiology of toxigenic microcystis blooms in Lake Taihu, China: Implications for water quality management. *Environ. Sci. Technol., 46*, 3480–3488.

Paerl, H. W., & Otten, T. G., (2013). Harmful cyanobacterial blooms: Causes, consequences, and controls. *Microb. Ecol., 65*, 995–1010.

Pikitch, E. K., (2018). A tool for finding rare marine species. *Science, 360*, 1180–1182.

Rees, H. C., Maddison, B. C., Middleditch, D. J., Patmore, J. R. M., & Gough, K. C., (2014). Review the detection of aquatic animal species using environmental DNA – a review of eDNA as a survey tool in ecology. *J. Appl. Ecol., 51*, 1450–1459.

Robson, H. L., Noble, T. H., Saunders, R. J., Robson, S. K., Burrows, D. W., & Jerry, D. R., (2016). Fine-tuning for the tropics: Application of eDNA technology for invasive fish detection in tropical freshwater ecosystems. *Mol. Ecol. Resour., 16*, 922–932.

Rudko, S. P., Turnbull, A., Reimink, R. L., Froelich, K., & Hanington, P. C., (2019). Species-specific qPCR assays allow for high-resolution population assessment of four species avian schistosome that cause swimmer's itch in recreational lakes. *Int. J. Parasitol.-Par., 9*, 122–129.

Schulz, R., (2004). Field studies on exposure, effects, and risk mitigation of aquatic nonpoint-source insecticide pollution: A review. *J. Environ. Qual., 33*, 419–448.

Schwartz, M. K., Luikart, G., & Waples, R. S., (2007). Genetic monitoring as a promising tool for conservation and management. *Trends Ecol. Evol., 22*, 25–33.

Smith, V. H., & Schindler, D. W., (2009). Eutrophication science: Where do we go from here? *Trends Ecol. Evol., 24*, 201–207.

Smith, V. H., (1998). Cultural eutrophication of inland, estuarine and coastal waters. In: *Successes, Limitations and Frontiers in Ecosystem Science* (pp. 7–49). Springer: New York, NY, USA.

Sogin, M. L., Morrison, H. G., Huber, J. A., MarkWelch, D., Huse, S. M., Neal, P. R., Arrieta, J. M., & Herndl, G. J., (2006). Microbial diversity in the deep sea and the underexplored "rare biosphere". *Proc. Natl. Acad. Sci. USA., 103*, 12115–12120.

Sondergaard, M., Bjerring, R., & Jeppesen, E., (2013). Persistent internal phosphorus loading during summer in shallow eutrophic lakes. *Hydrobiologia, 710*, 95–107.

Stoeckle, M. Y., Soboleva, L., & Charlop-Powers, Z., (2017). Aquatic environmental DNA detects seasonal fish abundance and habitat preference in an urban estuary. *PLoS One, 12*, 175–186.

Thomsen, P. F., Kielgast, J., Iversen, L. L., Moller, P. R., Rasmussen, M., & Willerslev, E., (2012). Detection of a diverse marine fish fauna using environmental DNA from seawater samples. *PLoS One, 7*, e41732.

Thomsen, P. F., Kielgast, J., Iversen, L. L., Wiuf, C., Rasmussen, M., Gilbert, M. T. P., Orlando, L., & Willerslev, E., (2012). Monitoring endangered freshwater biodiversity using environmental DNA. *Mol. Ecol., 21*, 2565–2573.

Vardaka, E., Moustaka-Gouni, M., Cook, C. M., & Lanaras, T., (2005). Cyanobacterial blooms and water quality in Greek waterbodies. *J. Appl. Phycol., 17*, 391–401.

Wagner, C., & Adrian, R., (2011). Consequences of changes in thermal regime for plankton diversity and trait composition in a polymictic lake: A matter of temporal scale. *Freshw. Biol., 56*, 1949–1961.

Wu, Z., Liu, Y., Liang, Z. Y., Wu, S. F., & Guo, H. C., (2017). Internal cycling, not external loading, decides the nutrient limitation in eutrophic lake: A dynamic model with temporal Bayesian hierarchical inference. *Water Res., 116*, 231–240.

Wurzbacher, C., Attermeyer, K., Kettner, M. T., Flintrop, C., Warthmann, N., Hilt, S., Grossart, H., & Monaghan, M. T., (2017). DNA metabarcoding of unfractionated water samples relates phyto-, zoo – and bacterioplankton dynamics and reveals a single-taxon bacterial bloom. *Environ. Microbiol. Rep., 9*, 383–388.

Xie, Y. W., Wang, J. Z., Yang, J. H., Giesy, J. P., Yu, H. X., & Zhang, X. W., (2017). Environmental DNA metabarcoding reveals primary chemical contaminants in freshwater sediments from different land-use types. *Chemosphere, 172*, 201–209.

Yamamoto, Y., (1988). Cynobacteria-lysing agents and their distribution patterns of Lakes and Rivers in Japan. *Bull. Jpn. Soc. Microb. Ecol., 2*, 77–88.

Yergeau, E., Lawrence, J. R., Sanschagrin, S., Waiser, M. J., Korber, D. R., & Greer, C. W., (2012). Next-generation sequencing of microbial communities in the Athabasca River and its tributaries in relation to oil sands mining activities. *Appl. Environ. Microb., 78*, 7626–7637.

Zeng, G. M., Wang, P., & Wang, Y., (2015). Algicidal efficiency and mechanism of *Phanerochaete chrysosporium* against harmful algal bloom species. *Algal Res., 12*, 182–190.

CHAPTER 14

Nutritional Composition of Coldwater Fishes

K. SRAVANI[1], FAISAL SOFI[2], TARIQ HUSSAIN[2], KAWKABUL SABA[3], S. VIJAY KUMAR REDDY[5], S. DEVADHARSHINI[1], P. GANESAN[1], K. DHANAPAL[4], and SHABIR A. DAR[6]

[1]*Department of Fish Processing Technology, Fisheries College and Research Institute, TNJFU, Thoothukudi, Tamil Nadu, India*

[2]*Division of Post-Harvest Technology, Faculty of Fisheries, SKUAST-K, Jammu and Kashmir, India*

[3]*Division of Fish Nutrition and Biochemistry, Faculty of Fisheries, SKUAST-K, Jammu and Kashmir, India*

[4]*Department of Fish Processing Technology, College of Fishery Science, SVVU, Muthukur, Andhra Pradesh, India*

[5]*Department of Harvest and Post-Harvest Technology, College of Fisheries, Gadvasu, Punjab, India*

[6]*Division of Fisheries Engineering, Faculty of Fisheries, SKUAST-K, Jammu and Kashmir, India*

ABSTRACT

Nowadays, the consumption of fish, as food has been increasing as fish contains almost all major and minor nutrients and show positive health benefits. Sensory attributes like texture, taste, flavor, appearance, and quality are the main concern for the consumers which are due to the net result of

Coldwater Fisheries and Aquaculture Management: Technology for Sustainable Food Production.
Mohd Ashraf Rather, Faisal Rashid Sofi, Adnan Amin, & Kawkabul Saba (Eds.)

complex reactions between major and minor nutrients. The geographical location, morphology, season, ecosystem, culture practices are the different components that play a significant role play in representing the quantity of these nutrients. When compared with amino acid (AA) composition, fatty acid composition is mostly affected by food. In one way or other, all the coldwater fishes are good in their nutrient composition. Most of the important cold water food fishes are superior in expressing the ratio of n–6/n–3 fatty acids. All the amino and fatty acid composition in coldwater fishes show good effect on the cardiovascular system of adults and prenatal development in the case of children. The present chapter is aimed to compile and complement to the available nutritional composition of coldwater fishes in view of proximate, fatty acid and AA phrases along with medicinal benefits.

14.1 INTRODUCTION

Coldwater fishes play a significant role as a part of the diet of the upland people of India and ensure good nutritional profile comprising of Essential amino and fatty acids, minerals, and vitamins. Coldwater fishes have immense prospects, a great potential in creating livelihood opportunities and occupy an important place in higher upland areas. Cold water fishery resources dispersed over the Himalayan and peninsular areas of India, where these are known as a candidate species for sports, ornamental, and food. In India, a number of Coldwater streams, rivers, lakes, and reservoirs are present. The overall lengths (OALs) of the upland rivers and coldwater streams are distributed to 10,000 km. Climate variations are seen in different parts of the country due to attribution, slopes of mountains, variation in geography, vegetation, and river valley expansion, on account of which the water resources harbor rich species diversity in upland areas. In total the coldwater fishes constitute 258 species, which are dispersed in 21 families as 76 genera. Out of which, 91 from the Deccan plateau, 203 from West and Central Himalaya and 255 species are from Northeast Himalayas. The variation in dispersion of hill stream fishes will vary depending on is primarily flow of water, characteristics of substratum, temperature profile, quality of water and the availability of food naturally. Salmons, trout, mahseers, and the mirror carp comprise the major group of the economically important coldwater fishes. As per economic importance, few of coldwater species are used for sports, some as food and a few others are as ornamental fishes.

14.1.1 SPORT FISHES

Tor tor (Golden mahseer), *Tor mossal* (Copper mahseer), *Mazirifor chelynoides* (Black mahseer), *Neolissochilus hexagonolepis* (Chocolate mahseer), *Raiamas bola* (Indian trout) among indigenous species and *Oncorhynchus mykiss* (rainbow trout) and *Salmo trutta fario* (brown trout) among exotics are considered as important sport fishes.

14.1.2 FOOD FISHES

The important genera among food fishes are–snow trout (*Schizothorax*), Garras (*Garra* sp.), minor carps (*Labeo* sp.), Bariels (*Barilius* sp.), exotic carps (*Ctenopharyngodon idella, Hypothtalmichthys molitrix, Cyprinus carpio*), eels (*Mastacembalus armatus*) are known as food fishes.

14.1.3 ORNAMENTAL FISHES

Colorful and captivating species have been recognized as ornamental fishes. The crucial ornamental fish species are *Carrassius auratus*, *Puntius gelius, Puntius conchonius, P. ticto, P. sophore, Badis badis, Barillus vagra, Brachydanio rerio.*

14.2 POTENTIAL BENEFITS OF FISH CONSUMPTION

Fish is being increasingly preferred due to health consciousness among the people all over the world and is playing a crucial role in human nutrition as low caloric foods. Fishes are best source of Omega-3 poly unsaturated fatty acids which proved to have beneficial results in reducing the chance of cardio-vascular diseases and are shown to have positive benefits in many other pathological situations especially, few types of cancers and arthritis.

The superiority of sea food is due to:

- More amount of poly unsaturated fatty acids;
- High amount of essential amino acids (EAAs);
- Easily digestible proteins;
- Abundance of several micro-nutrients.

14.2.1 NUTRIENT CONTENT

Seafoods reflect as an excellent source of major nutrients. Fish provides a significant range of nutrients on a caloric unit basis, though more intake of fish is compatible with a reduction of both saturated fatty acid and caloric intakes. Fish nutrients are having a lot of advantages and some curative benefits for contemporary health problems like osteoporosis, obesity, coronary heart disease (CHD), cancer, hypertension, iron, protein deficiency, and arthritis.

14.2.1.1 SIGNIFICANCE OF FISH PROTEINS AND AMINO ACIDS (AAS)

The major nutrient of fish muscle is protein, a 100 gr of cooked fish in most species serve approximately 11–24 gr of proteins, which is about one third of the average daily recommended protein intake, thus playing a beneficial role in eliminating the protein-calorie malnutrition. Fish protein digestibility is more due to the presence of very less stroma protein fraction and has an outstanding range of essential AAs as per the requirement of the body for growth and repair of the tissue. Some proteins are incomplete and must be supplemented with other protein foods, containing all most all AAs as per the requirement of the body. The protein from vegetable resources and fish will show a lesser burden in renal hemodynamics when compared with the animal protein sources.

There is proven evidence that the fish protein has the capacity to improve lipid profiles of blood in humans. Fish proteins also play a role as heart protective major nutrient as they are participating in reducing the cholesterol in plasma due to less fraction of lysine to arginine and methionine to glycine. The proteins in fish protein exerts beneficial effects on some of the nutritional abnormalities related to body weight.

The knowledge on composition of AAs of fish is important because of their role in setting up probable nourishment as they provide AAs for protein biosynthesis, which is important for growth of infants and children and also for the replacement of body proteins in adults, Work as precursors of hormones, other biomolecules, secondary metabolites and provide a significant fraction of the total daily energy requirement of the body.

Conventionally, the AAs are categorized as non-essential (NEAA), essential (EAA) and conditionally EAA (CEAA). Newly, the group called the functional amino acids (FAA) has been initiated. The group of AAs which regulate the crucial metabolic pathways for survival, improving health, development, growth, reproduction of an organisms and lactation are called functional AAs. Examples for these functional AAs are Arginine, methionine, leucine, cystine, proline, tyrosine, aspartic acid, tryptophan, glycine, glutamine, glutamic acid, and taurine. Arginine, methionine, cystine, histidine, isoleucine, leucine, lysine, proline, phenylalanine, threonine, tryptophan, tyrosine, valine are considered as EAA whereas alanine, asparagine, aspartic acid, serine as non-essential AAs and glycine, glutamic acid, glutamine, and taurine as CEAA (Wu, 2013).

Arginine plays a key role in healing of wounds, cell division, immune properties, hormone release and removal of ammonia. It works as a key signaling molecule in a various of biological processes because it is a precursor of nitric oxide (NO), it has a role in the treatment of conditions where vasodilatation is required. It can be given as a supplement for treatment of diseases like hypertension, pre-eclampsia, sepsis, erectile dysfunction, anxiety, etc. In case of coldwater fishes *O. mykiss* (6.5±0.3 gm 100 g^{-1} protein) *T. putitora, N. hexagonolepis* were reported to contain great amount of arginine, when compared with the other species studied and hence, these fishes can be suggested for arginine deficiency (Sarma et al., 2013).

Leucine, is the AA that plays a role in stimulation of protein synthesis by muscle (Etzel, 2004) and is also having curative role in stress conditions like sepsis, burn, and trauma (DeBandt & Cynober, 2006). Leucine content is found to be high marine fishes like *S. waitei* and *R. kanagurta* (10.4±0.4 and 10.3±0.410.4 g 100^{-1} g protein, respectively), carps *C. mrigala, L. rohita* and catfishes *C. batrachus* and *H. fossili*.

The AA Methionine finds it way for treatment of psychosis, depression, liver disorders, improving wound healing, dipsomaniac, hypersensitivity, branchial asthma, copper poisoning, radiation side effects, drug withdrawal, and Parkinson's disease (Mischoulon & Fava, 2002). Methionine content is higher in mutton when compared with fish. The marine fish *S. waitei* (4.0±0.4 100^{-1} g protein) and coldwater fish *T. putitora* (3.6±0.3 g 100^{-1} g protein) was reported to have good quantity of methionine than other fishes (Loest et al., 1997).

Glutamic acid is having a crucial role in the metabolism of AAs as it is involved in transamination reactions which are required for the synthesis of important molecules, called glutathione. Glutamic acid also plays an important role regulation of metabolism, tissue injury prevention, anti-oxidant activity enhancement, promotion of protein synthesis, wound healing, increasing immunity and in curing of obesity, cardio vascular disorders, certain types of cancers, and some inflammatory diseases (Wang et al., 2013).

Tryptophan is a precursor for serotonin, a brain neurotransmitter which plays a role in reducing pain (Segura & Ventura, 1988). It is also a precursor of kynurenine, tryptamine, and melatonin and has a major role in the functioning of neurotransmitters like nor-epinephrine and dopamine. Tryptophan can also be used as a supplement to get relief from pain, sleeplessness, dejection, seasonal affective disorder, bulimia, premenstrual dysphoric disorder, ADH disorder and chronic fatigue (Richard et al., 2009). *T. putitora* was found to contain the highest amount of tryptophan among the fishes studied coldwater fishes.

Histidine, plays a role as a precursor during the biosynthesis of histamine and carnosine. The stability of proteins has been found to be increasing during thermal stress due to the presence of histidine which is responsible for tissue repair and growth, myelin sheaths maintenance and in heavy metals removal from the body (Heimann, 1982).

Glycine is considered as an essential and functional AA for fish and plays key roles in a wide range of metabolic processes and chronic insufficiency of glycine leads to insignificant growth, immune response impairment and various bad effects on nutrient metabolism and health.

The essential AA that is broadly important for optimal growth is Lysine and its deficiency results in immunodeficiency (Chen et al., 2003). Lysine can be used as an oral supplement or directly to the skin for reducing or preventing cold sores. Lysine content is recorded to be very high in *S. commersoni* (16.1 ± 0.9 g 100^{-1} g protein) and *T. putitora* (9.4 ± 0.6 g 100^{-1} g protein).

Threonine finds its role in curing different disorders related to brain which includes spinal spasticity is used for to treating various nervous system disorders including spinal spasticity, weakness, amyotrophic lateral sclerosis and familial spastic paraparesis (Hyland, 2007). *S. waitei* can be suggested as source as natural supplement for threonine.

Isoleucine is the AA that is important for correct growth and formation of muscle (Charlton, 2006). Leucine, isoleucine, and valine are reported to be in low in plasma of patients with chronic renal failure. Alanine is a major fraction of the antifreeze proteins (AFPs) where it is present as alanine-alanine-threonine in the blood plasma of polar region fishes and animals to prevent crystallization in blood. Alanine content is observed high in patients with hypertension, more cholesterol and BMI.

Non-essential AAs are having role in gene expression regulation and animal cell metabolism. Aspartic acid which is considered as functional AA, works as a precursor to methionine, threonine, isoleucine, and lysine and also regulates the hormonal secretion. Similarly, for glycine, cysteine, and tryptophan, serine acts as a precursor and also important for cell signaling. Most commonly serine is used as a supplement in curing of schizophrenia.

Taurine, conditionally essential AA present in the most of the fishes is helpful for healing people having edema, high blood pressure, cardiac disorders, proper eye functioning, eagerness, and arthrosclerosis and is important for better usage of sodium and along with protective effect on brain.

Particularly some fishes are considered as good source of various AAs viz., *Oncorhynchus mykiss, Tor putitora, Neolissochilus hexagonolepis* are rich in Arginine, *Rastrellger kanagurta, Catla catla, Stolephorus waitei, Amblypharyngodon mola, Puntius sophore* are high source of Histidine, *Oncorhynchus mykiss, Labeo rohita, Stolephorus commersonii* are good sources of Isoleucine and Leucine, *Stolephorus commersonii, Thunnus albacares, Tor putitora* are considered as rich source of Lysine, *Stolephorus waitei, Tor putitora, Rastrelliger kanagurta* are source of Methionine, *Cirrihus mrigala, Catla catla, Labeo rohita* are nice source of Phenyl alanine, threonine content is more in *Thunnus albacares, Nemipterus japonicus, Stolephorus waitei, Stolephorus commersoni*, Tyrosine concentration is more in *Oncorhynchus mykiss, Tor putitora* whereas *Nemipterus japonicas, Cirrihus mrigala, Rastrelliger kanagurta* are good source of Valine, *Tor putitora, Cirrihus mrigala, Catla catla, Labeo rohita* are showing high volume of Tryptophan, Glutamic acid, Glycine, and Proline is present in *Oncorhynchus mykiss, Tor putitora.* The nonessential AAs like Alanine is present in *Nemipterus japonicus, Labeo rohita, Catla catla, Stoleophorus commersonii, Heteropneustes fossilis, Clarias batrachus, Nemipterus japonicas, Thunnus albacares* are a rich source of Aspartate and Serine (AICRPs/Network Projects/Outreach Activities, 2014; Mohanty et al., 2014).

14.2.1.2 LIPIDS

Most fish are considered excellent because of their low-fat content. The majority of fish consumed are low in fat, containing 1–2% fat present mostly as phospholipids components of muscle membranes. A few popular finfish contain more than 2–3% fat. A few species contain more than 10% fat, located mostly in the sub-epithelial layer. The fat in finfish is composed mostly of triglycerides which are fatty acid esters of the trihydric alcohol, glycerol. The composition of fatty acid in fish varies with species, season of catch and diet. The eicosapentenoic acid (EPA) and docosahexaenoic acid (DHA) are the two important fatty acids observed in fish oils out of seven omega-3 fatty acids.

14.2.1.2.1 Functions of Ω-3 Fatty Acids

The Ω-3 fatty acids play a major role in various body functions viz.:

- Notably decrease the chance of heart attack and stroke by:
  - o Reducing the hypertension by inhibiting the body's production of eicosanoids that cause inflammation in blood vessels.
  - o Reduces the levels of triglycerides in the blood, decreases the growth rate of artery narrowing plague.
- Reduction of rheumatoid arthritis when supplemented through diet, i.e., reduction of swelling, pain, and redness of joints.
- Help in curing people with psoriasis or eczema by blocking the action of arachidonic acid.
- Omega 3 fatty acid (DHA) helps to improve the mental health, especially depression and attention deficit/hyperactivity disorders (ADHD).
- Reduce the tumor growth in people with cancer.
- Thrombosis and ameliorate ischemic heart diseases can be lowered by consuming a good quantity of triglycerides.
- For the many years only Ω-6 fatty acids, arachidonic, and linoleic are considered necessary (at 1–2% of dietary calories) for normal growth and health, especially from the skin.
- Proofs shown that Docosahexaenoic acid (C22: 6n–3), is a component of phospholipid membrane in retinal receptors, cerebral gray matter and sperm and it is required for regular retinal functions. Comparatively more amount of DHA is required for normal brain development and learning ability in newborn. Hence, the mechanism of Selective incorporation is practiced supplying high levels of DHA to the fetal brain.
- The role of DHA is suggested to be one of regulation of membrane function. The fetus has specific requirements for Ω-3 fatty acids. DHA is preferentially taken by brain cell membrane phospholipids. Liver, forebrain, and retinal cells for the developing fetus show proportional increase of Ω-3 and decrease of Ω-6. A high Ω-3: Ω-6 ratio can be damaging.

14.2.1.2.2 Cholesterol

Seafood is not a major dietary source of cholesterol. Finfish generally contain modest amounts of cholesterol, i.e., 50–90 mg/100 g muscle. The cholesterol content of fish tends to increase with fat content. Shellfishes are reported to contain moderately high quantity of cholesterol; so, crustaceans (crabs, lobsters, and shrimps) are having 60–100 mg/100 g and mollusks (clams, oysters, and mussels), 40–110 mg/100 g. Squid and fish roe contain relatively high amount of cholesterol.

14.2.1.2.3 Squalene

Shark liver oils contain more amount of Squalene which is an isoprenoid molecular protein, which works as a lipid lowering agent (antilipidemic), reduces or prevents the rate of oxidation (antioxidant) and membrane stabilizing agent. Squalene is included in cholesterol metabolism as a precursor in the biosynthesis of cholesterol and inhibitor of HMG CoA reductase. It is also having the property of suppressing the action of carcinogenic tumor, on exposure to squalene over a period of time. Squalene improves the effect of cholesterol reducing drugs.

14.2.1.3 MINERALS

14.2.1.3.1 Sodium

There has been probably more publicity about this element in these past few years than virtually any other nutrient. This is largely because sodium is strongly linked to blood pressure or hypertension. Controlling the amount of sodium consumption can be helpful in lowering high blood pressure. In other words, one should consume mainly foods naturally low in sodium. Fresh seafood is one best choice for curbing sodium intake since all fish varieties are low in sodium. Most fresh fish range from about 60 to 110 mg of sodium per 100 g of raw muscle. Most shellfishes, on the other hand, have more sodium, usually in the range of 200–400 mg/100 g. Squid falls in between at about 160 mg. In case of processed fish like freezing, canning, smoking, and curing, sodium content is found to be high, which varies from 300 to 900 mg/100 g. The reason behind the more content of sodium is due to conventional processing and on-board handling procedures like brine treatment and stowage in refrigerated sea water system, treatment of filets with sodium tripolyphosphate to reduce drip loss, etc., which results in the infusion of sodium. Frozen battered seafood products contain an average of 400 mg/100 g. In the majority of these products, the battered/breaded coatings have more sodium than the meat portion. The increased popularity of surimi may require the use of salts other than sodium.

14.2.1.3.2 Potassium

Potassium is well-known as an electrolyte, one of the minerals that body's cells to function properly and to help regulate fluid balance. Potassium has other functions that help in critical activities like regulating heartbeat, muscle contractions and nerve signaling. It's required to help some enzymes to function, aids in bone health, lowers the occurrence of some heart related disorders, and can lower the risk of kidney stones, too. Normally, fresh fishes will have more amount of potassium, ranging from 250–320 mg/100 g. The safe daily intake of potassium is between 1.8 and 5.6 grams.

14.2.1.3.3 Calcium

Calcium is necessary for the formation of strong bones and bone strength. Proper absorption of calcium is associated with vitamin D. Calcium plays an important role in the proper functioning of muscles, nervous system, bold clotting process. Small fishes can be consumed along with bones, which forms a good source calcium rather than discarding. Calcium deficiency is associated with a condition called rickets in case of young children and in case of adults leads to osteomalacia where bones will become soft. Fluorine is also important for strong bones and teeth. Calcium is vital for the growth of children, women, especially during reproductive years to immense subsequent osteoporosis. Intakes of 1.25 g/day are recommended, where milk is the major source. Seafoods can also be useful as source of calcium, which are having a calcium range of 6 to 120 mg/100 g, depending on the species and processing.

14.2.1.3.4 Iron

This can be a limiting element in the diets of young women. The level found in seafood range from 0.3 to 7 mg/100 g. Seafoods, especially shellfish and dark fleshed fish (mackerel, tuna, and sardine) are reasonably good source of iron, supplying 1–2 mg/100 g muscle.

14.2.1.3.5 Zinc

This is an essential component of enzymes involved in the biosynthetic reactions. There is some indication that the dietary intake can fall below the daily recommended allowances of 15 mg per day. Seafoods, particularly shell fishes are major source of zinc, and also a probable source of copper and iodine. Seafoods are also rich source of selenium which, as an integral component of peroxidase, provides protection against the toxicity of heavy metals.

14.2.1.4 VITAMINS

Vitamin A is very important in the diet of man. It is required for vision, epithelial membrane integrity and perhaps, as an anticancer retinoid. It may also function on an anti-oxidant capacity as a free radical quencher, but it is toxic when consumed in large amounts. Fish liver oils contain more amounts of vitamin A, and small amounts occur in fish muscle. Hence, fishes are not considered as extensive source of this vitamin. Fatty fish contain a modest amount of vitamin A, in contrast to lean fish which contain very little of it. Fish liver oil contain vitamin which is associated with calcium transport and calcification of bones. In its active form, vitamin D promotes synthesis of the protein involved in the transport of calcium. The high fat species of fish such as mackerel and sardines are having high amount of D vitamin in the tissues. Vitamin E is a natural antioxidant that helps in protecting cell membranes and strengthening the immune system. Vitamin K plays a role in blood coagulation by preventing internal bleeding.

Fish contain varying concentration of B-vitamins, depending upon the specific vitamin and the fish species. Although a poor source of thiamine, fish is having moderate amount of vit B9 (folic acid), B7 (biotin), B3 (niacin) and B5 (pantothenic acid). Dark fleshed fishes are considered as moderate source of Riboflavin. Fish and shellfishes contain a reasonable quantity of Pyridoxine and are rich source of vitamin B12.

14.3 NUTRITIONAL COMPOSITION OF COLDWATER FISHES

Most of the population from hill stations consumes fish as daily food. Geographical location, seasons, ecosystem, culture practices show their effect on proximate composition. For example, when compared with the other freshwater fishes coldwater fishes are having high amount of PUFA.

Nutritional composition of golden Mahseer changes based on the seasonal changes and geographical changes (Sarma et al., 2015). The average protein value varied from 15.59 to 17.29 g/100 g fish, whereas ash, crude fat and moisture values observed to be varied as 1.23–1.55, 0.62–1.52, and 76.24–79.24 g/100 g fish, respectively. Seasonal changes influence on mean protein content of Mahseer where highest protein content and moderate fat content was reported during the months of June to September, as it is breeding season for the particular species. Mahseer is having high quantity of calcium, which is reported as 219.33 ± 9.018–1951.67 ±3.510 mg/100 g followed by potassium (755.33±14.740–1523.0±3.000 mg/100 g) but the average sodium concentration is low comparatively (99.67±2.082–253.33±4.163 mg/100 g). The average micro-mineral contents like iron, manganese, zinc, and selenium are 0.83 ±0.208, 0.17 ±0.036, 1.43 ±0.153, 0.87 ±0.306 mg/100 g, respectively.

In *Oncorhynchus mykiss* and *Schizothorax richardsonii* proximate composition also varied along with geographical locations and seasonal variations. In *O. mykiss*, the protein, ash, fat, and moisture content on dry matter basis are reported as 62–75.81%, 4.34–6.21%, 10.46–13.76%, 72.67–77.14% respectively and the macro-and micro-minerals as sodium (116.66–216.66 mg/100 g), Potassium (868–1538.67 mg/100 g), Calcium (299.66–488 mg/100 g), Iron (0.429–5.171 mg/100 g), Manganese (0.077–0.193 mg/100 g), Zinc (0.479–1.789 mg/100 g), Selenium (0.268–1.660 mg/100 g). Whereas *S. richardsonii* showed an average protein content of 56.96–72.45%, ash content of 5.49–10.68%, fat content of 11.18–20.59% and a moisture content of 70.12–77.34% and is abundant source of Potassium (608–1246.67 mg/100 g), followed by Calcium (208–417 mg/100 g), sodium (60–241 mg/100 g), Zinc (1.197–2.705 mg/100 g), Selenium (0.826–1.032 mg/100 g), Iron (0.461–0.892 mg/100 g) and Manganese (0.204–0.345 mg/100 g) (Das et al., 2012).

The habituated coldwater fishes of kosi river viz., *Labeo dero*, *Labeo dyocheilus*, *Labeo pangusia*, *Tor putitora* and *Schizothorax rechardsonii* forms a major contribution to total catch. The Proximate composition of the fish has been reported by Sarma et al. (2011). The Average moisture, crude protein, crude fat and ash of *Labeo dero* are noted as 76.92%, 70.24%, 32.53%, 8.86% respectively. The macro- and micro-mineral content is reported as sodium (138 mg/100 g), Potassium (1,233 mg/100 g), Calcium (342 mg/100 g), Iron (1.016 mg/100 g), Zinc (1.239 mg/100 g), magnesium

(0.213 mg/100 g) and selenium (0.265 mg/100 g) and in *Labeo dyocheilus* the average moisture was found to be 76.32%, crude protein as 68.53%, crude fat as 12.66% and ash content as 6.51% whereas the macro- and micro-mineral content was reported as sodium (100 mg/100 g), Potassium (925 mg/100 g), calcium (263 mg/100 g), Iron (0.984 mg/100 g), Zinc (0.963 mg/100 g), magnesium (0.076 mg/100 g) and selenium (0.442 mg/100 g).

In *Labeo pangusia* the values were observed as 81.54%, 73.01%, 24.19%, 7.39% for moisture, crude protein, crude fat and ah content, respectively and is a rich source potassium (784 mg/100 g) and calcium (270 mg/100 g) followed by low level of sodium (103 mg/100 g), iron (0.263 mg/100 g), zinc (0.769 mg/100 g), magnesium (0.164 mg/100 g) and selenium (0.565 mg/100 g). *Tor putitora* is having a moisture content of 72.68%, crude protein 52.43%, crude fat 19.25% and ash as 6.23%. The macro- and micro-mineral content is shown as sodium (241 mg/100 g), potassium (1,228 mg/100 g), calcium (397 mg/100 g), iron (0.628 mg/100 g), zinc (1.407 mg/100 g), magnesium (0.128 mg/100 g), and selenium (1.311 mg/100 g). Snow trout (*Schizothorax* spp.) are distributed from Jammu and Kashmir to Arunachal Pradesh. Of these *S. niger*, *S. progastus*, *S. plagiostomus*, *S. curvifrons,* and *S. esocinus* are important food fishes. In all the five species the protein content ranged from 15.43–17.17%, crude fat from 2.21 to 7.98%, Ash content from 0.89–1.18% and moisture content from 75.20 to 82.74% on wet weight basis and the macro- and micro-mineral content was reported as sodium (241 mg/100 g), potassium (1,246 mg/100 g), calcium (406 mg/100 g), Iron (0.464 mg/100 g), zinc (1.497 mg/100 g), magnesium (0.221 mg/100 g), and selenium (0.121 mg/100 g).

Some important coldwater fishes in Himalayas are *Semiplotus semiplotus, Labeo dero, Labeo dyocheilus, Sanguina sanguine, Barilius bendelisis, Garra mullya, L. pangusia, Tortor, Salmo trutta,* and *Neolissochilus hexagonolepis* (Sarma et al., 2014). These fishes are mainly consumed by the people in entire North Eastern India and among all the 10 species the moisture content ranged from 71.21 to 78.22%, crude protein content varied from 16.3 to 20.5%, fat content from 1.51 to 9.65% and ash content differed from 0.91 to 3.54%. *Labeo dero* is rich source of Potassium (1435.56 mg/100 g) and Calcium (1423.31 mg/100 g) whereas *Sanguina sanguine* is considered as a good source of Sodium (309.12 mg/100 g) and Selenium (177.52 mg/100 g). The other species *Barilius bendelisis* shown high amount of Iron (8.74 mg/100 g) and Manganese (1.24 mg/100 g) and *Semiplotus semiplotus* showed high amount of Zinc (1.72 mg/100 g).

14.4 AMINO ACID (AA) COMPOSITION OF COLDWATER FISHES

AAs are having good nutritive value and are having good health benefits (Chyun & Griminger, 1984). The reference AA score (mg/g protein) for histidine, Isoleucine, leucine, lysine, methionine-cysteine, phenylalanine-tyrosine, threonine, tryptophan, valine was 14, 28, 66, 58, 25, 63, 34, 11, 35, respectively.

Sarma et al. (2013) reported the AA composition of five coldwater fishes which are most important and the results were discussed below:

1. ***Tor putitora:*** The golden mahseer is a freshwater fish inhabiting along streams, rivers, and lakes and is benthopelagic in habitat. It had amino composition (mg/g protein) with more quantity of Tryptophan (618 mg/g protein) followed by phenylalanine-tyrosine (187 mg/g protein), lysine (170 mg/g protein), Methionine-cysteine (149.78 mg/g protein), Isoleucine (137.73 mg/g protein), leucine (120 mg/g protein), valine (114.15 mg/g protein) and Histidine (36.15 mg/g protein).
2. ***Neolissochilus hexagonolepis* (McClelland, 1839):** The chocolate or copper mahseer is having highest amount of histidine (183.43 mg/g protein) followed by tryptophan (149.98 mg/g protein), methionine-cysteine (143.30 mg/g protein), valine (126.69 mg/g protein), lysine (92.83 mg/g protein), leucine (77.34 mg/g protein), threonine (74.67 mg/g protein), isoleucine (72.84 mg/g protein), and phenylalanine-tyrosine (67.16 mg/g protein).
3. ***Oncorhynchus mykiss* (Walbaum, 1792):** i.e., rainbow trout had 17 AAs of which the most abundant AA (g/100 g of muscle) was Proline (1.67), Aspartic acid (1.48), Tyrosine (1.44), Glycine (1.21), Serine (1.16), Arginine (1.13), Isoleucine (1.12) and tryptophan (1.07). In total 20, AAs were identified with *Cyprinus carpio* (Linnaeus, 1758) with highest amount of Aspartic acid (2.721) followed by Serine (2.441), Glutamic acid (1.903), Alanine (1.553), Glycine (1.497), and Arginine (1.011).

The composition of AAs (mg/g protein) in trout were analyzed of which all the species showed that they are good sources of alanine, Leucine, and Phenyl alanine. *S. curvifrons* is the first highest source of Alanine (37.48 g/100 g protein) and second highest source of Leucine (10.38 g/100 g protein), whereas *S. esocinus* is first highest source of Leucine (12.51 g/100 g protein) and second highest source of Alanine (22.62 g/100 g protein). *S. labiatus* reflected with a high amount of Leucine (12.51 g/100 g protein) and

Phenyl Alanine (16.93 g/100 g protein) and *S. niger* showed large amount of Alanine (15.63 g/100 g protein) and Leucine (14.63 g/100 g protein).

14.5 COMPOSITION OF FATTY ACIDS IN COLDWATER FISHES

The composition of fatty acids in coldwater fishes is categorized under the groups as saturated fatty acids (SFAs), monounsaturated fatty acids (MUFAs) and polyunsaturated fatty acids (PUFAs) and the fatty acid composition of fishes varies depending on type of species, period of maturity, age of fish, size of fish, seasonal variations, and location.

The fatty acid profile for cold water fishes like *T. putitora, N. hexagonolepis, O. mykiss, S. richardsonii, C. carpio* were analyzed by Sarma et al. (2013). The total saturated fatty acid content, total monounsaturated fatty acid content, total n–3 PUFA and the total n–6 PUFA in *Tor putitora* were analyzed and the results were 52.97%, 28.06%, 8.69%, 9.67% whereas in *N. hexagonolepis* the fatty acid components were shown as 44.29%, 23.95%, 21.48%, 9.72% respectively. The ratio of n–3 to n–6 PUFA in golden Mahseer was 0.9% and in *N. hexagonolepis* was 2.21%. The saturated fatty acid content was the most dominant fatty acids in both species. From the SFAs, palmitic acid (C16:0) is major where stearic acid (C18:0) consists of 5.87–9.58% of the total fatty acids. Next to these MUFAs were high in quantity which are reported as 23.95% and 28.06% in *N. hexagonolepis* and *T. putitora*, respectively. The major contributions from poly unsaturated fatty acids includes Linolenic acid (C18:3n–3), Linoleic acid (C18:2n–6), EPA (C20:5n–3), DHA (C22:6n–3) and Arachidonic acid (C20:4n–6).

The total saturated (TSF) and monounsaturated fatty acid, total n–3 PUFA and n–6 PUFA content of *O. mykiss* was reported as 34.53%, 34.7%, as 13.61% and 17.77% respectively. The fraction of n–3 to n–6 PUFA was found to be 0.77%. In *S. richardsonii*, the TSF, total MUFA, total n–3 PUFA and the total n–6 PUFA were found to be 42.55%, 37.32%, 16.18%, 3.25% respectively and the ratio of n–3 to n–6 PUFA was 4.98%, whereas, in case of *C. carpio* the values are 46.19%, 31%, 9.81%, 13.91% respectively and the fraction of n–3 to n–6 PUFA was 0.7%.

In both rainbow trout and common carp species, MUFA was found to be dominant followed by SFAs and PUFAs. Among MUFA, Palmitic acid (C16:0) was abundant followed by stearic acid (C18:0). Oleic (C18:1) and Palmitoleic (C6:1) acids and from PUFA the highest to lowest quantity of acids are Linoleic acids (C18:2n–6), DHA (C22:6n–3), Linolenic (C18:3n–3), Arachidonic acid (C20:4n–6) and EPA (C20: %n–3) in both species.

14.6 FATTY ACID COMPOSITION OF SOME CULTURED AND NATURAL COLDWATER FISHES

14.6.1 SALMON

Atlantic salmon is the major consumed coldwater species by the people. The total lipids in farmed and wild Atlantic salmon were 7.4 and 7 g/100 g whereas the total fat content was about 4.1 g/100 g in farmed and 4 g/100 g in wild salmon, respectively. In farmed Atlantic salmon the total PUFAs were abundant (41%) later to that monosaturated fatty acids (33.4%) and SFAs (25.6%). The percentage of n–6 PUFA was 9.8%, n–6 HUFA was 1.7%, C18:2 (linoleic acid) was 7.4% and C20:4 (arachidonic acid) was 0.9%. The percentage of n–3 PUFA was 31.1% and n–3 HUFA was 28.3%, C 18:3 (α-linolenic acid) was 1.6%, C 20:5 (EPA) was 7.9% and C 22:6 (DHA) was 15.2% respectively. In case of wild Atlantic Salmon, the total MUFAs were more (53.7%) trended by PUFAs (27.3%) and SFAs (19%). The percentage of n–6 PUFA was 2.3%, n–6 HUFA was 0.8%, C18:2 (linoleic acid) was 1.2% and C20:4 (arachidonic acid) was 0.4%. The percentage of n–3 PUFA was 25% and n–3 HUFA was 23.8%, C 18:3 (α-linolenic acid) was 0.5%, C 20:5 (EPA) was 6.6% and C 22:6 (DHA) was 13.1% respectively. The ratio for n–3 to n–6 value was high in wild Atlantic salmon (11%) in contrast to Farmed Atlantic salmon (3.6%). Comparatively the fatty acid profile was high in farmed species to that of wild species (Blanchet et al., 2005).

14.6.2 RAINBOW TROUT

The total fatty acid and lipid in cultured and natural rainbow trout were 3.2 g/100 g, 0.6 g/100 g and 5.6 g/100 g, 1 g/100 g, respectively. Farmed rainbow trout has the highest amount of PUFA (40.6%) followed by MUFA (32.5%), Saturated fatty acid (26.9%). Wild rainbow trout has also contained the highest amount of PUFA (58.6%) followed by SFAs (24.4%), MUFAs (17%). In Farmed rainbow trout the percentage of n–6 PUFA, n–6 HUFA, C18:2 (linoleic acid), C20:4 (arachidonic acid) was 8.5%, 1.7%, 6.2% and 0.9% respectively. The percentage of n–3 PUFA, n–3 HUFA, C 18:3 (α-linolenic acid), C 20:5 (EPA), C 22:6 (DHA) was 32.2%, 30.1%, 1%, 7.3%, 18.7% respectively. Natural rainbow trout the percentage of n–6 PUFA, n–6 HUFA, C18:2 (linoleic acid), C20:4 (arachidonic acid) was 12.5%, 7.9%, 4.2% and 5.4% respectively. The percentage of n–3 PUFA, n–3 HUFA, C 18:3

(α-linolenic acid), C 20:5 (EPA), C 22:6 (DHA) was 46.2%, 44%, 1.7%, 8.1%, 32.2% respectively. Cultured rainbow trout contain high levels of n–3 HUFA and reported to show more health benefits to the consumers (Blanchet et al., 2005).

14.6.3 DICENTRARCHUS LABRAX

In case of wild *Dicentrarchus labrax,* the total SFAs, total monoenoic, n–6, n–3, total polyenoic and ratio of n3:n6 is 33.4, 19.4, 11.8, 35.6, 47.4, and 3.02% where as in case of Farmed species the values were noted as 29.3, 34.6, 9.3, 26.8, 36.1, and 2.88, respectively. In comparison between the wild and farmed one, earlier one is showing high fatty acid content than the later (Nettleton & Exler, 1992).

14.6.4 ONCORHYNCHUS KISUTCH

In this species the total SFAs, total monoenoic, n06, n–3, total polyenoic, n3: n6 and total PUFA are observed as 0.77, 1.02, 0.06, 0.9, 0.96, 14, and 2.9, where as in case of farmed one the values are 1.84, 2.88, 0.46, 1.42, 1.87, 3.1, and 6.88, respectively (Nettleton & Exler, 1992).

14.6.5 ICTALURUS PUNCTATUS

In case of wild *Ictalurus punctatus,* the total SFAs are 0.56%, total monoenoic acids are 0.82, n–6 series is 0.22%, n–3 series is 0.28%, total polyenoic acids are 0.5%, the ratio of n3:n6 is 1.3 and total PUFA are 1.98%, whereas in case of Farmed species the values are recorded as 2.48, 5.72, 1.52, 0.37, 2.02, 0.2, and 10.4% respectively and farmed species is showing good fatty acid profile (Nettleton & Exler, 1992).

14.6.6 SALMO GAIRDNERRI

Salmo gairdnerri, wild species is observed with 1.22% of total SFAs and farmed one is 1.5%, the total monoenoic acid content was found to be 1.27% in wild and 1.49% in farmed species. The total PUFA of wild and farmed species are recorded as 4.17 and 4.92% respectively and the ration of n3:n6 as 4.2 and 1.4% respectively in wild and farmed one (Nettleton & Exler, 1992).

14.7 MEDICINAL AND THERAPEUTIC VALUE OF FISHES

Fish is having great medicinal and therapeutic value and will be utilized as medication for preventing various disorders. Fish proteins are generally superior to other plant or animal proteins. Oils produced from fishes are having good therapeutic values, especially liver oils can extensively use for medical purposes. Oils from fish are great flavor enhancers and are a root of essential fatty acids (EFAs) such as omega-3 and omega-6. Most of the diseases like Alzheimer's, asthma, cancer, bronchitis, heart diseases, arterio-sclerosis, blood pressure can be cured with oils derived from seafood. These omega 3 oils play a vital role in brain and eye development of infants. Hence, pregnant women are prescribed to have omega 3 fatty acids available in fish.

The cartilage of shark has the capacity to fight with cancer. Halibut is a good source of vitamins (folic acid) and minerals (Magnesium) source and provides protection against ovarian and digestive tract cancers. Folic acid in halibut fishes will reduce the homocysteine compound, which is responsible for damaging artery walls. Magnesium increases blood flow in the body which indirectly helps to supply oxygen and nutrients. Consumption of fish roe is also shown to have medicinal properties as roe is a good source of vitamins, minerals, and long chain fatty acids. Salmon roe is prescribed for consumption to the people suffering with chronic liver diseases as it contains phosphatidylcholine. Haddock and cod, are low calorie foods that help in weight loss. Shellfishes are having more fraction of zinc, and hence can be used for healthy skin, good muscle and proper fertility. To supplement iodine deficiency, in treating the people with goiter marine sea foods like cod, sea bass, perch, and seabass are consumed in olden days. Chondroitin sulfate is known for its capacity to treat arthritis pain, which is abundantly present in sea cucumber. Along with chondroitin sulfate oils from sea cucumber consists of anti-inflammatory fractions, which can be used as an alternate to fish oil and as a potential inhibitor of 5-LOX (lipoxygenase) enzyme system. These lipoxygenase enzyme system inhibitors play an important role in modern drug development for treating asthma, ulcers, and arthritis. Bacteria present in sea cucumbers, sponges, and tunicates are able to produce some fatty acids that help in inhibiting cancer effects which have been reported in initial research with prostate cancer cells line and other human cancer cells.

14.8 CONCLUSION

In the nutrition of humans, fish occupies a unique position along with fish takes a vital role in the reduction of poverty and hunger. Fish as food, will nourish the households, especially for young children by reducing malnutrition and improving the overall growth, health, and maintenance of the body. All the coldwater fishes are best in their nutritive value in one or other way. Though the cold water fishes were widely distributed, their consumption is limited because of a lack of information about the nutritional profile. The coldwater fishes *Oncorhynchus mykiss, Tor putitora, Schizothorax richardsonii,* and *Neoliccochilush exagonolepsis* are very good sources of omega-3/6 PUFAs. Besides, these fishes are rich sources of quality proteins (16–18% of body weight) and functional AAs arginine, methionine, tyrosine, and proline.

KEYWORDS

- **antifreeze proteins**
- **coldwater fishes**
- **docosahexaenoic acid**
- **eicosapentaenoic acid**
- **functional amino acids**
- **tyrosine**

REFERENCES

Blanchet, C., Lucas, M., Julien, P., Morin, R., Gingras, S., & Dewailly, É., (2005). Fatty acid composition of wild and farmed Atlantic salmon (*Salmo salar*) and rainbow trout (*Oncorhynchus mykiss*). *Lipids, 40*(5), 529–531.

Charlton, M., (2005). Branched-chain amino acid enriched supplements as therapy for liver disease. *Journal of Nutrition, 136*(1), 295S–298.

Chen, C., Sander, J. E., & Dale, N. M., (2006). The effect of dietary lysine deficiency on the immune response to Newcastle disease vaccination in chickens. *Avian Diseases, 47*(4), 1346–1351.

Chyun, J. H., & Griminger, P., (1984). Improvement of nitrogen retention by arginine and glycine supplementation and its relation to collagen synthesis in traumatized mature and aged rats. *The Journal of Nutrition, 114*(9), 1697–1704.

Das, P., Sarma, D., Bisht, H. C. S., & Das, P., (2012). Nutritional quality of exotic rainbow trout (*Oncorhynchus mykiss*) and indigenous snow trout (*Schizothorax richardsonii*) in upland Himalayan region. *Indian Journal of Animal Nutrition, 29*(2), 121–126.

Debandt, J. P., & Cynober, L., (2006). Therapeutic use of branched chain amino acids in burn, trauma and sepsis. *Journal of Nutrition, 185*(1), 308S–313.

Etzel, M. R., (2004). Manufacture and use of dairy protein fractions. *Journal of Nutrition, 134*(4), 996S–1002.

Heimann, W., (1982). *Fundamental of Food Chemistry*. AVI Publishing Company, USA.

Hyland, K., (2007). Inherited disorders affecting dopamine and serotonin: Critical neurotransmitters derived from aromatic amino acids. *Journal of Nutrition, 137*(6), 1568S–1572.

Loest, C. A., Ferreira, A. V., Van, D. M. H. J., & Fair, M. D., (1997). Chemical and essential amino acid composition of South African mutton merino lamb carcasses. *South African Journal of Animal Sciences, 27*, 7–12.

Mischoulon, D., & Fava, M., (2002). Role of S-adenosyl-L-methionine in the treatment of depression: A review of the evidence. *American Journal of Clinical Nutrition, 76*(5), 11585–11615.

Mohanty, B. P., Mahanty, A., & Ganguly, S., (2014). Amino acid composition of 27 food fishes and their importance in clinical nutrition. *Journal of Amino Acids*. http:// dx.doi.org/10.1155/2014/269797.

Nettleton, J. A., & Exler, J., (1992). Nutrients in wild and farmed fish and shellfish. *Journal of Food Science, 57*(2), 257–260.

Richard, D. M., Dawes, M. A., Mathias, C. W., Acheson, A., Hill-Kapturczak, N., & Dougherty, D. M., (2009). L-Tryptophan: Basic metabolic functions, behavioral research and therapeutic indications. *International Journal of Tryptophan Research, 2*, 45–60.

Sarma, D., Akhtar, M. S., Das, P., Das, P., Shahi, N., Ciji, A., Mahanta, P. C., et al., (2013). Nutritional quality in terms of amino acid and fatty acid of five Coldwater fish species: Implications to human health. *National Academy Science Letters, 36*(4), 385–391.

Sarma, D., Akhtar, M. S., Das, P., Das, P.,Gadiya, G., Shahi, N., & Ciji, A., (2014). Proximate and mineral composition of some selected cold water fishes of upland Himalaya. *Nutrition & Food Science*.

Sarma, D., Das, P., Das, P., Sanwa, S., & Akhtar, M. S., (2015). Nutritional composition of golden mahseer (*Tor putitora*) in Coldwater Himalayan Region of India. *SKUAST Journal of Research, 17*(1), 23–28.

Sarma, D., Tiwari, T., Das, P., Jha, G., & Das, P., (2011). Proximate and mineral composition of indigenous hill stream fishes of Uttarakhand. *Indian Journal of Animal Nutrition, 28*(2), 203–206.

Segura, R., & Ventura, J., (1988). Effect of L-tryptophan supplementation on exercise performance. *International Journal of Sports Medicine, 9*(5), 301–305.

Sehgal, K. L., (1999). Coldwater fish and fisheries in the Indian Himalayas: Rivers and streams. In: Petr, T., (ed.) *Fish and Fisheries at Higher Altitudes: Asia* (p. 304). FAO Fisheries Technical Paper 385. FAO, Rome.

Singh, A. K., & Akhtar, M. S., (2015). *Coldwater Fish Diversity of India and its Sustainable Development* (pp. 97–105). Biodiversity for Sustainable development Editor-in-Chief Pratibha Singh Published by UP Biodiversity Board, Lucknow. International Day for Biological Diversity.

Wang, W., Wu, Z., Dai, Z., Yang, Y., Wang, J., & Wu, G., (2013). Glycine metabolism in animals and humans: Implications for nutrition and health. *Amino Acids, 45*(3), 463–477.

Wu, G., (2013). Functional amino acids in nutrition and health. *Amino Acids*, *45*(3), 407–411.

CHAPTER 15

Genotoxicity in Fishes: With Special Reference to Micronucleus Formation in Hematocytes

IMTIYAZ QAYOOM, SAMEENA KHAN, MONISA M. MALIK, ADNAN AMIN, and ADNAN ABUBAKR

Division of Aquatic Environmental Management, Faculty of Fisheries, Sher-e-Kashmir, University of Agricultural Sciences and Technology (SKUAST) of Kashmir, Rangil, Ganderbal, Jammu and Kashmir, India

ABSTRACT

Toxic insults of xenobiotic compounds pose a threat to non-target organisms in aquatic biota. Contaminants like pesticides, heavy metals, sewage, polycyclic hydrocarbons, plastics, and persistent organic pollutants are known to induce physiological dysfunctions in fishes under sub-lethal exposure. Among various injuries induced by them, environmental toxins are also potent genotoxins that damage, modify or transform the genetic makeup of fishes by altering the structure of DNA. Moreover, the formation of micronuclei in fish hematocytes has been a long and basic tool to quantify the level of genetic injuries caused in fish due to continuous xenobiotic exposure in nature. Hence this chapter identifies the mechanism of genotoxicity in fish and the role significance of micronuclei evaluation in fish blood cells for the quantification of genotoxic abuses.

15.1 INTRODUCTION

Fishes live in an environment that is susceptible to toxic insults by the contaminants. Waterbodies usually lie in the depression and act as ultimate

Coldwater Fisheries and Aquaculture Management: Technology for Sustainable Food Production.
Mohd Ashraf Rather, Faisal Rashid Sofi, Adnan Amin, & Kawkabul Saba (Eds.)

sinks of all the substances coming to from their watershed. Hence the environmental toxins, which include pesticides, heavy metals, plastics, persistent organic pollutants, and sewage coming as a part of municipal left over pose threat to fishes under natural conditions. These toxicants get washed away through leaching, drift, and percolation to aquatic ecosystems (Miyamoto et al., 1990) and induce toxicity in the non-target organisms like fishes which are badly hit by their exposure, even though the xenobiotics are bioavailable in minute quantities. Apart from various physiological dysfunctions, environmental toxins cause alterations in the structure and function of DNA resulting in genotoxicity and DNA damage.

Genotoxicity or genetic toxicity refers to some total cumulative damages or destructive changes in the genetic material (DNA and RNA) of fish cells distressing its integrity. Genotoxin, a substance inducing genotoxicity has an ability to prompt genetic damage in the cells by interacting with DNA. They have common physical and chemical properties that enable them to interact with nucleic acids and cause damage in them (Shugart, 1995; Sharma, 2016). Genotoxicity and mutagenicity are different terms. Every mutagen is necessarily a genotoxin but the reverse of it is not true. In definition, genotoxicity only refers to the potential of genotoxins to cause damage to DNA or RNA. They may be physical or chemical substances that have the ability to alter gene sequences and cause changes in the genetic information of fishes. If a genotoxin affects the genetic material of a somatic cell, it will not be transferred to the progeny, hence heredity will not be witnessed. Conversely, the effect of genotoxic substance on germ cells can get transferred to the progeny as hereditary. Moreover, the genotoxic effects can be diminished by DNA repairments, majorly by the enzyme activity of cells. Various contaminants such as polyaromatic hydrocarbons (PAHs), more particularly benzo-pyrenes and aldehydes are well known genotoxins and cause DNA adducts while binding with a nucleotide in DNA, more particularly guanosine base (Nikinmaa, 2014). An increase in the frequency of mutation is directly dependent on the type of DNA adduct and keeps on increasing if damaged DNA is not repaired by the replacement of adducted base by unmodified one.

Unlike a genotoxin, mutagen causes permanent and transmissible change in DNA, resulting in mutations. Mutagens can be physical which include different types of ionizing and non-ionizing radiations known to disrupt DNA structure. The transitions and transversions of the DNA sequence can be produced by chemical and biological mutagens. Base similarities, NO species, and intercalating agents are mediators of chemical mutagens that leads to the formation of apurinic and apyrimidinic sites, creating alterations in DNA. However, various viruses that cause host mutations by incorporating their DNA into the host cell, causing the change in the sequence of DNA of host cell are included in biological mutagens. There is a broad range of adverse effects on genetic components of the cell which may include disruption of genetic materials of cellular components that regulates the fidelity of the genome such as the spindle apparatus, topoisomerases, DNA repair systems and DNA polymerases WHICH can be used in Experimental structures that measures the capability of genotoxins in genotoxicity testing.

15.2 MECHANISM OF GENOTOXICITY IN FISHES

Various mechanisms for the induction of genotoxicity in fishes have been studied. Important among all are point mutations, clastogenicity, and aneuploidy. All these changes occur in DNA structure after genotoxin-DNA interactions. Genotoxins cause lesions, breakage, fusion, deletion, mis-segregation, or non-disjunction leading to damage and mutation (Anonymous). Genotoxins likely affect actively transcribed and readily dividing cells rather than uncoiled genomes. It manifests that toxic insult in embryos lead to teratogenic effects more quickly and profoundly than adults on account of their actively diving cells to form an organism. Exposure to genotoxins can cause change in nucleotide sequence either by addition of a base, deletion, insertion of a new gene sequence or replacement of one base by another. All these types are called point mutations. In addition to the point mutations, toxicants can be clastogenic, i.e., they lead to breakage of DNA strands which are inherited into daughter cells and cause genotoxicity. It is known that exposure of fishes to arsenic and benzene containing compounds lead to clastogenicity in them. Formation of micronuclei in fish blood cells is the best studied clastogenic effects till date.

15.3 MICRONUCLEUS AS AN INDICATOR OF GENOTOXICITY

During the process of cell division, the cytoplasmic chromatin-containing bodies which are formed when acentric chromosome fragments fail to get incorporated into daughter cell nuclei during anaphase are called micronuclei (Kirsch-Volders et al., 2011). The delay during anaphase arises due to genotoxic exposure leading to genetic damage either by the lack of centromeres or by abnormalities in the formation of mitotic spindle, which triggers and accomplishes the micronuclei formation (Schmid, 1975). During cell division, micronuclei are formed regardless of the type of DNA damage

(Salvadori et al., 2003; Nandanpawar et al., 2018). And the formation of micronucleus is related to the number of cells under n division. The major mechanism of MN formation resulting from chromosome mal-segregation is hypo-methylation of centromeric and para-centromeric regions satellite repeats However, the satellites are hyper-methylated, and loss of methylation lengthens repeat areas resulting of lowering tension in kinetochores and resulting in incorrect connections between mitotic spindle microtubules and chromosomes in most of the cases (Fenech et al., 2011). Due to tubulin depolymerization, chromatids/chromosomes are sometimes unable to segregate as the mitotic spindle cannot force them apart. Moreover, chromosomes lag behind at telophase due to the absence of kinetochore or centromeric abnormalities. A dicentric chromosome having two centromeres could get attached to two opposite spindle bodies thereby pulling chromatids in opposite directions. When breakage does not occur, the nuclear membrane surrounds both nuclei creating nucleoplasmic bridges between them. The nucleoplasmic bridge is eventually disrupted during cytokinesis, ending in the creation of a micronucleus. Boller & Schmidt (1970) proposed the term micronucleus test (MNT) in the early 1970s. MNT is used to identify the genomic damage induced by genotoxins in fishes and is the simplest of all techniques. This procedure targets interphase cells of any proliferating cell population regardless of its karyotype and is technically easier. Hence, this biomarker is widely utilized in environmental bio-monitoring programs. The potential toxicity of a compound in fishes is indicted by frequency of micronuclei formation which must meet the following criteria like the diameter of MN should be less than one-third of the main nucleus; there should be clear identification between the nuclear boundary of MN and the main nucleus and MN and the main nucleus should be identical in staining Fenech et al. (2003). As mentioned above, the micronuclei can be different as per their shape, like notched, blebbed, lobbed, budded, fragmented or bi-nucleated. All are considered as high quality indicators of cytotoxicity irrespective of their shapes observed in fish hematocytes (Ayllon & Vazquez, 2000, 2001). The health status index of different species cab be evaluated by Hematology as it provides a steadfast estimation via non-lethal means (Satheeshkumar et al., 2012). Though due to several internal and external factors, it can be quite challenging, yet hematological parameters serve as a valuable indicator to evaluate the health conditions of aquatic organisms. Blood parameters of fish are closely related to environmental and biological factors, also, the conditions like physiological changes could be a consequence of stress, and can be identified by using a number of hematological tools as studied by Fernandes & Mazon (2003). Establishment of hematological status based on data achieved from healthy animals is considered fundamental to illustrate a normal range variation. However, to establish a reference index before using hematological parameters is considered necessary to evaluate the physiological condition of different species of fish (Thrall et al., 2007; Leatherland et al., 1998).

15.4 GENOTOXICITY INDUCED DUE TO PESTICIDES

Pesticides induce the formation of micronuclei in fish hematocytes which indicates DNA damage on account of their exposure. A dose and time dependent increase of micronuclei is reported in almost all fishes such as *Cyprinus carpio specularis* and *Cyprinus carpio communis* (Al-Sabti, 1986; Nepomuceno et al., 1997; Gustavino et al., 2001; Buschini et al., 2004). Thiometon, an organophosphorus (OP) pesticide is also known to induce micronuclei in peripheral erythrocytes of fish along with other OPs. Similarly, chlorpyrifos is rated as a dangerous pesticide in India to induce micronuclei in peripheral erythrocytes are often dot-shaped and near to the major nucleus, with size and form varying from cell to cell (Bhatnagar et al., 2016). Various other classes of pesticides have been found to induce micronuclei in peripheral blood erythrocytes of different fishes such as *Oreochromis mossambicus* (Naqvi et al., 2016), *Chanos chanos* (Palanikumar et al., 2014), *Cirrhinus mrigala* (Anita et al., 2016), *Cyprinus carpio* var. *communis* and *Cyprinus carpio* var. *specularis* (Malik & Ganaie, 2011; Dar et al., 2014). Induction of micronuclei due to pesticides in various fishes is given in Table 15.1.

15.5 GENOTOXICITY INDUCED DUE TO HEAVY METALS

Heavy metals are potential water contaminants which includes lead (Pb), chromium (Cr), zinc (Zn), copper (Cu), cadmium (Cd), nickel (Ni), etc. (Patel et al., 2017; Kumar & Singh, 2018) and pose threat to fish life. They find their way into the water supply through industrial and consumer materials. Also, from acidic rain that breaks down soils thereby, releasing heavy metals into streams, lakes, rivers, and groundwater. They are point of concern worldwide because of their persistent, bio-accumulative, and toxic nature (Ali et al., 2019) the induction of morphological, behavioral, reproductive, genotoxic, and mutagenic damages to the aquatic biota is caused by Heavy metals (Pulley et al., 2016). When the heavy metals are taken up, they are stored at a faster rate than they are metabolized or excreted, hence, they get accumulated in living organisms. The three most dangerous heavy metals identified in fishes are lead (Pb), mercury (Hg) and cadmium (Cd). Several studies have shown that fishes exposed to water and sediments contaminated by heavy metals can exhibit genotoxic effects (Turan et al., 2020; Kontas & Bostanc, 2020) and DNA damage as reported in *Geophagus brasiliensis* exposed to Fe, Mn, Cd, Cu, and Pb (Gomes et al., 2019). Da Silva et al. (2020) reported the effect of heavy metals on *Astyanax lacustris* as a genotoxic effect due to various heavy metals in the order of Mn > Ba > Zn > Ni > Ti > Cr > Cu. There are possibilities of synergistic effects of heavy metals that lead to profound effects in fishes when compared with the individual toxic potential of any metal (Enserink et al., 1991). Some studies have shown that nanoparticles of Titanium dioxide exposure caused significant genotoxic changes in the blood when it reaches the blood stream, gets accumulated in the muscles and brain of fishes and also lead to oxidative stress by the generation of ROS (Carmo et al., 2019). Studies on *Oreochromis niloticus, Poronotus triacanthus* and *Puntius altus* lead (Pb), copper (Cu) and cadmium (Cd) induced MN formation and genetic damage in the order of Pb > Cu > Cd (Jiraungkoorskul et al., 2007). Induction of micronuclei due to pesticides in various fishes is given in Table 15.2.

TABLE 15.1 MN Nucleus Frequency Induced by Pesticides in Different Fish Species

SL. No.	Species	MN per 1,000 Cells	Causative Pesticide	References
1.	*Channa punctatus*	0.02	Dimethoate, dichlorvos, chlorpyriphos, and malathion, methyl parathion, fenvalerate, cypermethrin, and carbaryl	Kumar et al. (2010)
2.	*Channa punctatus*	0.028–0.048	Pesticide toxicity	Ali et al. (2008)
		0.72–4.11	Pesticide toxicity	Farah et al. (2002)
3.	*Carassius auratus*	3.17	Herbicide	Canvas & Konen (2007)
4.	*Cheirodon interruptus*	0.0–1.0	Pesticide toxicity	Campana et al. (1999)
5.	*Clarias batrachus*	0.47–3.95	Pesticide toxicity	Ateeq et al. (2002)

15.6 GENOTOXICITY INDUCED DUE TO OTHER SOURCES

Various other substances are also known to induce genotoxicity in fishes. Some chemicals like colchicine and radiations like X-rays are reported to induce clastogenic and mitoclastic (spindle poisoning) in *Cyprinus carpio* species (Gustavino et al., 2001). Moreover, discharge of armament factory wastes is also known to cause progressive genotoxic insults by micronuclei induction (Ayllon & Vazquez, 2001). Even intermittent heat shocks at variant temperature have also been reported as external genotoxins to induce nuclear anomalies like micronuclei formations, chromosomal aberrations, DNA damage and cell proliferation in *Carassius auratus* (Anitha et al., 2000). Induction of micronuclei due to pesticides in various fishes is given in Table 15.3.

TABLE 15.2 MN Nucleus Frequency Induced by Heavy Metals in Different Fish Species

SL. No.	Species	MN per 1,000 Cells	Causative Pesticide	References
1.	*Carassius auratus*	2.26	Metal toxicity	Cavas (2008)
2.	*Carassius auratus gibelio*	13; 10; 5.2	Metal toxicity (Cr)	Al-Sabti et al. (1994)
3.	*Carassius* sp.	1.8	Metal toxicity (Cu ions)	Hayashi et al. (1998)
4.	*Cyprinus carpio*	1.2	Metallic mercury	Nepomuceno et al. (1997)
5.	*Oncorhynchus mykiss*	0.1	Cadmium injections	Castano et al. (1998)
6.	*Oreochromis niloticus*	0.21	Heavy metals	Barbosa et al. (2010)
7.	*Phoxinus phoxinus*	0.3–0.7	Metal toxicity	Ayllon & Garcia-Vazquez (2000)
8.	*Poecilia latipinna*	0–0.4	Metal toxicity	Ayllon & Garcia-Vazquez (2000)

15.7 SIGNIFICANCE OF GENOTOXICITY TESTING

Genotoxicity testing in fishes provides information about damage caused to their DNA. The endpoints of genotoxic insults resulting due to unscheduled DNA synthesis, sister-chromatid exchange, adduct formation in DNA, mitotic recombination, chromosomal aberrations studies and micronuclei formation are evaluated and quantified by genotoxicity testing tools. Thus, toxic endpoints of all the substances, chemicals, pesticides, and drugs, etc., can be measured by toxicity testing. Safety evaluation of new chemical entities (NCE) and their efficiency are one of the mandates of toxicity testing. similarly, evaluating natural waterbodies for the presence of genotoxins in order to generate baseline data for evaluation of potable waters is mandatory. Moreover, its safety for commercially significant flora and fauna is need of an hour for human safety.

TABLE 15.3 MN Nucleus Frequency Induced Due to Other Sources in Different Fish Species

SL. No.	Species	MN per 1,000 Cells	Causative Pesticide	References
1.	*Carassius auratus*	0.18	Mutagenic chemicals (orange (AO) fluorescent staining)	Ueda et al. (1992)
2.	*Carassius auratus*	0.26	Heat shock	Anitha et al. (2000)
3.	*Clarias lazera*	0.31	Industrial waste ethyl methane sulfonate	Odeigah & Osanyinpeju (1995)
4.	*Cyprinus carpio*	0.02	X-rays and colchicine	Gustavino et al. (2001)
5.	*Cyprinus carpio*	0.52	Chemical industries and/or a nuclear power plant waste	Llorente et al. (2002)
6.	*Cyprinus carpio*	0.9	Sewage treatment plant discharges.	Grisolia & Starling (2001)
7.	*Odontobutis oscura* Obscura	0.181	Wastewater	Ayllon & Garcia-Vazquez (2001)
		0.33	River pollution	De Flora et al. (1993)
		1.0	Radiation	Schultz et al. (1993)
8.	*Oreochromis niloticus*	0.8	Sewage treatment plant discharges	Grisolia & Starling (2001)
9.	*Salmo trutta*	0.4–2.3	Waste discharge	Ayllon et al. (2000)
10.	*Pseudopleuronectes Americanus*	0–1.30	Wastewater	Hughes & Hebert (1991)

15.8 CONCLUSION

At different biodiversity levels, multiple consequences are produced by aquatic pollutants, the most dangerous being carcinogenic and mutagenic compounds as their effects may not be limited to that of the individual but can reach to the several generations. Pesticides, heavy metals, sewage, and PAHs are known genotoxins in fishes and are considered as environmental toxins. Moreover, some other sources like radiations, heat shocks, factory wastes and chemicals like colchicine are documented as genotoxins in vide rage of fish species leading to deleterious effects in them. The application of genotoxicity biomarkers allows assessing the mutagenic hazards in sentinel organisms and for the contaminants fate and sources identification.

KEYWORDS

- **genotoxicity**
- **micronucleus**
- **micronucleus test**
- **new chemical entities**
- **organophosphorus**
- **toxicants**

REFERENCES

Ali, D., Nagpure, N. S., Kumar, S., Kumar, R., & Kushwaha, B., (2008). Genotoxicity assessment of acute exposure of chlorpyrifos to freshwater fish *Channa punctatus* (Bloch) using micronucleus assay and alkaline single-cell gel electrophoresis. *Chemosphere, 71*, 1823–1831.

Ali, H., Khan, E., & Ilahi, I., (2019). Environmental chemistry and ecotoxicology of hazardous heavy metals: Environmental persistence, toxicity, and bioaccumulation. *Hindawi J. Chem.* Article ID 6730305.

Al-Sabti, K., (1986). Clastogenic effect of five carcinogenic-mutagenic chemicals on the cells of the common carp *Cyprinus carpio*. *Comp. Biochem. Physiol. C, 85*, 5–9.

Al-Sabti, K., Franko, M., Adrijanic, B., Knez, S., & Stegnar, P., (1994). Chromium induced micronuclei in fish. *J. Appl. Toxicol., 14*, 333–336.

Anita, B., Yadav, A. S., & Cheema, N., (2016). Genotoxic effects of chlorpyrifos in freshwater fish *Cirrhinus mrigala* using micronucleus assay. *Advances in Biology*, 1–6.

Anitha, B., Chandra, N., Gopinath, P. M., & Durairaj, G., (2000). Genotoxicity evaluation of heat shock in gold fish (*Carassius auratus*). *Mutat. Res., 469*, 1–8.

Anonymous: http://www.medscape.com/viewarticle/499098_4. WebMD, LLC. *This Week in Cardiology Podcast – Medscape* (accessed on 3 February 2023).

Ateeq, B., Abul, F. M., Niamat, A. M., & Ahmad, W., (2002). Induction of micronuclei and erythrocyte alterations in the catfish *Clarias batrachus* by 2,4-dichlorophenoxyacetic acid and butachlor. *Mutat. Res., 518*, 135–144.

Ayllon, F., & Garcia-Vaszquez, E., (2000). Induction of micronuclei and other nuclear abnormalities in European minnow *Phoxinus phoxinus* and *Mollie Poecilia latipinna*: An assessment of the fish micronucleus test. *Mutat. Res., 467*, 177–186.

Ayllon, F., & Garcia-Vazquez, E., (2001). Micronuclei and other nuclear lesions as genotoxicity indicators in rainbow trout *Oncorhynchus mykiss*. *Ecotoxicol. Environ. Saf., 49*, 221–225.

Ayllon, F., Suciu, R., Gephard, S., Juanes, F., & Garcia-Vazquez, E., (2000). Conventional armament wastes induce micronuclei in wild brown trout *Salmo trutta. Mutat. Res., 470*, 169–176.

Barbosa, J. S., Cabral, T. M., Ferreira, D. N., Agnez-Lima, L. F., & De Medeiros, S. R., (2010). Genotoxicity assessment in aquatic environment impacted by the presence of heavy metals. *Ecotoxicol. Environ. Saf., 73*, 320–325.

Bhatnagar, A., Yadav, A. S., & Cheema, N., (2016). Genotoxic effects of chlorpyrifos in freshwater fish *Cirrhinus mrigala* using micronucleus assay. *Advances in Biology*, Article ID 9276963.

Boller, K., & Schmid, W., (1970). Chemical mutagenesis in mammals: The Chinese hamster bone marrow as an *in vivo* test system. Hematological findings after treatment with trenimon. *Human Genetic, 11*, 35–54.

Buschini, A., Martino, A., Gustavino, B., Monfrinotti, M., Poli, P., Rossi, C., Santoro, M., et al., (2004). Comet assay and micronucleus test in circulating erythrocytes of *Cyprinus carpio* specimens exposed in situ to lake waters treated with disinfectants for potabilization. *Mutat. Res., 557*, 119–129.

Campana, M. A., Panzeri, A. M., Moreno, V. J., & Dulout, F. N., (1999). Genotoxic evaluation of the pyrethroid lambda-cyhalothrin using the micronucleus test in erythrocytes of the fish *Cheirodon interruptus. Mutat. Res., 438*, 155–161.

Carmo, T. L. L., Siqueira, P. R., Azevedo, V. C., Tavares, D., Pesenti, E. C., Cestari, M. M., Martinez, C. B. R., & Fernandes, M. N., (2019). Overview of the toxic effects of titanium dioxide nanoparticles in blood, liver, muscles, and brain of a neotropical detritivorous fish. *Environ. Toxicol., 34*(4), 457–468.

Castano, A., Carbonell, G., Carballo, M., Fernandez, C., Boleas, S., & Tarazona, J. V., (1998). Sublethal effects of repeated intraperitoneal cadmium injections on rainbow trout (*Oncorhynchus mykiss*). *Ecotoxicol. Environ. Saf., 41*, 29–35.

Cavas, T., & Ko¨nen, S., (2007). Detection of cytogenetic and DNA damage in peripheral erythrocytes of goldfish (*Carassius auratus*) exposed to a glyphosate formulation using the micronucleus test and the comet assay. *Mutagenesis, 22*, 263–268.

Cavas, T., (2008). *In vivo* genotoxicity of mercury chloride and lead acetate: Micronucleus test on acridine orange stained fish cells. *Food Chem. Toxicol., 46*, 352–358.

Da Silva, E. P., Benvindo-Souza, M., Cotrim, C. F. C., Motta, A. G. C., Lucena, M. M., Filho, N. R., Pereira, J., et al., (2020). Genotoxic effect of heavy metals on *Astyanax lacustris* in an urban stream. *Heliyon.*

Dar, S. A., Yousuf, A. R., Balkhi, M. H., Ganai, F. A., & Bhat, F. A., (2014). Assessment of endosulfan induced genotoxicity and mutagenicity manifested by oxidative stress pathways in freshwater cyprinid fish crucian carp (*Carassius carassius* L.). C*hemosphere, 120*, 273–283.

Enserink, E. L., Maas-Diepeveen, J. L., & Van, L. C. J., (1991). Combined effects of metals; an ecotoxicological evaluation. *Water Res., 25*(6), 679–687.

Farah, M. A., Ateeq, B., Ali, M. N., & Ahmad, W., (2002). Evaluation of genotoxicity of PCP and 2,4-D by micronucleus test in freshwater fish *Channa punctatus*. *Ecotoxicol. Environ. Saf., 54*, 25–29.

Fenech, M., Chang, W. P., Kirsch-Volders, M., Holland, N., Bonassi, S., & Zeiger, E., (2003). Human micronnucleus project. HUMN project: Detailed description of the scoring criteria for the cytokinesis-block micronucleus assay using isolated human lymphocyte cultures. *Mutat. Res. 534*(1, 2), 65–75.

Fenech, M., Kirsch-Volders, M., & Natarajan, A. T., (2011). Molecular mechanisms of micronucleus, nucleoplasmic bridge and nuclear bud formation in mammalian and human cells. *Mutagenesis, 26*, 125–132.

Fernandes, M. N., & Mazon, A. F., (2003). Environmental pollution and fish gill morphology. In: *Fish Adaptations* (pp. 203–231). Enfield, Sciences Publishers.

Gomes, L. C., Chippari-Gomes, A. R., Miranda, T. O., Pereira, T. M., Mercon, J., Davel, V. C., Barbosa, B. V., et al., (2019). Genotoxicity effects on *Geophagus brasiliensis* fish exposed to Doce River water after the environmental disaster in the city of Mariana, MG, Brazil. *Braz. J. Biol., 79*(4), 659–664.

Grisolia, C. K., & Starling, F. L., (2001). Micronuclei monitoring of fishes from Lake Paranoa, under the influence of sewage treatment plant discharges. *Mutat. Res., 491*, 39–44.

Gustavino, B., Scornajenghi, K. A., Minissi, S., & Ciccotti, E., (2001). Micronuclei induced in erythrocytes of *Cyprinus carpio* (*Teleostei,* Pisces) by X-rays and colchicines. *Mutat. Res., 494*, 151–159.

Hayashi, M., Ueda, T., & Uyeno, K., (1998). Development of genotoxicity assay systems that use aquatic organisms. *Mutat. Res., 399*, 125–133.

Hughes, J. B., & Hebert, A. T., (1991). Erythrocyte micronuclei in winter flounder (*Pseudopleuronectes americanus*): Results of fields surveys during 1980–1988 from Virginia to Nova Scotia and in long Island sound. *Arch. Environ. Contam. Toxicol., 20*, 474–479.

Jiraungkoorskul, W., Kosai, P., Sahaphong, S., Kirtputra, P., & Shanida, J. C., (2007). Evaluation of micronucleus test's sensitivity in freshwater fish species. *Research Journal of Environmental Sciences, 1*(2), 56–63.

Kirsch-Volders, M., Plas, G., & Elhajouji, A., (2011). The *in vitro* MN assay in 2011: Origin and fate, biological significance, protocols, high throughput methodologies and toxicological relevance. *Arch. Toxicol., 85*, 873–899.

Kontas, S., & Bostanc, D., (2020). Genotoxic effects of environmental pollutant heavy metals on *Alburnus chalcoides* (Pisces: *Cyprinidae*) inhabiting lower Melet river (Ordu, Turkey). *Bull. Environ. Contam. Toxicol., 104*, 763–769.

Kumar, B., & Singh, U. K., (2018). Source apportionment of heavy metals and their ecological risk in a tropical river basin system. *Environ. Sci. Pollut. Control Ser., 25*(25), 25443–25457.

Leatherland, J. F., Ballantyne, J. S., & Van, D. K. G., (1998). Diagnostic assessment of non-infectious disorders of captive and wild fish populations and the use of fish as sentinels organisms environmental studies. *Fish Diseases and Disorders* (pp. 335–366). Massachusetts, Cab International.

Llorente, M. T., Martos, A., & Castano, A., (2002). Detection of cytogenetic alterations and blood cell changes in natural populations of carp. *Ecotoxicology, 11*, 27–34.

Malik, M., & Ganie, F., (2011). Genotoxic effects of organophosphate insecticide thiometon in some exotic fishes of Kashmir. *International Journal of Pharmaceutical Sciences Review and Research, 8*(1), 19.

Miyamoto, J., Mikami, N., & Takimoto, Y., (1990). The fate of pesticides in aquatic ecosystems. In: Hutson, D. H., & Roberts, T. R., (eds.), *Environmental Fate of Pesticides* (pp. 123–476). Chichester, England: Wiley.

Naqvi, G. Z., Shoaib, N., & Ali, A. M., (2016). Genotoxic potential of pesticides in the peripheral blood erythrocytes of fish *Oreochromis mossambicus*. *Pakistan Journal of Zoology, 48*(6), 1643–1648.

Nepomuceno, J. C., Ferrari, I., Spano, M. A., & Centeno, A. J., (1997). Detection of micronuclei in peripheral erythrocytes of *Cyprinus carpio* exposed to metallic mercury. *Environ. Mol. Mutagen., 30*, 293–297.

Nikinma, M., (2014). *An Introduction to Aquatic Toxicology*. ISBN: 978-0-12-411574-3. Elsevier.

Odeigah, P. G., & Osanyinpeju, A. O., (1995). Genotoxic effects of two industrial effluents and ethyl methane sulfonate in *Clarias lazera*. *Food Chem. Toxicol., 33*, 501–505.

Palanikumar, L., Kumaraguru, A. K., Ramakritinan, C. M., & Anand, M., (2014). Toxicity, biochemical and clastogenic response of chlorpyrifos and carbendazim in milkfish *Chanos chanos*. *International Journal of Environment, Science and Technology, 11*, 765–774.

Patel, P., Raju, M. J., Reddy, B. C. S. R., Suresh, U., Sankar, D. B., & Reddy, T. V. K., (2017). Heavy metal contamination in river water and sediments of the Swarnamukhi River Basin, India: Risk assessment and environmental implications. *Environ. Geochem. Health, 40*(2), 609–623.

Pulley, S., Foster, I., & Antunes, P., (2016). The dynamics of sediment-associated contaminants over a transition from drought to multiple flood events in a low land UK catchment. *Hydrol. Process, 30*, 704–719.

Salvadori, D., Ribeiro, L., & Fenech, M., (2003). Micronuclei test in human cells *in vitro* In: Ribeiro, L., Salvadori, D., & Marques, E., (eds.), *Environmental Mutagenesis* (pp. 201–223). Canoas: ULBRA.

Satheeshkumar, P., Ananthan, G., Senthilkumar, D., Khan, A. B., & Jeevanantham, K., (2012). Comparative investigation on haematological and biochemical studies on wild marine teleost fishes from Vellar estuary, southeast coast of India. *Comparative Clinical Pathology, 21*, 275–281.

Schmid, W., (1975). The micronucleus test. *Mutat. Research, 31*, 9–15.

Shugart, L. R., (1995). environmental genotoxicology. In: Gary, M. R., (ed.), *Fundamentals of Aquatic Toxicology, Effects, Environmental Fate, and Risk Assessment* (IInd edn.). CRC press.

Thrall, M. A., Baker, D. C., Campbell, T. W., DeNicola, D., Fettman, M. J., Lassen, E. D., Rebar, A., & Weiser, G., (2007). *Hematologia e Bioquímica Clínica Veterinária* (P. 582). São Paulo, Roca.

Turan, F., Karan, S., & Ergenler, A., (2020). Effect of heavy metals on toxicogenetic damage of European eels *Anguilla anguilla*. *Environmental Science and Pollution Research, 27*, 38047–38055pp.

Ueda, T., Hayashi, M., Ohtsuka, Y., Nakamura, T., Kobayashi, J., & Sofuni, T., (1992). A preliminary study of the micronucleus test by acridine orange fluorescent staining compared with chromosomal aberration test using fish erythropoietic and embryonic cells. *Water Sci. Technol., 25*, 235–240.

CHAPTER 16

Infectious Diseases of Coldwater Fishes: Focus on Viral and Fungal Infections

SYED SHARIQ N. QADIRI, INAIN JAIES, FEROZ A. SHAH, SHABIR A. DAR, and ASIFA WALI

Division of Aquatic Animal Health Management, Faculty of Fisheries, Sher-e-Kashmir University of Agricultural Sciences and Technology (SKUAST) of Kashmir, Rangil, Ganderbal, Jammu and Kashmir, India

ABSTRACT

Intensification of aquacultural practices results in disease outbreaks of prevalent pathogens alongside new and emerging diseases. Water as a medium in an aquatic environment facilitates the spread of disease from infected specimens to naïve fish. The transmission of disease can be either horizontal (fish to fish) or vertical (parent to offspring). Stress is always a predisposing factor to a disease outbreak. The pathogenicity and virulence of the invading pathogen will determine the severity of infection. In addition, the immune status and genetic make-up of fish will define whether the host succumbs to the invading pathogen or not. It is to mention that the immune system of a coldwater fish is generally weak in comparison to a warm water fish given the low temperature which makes fish more susceptible to infectious diseases. Infectious diseases of coldwater fishes are either of bacterial, viral, parasitic, or fungal etiology. In the context of the present chapter, the focus here will only be on viral and fungal pathogens prevalent in cold water aquaculture.

Coldwater Fisheries and Aquaculture Management: Technology for Sustainable Food Production.
Mohd Ashraf Rather, Faisal Rashid Sofi, Adnan Amin, & Kawkabul Saba (Eds.)

16.1 INTRODUCTION

Any detrimental divergence from the standard structural and functional state of an organism is termed as disease. It is generally coupled with certain signs and symptoms which are indicative of the abnormal state of the host. Being always immersed in the aquatic environment, fish is very sensitive to its surroundings which contains myriad of pathogenic organisms (Rombout et al., 2014; Qadiri et al., 2019a). Under normal conditions, fish is in equilibrium with pathogen and once this equilibrium deviates, disease outbreak occurs. Although, the defense mechanisms in fish are well developed but whenever these pathogens breach the host defenses, fish gets infected. Study of disease (pathology) involves the identification of causative agent (etiology), understanding the mechanism of disease development (pathogenesis) and the changes associated with the disease process. The epidemiological triad (Figure 16.1) which involves the interaction of host, environment, and pathogen determine disease process and pathogen ecology (Kelly & Renukdas, 2020). The common viral and fungal diseases of finfish prevalent in cold water fish are enlisted in Table 16.1.

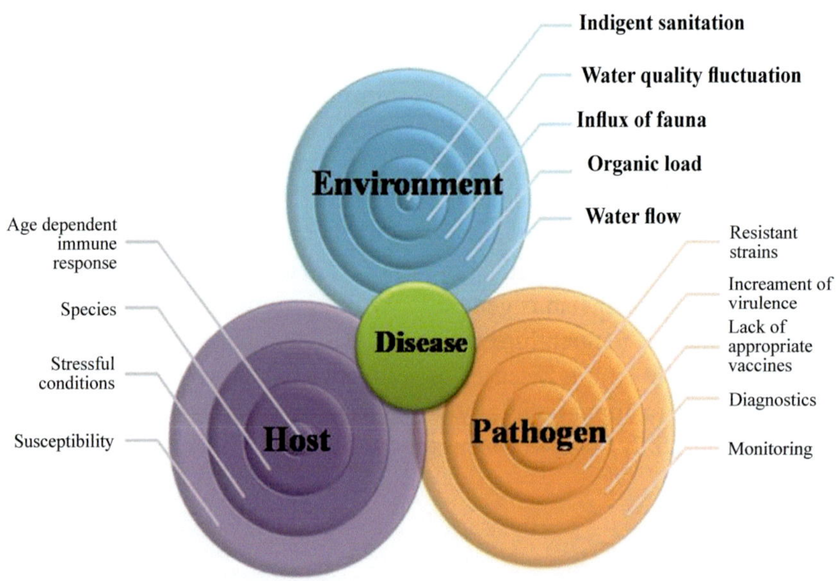

FIGURE 16.1 Epidemiological triad (host-pathogen-environment).

TABLE 16.1 Commonly Prevalent Viral and Fungal Diseases in Cold Water Fish

SL. No.	Disease	Type	Causative Agent	Clinical Signs
1.	Viral hemorrhagic septicemia (VHS)	Viral	VHSV	• Anorexia • Lethargy • Petechial hemorrhages • Dark color
2.	Spring viremia of carp (SVC)	Viral	SVCV	• Exophthalmia • Edema • Abdominal distension • Hemorrhages
3.	Infectious hematopoietic necrosis (IHN)	Viral	IHNV	• Ascites • Scoliosis • Dark coloration • Pale gills
4.	Infectious pancreatic necrosis (IPNV)	Viral	IPNV	• Anemic gills • Swelled abdomen • Pop eye
5.	Saprolegniasis	Fungal	*Saprolegnia* sp.	• Depigmented lesions • Whitish brown cotton-like patches • Lethargy • Bloating

TABLE 16.1 *(Continued)*

SL. No.	Disease	Type	Causative Agent	Clinical Signs
6.	Branchiomycosis	Fungal	*Branchiomyces* sp.	• Pale and rotten gills • Fishes become weak • Gasping for air • Mouth wide open
7.	Epizootic ulcerative syndrome (EUS)	Fungal	*Aphanomyces invadans*	• Lesions on the lateral sides • Protruding scales • Scale loss • Skin erosion • Reddened skin

16.2 VIRAL DISEASES OF COLD-WATER FISH

16.2.1 VIRAL HEMORRHAGIC SEPTICEMIA (VHS)

16.2.1.1 ETIOLOGICAL AGENT

The viral hemorrhagic septicemia virus (VHSV) causes viral hemorrhagic septicemia (VHS) infection in cultured rainbow trout (*O. mykiss*) and is one of the most lethal finfish viral diseases. VHSV is a member of the Novirhabdovirus genus in the Rhabdoviridae family (Ammayappan & Vakharia, 2009), with a single linear, negative sense ssRNA of roughly 11.1 kb as genetic material. The disease was first described by Schäperclaus (1938), and the etiology was later identified as Egtved virus (Jensen, 1963) as it initially occurred in rainbow trout near the Danish village of Egtved. Therefore, VHS disease is also known as "Egtved illness."

16.2.1.2 CLINICAL SIGNS AND PATHOLOGY

Affected fish show anorexia and are lethargic. Dark color and pale gills are common. Petechial hemorrhages are seen on the skin, fins, and internal organs. Infected fish shows irregular swimming and heavy mortality. In the liver, kidney, spleen, and skeletal muscle, the majority of tissue alterations are observed which usually indicate moderate to significant necrosis or cell death (Nishizawa et al., 2006). The primary site of infection is the hematopoietic (blood-forming) regions of the kidney and spleen which display severe necrosis. The virus shows a strong tropism for blood and is frequently localized in organs such as the heart due to high blood supply and likely gains entry into fish via gill (Qadiri et al., 2019b, 2020). Thickened lamellae in the gills and pyknotic nuclei in the liver are also apparently visible. Blood collects in skeletal muscle, although it is not severely damaged. Due to the tropism of virus for brain tissue, living fish infected with VHS may appear listless or limp, hang just beneath the surface, or swim very erratically, such as continual flashing circling.

16.2.1.3 DIAGNOSIS

Histopathological (HP) investigation and tissue observation under a microscope are used to get a preliminary diagnosis. To diagnose VHS according to OIE guidelines, the virus must first be isolated in cell culture and then neutralized using serum. Commonly used fish cell lines are BF-2, EPC, and FHM (OIE, 2019). The bullet-shaped rhabdovirus can be seen using electron microscopy, but it is insufficient for a definite diagnosis. RT-PCR (Snow et al., 2004) and standardization of various qRT-PCR protocols are in routine use for molecular diagnosis of VHSV (Garver et al., 201; Kim et al., 2014).

16.2.1.4 CONTROL AND TREATMENT

Utilization of virus-free water supplies (water from wells or spring sources that are separate from surface water sources), sourcing eggs from certified virus-free stocks, and disinfection of eggs prior to introduction to the production facility are all measures to prevent VHS outbreaks in cultured fish. Modern aquaculture operations have a facility biosecurity plan that defines site-specific equipment and disinfection procedures for workers and equipment that transit between sites. Disease outbreaks have been demonstrated to be reduced when processed (pelleted) feed is used instead of raw fish-derived feed. The use of all in-all out production systems is one of the management techniques for marine culture systems. Disinfection of hatchery, specific pathogen free stock and other management practices should be employed. Common disinfectants such as formalin, sodium hydroxide and sodium hypochlorite are to be used.

16.2.2 SPRING VIREMIA OF CARP (SVC)

16.2.2.1 ETIOLOGICAL AGENT

It is caused by SVC virus belonging to the genus Vesiculovirus and family Rhabdoviridae (Carstens, 2010). The name refers to a viremia that mostly affects common carp (*Cyprinus carpio*), particularly in the spring (Kibenge, 2016). The term was coined to distinguish virus-induced disease from infectious dropsy of carp which has a complex etiology (Fijan et al., 1971, 1984). Viral genome is a single strand of non-segmented, negative-sense RNA of 11,019 nucleotides that codes for five proteins: a nucleoprotein (N), a phosphoprotein (P), a matrix protein (M), a glycoprotein (G), and an RNA-dependent, RNA polymerase (L).

16.2.2.2 CLINICAL SIGNS AND PATHOLOGY

Exophthalmia, edema, petechial hemorrhages, and abdominal distension are commonly observed in SVCV infection. The virus quickly spreads to the liver, kidney, spleen, and alimentary tract, resulting in viremia and peritonitis (Dikkeboom et al., 2004). Edematous and enlarged kidneys with tubular degeneration and enlarged spleen are frequently observed with hemorrhages in the swim bladder. Vasculitis is frequently caused by SVCV multiplication in blood vessel endothelium and is the most likely source of both exterior and interior hemorrhages (Ahne et al., 2002). Diseased fish move slowly, gather at sides, and are sluggish to any stimulus. Moribund fish have been observed lying on their sides, frequently on the tank's bottom, and swimming up but then returning to the bottom when frightened.

16.2.2.3 DIAGNOSIS

The virus can be isolated using OIE recommended cell lines like FHM or EPC (Fijan et al., 1983; Clark et al., 1974). If the virus is present, the cells will degenerate and round out. One or more indirect test procedures should be done by a recognized laboratory to confirm that SVC virus is present in the cell line. Definite diagnosis of SVCV requires isolation of virus followed by neutralization assay, IFAT, ELISA, and RT-PCR (OIE, 2015).

16.2.2.4 CONTROL AND TREATMENT

SVC-free providers and farms should be used to obtain new fish. Surface water should be disinfected before being supplied to the farm. Maintain a separate set of SVC-susceptible species-specific equipment and disinfect it between ponds or tanks. Birds, livestock, pets, and other animals should not be allowed access to ponds or tanks.

16.2.3 INFECTIOUS HEMATOPOIETIC NECROSIS (IHN)

16.2.3.1 ETIOLOGICAL AGENT

This fish virus belongs to the genus Novirhabdovirus and family Rhabdoviridae and is negative sense single stranded RNA molecule. It is bullet shaped containing a unique non-virion protein gene. IHNV isolates are classified into three genetic categories based on their geographic distribution (Lakshmi et al., 2019). Isolates from Alaska, British Columbia, coastal Washington watersheds, and the Columbia River basin, as well as a few from Oregon, California, and Japan, belong to the U genogroup. Most viruses from California and Oregon's coast belong to the L genogroup. Isolates from Idaho, the Columbia River basin, and Europe, as well as a virus from the Washington coast, make up the M genogroup. The genetic diversity of the M genogroup is much higher than that of the L or U genogroups.

16.2.3.2 CLINICAL SIGNS AND PATHOLOGY

Ascites, dark color, exophthalmia, pale gills are seen in infested specimens. Petechial hemorrhages are seen near fins and mouth. Scoliosis is also seen in surviving specimens and infected specimens swim in a corkscrew manner. IHN may be suspected based on clinical indicators and a history of past outbreaks. Depending upon the viral strain, environmental conditions, and the condition of the host, it may cause 10% to 90% mortality in salmons. The main symptom in some cases is a large increase in mortality in juvenile fish with little clinical indications. The majority of outbreaks take place in the spring and early summer. With a broad viremia and concomitant necrosis in all tissues, this type of the illness leads to a deadly necrosis of the hematopoietic tissues of the kidney and spleen. Renal failure induced by an electrolyte imbalance can result in fish death (Rodriguez et al., 2003).

IHNV is spread by asymptomatic carriers and clinically unwell fish. This virus can be found in feces, urine, sexual fluids, and mucus. The disease is spread mostly from fish to fish, mostly through direct contact, but also through the water. IHNV can persist for at least a month in water, especially if it contains organic material. Although the gills and the digestive tract have been proposed as significant virus entry points, research reveals that IHNV may enter near the base of the fins (Harmache et al., 2006). There is also "egg-associated" (vertical) transmission; whether IHNV can be found inside the egg as well as on the surface is debatable. It is possible that invertebrate vectors exist.

16.2.3.3 DIAGNOSIS

To confirm a diagnosis, procedures such as staphylococcal agglutination, virus neutralization (VN), indirect fluorescent antibody detection, ELISA, RT-PCR, qRT-PCR, and DNA probe technology can be utilized (Dhar et al., 2008; Purcell et al., 2006; Barlic-Maganja et al., 2002). Alternatively, histology can be utilized to diagnose infection by identifying degeneration and necrosis of granular cells in the lamina propria, stratum compactum, and stratum granulosum of the gastrointestinal tract.

16.2.3.4 CONTROL AND TREATMENT

IHNV infection prevention relies on avoiding viral exposure through the establishment of stringent control rules and good hygiene measures. To prevent infection with IHNV at a fish production site, proper cleaning of fertilized eggs, the use of virus-free water supplies for incubation and rearing, and the operation of facilities under established biosecurity procedures are all essential. Disease incidence is reduced by good husbandry practices such as excellent water quality, moderate stocking density, and no batch mixing. To eradicate the disease, highly rigorous protocols regarding mobility, water supplies, and stock replacement must be in place which is still difficult to execute and comes at a hefty expense.

16.2.4 INFECTIOUS PANCREATIC NECROSIS (IPN)

16.2.4.1 ETIOLOGICAL AGENT

IPNV was the first isolated birnavirus in rainbow trout (Wolf et al., 1960). IPN virus, a non-enveloped, bisegmented, double-stranded RNA genome member of the genus Aquabirnavirus and family Birnaviridae, is the causative agent of IPN disease. Serologically, IPNV strains are separated into groups A and B, and further classified based on deduced amino acid (AA) similarities of VP2. Infectious pancreatic necrosis (IPN) is a highly contagious viral infection that primarily affects trout and salmon, although the virus has also been found in a range of other fish species. It is a contagious disease that mostly affects the Salmonidae family (Dopazo, 2020). IPN is a highly infectious viral disease with a significant mortality rate in adult fish and a 100% mortality rate in fry and fingerlings.

16.2.4.2 CLINICAL SIGNS AND PATHOLOGY

The virus enters the fish body through gills, intestine, and skin, then spreads through the bloodstream to the internal organs (Munang'andu et al., 2012).

The head kidney, spleen, and pancreas are predilection locations for primary IPNV replication in internal organs. Dark coloration, distended abdomen, exophthalmia, pale gills, petechial hemorrhages on the ventral surface and weak respiration are common signs. Very young farmed salmonids develop clinically acute or chronic illness conditions. Only a small percentage of older fish are harmed. Variations can be seen in the outbreak's clinical manifestations. Infected specimens show corkscrew manner of rotation and large mortality generally occurs. The loss of exocrine pancreatic acinar tissue is the only persistent HP hallmark of this illness. The presence of "McKnight cells" in the lamina propria of the gut in surviving fish is commonly observed in IPNV infection (McKnight & Roberts, 1976). The necrotic alterations are caused by infected cells rupturing and releasing zymogen granules. Internal organs including the spleen, heart, liver, and kidney can also seem unusually pale with empty stomach and white fecal casts from their anal orifice (Roberts, 2012).

16.2.4.3 DIAGNOSIS

Virus isolation in any of a variety of standard fish cell cultures confirms the diagnosis of infectious pancreatic necrosis in fish with characteristic clinical or microscopic lesions. Because substantial amounts of virus are present in the kidneys of fish with either clinical or subclinical infections, kidney is the tissue of choice for sampling. Plaque assay can be used to titrate virus in fish cells. The standard cell lines for isolation of aquatic birnavirus are FHM, RTG-2, CHSE-214, BF-2 and EPC. For direct detection of viral antigens in internal organs and confirmation of virus identity from cell cultures, immunofluorescence (frozen sections or tissue smears) using virus-specific monoclonal or polyclonal antibodies can be utilized. Neutralization, enzyme-linked immunosorbent assays, and RT-PCR assays can all be used to determine the identity of infectious pancreatic necrosis virus (IPNV). Antiviral antibodies are seen in fish with subclinical infections, although serology is not employed as a diagnostic method very often (Pérez-Prieto, 2003).

16.2.4.4 CONTROL AND TREATMENT

The most effective control measure is avoidance. In order to do so, virus-free eggs must be incubated, and IPN-free stock must be propagated in an uncontaminated water supply and this approach is the preferred method. To avoid the introduction or unintentional spread of IPN, a thorough fish health inspection program is essential. Good sanitation and proper management techniques reduce the disease incidence.

16.3 FUNGAL DISEASES OF COLD-WATER FISH

16.3.1 SAPROLEGNIASIS

16.3.1.1 ETIOLOGICAL AGENT

The disease saprolegniasis is caused by water mold (oomycetes) mostly belonging to the genus *Saprolegnia* (Van West, 2006). This fungus is characterized by having non-septate multinucleate hyphae and branched mycelium can be easily identified on substratum. Morphological, physiological, and molecular analysis of *Saprolegnia* isolates has shown several species misassignments apparent for *S. diclina, S. australis, S. litoralis* or *S. ferax* due to similar oogonial morphology (Diéguez-Uribeondo *et al.*, 2007).

16.3.1.2 CLINICAL SIGNS AND PATHOLOGY

Small, round, depigmented lesions appear first, sometimes with hemorrhagic edges. Lesions can become ulcerative in advanced cases, extending through the skin and into muscular tissue (Willoughby et al., 1983). Whitish brown cotton-like patches appear on the surface of the skin and/or on the gills. Hemorrhage, necrosis, exophthalmia, lethargy are commonly seen in affected specimens. Bloating caused by gut obstruction may progress to perforation of the abdominal wall in small juvenile fishes (Jaies et al., 2020).

Infectious biflagellated zoospores secreted from hyphal sporangia transmit external mold infections through ambient water. *S. parasitica* prefers skin and gill as anatomical predilection sites in rainbow trout and any damage to the structural base of these organs results in loss of body fluids and imbalance in osmoregulation (Jaies et al., 2021). In cultured fish, systemic infections are caused by the intake of uneaten food that has been colonized by mold hyphae. Environmental stress is a significant contributor to the etiology of external disease. Outbreaks are most common after minor injuries from handling or in crowded environments with poor environmental quality. Adult salmon traveling to spawning grounds have compromised immune systems and are frequently infected with Saprolegnia on the outside.

Cold water temperatures also make fish more susceptible to mold sickness because zoospores and sexual stages thrive in cold water, while host tissue healing and the inflammatory response are hampered by the lower host metabolism.

16.3.1.3 DIAGNOSIS

Gross clinical signs of white, cottony tufts of hyphae on the skin, gills, and other surfaces of infected fish or eggs are common diagnostic measures. Branching and non-septate hyphae are seen in the wet mounts of fungal mycelium. Polymerase chain reaction (PCR) is essential for species identification and can be accomplished by using published protocols (Eissa et al., 2013; Touhali et al., 2018).

16.3.1.4 CONTROL AND TREATMENT

The easiest way to avoid Saprolegnia is to keep your water clean and feed your fish a healthy diet. This is the finest way to boost their immune system, which is the best Saprolegnia deterrent. Maintaining a proper maintenance schedule, removing of trash from the gravel, and maintenance of a hospital tank on hand to confine any potentially sick fish are some measures to reduce fish saprolegniasis. The effect of this disease can be further reduced by improving water quality and minimizing stress, especially in the late summer and fall. Saprolegniasis can be treated with saltwater baths (10–25 g/l for 5–30 min SID), benzalkonium chloride (2 mg/l bath for 10–60 min), malachite green (1–3.3 mg/l for 1 hour or 0.15–0.2 mg/l aap prolonged bath), methylene blue (10–20 mg/l for 15 min), formalin, potassium permanganate, and copper sulfate.

16.3.2 BRANCHIOMYCOSIS

16.3.2.1 ETIOLOGICAL AGENT

It is also known as "gill rot" and was first reported in striped bass (Meyer & Robinson, 1973). It is caused by two *Branchiomyces* species, namely, *B. sanguinis* and *B. demigrans.* Carps are the fish that are most afflicted. Branchiomycosis is most commonly found in fish in waters with a high organic content and at temperatures of 20°C or above. This deadly disease can cause up to 50% morbidity in infected fish, and they can die quickly (Harrell, 1997).

16.3.2.2 CLINICAL SIGNS AND PATHOLOGY

Gills appear pale and anemic, while as fungus development can be seen in gill tissue with naked eyes. Fishes become weak and gather in groups at the inlet/outlet of the water body. Infected specimens are seen gulping oxygen at the water surface with wide open mouth. Infected fish have respiratory symptoms as well as a loss of balance. Histologically, hyperplasia, fusion of gill lamellae, and areas of massive necrosis due to thrombosis of vessels by fungal hyphae are apparent. Affected fish may succumb to death within two days after infection, and morbidity can reach to 50% (Devashish, 2016).

16.3.2.3 DIAGNOSIS

Gross examination of clinical signs, wet mount preparation and examination and the molecular identification of pathogen are important for disease diagnosis. The hemorrhages and necrosis result in a patchy marbled look of the gills. A diagnosis can be made using wet mounts or histology of the lesions. Typical hyphae will be evident within the gill vessels or piercing the gill tissue. Light brown, slightly refractile, branching, and non-septate hyphae are routinely observed. To identify the fungal elements, special stains such as the Gomori methenamine silver (GMS) can be used. *B. denigrans* appears to impact the entire gill, whereas *B. sanguinis* appears to harm only the blood vessels of the gill. Strict sanitation, disinfection is essential for controlling the disease.

16.3.2.4 CONTROL AND TREATMENT

For damaged fish, greater water flow and adequate hygiene are recommended which includes removing dead fish, boosting water supply, and avoiding overfeeding, especially at high water temperatures. Branchiomycosis can be prevented using good water quality management practices. However, keeping fish in a malachite green solution for 24 hours has also been reported to be effective. Infected specimens are treated with malachite green @ 0.1 mg/l for extended time or 0.3 mg/l for 12 hours.

16.3.3 EPIZOOTIC ULCERATIVE SYNDROME (EUS)

16.3.3.1 ETIOLOGICAL AGENT

Aphanomyces invadans is the causative agent of EUS. Aphanomyces species have typhoid mycelium that is cylindrical and coenocytic. EUS is also known as red spot disease (RSD), mycotic granulomatosis (MG) and ulcerative mycosis (UM). It was first reported in Japan in goldfish and ayu in the year 1971 (Egusa & Masuda, 1971). Hyphae in infected fish have minimal branching and a diameter ranging from 6 to 27 mm, but they are much narrower in culture (Roberts et al., 1993).

16.3.3.2 CLINICAL SIGNS AND PATHOLOGY

Infested specimens have lesions on their body surface. Lesions are most commonly found on the lateral sides, however they can develop elsewhere on the body. Pinpoint red spots or hemorrhagic spots or localized inflammation is common. Protruding scales, scale loss, skin erosion, reddened skin and ulceration are the common clinical signs (Roberts et al., 1993). Because of the blood loss induced by the hemorrhagic lesions, fish with EUS have a considerable increase in white blood cells (particularly neutrophils) due to a local inflammatory response, while the quantity of red blood cells, hemoglobin concentration, and hematocrit level are all reduced.

16.3.3.3 DIAGNOSIS

Speculative diagnosis is based on sickening appearance and aseptate hyphae in squashed preparations. For confirmatory diagnosis, histological analysis is a must approach. There are no viable serological approaches for detecting and identifying *A. invadans* in EUS specimens.

16.3.3.4 CONTROL AND TREATMENT

Stocking numbers should be kept as low as possible during high-risk periods, such as when EUS prevalence is high in nearby wild fish populations, and farmed populations should be subjected to minimal stress. Similarly, the pond environment might be monitored to ensure that abiotic elements like as low DO concentrations, which can cause skin injury, are kept within safe ranges (Devashish, 2016). Disinfection of fish eggs and larvae against water molds is effective. Disease eradication, good husbandry practices, proper surveillance and biosecurity measures are necessary. Because mycelium and zoospores of *A. invadans* are killed at very low concentrations, treatment with Malachite green and formalin is currently the most effective strategy to prevent EUS.

16.4 CONCLUSION

Disease can be a major threat to cold water aquaculture in terms of economic loss to fish farmers. The focus should always be on prophylaxis rather than therapeutics to prevent the disease outbreak at a fish farm. Ensuring a water source that is free of pathogens, maintenance of proper hygiene at farm, disinfection of nets and equipment at the farm site, vaccination of the stock, procuring of the disease-free seed, proper stocking density, and use of immunostimulants are some of the reliable measure to reduce disease outbreaks in cold water aquaculture.

KEYWORDS

- **cold water aquaculture**
- **cold water fishes**
- **fungal and viral diseases**
- **infectious hematopoietic necrosis**
- **infectious pancreatic necrosis**
- **stocking density**

REFERENCES

Ahne, W., Bjorklund, H. V., Essbauer, S., Fijan, N., Kurath, G., & Winton, J. R., (2002). Spring viremia of carp (SVC). *Diseases of Aquatic Organisms, 52*(3), 261–272.

Ammayappan, A., & Vakharia, V. N., (2009). Molecular characterization of the great lakes viral hemorrhagic septicemia virus (VHSV) isolate from USA. *Virology Journal, 6*(1), 1–16.

Barlič-Maganja, D., Štrancar, M., Hostnik, P., Jenčič, V., & Grom, J., (2002). Comparison of the efficiency and sensitivity of virus isolation and molecular methods for routine diagnosis of infectious haematopoietic necrosis virus and infectious pancreatic necrosis virus. *Journal of Fish Diseases, 25*(2), 73–80.

Carstens, E. B., (2010). Ratification vote on taxonomic proposals to the international committee on taxonomy of viruses (2009). *Archives of Virology, 155*(1), 133–146.

Clark, H. F., & Soriano, E. Z., (1974). Fish rhabdovirus replication in non-piscine cell culture: New system for the study of rhabdovirus-cell interaction in which the virus and cell have different temperature optima. *Infection and Immunity, 10*(1), 180–188.

Devashish, K., (2016). *Epizootic Ulcerative Fish Disease Syndrome* (Vol. 294). Department of life Science Assam (central) University Silchar, India.

Dhar, A. K., Bowers, R. M., Licon, K. S., & Lapatra, S. E., (2008). Detection and quantification of infectious hematopoietic necrosis virus in rainbow trout (*Oncorhynchus mykiss*) by SYBR green real-time reverse transcriptase-polymerase chain reaction. *J. Virol. Methods, 147*, 157–166.

Diéguez-Uribeondo, J., Fregeneda-Grandes, J. M., Cerenius, L., Pérez-Iniesta, E., Aller-Gancedo, J. M., Tellería, M. T., & Martín, M. P., (2007). Re-evaluation of the enigmatic species complex *Saprolegnia diclina–Saprolegnia parasitica* based on morphological, physiological and molecular data. *Fungal Genetics and Biology, 44*(7), 585–601.

Dikkeboom, A. L., Radi, C., Toohey-Kurth, K., Marcquenski, S., Engel, M., Goodwin, A. E., & Longshaw, C., (2004). First report of spring viremia of carp virus (SVCV) in wild common carp in North America. *Journal of Aquatic Animal Health, 16*(4), 169–178.

Dopazo, C. P., (2020). The infectious pancreatic necrosis virus (IPNV) and its virulence determinants: What is known and what should be known. *Pathogens, 9*(2), 94.

Egusa, S., & Masuda, N., (1971). A new fungal of *Plecoglossus altivelis*. *Fish Pathology, 6*(1), 41–46.

Eissa, A. E., Abdelsalam, M., Tharwat, N., & Zaki, M., (2013). Detection of *Saprolegnia parasitica* in eggs of angelfish *Pterophyllum scalare* (Cuvier–Valenciennes) with a history of decreased hatchability. *International Journal of Veterinary Science and Medicine, 1*(1), 7–14.

Fijan, N., Matasin, Z., Jeney, Z., Olah, J. J., & Zwillenberg, L. O., (1984). Isolation of *Rhabdovirus carpio* from sheatfish (*Silurus glanis*) fry. In: *Symposia Biologica Hungarica*. 1984.

Fijan, N., Petrinec, Z., Sulimanovic, D., & Zwillenberg, L. O., (1971). Isolation of the viral causative agent from the acute form of infectious dropsy of carp. *Veterinarski Arhiv, 41*(5, 6), 125–138.

Fijan, N., Sulimanović, D., Bearzotti, M., Muzinić, D., Zwillenberg, L. O., Chilmonczyk, S., & De Kinkelin, P., (1983). Some properties of the epithelioma papulosum cyprini (EPC) cell line from carp *Cyprinus carpio*. In: *Annales de L'institut Pasteur/Virologie* (Vol. 134, No. 2, pp. 207–220). Elsevier Masson.

Garver, K. A., Hawley, L. M., McClure, C. A., Schroeder, T., Aldous, S., Doig, F., & Richard, J., (2011). Development and validation of a reverse transcription quantitative PCR for universal detection of viral hemorrhagic septicemia virus. *Diseases of Aquatic Organisms, 95*(2), 97–112.

Harmache, A., LeBerre, M., Droineau, S., Giovannini, M., & Brémont, M., (2006). Bioluminescence imaging of live infected salmonids reveals that the fin bases are the major portal of entry for novirhabdovirus. *Journal of Virology, 80*(7), 3655–3659.

Harrell, R. M., (1997). *Striped Bass and Other Morone Culture*. Elsevier.

Jaies, I., Shah, F. A., Abubakr, A., Asmi, O., & Bhat, B. A., (2020). PCR based identification of *Saprolegnia parasitica* affecting rainbow trout aquaculture in Kashmir region of Indian Himalayas. *Indian Journal of Animal Research, 54*(9), 1149–1154.

Jaies, I., Shah, F. A., Qadiri, S. S. N., Dar, S. A., Wali, A., & Shah, T. H., (2021). Pathological alterations induced by *Saprolegnia parasitica* in the skin and gills of rainbow trout of Kashmir Himalayas, India: A scanning electron microscopic study. *Aquaculture Research, 53*(5), 1–6.

Jensen, M. H., (1963). Preparation of fish tissue cultures for virus research. *Bull. WCE Intern. Epizooties, 59*, 131–134.

Kelly, A. M., & Renukdas, N. N., (2020). Disease management of aquatic animals. In: *Aquaculture Health Management* (pp. 137–161). Academic Press.

Kibenge, F. S., (2016). Rhabdoviruses of fish. In: *Aquaculture Virology* (pp. 267–297). Academic Press.

Kim, J. O., Kim, W. S., Kim, S. W., Han, H. J., Kim, J. W., Park, M., & Oh, M. J., (2014). Development and application of quantitative detection method for viral hemorrhagic septicemia virus (VHSV) genogroup IVa. *Viruses, 6*(5), 2204–2213.

Lakshmi, B., Syed, S., & Buddolla, V., (2019). Current advances in the protection of viral diseases in aquaculture with special reference to vaccination. In: *Recent Developments in Applied Microbiology and Biochemistry* (pp. 127–146). Academic Press.

McKnight, I. J., & Roberts, R. J., (1976). The pathology of infectious pancreatic necrosis. I. The sequential histopathology of the naturally occurring condition. *British Veterinary Journal, 132*(1), 76–85.

Meyer, F. P., & Robinson, J. A., (1973). Branchiomycosis: A new fungal disease of North American fishes. *The Progressive Fish-Culturist, 35*(2), 74–77.

Munang'andu, H. M., Fredriksen, B. N., Mutoloki, S., Brudeseth, B., Kuo, T. Y., Marjara, I. S., & Evensen, Ø., (2012). Comparison of vaccine efficacy for different antigen delivery systems for infectious pancreatic necrosis virus vaccines in Atlantic salmon (*Salmo salar* L.) in a cohabitation challenge model. *Vaccine, 30*(27), 4007–4016.

Nishizawa, T., Savas, H., Isıdan, H., Üstündağ, C., Iwamoto, H., & Yoshimizu, M., (2006). Genotyping and pathogenicity of viral hemorrhagic septicemia virus from free-living turbot (*Psetta maxima*) in a Turkish coastal area of the Black Sea. *Applied and Environmental Microbiology, 72*(4), 2373–2378.

OIE, (2015). Spring viraemia of carp. In: World Organization for Animal Health, (ed.), *Manual of Diagnostic Tests for Aquatic Animals Diseases* (7th edn., pp. 357–373). OIE, Paris.

OIE, (2019). Viral hemorrhagic septicemia. In: *World Organization for Animal Health, Manual of Diagnostic Tests for Aquatic Animals* (pp. 12–14). Paris, France. Chapter 2.3.10.

Pérez-Prieto, S. I., (2003). Infectious pancreatic necrosis virus: Biology, pathogenesis, and diagnostic methods. *Advances in Virus Research, 62*, 113–135.

Purcell, M. K., Hart, S. A., Kurath, G., & Winton, J. R., (2006). Strand-specific, real-time RT-PCR assays for quantification of genomic and positive-sense RNAs of the fish rhabdovirus, infectious hematopoietic necrosis virus. *J. Virol. Methods, 132*, 18–24.

Qadiri, S. S. N., Kim, S. J., Krishnan, R., Kim, J. O., Kole, S., Kim, W. S., & Oh, M. J., (2019b). Localization and tissue tropism of viral haemorrhagic septicemia virus (VHSV) in experimentally infected juvenile olive flounder, *Paralichthys olivaceus*: An in-situ hybridization and immunohistochemical study. *Aquaculture, 505*, 242–252.

Qadiri, S. S. N., Kim, S. J., Krishnan, R., Kole, S., & Oh, M. J., (2020). Early detection and localization of viral haemorrhagic septicemia virus (VHSV) genomic mRNAs in external tissues of juvenile olive flounder (*Paralichthys olivaceus*) following an immersion challenge. *Aquaculture, 518*, 734859.

Qadiri, S. S. N., Makesh, M., Rajendran, K. V., Rathore, G., & Purushothaman, C. S., (2019a). Specific immune response in mucosal and systemic compartments of *Cirrhinus mrigala* vaccinated against *Edwardsiella tarda*: *In vivo* kinetics using different antigen delivery routes. *Journal of the World Aquaculture Society, 50*(4), 856–865.

Roberts, R. J., (2012). The virology of teleost. In: *Fish pathology* (pp. 246–48). John Wiley & Sons.

Roberts, R. J., Willoughby, L. J., & Chinabut, S., (1993). Mycotic aspects of epizootic ulcerative syndrome (EUS) of Asian fishes. *J. Fish. Dis., 16*, 169–183.

Rodriguez Saint-Jean, S., Borrego, J. J., & Perez-Prieto, S. I., (2003). Infectious pancreatic necrosis virus: Biology, pathogenesis, and diagnostic methods. *Adv. Virus Res., 62*, 113–165.

Rombout, J. H., Yang, G., & Kiron, V., (2014). Adaptive immune responses at mucosal surfaces of teleost fish. *Fish & Shellfish Immunology, 40*(2), 634–643.

Schäperclaus, W., (1938). Die schädigungen der deutschen fischerei durch fischparasiten und fischkrankheiten. *Fischerei-Zeitung, 41*, 267–270.

Snow, M., Bain, N., Black, J., Taupin, V., Cunningham, C. O., King, J. A., et al., (2004). Genetic population structure of marine viral haemorrhagic septicaemia virus (VHSV). *Dis. Aquat. Organ, 61*, 11–21.

Touhali, I. S., & Al-uguali, I. S., (2018). Isolation and identification of *Saprolegnia parasitica* and other fungi from farms fishes in the province of Wasit/Iraq. *Journal of Global Pharma Technology, 10*, 135–142.

Van, W. P., (2006). *Saprolegnia parasitica*, an oomycete pathogen with a fishy appetite: New challenges for an old problem. *Mycologist, 20*(3), 99–104.

Willoughby, L. G., McGrory, C. B., & Pickering, A. D., (1983). Zoospore germination of *Saprolegnia* pathogenic to fish. *Transactions of the British Mycological Society, 80*(3), 421–435.

Wolf, K., Snieszko, S. F., Dunbar, C. E., & Pyle, E., (1960). Virus nature of infectious pancreatic necrosis in trout. *Proceedings of the Society for Experimental Biology and Medicine, 104*(1), 105–108.

CHAPTER 17

Nutrigenomics: Boost for Aquaculture Research and Development

KAWKABUL SABA[1], FAISAL SOFI[2], K. SRAVANI[3], S. VIJAY KUMAR REDDY[4], OYAIS AISMI[1], ASHWINI KUMAR[1], TARIQ HUSSAIN[2], S. DEVADHARSHINI[3], P. GANESAN[3], SHABIR A. DAR[5], and NEERAJ PATHAK[6]

[1]*Division of Fish Nutrition and Biochemistry, Faculty of Fisheries, SKUAST-K, Jammu and Kashmir, India*

[2]*Division of Post-Harvest Technology, Faculty of Fisheries, SKUAST-K, Jammu and Kashmir, India*

[3]*Department of Fish Processing Technology, Fisheries College and Research Institute, TNJFU, Thoothukudi, Tamil Nadu, India*

[4]*Department of Fish Processing Technology, College of Fisheries, Gadvasu, Ludhiana, Punjab, India*

[5]*Division of Fishery Engineering, Faculty of Fisheries, SKUAST-K, Jammu and Kashmir, India*

[6]*Department of Quality Assurance and Management, College of Fisheries, Thoothukudi, Tamil Nadu, India*

ABSTRACT

Nutrigenomics is a broad field of science that focuses on the interaction between genes and the activity of common dietary nutrients, as well as their function in achieving a balance between disease and health by altering the structure or expression of a person's genetic composition. Nutrigenomics is a field that uses postgenomic and related technologies to study an individual

Coldwater Fisheries and Aquaculture Management: Technology for Sustainable Food Production.
Mohd Ashraf Rather, Faisal Rashid Sofi, Adnan Amin, & Kawkabul Saba (Eds.)

organism's reaction to food contents. Nutrigenomics is an area that requires more rigorous investigation and evaluation, and it can be thought of as a new tool for nutrition research in understanding concerns related to animal production and health. Even while more research is being done, public awareness is not being raised to the point where science can be used to its full potential. As a result, efforts must be made to investigate the existing state of knowledge, practices, and abilities linked to the use of nutrigenomics to improve animal health and production. So, the purpose of this chapter is to explore the importance of nutrigenomics, how it works, and what its applications are in the animal sciences field.

17.1 INTRODUCTION

Scientific interventions are playing a major role in fishery because of which aquaculture industry in India is showing enormous growth for the last 40 years with 10–11 folds increase in fish production. It is targeted to produce 10 million metric tons (MT) by the end of 2030 year, for which more research has to be planned for understanding the basic process of growth necessary for improving the aquaculture productivity (Anonymous, 2013). Growth is considered as the conversion of feed into flesh, which is commonly calculated as an increase in biomass production, i.e., the variation between the initial and final biomass of fish, which is nothing but the accumulation of muscle mass, especially the accumulation of protein. Many complex reactions are involved in individuals' bodies for the transformation of protein in the diet to tissue protein and revealing the molecular mechanisms of these complex reactions will help to provide a picture of the growth phenomenon taking place inside the body.

In the nutrition of fish, especially those that are maintained in aquaculture, approximately 31.5 MT of cultured fish are depending on the external feeds that may be farm-made, fresh materials, or commercially manufactured feeds. The aquaculture feed production has increased to 39.9 million tons in the year 2016, when compared to 7.6 million tons in 1995, which is showing a growth of more than three folds, with an average production rate of 12% per annum (All Tech, 2016). The formulated feeds are notably different from naturally available diets, as the nutrients in formulated feeds, particularly oil and protein sources are dominated by land-based ingredients. Nowadays, new feeds, known as functional feeds, are prepared by incorporating nutrients, believing that the nutrients will help in enhancing the health of fish, reducing outbreaks of diseases, or increasing post-infection recovery.

By knowing the new technologies like nutrigenomics, nutritranscriptomics, epigenetic interactions, neuroproteomics, and nutrimetabolomics the level of fish nutrition can be increased which boosts the production and health of farmed fishes (Figure 17.1).

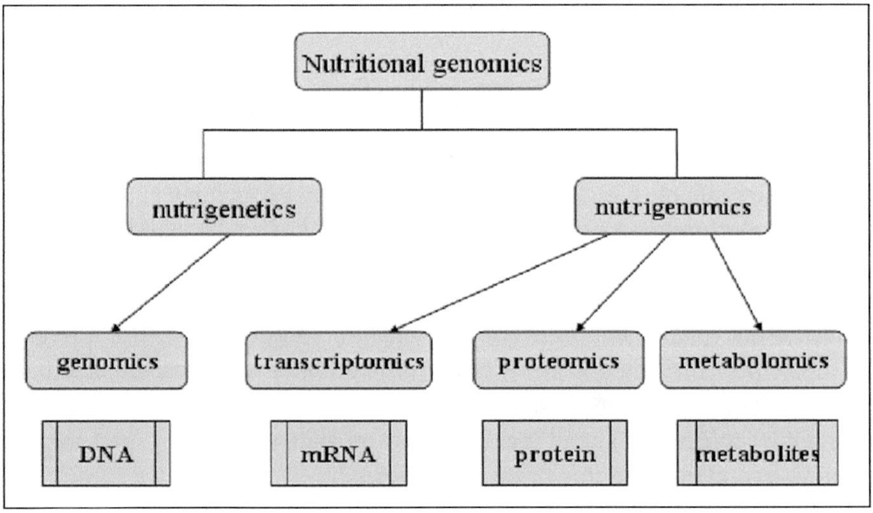

FIGURE 17.1 Nutritional genomics in association with fish nutrition

17.2 NUTRIGENOMICS

Nutrigenomics is a science that deals with the relation between diet, nutrients, and the expression of genes. It also deals with the association between human nutrition, genome, and health. The researcher in the field of nutrigenomics works on the responses of the entire body towards a single nutrient or particular food component as well as a gene. Nutrigenomics or Nutritional genomics deals with the occurrence of the role of natural ingredients or chemicals present in food that will alter the expression of individual genetic information. The part of nutrients on the expression of genes and regulation was highlighted by nutritional studies. There are so many advanced techniques like genomics, metabolomics, and proteomic techniques that provide a facility to know about the constituents of the diet, and the crucial factors involved in gene-nutrient interactions at the cellular level. During gene expression or activation, that particular gene will have some biochemical and physiological function at the cell level as a result of which proteins will be produced. With the advent of nutrigenomics, it is possible to identify that particular gene responsible for producing nutritionally required proteins and cofactors at the site of use and enzymes required for digestion. Thus, genetic interactions include both the effect of genes and the effect of genetics on genetic metabolism.

The field of Nutrigenomics is popularly identified as a tool, for the function of diet which is having an important role in the prevention of cancer, balancing homeostasis, avoiding the risk of chronic disease, or reduction of disease progression. The outlook of nutrition research may currently be unrewarding.

In today's world, nutrition research has focused on health problems and malnutrition. The beginnings of genomics can be described as a collection of advanced technologies for the use, processing, and production of scientific information about the functions and structure of genomes, leading to unique opportunities to successfully understand how genes modify genetic expression and protein, ultimately affecting the environment and cellular metabolism. Nutrigenomics, often known as nutritional genomics, is the study of the relationship between diet, health, and genomics. It's a hybrid of molecular nutrition and genetics, to put it another way. Post-translational modifications (proteome), metabolite profiles (metabolome), chromatin structure (epigenome), and gene-expression patterns are all part of the protein-expression patterns (transcriptome).

17.3 NUTRITRANSCRIPTOMICS

Sometimes, nutrients behave as transcription factors during the interaction with a receptor, which can induce gene expression after binding to DNA. In the recent past, various studies were conducted to demonstrate the interaction of nutrients with genes in cultured fish species. Others have reported the influence of nutrients in genetic expression through written features (Afman & Muller, 2006; Fuks, 2005) – called "nutritional nerves." The nuclear receptor superfamily of transcription factors binds to many receptors, fatty-binding receptors (PPAR) or cholesterol, and metabolites that bind to receptors (X receptors), which bind directly to the nucleotide sequence (reactions elements) to the facilitator of various genes (Afman & Muller, 2006). Nuclear receptors begin to function in the binding of ligands due to coherent flexibility driven by the systematic division of co-activators and the recruitment of co-activator proteins. As a result, these factors act as nutrient sensors by altering the genetic code of certain genes in response to changes in nutrition.

17.4 EPIGENETIC INTERACTIONS

Gene expression is altered due to the alteration in the DNA or histone protein structure by the nutrients. The process of DNA methylation or methylation and acetylation of histones or by both ways mediates the epigenetic effects (Holiday & Grigg, 1993). These mutations result in genetic mutations that last throughout the life of an animal or even for generations. DNA methylation on cytosine bases influences gene registration and genomic stability, and these mutations in genetic modifier regions modify genetic expression (Kersten, 2002).

Methylated 5′ CpG 3′ islands attract capping proteins, which prevent transcription factors from gaining access because of the increased methylation. Once these islands in genes have been methylated, the methylation is replicated each time the gene is copied. The methylation effects are maintained in this manner. When DNA is wrapped around proteins, it inhibits access to gene promoter regions when they are firmly packed. The channels will be produced by the methylation and acetylation of histones, which will allow transcription factors to pass through and activate gene promoters. Some nutrients, including vitamin B6, methionine, vitamin B12, folate, and choline, operate as methyl donors in one-carbon metabolism. Numerous research and publications show that nutrients interact epigenetically with genes to alter biological processes including growth, breeding, and immune response.

17.5 NUTRIPROTEOMICS

Nutriproteomics is the study of molecular and cellular changes in protein structure and expression in response to food intervention using the dynamics of proteomic methods. Proteins play a crucial part in a cell's structure and function. The proteome is the cell's large and constantly changing network of proteins. The proteome changes depending on the cell type, while the genome stays the same. The large amount of proteins derived from the genome is attributable to the following factors:

- Splicing of mRNA;
- Protein processing;
- Post-translation editing.

Dietary supplements and non-nutritious food components act as signals that the written material and nutrients are beginning to emerge. The sensors of nutrition change the gene. However, there is another way the dietary component can affect the proteome: disruption at the level of posttranslational transformation. Nutritional interventions cause two types of changes in the proteome: changes in quantity and function. Various tools for detecting these changes are also being developed. However, there are a few common types of protein separation and purification processes. Mass spectroscopy, peptide digestion, and isotope labeling are some of the methods used in quantitative proteomics. 2D PAGE, mass spectroscopy, and antibody-based methods are all used in active proteomics. The structure and function of proteins should be determined by active proteomics. Proteins do not move to the immune system and are acquired after the release of 2D PAGE in this area. Allows precision specification of protein micro-heterogeneity. Active proteomics can also be illuminated by real-time protein kinetics and protein sequence.

17.6 NUTRIMETABOLOMICS

The study of the distinct chemical fingerprints created by specific biological processes as a function of the animal's nutritional condition or in response to nutritional intervention. Oliver coined the word "metabolome" in 1998 to describe a collection of low-molecular-mass chemicals found in cells. Genetics, environmental variation, and food all have an impact on a cell's or organism's metabolome. Nutrimetabolomics testing is usually done on bodily fluids like blood, urine, or feces, but it can also be done on other tissues. The following approaches are being used in nutrimetabolomic studies:

i. **Metabolite Fingerprinting:** It is used to identify not only the metabolites but the identification of overall nature. It involves the physiochemical characteristic too.
ii. **Fingerprinting:** It is known as the output of the sensors in response to the sample.

Research on different chemical fingerprints created by specific biological processes such as the function of an animal's nutritional status or in response to a nutritious food intervention. Oliver coined the term "metabolome" in 1998 to describe a set of low-cell chemicals found in cells. Genetics, biological diversity, and diet all contribute to the cell or biological metabolome. Nutrimetabolomics tests are usually done on body fluids such as blood, urine, or feces, but can also be done on other tissues. The following methods are used in nutrimetabolomic studies:

1. **Metabolite Fingerprints:** It is used to identify not only metabolites but to identify the whole environment. It includes the physiochemical

aspect as well. Fingerprints are known as nerve output in response to a sample.

2. **Metabolite Profiling:** The analysis identifies the metabolites using chromatographic and spectrophotometric methods.
3. **Metabolome Analysis:** It is defined as a time-dependent resolution-based analysis of metabolite molecules. It entails using a combination of analytical tools to investigate the full range of metabolites present in a sample.
4. **Systems Biology:** It performs partial or complete integration of transcriptomics, proteomics, and metabolomics info.

There are two sorts of tools utilized in nutrimetabolomic investigations:

- Metabolites were separated using the following tools: Thin layer chromatography, gas chromatography, and capillary electrophoresis are some of the techniques used; and
- Metabolites are detected using the following tools: It involves nuclear magnetic resonance and mass spectroscopy.

17.7 GENERAL OVERVIEW OF NUTRIGENOMICS

Conventionally more attention is towards the effect of macro- and micronutrients like proteins, lipids, carbohydrates, vitamins, and minerals. The main objective of nutritional research is mainly based on their requirements, presence in excess or deficiency. Modern-day research is incomprehensible that most people have identical dietary requirements. Mostly every individual will not get benefit from the same dietary regime as they possess different nutritional phenotypes. The human genome is 99.9% identical with some 0.1% variations and this variation is due to their responsibilities towards their diet or food. In the case of having a meal, a few dietary compounds may get metabolized whereas some don't as they get attached to the cellular receptors by acting as a ligand. For an instance, the meal contains some ligand it can able to turn on disease-fighting genes or turn off the genes capable of causing disease in future nutritional research. According to reports, the human genome project began in the 1990s and finished in 2003, with the primary goal of identifying the human genetic code by mapping out every single gene. Nowadays, a new field of research has emerged in relation to the human genome project, which comprises nutrigenetics and nutrigenomics, both of which are concerned with the interaction between human genes and nutrition. Similarly, Nutrigenetics scientifically describes about the result of genetic differences in dietary response, ensuring differences in nutrients metabolism. Although the term Nutrigenomics,' is defined as the technologically how the nutrients adjust the gene expression, it is often measured as a critical excessive-throughput genetic tool inside the field of nutrition studies. Excessive throughput is applied to display nutrients research that enables us to find in what manner nutrients have an effect on the lots of genes expressed in the human genome. This chapter focuses more on Nutrigenomics and the many aspects of nutrigenomics and nutrition research, as well as their overall importance.

Nutrigenomics adheres to the following principles:

i. Nutrition may directly affect gene expression or indirectly induce genetic mutations that alter gene expression or gene silencing in metabolic pathways.
ii. Depending on inherited genetic variants, nutrient outcomes in health can also be affected by the biochemical reaction involving enzymes and their cofactors as well as the metabolism of nutrients.
iii. Health outcomes can be improved by taking into account inherited and acquired genetic characteristics based on age, gender, and health status.

Nutrigenomics explains how certain foods affect the genes in our body. Our diet determines what genetic messages our bodies receive. These messages, in turn, regulate the molecules that make up our metabolism: genes are molecules that control the body to burn or store calories. We can take control of our bodies and digest our food if we learn to understand our genes and control their messages and instructions. By changing the way food interacts with our bodies, we can lose weight and improve our health (Mark Hyman, 2006). Nutrigenomics began with the idea that food affects health. It includes the knowledge that food affects genes, which have long been used in dietary modification. Phenylketonuria (PKU) is one of these, and it is caused by a single genetic mutation. People with this condition should avoid foods that contain the amino acid (AA) phenylalanine. Because of a gene that encapsulates lactase, an enzyme that breaks down lactose and is often eliminated after weaning, most adults in the world suffer from lactose intolerance, which means they are unable to digest milk proteins. About 10,000 to 12,000 years ago, the people of Northern Europe developed a polymorphism from a single nucleotide of DNA. This polymorphism caused the lactase gene to continue maturing. The genetic makeup of cells helped scientists to identify other genes that interact with genes in the late 20th century, making this SNP beneficial to people living in areas with shorter growth periods.

Nutrigenomics was marketed in the 1980s. The science of nutrigenomics was started in the 1990s by a human genetic project, which traced all the DNA bases to the human genome. In 2007, a number of genetic, nutritional, and disease-related interactions were identified. Nutrigenomics has introduced many new ideas, discourses, and experimental processes in healthy food research, such as high-throughput technology that allows global research into genetic expression in a cell or organism. Geneticists, public health professionals, nutritionists, and cooks will need to work together on nutrigenomics. It is not difficult to prepare delicious food. If you add butter or butter to it, it will taste better. We need to find a way to cut fat while making nutritious and delicious dishes. Considering nutrigenomics, as well as the increase in obesity and chronic diseases such as type 2 diabetes, this could be a panacea for the future.

Chronic diseases are largely ignored in the global health agenda, despite rising rates of chronic diseases worldwide. As a result of dietary and lifestyle changes, developing countries are now facing an epidemic of non-communicable diseases (NCDs) linked to global trade, as well as competing for health problems. Globalization has led to something that can be linked to global health issues and must compete with competing health priorities. In 2005, NCDs killed an estimated 35 million people worldwide, mainly cancer, respiratory disease, heart disease, and diabetes. Over the next decade, deaths from NCDs are expected to increase by another 17%. NCDs are expected to deal with 80% of the global disease burden by 2020 and will be responsible for seven out of 10 deaths in developing countries and 60% of global deaths by 2005 (approximately 35 million deaths). As a result, limited healthcare funds, especially in less developed countries, face a (double) burden. Therefore, the promises of nutrigenomics should be considered in view of the increasing epidemic in developed and developing countries (Gobard & Hurlimann, 2009). In India, the death toll from deadly diseases is lower than in Western lands. These rates, however, are rising as a result of urbanization and lifestyle changes (Rao, 2001; Shetty, 2002; Sharma & Majumdar, 2009). The consumption of rice and wheat has changed from rough grains to refined grains in recent decades. In India, there are very high rates of coronary heart disease (CHD), with urban rates three times higher than in rural areas. Obesity and diabetes are also increasing in both remote and high-income sub-regions (Sinha et al., 2003). Many chronic illnesses, such as type 2 diabetes, high blood pressure, cardiovascular disease, cancer, and mental health problems are associated with obesity. Lifestyle changes and eating habits are the major contributors to its development. Diet appears to be associated with higher levels of diabetes, heart disease, and obesity

(Kaput et al., 2007; Hossain et al., 2007), although genetic factors may play a role. Therefore, general awareness of food-related problems that lead to genetic mutations is needed, where nutrigenomics should be studied in more detail.

17.8 BENEFITS OF NUTRIGENOMICS TO AQUACULTURE INDUSTRY

Research on nutrigenomics will benefit aquaculture industry in following modes:

1. To assess the reaction of an animal to nutrients. Nutrigenomics research will help to assess the different reactions between the animals and particular nutrients. For example, some animals will show particular reactions to carbohydrates, whereas other animals will not be affected by that. Nutrigenomics will explore the knowledge about the response between animals and nutrients. This can be explained in a better way as more dietary carbohydrates in mammals generally decrease gluco-neogenic activity whereas the same reaction is not seen in fishes.
2. For enhancing the diet. Nutrigenomics will inform the researcher about the potential diet requirements of a particular animal and their interactions with that particular nutrient, which helps in formulating or enhancing a diet as per the dietary requirement of the animal. Dietary carbohydrates, in the case of herbivorous fish, will maintain the activity and appearance of digestive enzymes of carbohydrates whereas in the case of carnivorous fish, its activity is less (Polakof et al., 2012).
3. Nutrigenomics will provide a better idea regarding the response of nutrients in cell.
4. Nutrigenomics research will help to synchronize the research results for improving husbandry practices (Volkoff et al., 2005).
5. Nutrigenomics studies will help to recognize the aspects that influence energy metabolism. In fish, the increase in lipase gene expression of lipoprotein of adipose tissue will initiate catabolism of fat, and released energy will be utilized for growth. And hence, the amount of protein and lipids in the feed has to be increased during this phase (Mead et al., 2002; Nurnberg et al., 1998; Hocquette et al., 2010; Caballero et al., 2002).
6. It enables to know the organ and tissue specificity for utilizing various nutrients. In the fish, Glucokinase is different from the other

three hexokinase enzymes (Moon, 2001; Hemre et al., 2002). The nutrigenomic studies confirmed that the expression of glucokinase is not inhibited by glucose-6-phosphate but it is able to restrict the expression of hexokinase in other tissues (Rocha et al., 2015). Insulin causes an increase in adipose lipoprotein lipase content but the same reduces the muscle LPL and fat deposition in the tissue (Bouraoui et al., 2011).

7. Specific changes in metabolism. In the adult stage, the amount of beta-oxidation in the red muscle of salmon is 10% but in the smolt stage, it is 60% (Jordl et al., 2006). Nutrigenomics will elucidate the reasons behind this change.
8. Regulation of basic metabolism by dietary factors. Zambonino & Cahu (2010) cite the retro-inhibition of delta-6 desaturase by DHA as an example of metabolic regulation by dietary variables. Nutrigenomics contributes greatly to a greater comprehension of metabolic regulation. Studies of nutrigenomics in the transportation of nutrients.

The development of different transporters in response to diverse forms of the same nutrients improves nutrition utilization. Hexokinases, for example, respond to dietary glucose levels, whereas glucokinase does not (Wilson, 1994). Furthermore, when dietary phosphate is limited, Na-PO_4 co-transporter mRNA expression is up-regulated, and when dietary phosphate is supplemented, it is down-regulated (Reed, 2014). The nutrigenomic technique can be used to confirm the type of phosphate that best up-regulates activity.

KEYWORDS

- **coronary heart disease**
- **co-transporter mRNA**
- **fatty-binding receptors**
- **non-communicable diseases**
- **nutrigenomic technique**
- **nutrigenomics**
- **phenylketonuria**

REFERENCES

Afman, L., & Müller, M., (2006). Nutrigenomics: From molecular nutrition to prevention of disease. *Journal of the American Dietetic Association, 106*(4), 569–576.

All Tech, (2017). Alltech Global Feed Survey http://www.alltech.com/news/newsarticles/2016/04/08/alltech-releases-first-globalaquaculture-feed-production-survey (accessed on 3 February 2023).

Anonymous, (2013). *Fish to 2030, Prospects for Fisheries and Aquaculture, FAO*, World Bank Report number 83177-GLB.

Bouraoui, L., Sánchez-Gurmaches, J., Cruz-Garcia, L., Gutiérrez, J., Benedito-Palos, L., & Pérez-Sánchez, J., (2011). Effect of dietary fish meal and fish oil replacement on lipogenic and lipoprotein lipase activities and plasma insulin in gilthead sea bream (*Sparus aurata*). *Aquaculture Nutrition, 17*(1), 54–63.

Caballero, M. J., Obach, A., Rosenlund, G., Montero, D., Gisvold, M., & Izquierdo, M. S., (2002). Impact of different dietary lipid sources on growth, lipid digestibility, tissue fatty acid composition and histology of rainbow trout, *Oncorhynchus mykiss*. *Aquaculture, 214*(1–4), 253–271.

Fuks, F., (2005). DNA methylation and histone modifications: Teaming up to silence genes. *Current Opinion in Genetics & Development, 15*(5), 490–495.

Gobard, B., & Hurlimann, T., (2009). Nutrigenomics for global health: Ethical challenges for underserved populations. *Current Pharmacogenomics and Personalized Medicine, 7*(3), 205–214.

Hemre, G. I., Mommsen, T. P., & Krogdahl, Å., (2002). Carbohydrates in fish nutrition: Effects on growth, glucose metabolism and hepatic enzymes. *Aquaculture Nutrition, 8*(3), 175–194.

Hocquette, J. F., Gondret, F., Baéza, E., Médale, F., Jurie, C., & Pethick, D. W., (2010). Intramuscular fat content in meat-producing animals: Development, genetic and nutritional control, and identification of putative markers. *Animal, 4*(2), 303–319.

Holliday, R., & Grigg, G. W., (1993). DNA methylation and mutation. *Mutation Research, 285*, 61–67.

Hossain, P., Kawar, B., & Nahas, M. E., (2007). Obesity and diabetics in the developing world—A growing challenge. *New England Journal of Medicine, 356*, 213–215.

Jordal, A. E., Hordvik, I., Pelsers, M., Bernlohr, D. A., & Torstensen, B. E., (2006). FABP3 and FABP10 in Atlantic salmon (*Salmo salar* L.)-general effects of dietary fatty acid composition and life cycle variations. *Comparative Biochemistry and Physiology Part B: Biochemistry and Molecular Biology, 145*(2), 147–158.

Kaput, J., Noble, J., Hatipoglu, B., & Kohrs, K., (2007). Application of nutrigenomic concepts to type 2 diabetes mellitus. *Nutrition, Metabolism and Cardiovascular Diseases, 17*, 89–103.

Kersten, S., (2002). Effects of fatty acids on gene expression: Role of peroxisome proliferator-activated receptor α, liver X receptor α and sterol regulatory element-binding protein. *Proceedings of the Nutrition Society, 61*(3), 371–374.

Mark, H., (2006). *Book on Ultra-Metabolism: The Simple Plan for Automatic Weight Loss* (p. 24). Atria Books, New York.

Mead, J. R., Irvine, S. A., & Ramji, D. P., (2002). Lipoprotein lipase: Structure, function, regulation, and role in disease. *Journal of Molecular Medicine, 80*(12), 753–769.

Moon, T. W., (2001). Glucose intolerance in teleost fish: Fact or fiction? Comparative *Biochemistry and Physiology Part B: Biochemistry and Molecular Biology, 129*(2, 3), 243–249.

Nurnberg, K., Wegner, J., & Ender, K., (1998). Factors influencing fat composition in muscle and adipose tissue of farm animals. *Livestock Production Science, 56*(2), 145–156.

Polakof, S., Panserat, S., Soengas, J. L., & Moon, T. W., (2012). Glucose metabolism in fish: A review. *Journal of Comparative Physiology, 182*(8), 1015–1045.

Rao, S., (2001). Nutritional status of Indian population. *Journal of Bioscience, 26*(4), 481–489.

Reed, M., (2014). *Transcriptional Adaptations to Low Salinity in Euryhaline Crustaceans (Doctoral dissertation).* Degree of Master of Science. Graduate Faculty of Auburn University.

Rocha, F., Dias, J., Engrola, S., Gavaia, P., Geurden, I., & Dinis, M. T., (2015). Glucose metabolism and gene expression in juvenile zebrafish (*Danio rerio*) challenged with a high carbohydrate diet: Effects of an acute glucose stimulus during late embryonic life. *British Journal of Nutrition, 113*(3), 403–413.

Sharma, M., & Majumdar, P. K., (2009). *Occupational Lifestyle Disease: An Emerging Issue, 3*(13), 109–112.

Shetty, P. S., (2002). Nutrition transition in India. *Public Health Nutrition, 5*(1A), 175–182.

Sinha, R., Anderson, D. E., & McDonald, S. S., (2003). Cancer risk and diet in India. *Journal of Postgraduate Medicine, 49*, 222–228.

Volkoff, H., Canosa, L. F., Unniappan, S., Cerda-Reverter, J. M., Bernier, N. J., & Kelly, S. P., (2005). Neuropeptides and the control of food intake in fish. *General and Comparative Endocrinology, 142*(1, 2), 3–19.

Wilson, R. P., (1994). Utilization of dietary carbohydrate by fish. *Aquaculture, 124*(1–4), 67–80.

Zambonino, J. L., & Cahu, C., (2010). Effect of nutrition on marine fish development and quality. In: *Recent Advances in Aquaculture Research* (pp. 103–124).

CHAPTER 18

Immune Components and Defense Mechanism in Fish: An Overview

SALIK NAZKI[1], MUNAZAH SHAHZAD[2], INAIN JAIES[3], SYED SHARIQ N. QADIRI3, and FEROZ A. SHAH[3]

[1]*The Pirbright Institute, United Kingdom*

[2]*Department of Veterinary Epidemiology and Public Health, University of Surrey, United Kingdom*

[3]*Division of Aquatic Animal Health Management, Faculty of Fisheries, Sher-e-Kashmir University of Agricultural Sciences and Technology (SKUAST) of Kashmir, Rangil, Ganderbal, Jammu and Kashmir, India*

ABSTRACT

During microbial invasion, an effective clearance and degradation system may be required to prevent severe manifestations and, eventually, mortality. Surface barriers, lysozyme, complement, and lectins are non-specific defense components that operate as the first line of defense and are critical at all stages of infection. During microbial invasion, non-specific defense systems may be potentiated, resulting in more efficient clearance and elimination of pathogens or other hazardous chemicals. Humoral immunity, cell-mediated immunity, and memory are three crucial characteristics of a specific defense system. Each component has its own inherent protective value, and the final combination is likely to be linked to the immunological history and evolutionary development of the fish. The present chapter will present a brief overview of innate and adaptive components of fish defense system.

Coldwater Fisheries and Aquaculture Management: Technology for Sustainable Food Production.
Mohd Ashraf Rather, Faisal Rashid Sofi, Adnan Amin, & Kawkabul Saba (Eds.)

18.1 INTRODUCTION

Network of cells, tissues, and organs make up the immune system. The innate and adaptive immune system are two branches of the fish defense system. The innate immune system is the first line of defense against infections, and it has no memory of prior reactions. If a pathogen overcomes the innate immune response, the adaptive immune system will target it with precision and memory. Immunoglobulins (Igs) and T-cell receptors (TCR) in conjunction with major histocompatibility complex (MHC) are components of the adaptive immune system. The adaptive immune system is highly specific to a specific antigen and can provide long-term immunity (Alberts et al., 2002). The lymphoid organs in the vertebrate immune system are classified as main or secondary based on their ontogeny and functional properties. As the majority of the infectious agents affect or initiates the process of infection in the mucous surfaces, the mucosal immune response plays a crucial role in the course of the infection (McNeilly et al., 2008). Despite certain differences, the immune systems of fish and higher vertebrates are biologically comparable. Fish is in close contact with water which can contain high quantities of bacteria, viruses, and other pathogens. During this phase, they are primarily reliant on a non-specific immune system to survive (Sahoo et al., 2021). While the non-specific immunity is the more primitive defense mechanism, acquired immunity is also important for maintaining homeostasis. The various innate and adaptive immune components in fish are presented in Figure 18.1.

18.2 NON-SPECIFIC DEFENSE MECHANISM IN FISH

18.2.1 SURFACE BARRIERS

The first line of defense against infection in fish includes skin mucus, gills, epidermis, and gastrointestinal tract. The constant sloughing of skin mucous prevents microbial invasion. In addition, fish mucus is constantly secreted and renewed, preventing the steady colonization of potentially pathogenic microbes as well as metazoan parasite invasion (Ángeles, 2012). Hyperplasia of Malpighian cells is the first response shown by non-keratinized epidermis against pathogens. Mucus production and a highly proliferative epithelium protect the gills.

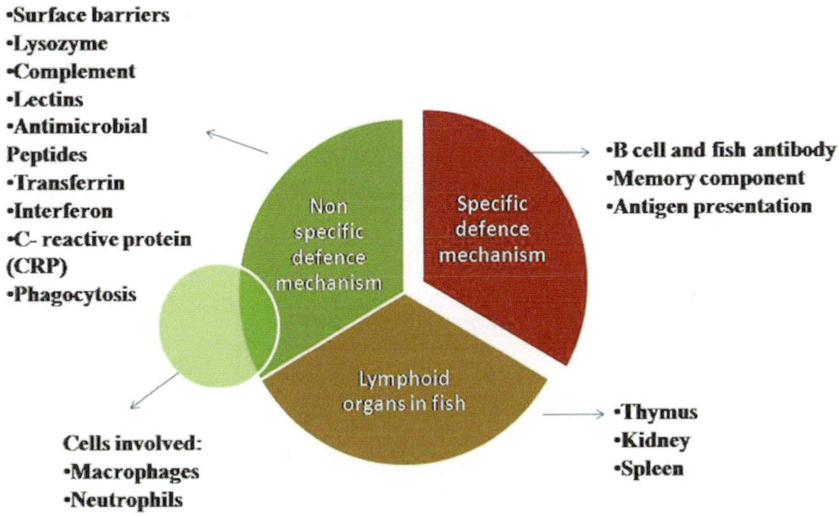

FIGURE 18.1 Different components of fish defense system.

18.2.2 LYSOZYME

Lysozyme is an enzyme that can lyse bacterial cell membranes, making it an effective antibacterial agent. The level or activity of lysozyme is a significant indicator of innate immunity in fish, and it is found in all living species. The lytic activity of fish lysozyme against Gram-positive and Gram-negative bacteria has been thoroughly characterized. Most fish species have lysozyme in their mucous, lymphoid tissue, and serum (Nigam et al., 2012). Lysozyme bacteriolytic action in fish skin mucus and other tissues aids in the host defense against bacterial infection (Subramanian et al., 2007).

18.2.3 COMPLEMENT

Complement is a group of roughly 35 proteins that make up an important part of the innate immune system. Fish complement proteins have many similarities to their mammalian counterparts, and their activation pathways appear to be similar to those found in mammals. Fish complement can lyse foreign cells and opsonize alien organisms for phagocytic destruction. Complement fragments may possibly play a role in inflammatory reactions, according to some evidence. Several complement proteins, such as C3 and factor B, have numerous isoforms in fish (Holland & Lambris, 2002).

18.2.4 LECTINS

Non-immune carbohydrate binding proteins are known as lectins. Lectins are components of the innate immune system that have an affinity for carbohydrate moieties and can cause cell agglutination and/or glycol-conjugate precipitation. Lectins bind to pathogenic surface features, causing opsonization, increased phagocytic activity or complement pathway activation (Matsushita et al., 2004; Ottinger et al., 1999).

18.2.5 ANTIMICROBIAL PEPTIDES (AMPS)

Antimicrobial peptides (AMPs) are becoming more well recognized as an important component of the host defense. They are effective against a wide range of pathogenic organisms, including Gram-positive and Gram-negative bacteria, yeast, fungus, enveloped viruses, and parasites, with little or no damage to host cells (Ángeles, 2012). Fish have a lot of alpha-helical amphipathic peptides (Smith & Fernandes, 2009). The piscidin family, which comprises the pleurocidins and piscidins, contains the majority of fish-helical peptides. Pardaxin, parasin 1, hipposin, oncorhyncin III, oncorhyncin II, SAMP-H1 are a few examples of AMPs found in fish epidermal mucus (Smith & Fernandes, 2009; Fernandes et al., 2003; Thompson et al., 1986).

18.2.6 TRANSFERRIN

Transferrin is made up of a single polypeptide chain with roughly 700 amino acid (AA) residues that is divided into two lobes, each with two domains where iron is coordinated (Reyes-López et al., 2015). Iron is required for the formation of infection by many pathogens, but due to its strong affinity for the blood protein transferrin, iron availability in vertebrate tissue fluids is exceedingly low. Importance of transferrin in the immune system stems from its ability to control serum free-iron levels, resulting in low-iron conditions wherein the potential of pathogenic bacteria to cause infection is limited (Chen et al., 1999).

18.2.7 INTERFERON

Interferons (IFNs) are proinflammatory, antiviral, and immunomodulatory cytokines that play a key role in the host response to infection (Martin et al., 2007). Two of the three IFN families found in higher vertebrates are now known to play a role in teleost fish antiviral defense (Zou & Secombes, 2011). Once IFN binds to the receptor, signal transduction activates IFN-responsive genes through conserved signal transduction pathways, causing the cell to change function (Schroder et al., 2004).

18.2.8 C-REACTIVE PROTEIN (CRP)

C-reactive protein (CRP) was discovered in the serum of a marine teleost fish, *Pleuronectes platessa* (White & Fletcher, 1985), and has since been isolated and described primarily from teleost fish. It binds to the phosphoryl choline found on the cell walls of microorganisms. Because its concentration rises in reaction to harmful substances and bacterial infections, CRP in trout is classified as an acute phase protein (Kodama et al., 2004). Salmon CRP is a non-acute phase protein of 208 AAs, and only one of five CRP types is activated by cytokines (Pathak & Agrawal, 2019). When tissue is injured, traumatized, or infected, the levels of these proteins rise. These proteins are involved in the immune system (Cook et al., 2003), the classical complement pathway (De Haas et al., 2000), and in the clearance of apoptotic cells (Nauta et al., 2003).

18.2.9 PHAGOCYTOSIS

In fish, neutrophils and macrophages are the major cells involved in phagocytosis (Secombes & Fletcher, 1992). Because it is the process that is least affected by temperature, phagocytosis is one of the most critical processes in poikilothermic species (Magnadottir et al., 2005). During a respiratory burst, these cells primarily eliminate microorganisms by producing reactive oxygen species (ROS).

18.3 CELLS INVOLVED IN NON-SPECIFIC DEFENSE MECHANISM

18.3.1 MACROPHAGES

The best-studied macrophage phenotype in teleosts is one that is similar to the M1 activation state, which plays an important role in host defense. M1 macrophages produce a wide range of cytokines, chemokines, and lipid mediators that help to enhance and fine-tune inflammatory and adaptive immune responses. These cells can kill pathogens quickly by engulfing them and producing toxic reactive intermediates, phagolysosomal acidification, and nutrition limitation (Neumann et al., 2000; Rieger et al., 2010; Grayfer et al., 2014). The synthesis of multicomponent enzyme nicotinamide adenine dinucleotide phosphate (NADPH) subunits at the plasma membrane during the macrophage respiratory burst results in the transfer of electrons from NADPH to molecular oxygen, culminating in the superoxide anion (Briggs et al., 1975). Superoxide is quickly converted to ROS such as hydrogen peroxide (H_2O_2), hydroxyl radical (OH), and hyperchlorous acid (HOCl) after it is formed (El-Benna et al., 2008). In comparison to those described in mammals, the inducible nitric oxide (NO) system of teleost macrophages is largely preserved. The activation of inducible nitric oxide synthase (iNOS/NOS_2), which catalyzes the conversion of l-arginine to l-citruline, culminating in the generation of NO, a strong antibacterial molecule, distinguishes classically activated macrophages (Nathan & Xie, 1994).

18.3.2 NEUTROPHILS

Teleost neutrophils are terminally differentiated leukocytes that defend the host and mount rapid and effective antimicrobial responses against infections. They are usually the first leukocytes to arrive at a location of inflammation (Havixbeck et al., 2015). Using toxic intracellular granules (Flerova & Balabanova, 2013), the generation of ROS (Katzenback & Belosevic, 2009), and the deployment of extracellular traps (Pijanowski et al., 2013), neutrophils become powerful killers. NO is required for neural transmission, cell growth inhibition, vasodilation, and intracellular signaling (Napoli et al., 2013). It also has strong toxic properties and is an important component of antimicrobial defenses. NO has been proven to have powerful antibacterial properties against a variety of diseases found in fish (Wiegertjes & Forlenza, 2010). *Trypanoplasma borreli*, a parasite hemoflagellate, has similarly been demonstrated to promote NO generation in carp head kidney neutrophils (Scharsack et al., 2003).

18.4 SPECIFIC DEFENSE MECHANISM IN FISH

18.4.1 B-CELL AND ANTIBODY

B1-B cells, which are implicated in innate activities, respond to stimulation *in-vitro* via toll-like receptors (TLR) (Rawlings et al., 2012), causing proliferation and differentiation of B1-B cells into Ig-secreting cells. After innate activation, B1-B cells also exhibit a quick capacity to release large levels of the immunomodulatory cytokine interleukin-10. Some fish species exhibit interesting B cell characteristics, such as the lack of pathogen-specific IgM in gadoids following successful pathogen immunization (Magnadottir et al., 2009), and the absence of the entire IgM gene in a coelacanth species (Amemiya et al., 2013).

Plasma blasts and plasma cells produce secretory Igs, which are important for maintaining mucosal homeostasis. Until recently, it was thought that teleost fish B cells only expressed two types of Igs, IgM, and IgD (Hordvik et al., 1999), with IgM being the only one that responded to pathogens in both the systemic and mucosal compartments. However, a third teleost Ig class, IgT/IgZ, was recently found to be the most common Ig in gut mucosal immune responses (Salinas et al., 2011). IgM, IgD, and IgT are the three Ig isotypes produced by teleost B cells. Antibodies found in teleost cutaneous mucus and skin play an important part in the host preventive defense against surface infections (Dickerson et al., 1998). The primary isotype in teleost fishes is IgM, which has one variable and four constant domains and is detected in plasma, bile, and skin mucus (Flajnik et al., 2003). The most common Ig found in teleosts is an IgM tetramer with eight antigenic combining sites. Salmon and Atlantic cod have different amounts of IgM depending on their size, temperature, and water quality season (Uribe et al., 2011). Ig has two identical heavy (H) chains and two identical light (L) chains as its basic structure (Mashoof & Criscitiello, 2016). The constant section of both heavy and light chains consists of one N-terminal variable domain (VH or VL) and one or more C-terminal constant domains (CH or CL). Complementarity-determining regions are three hyper variable regions found within each VL and VH domain. The antigen binding site of the Ig molecule is made up mostly of the three CDR segments of the VL domain and the three CDR segments of the VH domain (Abbas et al., 2010). Typical structure of a monomeric antibody is shown in Figure 18.2.

18.4.2 MEMORY COMPONENT

Before a second antigen encounter, fish acquire a memory response (Whittington et al., 1994). The memory mechanism is based on the generation of long-lived T and B memory cells, as well as alterations in the nature of reacting cells, resulting in the selection of clones with high antigen affinity. Both humoral and cell-mediated immunity are present in fish. Memory formation in carp appears to take longer than in mammals, roughly 3–6 months and lasts for 8–12 months after initial stimulation. Increased antigen sensitivity is another sign of memory (Lamers et al., 1985). For example, trout that had been primed with an ideal dose of antigen and then boosted with a sub immunogenic dosage had a larger antibody titer than those that had been primed with the optimal dose. T-dependent antigens require two exposures before the fish responds to the second dose, but T-independent antigens just require one exposure (Uribe et al., 2011).

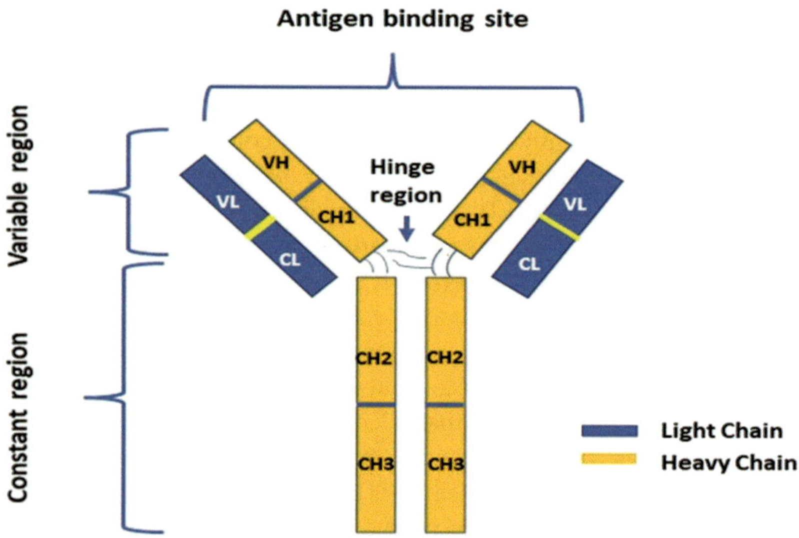

FIGURE 18.2 Structure of a monomeric antibody.

18.4.3 ANTIGEN PRESENTATION

Professional antigen presenting cells (APCs) play an important role in pathogen identification by PRRs and adaptive immunity activation by antigen presentation to naïve T cells. Dendritic cells (DCs), monocyte/macrophages, and B cells ingest pathogens and convert antigen to peptides as professional APCs. MHC class I molecules present intracellular antigens to cytotoxic T cells (CD8+ T-cells), while MHC class I molecules present extracellular antigens to helper T cells (CD4+ T-cells) (Keech et al., 2010). DCs are the most common and strong APCs in vertebrates, and they play an important role in bridging and modifying both innate and adaptive immune responses (Mildner & Jung, 2014). Macrophages are massive mononuclear cells found in nearly every animal tissue. Macrophages are phagocytes with germline-encoded PRRs that recognize pathogen-associated molecular patterns (PAMPs) like lipopolysaccharide, peptidoglycan, and LTA from Gram-positive bacteria's cell wall (Silva & Correia-Neves, 2012). Teleost B cells, like B-1 cells in mammals, can swallow particles and eliminate ingested microorganisms intracellularly (Sunyer, 2013). B lymphocytes in zebrafish, for example, can phagocytize both soluble and particulate antigens (Zhu et al., 2014).

18.5 LYMPHOID ORGANS IN FISH

18.5.1 THYMUS

The thymus is found near the gill cavity in most teleosts and is present even in adult fish, however its volume decreases with age or sexual development. The thymus is the first lymphoid organ to grow and become lymphoid in many teleost species throughout development. The thymus can be thought of as a collection of macrophages that stimulate T cell encapsulation and proliferation (Davis et al., 2002). The thymus is thought to play a crucial role in the immunological response of higher vertebrates in fish. Fish have been shown to exhibit immunological reactions that are dependent on thymus-derived cells. T-like cells are the cells that have a role in these tasks (Chilmonczyk, 1992). In teleosts, the differentiation of the thymic structure is exceedingly diverse, and in many species, there is no apparent distinction between the cortex and medulla as seen in higher vertebrates (Bowden et al., 2005).

18.5.2 KIDNEY

The teleost kidney is divided into two sections: the head kidney, which makes up about 20% of the anterior region of the kidney, and the trunk kidney, which makes up the remaining 80% and runs down the dorsal wall. The head kidney is mostly made up of steroidogenic interrenal cells, chromaffin cells, and hematopoietic tissue, whereas the trunk kidney is primarily made up of filtering nephrons, as well as hematopoietic and pigmented cells. The head kidney is a secondary lymphoid organ rich in monocytes and macrophages that are involved in phagocytosis of invading pathogens and secretion of cytokines such as the pro-inflammatory cytokines interleukin-1 and tumor necrosis factor, thus playing a key role in the initiation of the immune response in fish (Soulliere & Dixon, 2017).

18.5.3 SPLEEN

The spleen is the fundamental secondary lymphoid organ, and it is found in almost all gnathostomes, where adaptive immune responses are formed (Flajnik, 2018). Fish have melano-macrophages in their spleens, which can be grouped in clusters or more loosely diffused inside the white pulp, depending on the species. Several studies on antigen retention have suggested that such areas function as "dumping grounds" for various types of waste material (Agius & Roberts, 2003).

18.6 CONCLUSION

Fish make up a large part of the aquatic biodiversity. The variation in immune system adaptability in different fish species may have resulted from evolutionary pressure from pathogens. Immunological parameters have equivocal functions at the individual, species, and population levels, and the interactions between immune parameters are poorly understood. Beyond the field of immune evolution, a better understanding of fish immune components could be beneficial in fish farming and possibly human health. Innate protective systems are the first line of defense against disease, and they provide adequate disease protection in healthy populations of fish. Moribund and dead fish emit significant amounts of germs into the water, putting the remainder of the population in danger. As a result, the unique defense mechanism counteracts the increased infection pressure.

KEYWORDS

- **adaptive immunity**
- **antibody**
- **antimicrobial peptides**
- **defense system**
- **immune organs**
- **immunoglobulins**
- **innate immunity**
- **interferons**

REFERENCES

Abbas, A. K., Lichtman, A. H., & Pillai, S., (2010). *Cellular and Molecular Immunology* (6th edn., p. 566). Saunders/Elsevier; Philadelphia, PA, USA.

Agius, C., & Roberts, R., (2003). Melano-macrophage centers and their role in fish pathology. *J. Fish Dis., 26*(9), 499–509.

Alberts, B., Johnson, A., & Lewis, J., (2002). Chapter 24: The adaptive immune system. In: *Molecular Biology of the Cell* (4th edn.). New York, NY: Garland Science.

Amemiya, C. T., Alföldi, J., Lee, A. P., Fan, S., Philippe, H., Scapigliati, G., et al., (2013). Analysis of the African coelacanth genome sheds light on tetrapod evolution. *Nature, 496*, 311–316.

Ángeles, E. M., (2012). An overview of the immunological defenses in fish skin. *International Scholarly Research Notices*.

Ángeles, E. M., (2012). An overview of the immunological defenses in fish skin. *International Scholarly Research Notices, 2012*.

Bowden, T. J., Cook, P., & Rombout, J. H. W. M., (2005). Development and function of the thymus in teleosts. *Fish and Shellfish Immunology, 19*, 413–427.

Briggs, R. T., Drath, D. B., Karnovsky, M. L., & Karnovsky, M. J., (1975). Localization of NADH oxidase on the surface of human polymorphonuclear leukocytes by a new cytochemical method. *J. Cell Biol., 67*, 566–586.

Chen, D., McMichael, J. C., VanDerMeid, K. R., Masi, A. W., Bortell, E., Caplan, J. D., et al., (1992). Evaluation of a 74-kDa transferrin-binding protein from *Moraxella* (Branhamella) catarrhalis as a vaccine candidate. *Vaccine, 18*, 109–118.

Chilmonczyk, S., (1992). The thymus in fish: Development and possible function in the immune response. *Annual Review of Fish Diseases, 2*, 181–200.

Cook, M. T., Hayball, P. J., Birdseye, L., Bagley, C., Nowak, B. F., & Hayball, J. D., (2003). Isolation and partial characterization of a pentraxin-like protein with complement-fixing activity from snapper (*Pagrus auratus, Sparidae*) serum. *Developmental and Comparative Immunology, 27*, 579–588.

Davis, J. M., Clay, H., Lewis, J. L., Ghori, N., Herbomel, P., & Ramakrishnan, L., (2002). Real-time visualization of mycobacterium-macrophage interactions leading initiation of granuloma formation in zebrafish embryos. *Immunity, 17*, 693–702.

De Haas, C. J., Van, L. E. M. M., Van, B. T., Verhoef, J., Van, K. K. P. M., & Van, S. J. A. G., (2000). Serum amyloid P component bound to gram-negative bacteria prevents lipopolysaccharide-mediated classical pathway complement activation. *Infection and Immunity, 68*, 1753–1759.

Dickerson, H., & Clark, T., (1998). *Ichthyophthirius multifiliis*: A model of cutaneous infection and immunity in fishes. *Immunological Reviews, 166*, 377–384.

El-Benna, J., Dang, P. M. C., & Gougerot-Pocidalo, M. A., (2008). Priming of the neutrophil NADPH oxidase activation: Role of p47phox phosphorylation and NOX2 mobilization to the plasma membrane. *Semin. Immunopathol., 30*, 279–289.

Fernandes, J. M. O., Saint, N., Kemp, G. D., & Smith, V. J., (2003). Oncorhyncin III: A potent antimicrobial peptide derived from the non-histone chromosomal protein H6 of rainbow trout, *Oncorhynchus mykiss*. *Biochemical Journal, 373*(2), 621–628.

Flajnik, M. F., (2018). A cold-blooded view of adaptive immunity. *Nature Rev. Immunol., 18*(7), 438–453.

Flajnik, M. F., Miller, K., & Du Pasquier, L., (2003). Evolution of the immune system. In: Paul, W. E., (ed.), *Fundamental Immunology* (5th edn., pp. 519–570). Lippincott Williams & Wilkins, Philadelphia, Pa, USA.

Flerova, E. A., & Balabanova, L. V., (2013). Ultrastructure of granulocytes of teleost fish (*Salmoniformes*, *Cypriniformes*, *Perciformes*). *J. Evol. Biochem. Physiol., 49*, 223–233.

Grayfer, L., Hodgkinson, J. W., & Belosevic, M., (2014). Antimicrobial responses of teleost phagocytes and innate immune evasion strategies of intracellular bacteria. *Dev. Comp. Immunol., 43*, 223–242.

Havixbeck, J. J., Rieger, A. M., Wong, M. E., Hodgkinson, J. W., & Barreda, D. R., (2015). Neutrophil contributions to the induction and regulation of the acute inflammatory response in teleost fish. *Journal of Leukocyte Biology, 99*(2), 241–252.

Holland, M. C. H., & Lambris, J. D., (2002). The complement system in teleosts. *Fish & Shellfish Immunology, 12*(5), 399–420.

Hordvik, I., Thevarajan, J., Samdal, I., Bastani, N., & Krossøy, B., (1999). Molecular cloning and phylogenetic analysis of the Atlantic salmon immunoglobulin D gene. *Scandinavian Journal of Immunology, 50*(2), 202–210.

Katzenback, B. A., & Belosevic, M., (2009). Isolation and functional characterization of neutrophil-like cells, from goldfish (*Carassius auratus* L.) kidney. *Dev. Comp. Immunol., 33*, 601–611.

Keech, C. L., Pang, K. C., McCluskey, J., & Chen, W., (2010). Direct antigen presentation by DC shapes the functional CD8+ T-cell repertoire against the nuclear self-antigen La-SSB. *European Journal of Immunology, 40*(2), 330–338.

Kodama, H., Matsuoka, Y., Tanaka, Y., Liu, Y., Iwasaki, T., & Watarai, S., (2004). Changes of C-reactive protein levels in rainbow trout (*Oncorhynchus mykiss*) sera after exposure to anti-ectoparasitic chemicals used in aquaculture. *Fish Shellfish Immunol., 16*(5), 589–597.

Lamers, C. H. J., De Haas, M. J. H., & Van, M. W. B., (1985). The reaction of the immune system of fish to vaccination: Development of immunological memory in carp, *Cyprinus carpio* L., following direct immersion in *Aeromonas hydrophila* bacterin. *Journal of Fish Diseases, 8*(3), 253–262.

Magnadottir, B., Gudmundsdottir, S., Gudmundsdottir, B. K., & Helgason, S., (2009). Natural antibodies of cod (*Gadus morhua* L.): Specificity, activity and affinity. *Comp. Biochem. Physiol. B Biochem. Mol. Biol., 154*, 309–316.

Magnadottir, B., Lange, S., Gudmundsdottir, S., Bogwald, J., & Dalmo, R. A., (2005). Ontogeny of humoral immune parameters in fish. *Fish and Shellfish Immunology, 19*, 429–439.

Martin, S. A., Taggart, J. B., Seear, P., Bron, J. E., Talbot, R., Teale, A. J., & Secombes, C. J., (2007). Interferon type I & type II responses in an Atlantic salmon (*Salmo salar*) SHK-1 cell line by the salmon TRAITS/SGP microarray. *Physiological Genomics, 32*(1), 33–44.

Mashoof, S., & Criscitiello, M. F., (2016). Fish immunoglobulins. *Biology, 5*(4), 45.

Matsushita, M., Matsushita, A., Endo, Y., et al., (2004). Origin of the classical complement pathway: Lamprey orthologue of mammalian C1q acts as a lectin. *Proceedings of the National Academy of Sciences of the United States of America, 101*(27), 10127–10131.

McNeilly, T. N., Naylor, S. W., Mahajan, A., et al., (2008). *Escherichia coli* O157:H7 colonization in cattle following systemic and mucosal immunization with purified H7 flagellin. *Infection and Immunity, 76*(6), 2594–2602.

Mildner, A., & Jung, S., (2014). Development and function of dendritic cell subsets. *Immunity, 40*(5), 642–656.

Napoli, C., Paolisso, G., Casamassimi, A., Al-Omran, M., Barbieri, M., Sommese, L., Infante, T., & Ignarro, L. J., (2013). Effects of nitric oxide on cell proliferation: Novel insights. *J. Am. Coll. Cardiol., 62*, 89–95.

Nathan, C., & Xie, Q. W., (1994). Nitric oxide synthases: Roles, tolls, and controls. *Cell, 78*, 915–918.

Nauta, A. J., Daha, M., Van, K. C., & Roos, A., (2003). Recognition and clearance of apoptotic cells: A role for complement and pentraxins. *Trends in Immunology, 24*, 148–154.

Neumann, N. F., Stafford, J. L., & Belosevic, M., (2000). Biochemical and functional characterization of macrophage stimulating factors secreted by mitogen-induced goldfish kidney leucocytes. *Fish Shellfish Immunol., 10*, 167–186.

Nigam, A. K., Kumari, U., Mittal, S., & Mittal, A. K., (2012). Comparative analysis of innate immune parameters of the skin mucous secretions from certain freshwater teleosts, inhabiting different ecological niches. *Fish Physiology and Biochemistry, 38*(5), 1245–1256.

Ottinger, C. A., Johnson, S. C., Ewart, K. V., Brown, L. L., & Ross, N. W., (1999). Enhancement of anti-*Aeromonas salmonicida* activity in Atlantic salmon (*Salmo salar*) macrophages by a mannose-binding lectin. *Comparative Biochemistry and Physiology C., 123*(1), 53–59.

Pathak, A., & Agrawal, A., (2019). Evolution of C-reactive protein. *Frontiers in Immunology, 10*, 943.

Pijanowski, L., Golbach, L., Kolaczkowska, E., Scheer, M., Verburg-Van, K. B. M. L., & Chadzinska, M., (2013). Carp neutrophilic granulocytes form extracellular traps via ROS-dependent and independent pathways. *Fish Shellfish Immunol., 34*, 1244–1252.

Rawlings, D. J., Schwartz, M. A., Jackson, S. W., & Meyer-Bahlburg, A., (2012). Integration of B cell responses through toll-like receptors and antigen receptors. *Nat. Rev. Immunol., 12*, 282–294.

Reyes-López, M., Piña-Vázquez, C., & Serrano-Luna, J., (2015). Transferrin: Endocytosis and cell signaling in parasitic protozoa. *Biomed Res. Int.*, 641392.

Rieger, A. M., Hall, B. E., & Barreda, D. R., (2010). Macrophage activation differentially modulates particle binding, phagocytosis and downstream antimicrobial mechanisms. *Dev. Comp. Immunol., 34*, 1144–1159.

Sahoo, S., Banu, H., Prakash, A., & Tripathi, G., (2021). Immune system of fish: An evolutionary perspective. In: *Antimicrobial Immune Response*. IntechOpen.

Salinas, I., Zhang, Y. A., & Sunyer, J. O., (2011). Mucosal immunoglobulins and B cells of teleost fish. *Developmental and Comparative Immunology, 35*(12), 1346–1365.

Scharsack, J. P., Steinhagen, D., Kleczka, C., Schmidt, J. O., Körting, W., Michael, R. D., Leibold, W., & Schuberth, H. J., (2003). Head kidney neutrophils of carp (*Cyprinus carpio* L.) are functionally modulated by the haemoflagellate *Trypanoplasma borreli*. *Fish Shellfish Immunol., 14*, 389–403.

Schroder, K., Hertzog, P. J., Ravasi, T., &Hume, D. A., (2004). Interferon-gamma: An overview of signals, mechanisms and functions. *J. Leukoc. Biol., 75*, 163–189.

Secombes, C. J., & Fletcher, T. C., (1992). The role of phagocytes in the protective mechanisms of fish. *Annual Review of Fish Disease, 2*, 53–71.

Silva, M., & Correia-Neves, M., (2012). Neutrophils and macrophages: The main partners of phagocyte cell systems. *Frontiers in Immunology, 3*(174).

Smith, V. J., & Fernandes, J. M. O., (2009). Non-specific antimicrobial proteins of the innate system. In: Zaccone, G., Meseguer, J., García-Ayala, A., & Kapoor, B. G., (eds.), *Fish Defenses* (Vol. 1, pp. 241–275). Science Publishers, Enfield, NH, USA.

Soulliere, C., & Dixon, B. (2017). Immune system organs of bony fishes. *Reference Module in Life Sciences* Amsterdam: Elsevier Ltd. https://doi.org/10.1016/B978-0-12 809633-8. 12179-X.

Subramanian, S., MacKinnon, S. L., & Ross, N. W., (2007). A comparative study on innate immune parameters in the epidermal mucus of various fish species. *Comparative Biochemistry and Physiology B., 148*(3), 256–263.

Sunyer, J. O., (2013). Fishing for mammalian paradigms in the teleost immune system. *Nature Immunology, 14*(4), 320–326.

Thompson, S. A., Tachibana, K., Nakanishi, K., & Kubota, I., (1986). Melittin-like peptides from the shark-repelling defense secretion of the sole *Pardachirus pavoninus*. *Science, 233*(4761), 341–343.

Uribe, C., Folch, H., Enríquez, R., & Moran, G. J. V. M., (2011). Innate and adaptive immunity in teleost fish: A review. *Veterinarni Medicina, 56*(10), 486.

White, A., & Fletcher, T. C., (1985). The influence of hormones and inflammatory agents on C-reactive protein, cortisol and alanine aminotransferase in the plaice (*Pleuronectes platessa* L). *Comp. Biochem. Physiol., 80*(1), 99–104.

Wiegertjes, G. F., & Forlenza, M., (2010). Nitrosative stress during infection-induced inflammation in fish: Lessons from a host-parasite infection model. *Curr. Pharm. Des., 16*, 4194–4202.

Zhu, L., Lin, A., Shao, T., Nie, L., Dong, W., Xiang, L., & Shao, J., (2014). B cells in teleost fish act as pivotal initiating APCs in priming adaptive immunity: An evolutionary perspective on the origin of the B-1 cell subset and B7 molecules. *The Journal of Immunology, 192*(6), 2699–2714.

Zou, J., & Secombes, C. J., (2011). Teleost fish interferons and their role in immunity. *Developmental & Comparative Immunology, 35*(12), 1376–1387.

CHAPTER 19

Comprehensive Transcriptomics Analysis of Coldwater Fish

DEEPAK AGARWAL[1] and MOHD. ASHRAF RATHER[2]

[1]*Tamil Nadu Dr. J. Jayalalithaa Fisheries University, IFPGS, OMR Campus, Chennai, Tamil Nadu, India*

[2]*Faculty of Fisheries, Rangil-Ganderbal SKUAST, Jammu and Kashmir, India*

ABSTRACT

Transcriptomics is a highly performed experimental technique recently used in fishes to enrich the molecular data associated with several vital physiology such as reproduction, immune system, growth, and adaption to different environmental conditions. Transcriptomics has been transformed into a powerful, cost-effective, and user-friendly tool with the advent of technology. The study starts with RNA isolation which represents the complete set of genes expressed at the point in time. Transcriptome analysis help to aid the differential expression of important genes at various developmental stages and time points before or after some pathogenic or immunogenic stimulation. Transcriptomics generates a huge number of biological data which demand computational systems and various bioinformatics tools to process, analyze, and store the data for further deep analysis. Several applications of transcriptomics have been done in freshwater aquaculture but very few in cold water fisheries. Transcriptomics is a recent state-of-the-art technique to develop molecular markers in huge numbers for important commercial fish species. Transcriptomics helps in enriching the genetic resources of candidate and novel fish species to make better fruitful breeding strategies of fishes.

Coldwater Fisheries and Aquaculture Management: Technology for Sustainable Food Production. Mohd Ashraf Rather, Faisal Rashid Sofi, Adnan Amin, & Kawkabul Saba (Eds.)

19.1 INTRODUCTION

In molecular biology, transcripts expression is one of the most frequently performed experimental techniques in identifying a cell or an organism at the molecular level and studying physiology before and after some stimulation. Transcriptomics is defined as the complete set of complementary mRNAs produced in a cell or an organism. Transcript sequencing is done by RNA-sequencing (RNA-Seq) in which the RNA from different tissues at different time points are first isolated and then analyzed on their expression level. The development of the RNA-Seq method allows an unprecedented opportunity to analyze the expression of protein-coding, noncoding RNA and also *de novo* transcript assembly of a new species or organism (Chatterjee et al., 2018). Comprehensive transcriptome analysis using high throughput sequencing techniques is believed to provide a resource for genome annotation, candidate gene identification and molecular marker development (Agarwal et al., 2016). Recent advancements in next-generation sequencing technologies like high-throughput mRNA sequencing (RNA-seq) facilitate deciphering the functional complexity of the whole transcriptome of an organism. RNA-Seq generates a huge volume of data and accurate analysis of this huge data is always challenging and involves different steps and several bioinformatics tools. In the recent era, we are blessed with some advanced technologies and supercomputers. By using these supercomputers and advanced bioinformatics tools, the data generated from NGS platforms can reveal several hidden and significant information regarding the physiology and biology of the fish at the micro-level.

Temperature is a crucial environmental factor affecting the survival and growth of aquatic organisms. Although cold water fish habitat to low-temperature water, however, the global warming caused by the growth of greenhouse gas emission has severely changed the climate which ultimately exerts varying degrees of physiological stress to the fishes through knocking the metabolic pathways closely associated with homeostasis and physiological adaptation to the changing environment. Several endocrine receptors have been reported which are regulated at the transcription level in the cell and control the reproduction in fish (Sundaray et al., 2021). Such a mechanism of adaptation brings a substantial change in the transcriptome profile of the different organs/tissues which can be identified through RNA sequencing. The genes associated with changing temperature and salinity conditions have been identified in many fish including cold water species. The transcriptome profiling provides a better understanding of the adaptation.

Exploration of such differentially expressed genes and associated pathways helps to understand the adaptation at the micro-level.

19.2 THE WORKFLOW OF TRANSCRIPTOMICS THROUGH ILLUMINA SEQUENCING

The RNA sequencing starts from sample collection which is a vital step in transcriptome analysis. Sample collection is the most significant step of any kind of research so needs high level precautions. Samples should be stored in RNA later and properly preserved in liquid nitrogen. To decrease the number of samples, the tissues from different fishes of the same experiment are pooled in equimolar concentration before the sequencing. Transcripts represent the RNA expressed inside the cell which needs to be extracted from the sample through the most appropriate and suitable method such as Trizol method. The integrity of purified RNA is a significant factor to assess the quality of RNA which is measured using Bioanalyzer. The samples that have RIN (RNA Integrity Number) ≥8 is considered best for library preparation. RNA samples are stored at – 80°C until further use. After purification and fragmentation, the mRNA is converted into cDNA using random hexamers. Following the remaining steps of Transcriptomics as described in Figure 19.1, the library is prepared and validated using the same Bioanalyzer instrument. The library can be sequenced on any high throughput sequencer, but for the Transcriptomics study, Illumina is the most preferred and recommended platform.

19.3 BIOINFORMATICS ANALYSIS

After sequencing, a huge data is received in raw form which needs accurate and reliable scanning. Numerous bioinformatics tools are available in the public domain to analyze the data. Some of the tools are freely available and some of the tools need subscription charges. In this section, the primary tools required for transcriptome data analysis are discussed. The raw reads generated in the FASTQ file format are checked using FastQC software which determines the quality of reads like quality score, per base sequence content, per base GC content, sequence length distribution and duplicate sequences. Then raw reads are subjected to a quality filter to trim and remove unwanted or low-quality sequences using different software such as PRINSEQ and Kraken. The short quality reads are then assembled

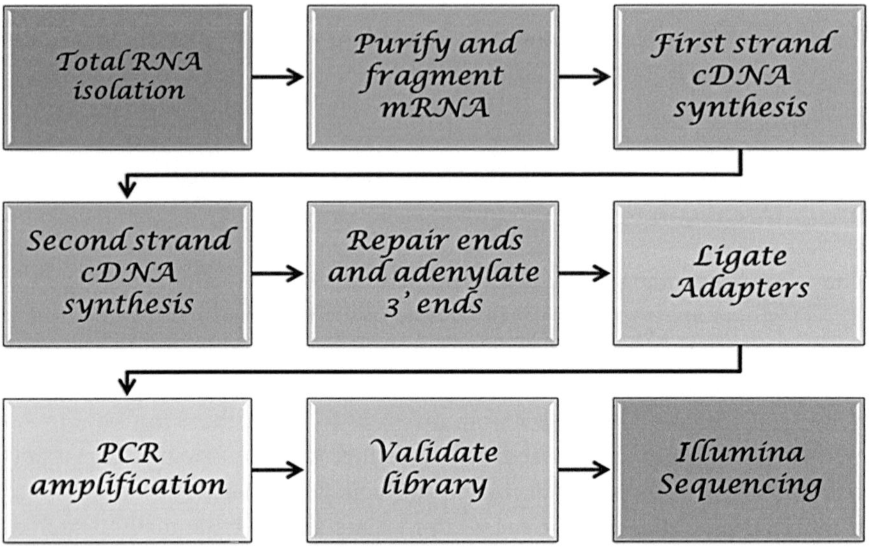

FIGURE 19.1 Flow chart of Illumina RNA-Seq.

into a reference transcriptome (Contigs). Two approaches can be followed depending on the availability of reference genome availability. Reference-guided assembly if a reference genome sequence is available otherwise, *de novo* RNA-Seq assembly in the absence of a reference genome sequence. Different software and packages are available like Trinity, Velvet, and CLC genomics workbench, TopHat-Cufflinks, etc. Once the assembly is completed, the next task is to find out the homology to determine putative gene descriptions of all the *de novo* assembled contigs. Often, local BLAST is used to perform the homology search which provides all needed details of the contigs such as gene id, protein id, coordinates of genes, hit numbers, expectation (E)-value, etc. For local blast, it is best to download the database first from the NCBI and then using the Linux operating system, BLASTX of the query sequences is done against the local nr (non-redundant) database. All contigs are also essential to be annotated functionally. GO (Gene Ontology) assignments were used to classify the predicted coding sequences based on their functions: Biological process, molecular function, and Cellular component. Usually, Blast2GO is the most common and popular tool for functional annotation. Several other tools have been developed for GO analysis like GOstat, EasyGO, Gorilla, etc. For the prediction of gene pathways, the assembled annotated contigs are also compared to the Kyoto

Encyclopedia of Genes and Genomes (KEGG) database using the Blast2GO program. As in transcriptome data, all the contigs are expressed genes, their EC (Enzyme Commission) number is often given with the help of several EC prediction tools such as E-zyme, ECOH, ECPred, etc.

19.4 TRANSCRIPTOMICS APPROACH FOR COLD-WATER FISH

Transcriptome profiling has been used to decipher the novel mechanisms behind several physiological processes of the fishes. Although the application of RNA-seq in fish Transcriptomics is at the nascent stage, the results of a PubMed literature search (with keywords, "fish" and "RNA-seq") indicated that the number of publications in this field has increased considerably in the last five years. The transcriptome analysis of several cold water fish species has not been done so far. The transcriptome can be used in a wide range of applications such as identifying a collection of protein-encoding genes with alternative splicing and post-translation modifications, molecular markers, generation of genetic resources and importantly measuring differential expression of thousands of genes within an individual at different environmental conditions or before and after the stimulation (Chandhini & Rejish, 2019).

The most basic application of transcriptomics is to enrich genetic resources and molecular data. Several researchers have been utilized this technique to enhance the data availability of many cold water fishes to understand the molecular mechanisms associated with important traits such as growth, immune response, reproduction, and sex determination. Lau et al. (2021) performed transcriptome analysis of Javan mahseer (*Tor tambra*) for the first time and suggested the data will be useful to gain insight into transcription regulation and biomarker discovery for the subsequent improvement of this species for aquaculture purposes. Barat et al. (2016) generated organ-specific transcriptome profiles of the golden mahseer (*Tor putitora*) using next-generation sequencing and suggested that the database will be helpful to understand the local adaptation, genome evolution, and also future functional studies on the immune system of the golden mahseer. Efforts have been made to obtain a comprehensive genome sequence for rainbow trout. Transcriptome sequencing was done to generate comprehensive de novo transcriptome information of rainbow trout suitable for functional and comparative genomics studies (Saleem et al., 2010, 2015).

One of the important objectives of transcriptome analysis is the identification of genes regulated during reproductive development and spawning in fish (Yang et al., 2019). Several genes and hormones are up and down-regulated during fish maturation and spawning. These genes are essential for inducing oocyte maturation, ovulation, and spermiation in domesticated fishes (Agarwal et al., 2020; Chutia et al., 2021). Numerous studies have been conducted to determine the transcriptional profile of gene programming in GnRH or GnRHa treated fish to provide better strategies for inducing breeding (Agarwal et al., 2020; Ahmad et al., 2108). RNA sequencing of gill and liver tissues of wild migrating sockeye salmon (*Oncorhynchus nerka*) was done to identify differentially expressed genes associated with changing salinity, temperature, pathogen exposure and dissolved oxygen indicate that these environmental variables most strongly impact physiology during spawning migrations (Evans et al., 2011). A similar kind of study responding to climate change with shifts in migration timing and habitat was done in sympatric sister species of salmon species (McKenzie et al., 2021).

Cold water fishes can survive at low temperatures due to the adaptation of fish to low temperatures which is the result of long-term evolution. The tolerance to temperature determines the habitat, distribution, and migration of fish and affects its reproduction, growth, and survival. Comparison of transcriptome profiles of fishes in different tissues at different temperatures may help to understand the physiology more precisely and will aid in deciphering the genetic basis of ecological and environmental adaptations of fishes. In Amur carp, cold-adaptive responses were identified through transcriptome profiling using high-throughput sequencing technology and suggested the strategies of gene induction during their six-month overwintering period (Liang et al., 2015). The molecular mechanisms of temperature adaptability of old-freshwater codfish (*Lota lota* lota) were studied and gene expression at the transcriptome level was revealed to decipher the response to acute temperature acclimation (Yang et al., 2021).

Change in seasons modulates the fish growth through the circannual rhythms in which the expression of genes associated with breeding show fluctuation in the brain and gonads. Danzmann et al. (2016) conducted the muscle transcriptome analysis of rainbow trout during different seasons unveiled that seasonal changes are correlated with a change in gene expression. They investigated the effects of a declining photoperiod regime on white muscle compared to an increasing photoperiod regime in fast and slow-growing rainbow trout through transcriptomics.

Recently transcriptome approaches are being widely utilized in the nutrigenomics field of the fisheries sector. The gene expression profile of genes involved in metabolism aid to determine the effect of feeding on growth, survival, and productivity. The ingredients given in diet play a major role in controlling physiology at the molecular level (Gora et al., 2018; Mir et al., 2020). Nowadays, studies are being done to replace the non-active or less available feed ingredients with another more active and easily available feed ingredient. Callet et al. (2018) performed the whole transcriptome of rainbow trout fry and detected the new pathways involved in the acceptance and the utilization of a plant-based diet in isogenic lines (Quillet et al., 2007). The whole-body transcriptome of trout alevins was done to determine the effects of long-term feeding of fish meal (FM) and fish oil free diet to the brooder females of rainbow trout and revealed the metabolic capacities of progeny (Lazzarotto et al., 2016). The level of micro- and macro-nutrients present in diet also influences fish performance and reproduction. Further transcriptomics studies have been provided valuable information for understanding the molecular mechanism and potential physiological pathway of dietary nutrients. Nazari et al. (2021), conducted the transcriptome profile of rainbow trout fed different zinc sources for the first time. Their findings provide molecular support to the notion and suggested that dietary zinc is useful material in commercial diets for brood stock rainbow trout.

Transcriptome profiling has been used to understand the novel mechanisms behind the immune responses of the fishes. Several studies have been done to decipher the molecular mechanism underlining immune response in fishes through a transcriptome approach. In one of the commercially important cold water fish golden mahseer (*Tor putitora*), infected with *A. hydrophila,* the transcriptome responses were recorded to understand the immune response of the fish (Kumar et al., 2017). Several studies have been conducted to enrich the transcriptome data and to determine the involvement of the immune system in response to bacterial infection in rainbow trout (Ali et al., 2014; Wang et al., 2021; Syahputra et al., 2019). The Immunogenomics data produced through several studies can be used in clarifying the origin and evolution of the immune systems (Kaiser et al., 2008). Transcriptome analysis studies have also been performed to better understand the immune response in fishes immunized with some widely used vaccines (Lim & Hong, 2021; Chinchilla et al., 2020).

19.5 CONCLUSION

Sequencing technologies have a tremendous impact on life science research and development work. The technology has evolved a lot over the period starting from classical first-generation Sanger's sequencing to high throughput and technologically advanced third and fourth generation sequencing, making it better than earlier in terms of the ability of throughput, error rate, long read lengths and of course the cost of sequencing. The availability of these technologies has enabled a more complete understanding of whole-genome and transcriptome in shorter time intervals with high accuracy. Transcriptomics has better advantages over other gene expression approaches such as microarray. Although, the transcriptome data have been generated for many commercially important fishes in the case of cold water fisheries the research progress is still in its initial stage. Very few cold water fishes have been sequenced and the data available in the public domain need more attention to be focused on. Several physiological activities have not been studied using high throughput sequencing so far. As transcriptomics expands endlessly and its cost keeps decreasing, within the next few years transcriptomics will be exploited to a larger extent without a doubt and lead to many more exciting discoveries in the cold water fisheries sector.

KEYWORDS

- **Kyoto Encyclopedia of Genes and Genomes**
- **mRNA sequencing**
- **RNA integrity number**
- **RNA isolation**
- **RNA-sequencing**
- **transcriptomics**

REFERENCES

Agarwal, D., Aich, N., Pavan-Kumar, A., Kumar, S., Sabnis, S., Joshi, C. G., & Chaudhari, A., (2016). SNP mining in transcripts and concomitant estimation of genetic variation in *Macrobrachium rosenbergii* stocks. *Conservation Genetics Resources, 8*(2), 159–168.

Agarwal, D., Gireesh-Babu, P., Pavan-Kumar, A., Koringa, P., Joshi, C. G., Gora, A., Bhat, I. A., & Chaudhari, A., (2020). Molecular characterization and expression profiling of 17-beta-hydroxysteroid dehydrogenase 2 and spermatogenesis associated protein 2 genes in endangered catfish, *Clarias magur* (Hamilton, 1822). *Animal Biotechnology, 31*(2), 93–106. https://doi.org/10.1080/10495398.2018.1545663.

Agarwal, D., Gireesh-Babu, P., Pavan-Kumar, A., Koringa, P., Joshi, C. G., & Chaudhari, A., (2020). Transcriptome analysis of *Clarias magur* brain and gonads suggests neuro-endocrine inhibition of milt release from captive GnRH-induced males. *Genomics, 112*(6), 4041–4052.

Ahmad, I., Jagtap, D. D., Kumar, C. S., Balasinor, N. H., Rani, A. B., Agarwal, D., & Saharan, N., (2018). Molecular characterization of inhibin-A: Structure and expression analysis in *Clarias batrachus*. *General and Comparative Endocrinology, 261*, 104–114.

Ali, A., Rexroad, C. E., Thorgaard, G. H., Yao, J., & Salem, M., (2014). Characterization of the rainbow trout spleen transcriptome and identification of immune-related genes. *Frontiers in Genetics, 5*, 348.

Barat, A., Kumar, R., Goel, C., Singh, A. K., & Sahoo, P. K., (2016). De novo assembly and characterization of tissue-specific transcriptome in the endangered golden mahseer, *Tor putitora*. *Meta Gene, 7*, 28–33.

Callet, T., Dupont-Nivet, M., Cluzeaud, M., Jaffrezic, F., Laloë, D., Kerneis, T., & Médale, F., (2018). Detection of new pathways involved in the acceptance and the utilization of a plant-based diet in isogenic lines of rainbow trout fry. *Plos One, 13*(7), e0201462.

Chandhini, S., & Rejish, K. V. J., (2019). Transcriptomics in aquaculture: Current status and applications. *Reviews in Aquaculture, 11*(4), 1379–1397.

Chatterjee, A., Ahn, A., Rodger, E. J., Stockwell, P. A., & Eccles, M. R., (2018). A guide for designing and analyzing RNA-seq data. In: *Gene Expression Analysis* (pp. 35–80). Humana Press, New York, NY.

Chinchilla, B., Encinas, P., Coll, J. M., & Gomez-Casado, E., (2020). Differential immune transcriptome and modulated signaling pathways in rainbow trout infected with viral haemorrhagic septicaemia virus (VHSV) and its derivative non-virion (NV) gene deleted. *Vaccines, 8*(1), 58.

Chutia, S. J., Bora, G., Nath, R. J., Yashwanth, B. S., Agarwal, D., Bhagawati, K., & Sarma, D. K., (2021). Gene regulation on spermiation of catfish. In: *Recent updates in molecular Endocrinology and Reproductive Physiology of Fish* (pp. 161–169). Springer, Singapore.

Evans, T. G., Hammill, E. D. D., Kaukinen, K., Schulze, A. D., Patterson, D. A., English, K. K., & Miller, K. M., (2011). Transcriptomics of environmental acclimatization and survival in wild adult Pacific sockeye salmon (*Oncorhynchus nerka*) during spawning migration. *Molecular Ecology, 20*(21), 4472–4489.

Gora, A. H., Sahu, N. P., Sahoo, S., Rehman, S., Ahmad, I., Agarwal, D., & Rasool, S. I., (2018). Metabolic and haematological responses of *Labeo rohita* to dietary fucoidan. *Journal of Applied Animal Research, 46*(1), 1042–1050.

Kumar, R., Sahoo, P. K., & Barat, A., (2017). Transcriptome profiling and expression analysis of immune responsive genes in the liver of golden mahseer (*Tor putitora*) challenged with *Aeromonas hydrophila*. *Fish & Shellfish Immunology, 67*, 655–666.

Lau, M. M. L., Lim, L. W. K., Chung, H. H., & Gan, H. M., (2021). The first transcriptome sequencing and data analysis of the Javan mahseer (*Tor tambra*). *Data in Brief, 39*, 107481.

Lazzarotto, V., Corraze, G., Larroquet, L., Mazurais, D., & Médale, F., (2016). Does broodstock nutritional history affect the response of progeny to different first-feeding diets? A whole-body transcriptomic study of rainbow trout alevins. *British Journal of Nutrition, 115*(12), 2079–2092.

Liang, L., Chang, Y., He, X., & Tang, R., (2015). Transcriptome analysis to identify cold-responsive genes in amur carp (*Cyprinus carpio haematopterus*). *PLoS One, 10*(6), e0130526.

Lim, J., & Hong, S., (2021). Transcriptome analysis in the head kidney of rainbow trout (*Oncorhynchus mykiss*) immunized with a combined vaccine of formalin-inactivated *Aeromonas salmonicida* and *Vibrio anguillarum*. *Vaccines, 9*(11), 1234.

McKenzie, J. L., Araújo, H. A., Smith, J. L., Schluter, D., & Devlin, R. H., (2021). Incomplete reproductive isolation and strong transcriptomic response to hybridization between sympatric sister species of salmon. *Proceedings of the Royal Society B, 288*(1950), 20203020.

Mir, I. N., Srivastava, P. P., Bhat, I. A., Dar, J. Y., & Agarwal, D., (2020). Expression profiling of lipolytic and long-chain fatty acid biosynthesis genes in catfish larvae fed with graded lipid levels. *Frontiers in Marine Science, 791.*

Nazari, S., Pourkazemi, M., Paknejad, H., Kazemi, E., Ghaderi, M., & Eslamloo, K., (2021). Transcriptome profiling of farmed rainbow trout (*Oncorhynchus mykiss*) liver from different sources of dietary zinc. *Aquaculture, 543*, 737017.

Quillet, E., Dorson, M., Le Guillou, S., Benmansour, A., & Boudinot, P., (2007). Wide range of susceptibility to rhabdoviruses in homozygous clones of rainbow trout. *Fish & shellfish Immunology, 22*(5), 510–519.

Salem, M., Paneru, B., Al-Tobasei, R., Abdouni, F., Thorgaard, G. H., Rexroad, C. E., & Yao, J., (2015). Transcriptome assembly, gene annotation and tissue gene expression atlas of the rainbow trout. *PloS one, 10*(3), e0121778.

Salem, M., Rexroad, C. E., Wang, J., Thorgaard, G. H., & Yao, J., (2010). Characterization of the rainbow trout transcriptome using sanger and 454-pyrosequencing approaches. *BMC Genomics, 11*(1), 1–10.

Sundaray, J. K., Rather, M. A., Kumar, S., & Agarwal, D., (2021). *Recent Updates in Molecular Endocrinology and Reproductive Physiology of Fish*. Springer Singapore.

Syahputra, K., Kania, P. W., Al-Jubury, A., Jafaar, R. M., Dirks, R. P., & Buchmann, K., (2019). Transcriptomic analysis of immunity in rainbow trout (*Oncorhynchus mykiss*) gills infected by *Ichthyophthirius multifiliis*. *Fish & Shellfish Immunology, 86*, 486–496.

Wang, D., Sun, S., Li, S., Lu, T., & Shi, D., (2021). Transcriptome profiling of immune response to *Yersinia ruckeri* in spleen of rainbow trout (*Oncorhynchus mykiss*). *BMC Genomics, 22*(1), 1–11.

Yang, T., Zhang, Y., Meng, W., Zhong, X., Shan, Y., & Gao, T., (2021). Comparative transcriptomic analysis brings new insights into the response to acute temperature acclimation in burbot (*Lota lota* lota). *Aquaculture Reports, 20*, 100657.

Yang, W., Lin, B., Li, G., Chen, H., & Liu, M., (2019). Sequencing and transcriptome analysis for reproduction-related genes identification and SSRs discovery in sequential hermaphrodite *Amphiprion ocellaris*. *Turkish Journal of Fisheries and Aquatic Sciences, 19*(12), 1049–1059.

CHAPTER 20

Predictable Threats to Coldwater Fisheries from the Unpredictable Nature of Climate Change: An Indian Perspective

ANKUR JAMWAL[1], VIKAS PHULIA[2], and SYED TALIA MUSHTAQ[3]

[1]*College of Fisheries and Center of Excellence on Water Management, Dr. Rajendra Prasad Central Agricultural University, Pusa, Samastipur, Bihar, India*

[2]*Krishi Vigyan Kendra, SAS Nagar, (HQ: Guru Angad Dev Veterinary and Animal Sciences University, Ludhiana), Punjab, India*

[3]*Fisheries Resource Management, Faculty of Fisheries, Sher-e-Kashmir University of Agricultural Sciences and Technology, Jammu and Kashmir, India*

ABSTRACT

The coldwater fishery of India contributes to only 2% of the total fisheries of the nation; however, it plays a vital role in food security and nutrition of people living in the remote Himalayan and peninsular regions. Since the Himalayas and the Peninsular Ghats of India are sensitive to the vagaries of climate change, the fishery and aquaculture practices along these regions are also vulnerable. The coldwater fisheries of India is susceptible to the hydrological and rheological changes caused by receding glacial headwaters, altered habitat range and increased severity of natural disasters. Changing climate is also expected to favor invasive species that can replace the indigenous fauna upon which the local populations are dependent culturally and for their nutritional requirements. Warmer temperatures and environmental stress also

Coldwater Fisheries and Aquaculture Management: Technology for Sustainable Food Production.
Mohd Ashraf Rather, Faisal Rashid Sofi, Adnan Amin, & Kawkabul Saba (Eds.)

pose a risk to the fishery and aquaculture industry from increased disease virulence. A thorough assessment of risks to the coldwater fisheries of India from climate change is required. Policy intervention at both state and national levels can help in the mitigation of some risks from climate change. Improved disaster management infrastructure and better engineering interventions in aquaculture can also help in the prevention of loss of human lives and aquaculture infrastructure from flooding and other natural disasters.

20.1 INTRODUCTION

The world population is estimated to cross 9.7 billion by 2050 and providing them with nutritious food will be one of the biggest challenges before humankind (FAO, 2018). At the present rate of agricultural growth and plateauing arable land area, it would be difficult to meet the growing demand for nutritious food (Bommarco et al., 2013; Hunter et al., 2017). Therefore, necessitating the development of other resources, such as meat and dairy. Further, it is estimated that improvement in economic status of Asian and Latin American countries will induce a shift in dietary pattern towards inclusion of more meat and other animal products in the diet increasing the demand for meat and dairy products in the near future (Gerbens-Leenes et al., 2010; Lange et al., 2018). In this context, there lies an opportunity for the fisheries sector to grow and play an important role in meeting the global demand for nutritious food.

Currently, fisheries have outpaced all other meat-producing sectors and is expected to remain the leading supplier of animal meat even in the future (FAO, 2018). People all over the world have started to recognize fish as an essential component of a healthy diet. This perception is also substantiated by a growing body of scientific literature that suggests multiple health benefits of fish in the human diet. For example, lower collagen content in fish meat makes it easy to digest and thus increases its nutritional value (Listrat et al., 2016). Fish meat is also rich in various other essential nutrients such as ω-3 fatty acids, vitamins, and minerals (Domingo, 2016). It is because of the superior nutritional quality of fish meat, its inclusion in the human diet has shown to help against various modern lifestyle ailments such as coronary heart disease (CHD), diabetes, and obesity (Domingo, 2016). The demand for fish is therefore expected to rise further as the awareness about the health benefits of eating fish increases. Although the future holds immense opportunities for this sunrise sector, it is not insulated from various threats such as climate change.

20.1.1 UNDERSTANDING CLIMATE CHANGE

Climate is the average of weather conditions prevailing over a long period of time and all the biological processes depend upon a delicate balance of temperature, rainfall, light, wind pattern, humidity, and many other abiotic factors that exist during a season in a specific climatic condition. It is the rhythmicity of the combination of abiotic factors that determine the growth and reproductive cycle of animals and plants. For example, Indian major carps breed primarily between June and August; when the subcontinent experiences monsoon-like conditions. Similarly, trees blossom and bear fruits in specific months only. Further, on a much larger and complex scale, all biological processes also depend on each other for the flow of energy and form a trophic hierarchy – breeding of fish mostly coincides with abundance of phyto and zooplankton (base of the aquatic trophic chain), which in turn depends on the availability of nutrients through upwelling, floods, and rain in a season (rhythmicity of abiotic factors). Interestingly, and contrary to the common belief, the climatic conditions on this planet have never remained stationary. For example, the Earth has transitioned from colder Pleistocene (Ice Age) to warmer Holocene epoch. Such transitions result from natural factors and processes such as the changes in Sun's intensity, volcanic eruptions, change in the Earth's orbit around the Sun, and change in ocean current circulation. However, the change in climate due to natural factors is slow and subtle, which allows enough time for the organisms to adapt. In contrast, climate change induced by anthropogenic activities leading to CO_2 accumulation, aerosol emission, and deforestation resulted in a sudden change in global climatic conditions. The adaptive processes, which is slow, cannot occur in response to drastic and sudden changes in climate. As per an estimate, the period after the Industrial age has resulted in increased global temperatures and more instances of extreme weather conditions leading to "exceptionally rapid loss of biodiversity over the last few centuries," indicating the sixth mass extinction event (Ceballos et al., 2015). A substantial amount of scientific evidence suggests that the global average temperatures will increase by 1.5°C before 2040 if serious interventions are not made to stop it (IPCC, 2018).

The effects of climate change are manifold, and their comprehensive discussion is beyond the scope of this review; however, some of the most apparent effects being currently experienced are altered precipitation patterns, increasing number of hot days, severe cyclones, extreme heatwaves, flooding, and desertification. As far as fisheries is concerned, it is mostly threatened by:

- Acidification of oceans;
- Coral bleaching;
- Reduced glacial cover;
- Loss of habitat to invasive species;
- Oceanic cyclones;
- Floods;
- Drought;
- More instance of disease and stress;
- Poor agricultural output resulting in scarcity of ingredients for fish feed.

The above-mentioned effects of climate change, in the context of cold-water fisheries of India, will be discussed in this review.

20.2 EFFECTS OF CLIMATE CHANGE ON COLD-WATER FISHERIES

The coldwater fishes of India live in some of the most vulnerable habitats on this planet – the Himalayas and the Nilgiris. Both these ecosystems are threatened by: (i) changes on account of natural causes; (ii) climate change due to anthropogenic factors; and (iii) exploitation of resources to sustain modern civilization. The recession of glaciers and altered flow of Himalayan rivers for hydroelectricity projects or other developmental activities pose the biggest threat to both capture and culture fisheries of coldwater species.

20.2.1 RECEDING GLACIERS AND ALTERED HABITAT RANGE

Geo-spatial studies indicate that almost all the glaciers outside the Karakoram range in the Himalayas are shrinking at a much faster rate over the last two decades (Bolch et al., 2012; Scherler et al., 2011). Glaciers form the headwaters for most coldwater streams and rivers, and their recession by excessive melting can result in floods in the short run and drought over the extended period. Unfortunately, the effects of receding glaciers have already become visible in India. As per the report published by NITI Aayog, nearly 50% of springs in the Indian Himalayan Region have either dried up or have become seasonal (NITI Aayog, 2018). In contrast, the population dependent on these springs is increasing exponentially. Therefore, there is extreme exploitation pressure on the cold freshwater resources for domestic consumption, electricity generation, fishery, and industrial purpose. Since the exploitation of water resources for domestic consumption and industrial applications takes precedence, the fisheries sector is left starving. Moreover, the detrimental effects of climate change can be accentuated by altered rheological features, habitat fragmentation from constructions, illegal mining activities, and unsustainable tourism activities leading to reduced carrying capacity of rivers and severe food security concerns in regions where fish forms a part of the diet (Sarkar et al., 2012, 2015; Ziv et al., 2012). Lack of in-depth and systematic studies on the dynamics of coldwater fish populations of India remains a significant bottleneck in assessing the impact of climate change and needs to be addressed with urgency.

Climate change is not only reducing glacial volume but with rising global temperatures, the habitat of native coldwater species is also receding and moving upwards (Shepard et al., 2016). Glacial melt in summers provides a thermal buffer to streams and rivers from a drastic change in water temperature. However, with reduced glacial volumes and higher atmospheric variation, the rivers are experiencing high variability in temperature – resulting in native species being outcompeted by other fish that are better suited for changed conditions (Shepard et al., 2016). For example, mahseer and snow trout (*Schizothorax* spp.), which prefer colder waters with high dissolved oxygen content, are being outcompeted by voracious and fastidious catfish that can survive better in warmer waters with slightly lower dissolved oxygen content (Sarkar et al., 2015).

In addition to the fast-flowing rivers, the Himalayan states of Himachal Pradesh, Sikkim, Uttarakhand, Arunachal Pradesh, and the union territories of Ladakh and Jammu and Kashmir also have 4,699 lakes (*see* Table 20.1) that are often ignored from the fisheries point of view (Panigrahy et al., 2012). Scientific literature also suggests that high-altitude lakes are shrinking due to increased evaporation and sedimentation from melting glaciers (Philip & Mazari, 2000). Since lakes are enclosed waterbodies, with no escape routes for fish, change in thermal regime can be disastrous for its inhabitants. The ichthyofaunal status of most high-altitude lakes in the Indian Himalayan Region is still unknown and is an impediment in assessing the effects of climate change on their fish biodiversity.

20.2.2 CLIMATE CHANGE AND NATURAL DISASTERS

Recent devastation due to floods, cloud bursts and landslides in Indian Himalayan Regions and the Western Ghats are not some isolated events that could be explained merely by chance. In contrast, various climatological models had already predicted increased penetration of monsoon into the

Himalayas with consequences such as higher incidences of floods, cloud bursts and increased riverine discharge (Wasson et al., 2013). These events negatively affect the fisheries industry. Flash floods and natural disasters not only keep fishers from going out to fish, but the organized coldwater aquaculture is also affected by such disasters. Most coldwater fishery infrastructure, such as hatcheries and farms, are situated along the banks of rivers/rivulets because these water bodies provide a continuous flow of water for raceways. Therefore, flash floods or abnormal discharge in these waterbodies can damage or wash-out the infrastructure – directly affecting the coldwater aquaculture industry. Although the Governments (e.g., Government of Himachal Pradesh) provide insurance coverage for the infrastructure but the rebuilding of structures and establishment of fish stocks can take years. Even worse, the cultured gene stock can contaminate the wild population if escaped during such disasters.

TABLE 20.1 List of High-Altitude Lakes in the Indian Himalayan Region Classified based on Their Area

SL. No.	Class	Range	No of Lakes	Area (ha)
1.	Very large	> 500 ha	12	95,499
2.	Large	100–500 ha	31	4,993
3.	Medium	25–100 ha	177	7,366
4.	Small	10–25 ha	498	7,679
5.	Very small	2.25–10 ha	1,985	8,592
6.	< 2.25 ha	< 2.25 ha	1,996	1,996
Total			**4,699**	**1,26,125**

Source: Reprinted from Panigrahy et al., 2012. © Space Applications Centre, ISRO, 2012.

20.2.3 LOSS OF HABITAT TO INVASIVE SPECIES

It is predicted that coldwater fish could lose about 31% of their habitat to warm water fish species in the US because of climate change (Mohseni et al., 2003). A similar situation is not out of the picture for India. The scientific community agrees that warmer water temperatures, shorter winters, altered rheological patterns, and increased demand for water can result in the establishment of non-native species by eliminating native coldwater fish (Rahel & Olden, 2008). Since the reservoirs will be used for the introduction of novel fish species for aquaculture practices, they could serve as hotspots for entry of invasive species. The introduction of silver and mirror carp in Indian reservoirs is an example of how the introduction of exotic species can affect the survival of native fish species. Silver carp (*Hypophthalmichthys molitrix*), after its introduction into the Gobind Sagar reservoir (Himachal Pradesh), has formed a breeding population that has outcompeted native mahseer by completely dominating the lower trophic food chain. Silver carp is a coldwater fish but grows at a much faster rate and matures much earlier (within one year) than the native fish species – thereby replacing native fish stocks. Similarly, mirror carp, which is also a coldwater fish, was introduced in various Himalayan lakes and reservoirs to augment fisheries. However, the introduction of mirror carp has again proven to be destructive for native fish species in Gobind Sagar reservoir, upland lakes of Kashmir and Kumaon, and Loktak Lake in the northeast (Sugunan, 2000).

Increasing demand for fish can motivate various resource managers to introduce fish species for outdoor culture in pens and cages. Warm water fish species have better growth rates and are often preferred over native species for such aquaculture operations. For example, the National Fisheries Development Board (NFDB) of India recommends culture of exotic cichlids (Tilapia) and catfish (*Pangasianodon* spp.) for reservoir-based aquaculture, especially in the scenario where warmers temperature may favor such species. Although the regulatory guidelines prohibit the culture of individual fish species that may establish a reproducing population, accidental introduction (as in the case of silver carp in Himachal) of a breeding pair can be detrimental and difficult to contain.

20.2.4 CLIMATE CHANGE AND INCREASED DISEASE VIRULENCE

Alien species or their invasion can bring with them disease and parasites not previously known to coldwater species–resulting in an epizootic outbreak in native species (Marcogliese, 2001a). Increased instances of whirling disease in North American and European salmonid species due to higher virulence of myxosporean parasite, *Myxobolus cerebralis*, has been observed and linked with climate change (Marcogliese, 2001b). Infection from *Streptococcus* is observed in temperatures above 30°C; therefore, a rise in global temperatures or increased number of warmer days in a year can result in *Streptococcal* outbreaks. Similar, threats to Indian coldwater fish stocks cannot be ruled out. Moreover, stress due to sudden changes in environmental variables is expected to reduced disease resistance capabilities in fish, hence increasing their vulnerability. It has been shown experimentally that finfish undergo metabolic depression when reared at temperatures close to their thermal maxima. Metabolic depression often results in suppression of disease response in fish.

20.3 MITIGATION AND ADAPTATION STRATEGIES

20.3.1 RISK ASSESSMENT

Climate change presents a situation of uncertainty and making decisions in the face of this ambiguity is the biggest challenge. However, accumulation and analysis of time-series data on the changing environmental variables, its effects on water bodies and fish caught in the past can help in risk assessment. Lack of baseline studies and continuous use of rivulets and rivers for developmental projects in ecologically fragile Himalayas and peninsular ghats puts the coldwater fisheries of India in a highly precarious situation. Therefore, the first steps towards mitigation of risks from climate change should begin with the initiation of cross-disciplinary risk assessment. Development of computer-based models to predict the effects of climate change on fish production, understanding of interactions between human activities and fish production, and assessment of losses due to inaction should also form part of mitigation policies.

Robust fish-disease monitoring and reporting system, based on the principles of one-health, can go a long way in preventing loss of fish in culture systems and natural waters.

20.3.2 CREATE A POLITICAL WILL AND ENGAGE SCIENTISTS TO PREPARE POLICIES

Since it is apparent that climate change and its consequences are inevitable, the governments will need to fund scientific data gathering and constitute policies to adapt coldwater fisheries infrastructure. The policies can be reactive or anticipatory (Smith & Lenhart, 1996). The reactive policies propose remedial measures in response to the changes and effects already visible. For example, relocation of fish hatcheries and farms constructed in sites vulnerable to flooding and landslides. Ranching operations to replenish the wild stocks can also be undertaken. Responsible tourism and strict policy-based development activities to restrict the emission of greenhouse gases should also be incorporated into the reactive policy measures. The coldwater fisheries of India on its own has a very little carbon footprint. Hence policies should be drafted to keep the fisheries clean in terms of fossil fuel use. The anticipatory policies, on the other hand, include preemptive measures based on sound scientific principles and predictions using computer-based models. Regulation of fishing with only selective, non-destructive gears is one such example of anticipatory measure. Currently, various states have their laws to regulate capture fishing activities; however, the regulations on the aquaculture industry are still missing, especially in the states along the Indian Himalayan region. Inclusion of the aquaculture industry, in consultation with stakeholders, for the development of policies towards climate change resilient coldwater fishery should be promoted.

20.3.3 DIVERSIFICATION OF AQUACULTURE INDUSTRY

While the coldwater capture fisheries of India is primarily dependent upon three groups of fishes, the mahseers, snow trout (*Schizothorax* spp.), and the rainbow trout (*Oncorhynchus mykiss*), the culture of rainbow trout forms the mainstay of coldwater aquaculture. Therefore, any detrimental effects of climate change on the survival or wild recruitment of rainbow trout, mahseer, and snow trout can crash the entire coldwater fishery of India. Efforts are required to diversify the species of the profile of coldwater aquaculture in India. The captive breeding and culture techniques of mahseer and snow trout have already been standardized but have yet to be popularized.

Efforts for selection fish strains that are resilient to disease and sudden change in climate should be selected and used for aquaculture and ranching purposes. Research into the diversification of coldwater aquaculture by introducing more fish species should also be explored. However, stringent measures to prevent escapement of exotic fish species and their reproductive proliferation in the wild must be formulated. Monosex culture or culture of sterile fish can prevent the indigenous gene pool from contamination. Research into improved techniques for fish cultured is also required. Currently, trout is primarily cultured in the raceways system that requires a continuous flow of water. Consequently, trout culture has a high-water requirement. Adoption of alternate fish species or fish culture techniques, such as the recirculatory aquaculture system that is less water-intensive is required. Research efforts are also required to reduce the dependency of aquaculture on FM as a protein source. Even better, fish species that feed on lower trophic levels should be preferred.

20.3.4 DISASTER MANAGEMENT

Loss of aquaculture infrastructure built along the banks of rivers or rivulets can be prevented by the construction of flood-control measures such as embankments. The installation of early warning systems can also help in the prevention of loss of human lives. Aquaculture techniques that are less water-intensive or do not require a continuous water supply offer the flexibility to move aquaculture infrastructure away from flood-prone regions.

20.4 CONCLUSION

Climate change is expected to affect the aquatic system the most. Altered environmental variables are predicted to shift the ecological balance against the native species and will favor natural disasters, disease outbreaks and invasion of exotic fish species. The threat from climate change is particularly imminent for coldwater fishes that are stenothermal and stenohaline. Disasters associated with climate change are expected to destroy natural fish breeding grounds and manmade aquaculture infrastructure. Additionally, altered thermal regime, concomitant with an invasion of exotic species, may result in loss of habitat range for cold water fishes. Unfortunately, there is a severe dearth of scientific literature on the vulnerability of Indian coldwater fisheries to climate change and there is an urgent need to address this lacuna.

KEYWORDS

- **aquaculture industry**
- **climate change**
- **coldwater fisheries**
- **disaster management**
- **mitigation strategies**
- **natural disasters**

REFERENCES

Bolch, T., Kulkarni, A., Kääb, A., Huggel, C., Paul, F., Cogley, J. G., Frey, H., et al., (2012). The State and fate of Himalayan glaciers. *Science*, *80*, 336, 310LP–314. https://doi.org/10.1126/science.1215828.

Bommarco, R., Kleijn, D., & Potts, S. G., (2013). Ecological intensification: Harnessing ecosystem services for food security. *Trends Ecol. Evol.*, *28*, 230–238. https://doi.org/https://doi.org/10.1016/j.tree.2012.10.012.

Ceballos, G., Ehrlich, P. R., Barnosky, A. D., García, A., Pringle, R. M., & Palmer, T. M., (2015). Accelerated modern human–induced species losses: Entering the sixth mass extinction. *Sci. Adv., 1,* e1400253. https://doi.org/10.1126/sciadv.1400253.

Domingo, J. L., (2016). Nutrients and chemical pollutants in fish and shellfish. Balancing health benefits and risks of regular fish consumption. *Crit. Rev. Food Sci. Nutr., 56*, 979–988. https://doi.org/10.1080/10408398.2012.742985.

FAO, (2018). *The State of World Fisheries and Aquaculture 2018 – Meeting the Sustainable Development Goals*. Rome.

Gerbens-Leenes, P. W., Nonhebel, S., & Krol, M. S., (2010). Food consumption patterns and economic growth. Increasing affluence and the use of natural resources. *Appetite 55*, 597–608. https://doi.org/https://doi.org/10.1016/j.appet.2010.09.013.

Hunter, M. C., Smith, R. G., Schipanski, M. E., Atwood, L. W., & Mortensen, D. A., (2017). Agriculture in 2050: Recalibrating targets for sustainable intensification. *Bioscience, 67*, 386–391. https://doi.org/10.1093/biosci/bix010.

IPCC, (2018). Global warming of 1.5°C. An IPCC special report on the impacts of global warming of 1.5°C above pre-industrial levels and related global greenhouse gas emission pathways, in the context of strengthening the global response to the threat of climate change. *Intergovernmental Panel on Climate Change*.

Lange, G. M., Naikal, E., & Wodon, Q., (2018). Richer or poorer?: Global and regional trends in wealth from 1995 to 2014. In: Lange, G. M., Wodon, Q., & Carey, K., (eds.), *The Changing Wealth of Nations 2018: Building a Sustainable Future* (pp. 43–68). World Bank, Washington, DC. https://doi.org/10.1596/978-1-4648-1046-6_ch2.

Listrat, A., Lebret, B., Louveau, I., Astruc, T., Bonnet, M., Lefaucheur, L., Picard, B., & Bugeon, J., (2016). How muscle structure and composition influence meat and flesh quality. *The Sci. World J. 2016*, 14.

Marcogliese, D. J., (2001a). Implications of climate change for parasitism of animals in the aquatic environment. *Can. J. Zool., 79*, 1331–1352. https://doi.org/10.1139/z01-067.

Marcogliese, D. J., (2001b). Implications of climate change for parasitism of animals in the aquatic environment. *Can. J. Zool., 79*, 1331–1352. https://doi.org/10.1139/z01-067.

Mohseni, O., Stefan, H. G., & Eaton, J. G., (2003). Global warming and potential changes in fish habitat in U.S. streams. *Clim. Change, 59*, 389–409. https://doi.org/10.1023/A:102484 7723344.

NITI Aayog, (2018). Report of Working Group 1: Inventory and Revival of Springs in the Himalayas for Water Security. *21*(3), 65–89.

Panigrahy, S., Patel, J. G., & Parihar, J. S., (2012). National wetland Atlas: High altitude lakes of India. Ahmedabad, India.

Philip, G., & Mazari, R. K., (2000). Shrinking lake basins in the proximity of the Indus suture zone of northwestern Himalaya: A case study of Tso Kar and Startsapuk Tso, using IRS-1C data. *Int. J. Remote Sens., 21*, 2973–2984. https://doi.org/10.1080/01431160050144901.

Rahel, F. J., & Olden, J. D., (2008). Assessing the effects of climate change on aquatic invasive species. *Conserv. Biol., 22*, 521–533. https://doi.org/10.1111/j.1523-1739.2008.00950.x.

Sarkar, U. K., Pathak, A. K., Sinha, R. K., Sivakumar, K., Pandian, A. K., Pandey, A., Dubey, V. K., & Lakra, W. S., (2012). Freshwater fish biodiversity in the River Ganga (India): Changing pattern, threats and conservation perspectives. *Rev. Fish Biol. Fish., 22,* 251–272. https://doi.org/10.1007/s11160-011-9218-6.

Sarkar, U. K., Sharma, J., & Mahapatra, B. K., (2015). A review on the fish communities in the Indian reservoirs and enhancement of fisheries and aquatic environment. *J. Aquac. Res. Dev., 6*, 297.

Scherler, D., Bookhagen, B., & Strecker, M. R., (2011). Spatially variable response of Himalayan glaciers to climate change affected by debris cover. *Nat. Geosci., 4*, 156.

Shepard, B. B., Al-Chokhachy, R., Koel, T., Kulp, M. A., & Hitt, N., (2016). Likely responses of native and invasive salmonid fishes to climate change in the rocky mountains and Appalachian mountains. In: Hansen, A. J., Monahan, W. B., Olliff, S. T., & Theobald, D. M., (eds.), *BT – Climate Change in Wildlands: Pioneering Approaches to Science and Management* (pp. 234–255). Island Press/Center for Resource Economics, Washington, DC. https://doi.org/10.5822/978-1-61091-713-1_12.

Smith, J. B., & Lenhart, S. S., (1996). Climate change adaptation policy options. *Clim. Res., 6*, 193–201. https://doi.org/10.3354/cr006193.

Sugunan, V. V., (2000). Ecology and fishery management of reservoirs in India. *Hydrobiologia, 430*, 121–147. https://doi.org/10.1023/A:1004081316185.

Wasson, R. J., Sundriyal, Y. P., Chaudhary, S., Jaiswal, M. K., Morthekai, P., Sati, S. P., & Juyal, N., (2013). A 1000-year history of large floods in the upper ganga catchment, Central Himalaya, India. *Quat. Sci. Rev., 77*, 156–166. https://doi.org/https://doi.org/10.1016/j.quascirev.2013.07.022.

Ziv, G., Baran, E., Nam, S., Rodríguez-Iturbe, I., & Levin, S. A., (2012). Trading-off fish biodiversity, food security, and hydropower in the Mekong River basin. *Proc. Natl. Acad. Sci., 109,* 5609 LP – 5614. https://doi.org/10.1073/pnas.1201423109.

Index

B

D

E

F

G

H

I

J

K

L

O

P

Q

U

V